Forensic DNA Trace Evidence Interpretation

Forensic DNA Trace Evidence Interpretation: Activity Level Propositions and Likelihood Ratios provides all foundational information required for a reader to understand the practice of evaluating forensic biology evidence given activity level propositions and to implement the practice into active casework within a forensic institution. The book begins by explaining basic concepts and foundational theory, pulling together research and studies that have accumulated in forensic journal literature over the last 20 years.

The book explains the laws of probability – showing how they can be used to derive, from first principles, the likelihood ratio – used throughout the book to express the strength of evidence for any evaluation. Concepts such as the hierarchy of propositions, the difference between experts working in an investigative or evaluative mode and the practice of case assessment and interpretation are explained to provide the reader with a broad grounding in the topics that are important to understanding evaluation of evidence. Activity level evaluations are discussed in relation to biological material transferred from one object to another, the ability for biological material to persist on an item for a period of time or through an event, the ability to recover the biological material from the object when sampled for forensic testing and the expectations of the prevalence of biological material on objects in our environment. These concepts of transfer, persistence, prevalence and recovery are discussed in detail in addition to the factors that affect each of them.

The authors go on to explain the evaluation process: how to structure case information and formulate propositions. This includes how a likelihood ratio formula can be derived to evaluate the forensic findings, introducing Bayesian networks and explaining what they represent and how they can be used in evaluations and showing how evaluation can be tested for robustness. Using these tools, the authors also demonstrate the ways that the methods used in activity level evaluations are applied to questions about body fluids. There are also chapters dedicated to reporting of results and implementation of activity level evaluation in a working forensic laboratory. Throughout the book, four cases are used as examples to demonstrate how to relate the theory to practice and detail how laboratories can integrate and implement activity level evaluation into their active casework.

Forensic DNA Trace Evidence Interpretation

Activity Level Propositions and Likelihood Ratios

Duncan Taylor
and Bas Kokshoorn

CRC Press
Taylor & Francis Group
Boca Raton London New York

CRC Press is an imprint of the
Taylor & Francis Group, an **informa** business

Cover image: "Through the Bayesian Lens," by Jasmin Elise Taylor.

First edition published 2023
by CRC Press
6000 Broken Sound Parkway NW, Suite 300, Boca Raton, FL 33487-2742

and by CRC Press
4 Park Square, Milton Park, Abingdon, Oxon, OX14 4RN

CRC Press is an imprint of Taylor & Francis Group, LLC

© 2023 Duncan Taylor and Bas Kokshoorn

Library of Congress Cataloging-in-Publication Data

Names: Taylor, Duncan, author. | Kokshoorn, Bas, author.
Title: Forensic DNA trace evidence interpretation : activity level propositions and likelihood ratios / Duncan Taylor, Bas Kokshoorn.
Identifiers: LCCN 2022027873 (print) | LCCN 2022027874 (ebook) | ISBN 9781032225821 (hardback) | ISBN 9781032225814 (paperback) | ISBN 9781003273189 (ebook)
Subjects: LCSH: DNA fingerprinting. | Forensic genetics. | Forensic biology.
Classification: LCC RA1057.55 .T39 2022 (print) | LCC RA1057.55 (ebook) | DDC 614/.1--dc23/eng/20220928
LC record available at https://lccn.loc.gov/2022027873
LC ebook record available at https://lccn.loc.gov/2022027874

ISBN: 978-1-032-22582-1 (hbk)
ISBN: 978-1-032-22581-4 (pbk)
ISBN: 978-1-003-27318-9 (ebk)

DOI: 10.4324/9781003273189

Typeset in Times New Roman
by Deanta Global Publishing Services, Chennai, India

Contents

Chapter 12 Beyond Forensic Biology 435
Bas Kokshoorn

Chapter 13 Looking to the Future 469
Duncan Taylor and Bas Kokshoorn

Chapter 14 Answers to Practice Questions 475
Duncan Taylor and Bas Kokshoorn

Index 561

Preface

Forensic science applies scientific technique to matters of criminal investigation, prosecution and defence. Prior to the use of any scientific method, trials and investigations relied mostly on confessions and eyewitness accounts. However, there are very early examples of what could be considered forerunners of modern forensic science. Perhaps the most famous anecdote of Archimedes (287–212 BC) is his 'forensic investigation' of a votive crown that had been commissioned by King Hiero II of Syracuse. Supposing the crown to be made of pure gold, the King asked Archimedes to devise a non-destructive test to ensure that the goldsmith had not substituted any of the pure gold he had been given for its construction. Archimedes determined (in what is known as the Archimedes' principle) that by submerging the crown in water its volume could be measured and comparing this with the weight would give the density of the material the crown was made of. From here, it was simply a matter of comparing the density of the crown to the density of pure gold to determine whether the goldsmith had indeed made any substitutions. This anecdote is an example of science (albeit a basic measurement) to investigate a potential crime.

Medical autopsies also have ancient roots. Ancient Greek physicians sought cause of death, and as early as the 4th century BC, Hippocrates recommended that physicians learn the signs of injuries inflicted during criminal offences. Famously, the autopsy of Julius Caesar in 44 BC identified the fatal stabbing blow, and some historians even attribute the origin of the word 'forensic' (Latin for 'before the forum') to this autopsy.

While these early examples exist, it is some time before forensic science as we know it now would evolve. In many courts, the likes of a scientific experiment would have been held in equal regard with the likes of divine instruction, witchcraft or possessions. While the scientific method of using observations to test and refine hypotheses that are then used to make predictions is generally thought of as arising in the 16th century, there are earlier examples. One example is the use of entomology by 11th-century Chinese physician and later judge Song Ci [1], who made many advances in the conduct of autopsies. In one instance, after identifying a sickle as the murder weapon, Song Ci called all the nearby suspects to bring their sickles and the murder weapon was identified as the one to which flies were attracted. Confronted with this evidence, the owner of the be-flied sickle confessed to the crime.

Through the 16th, 17th and 18th centuries, many advances were made and scientific disciplines that would be recognisable today were introduced, such as anthropometrics by Alphonse Bertillon and fingerprint comparison championed by Sir William Herschel. In the late 19th century, the idea of forensic-evidence-based investigation and deductive reasoning was immortalised in Sir Arthur Conan Doyle's character Sherlock Holmes. By the early 20th century, the first school of forensics was opened by Archibald Reiss (the Institut de police scientifique of the University of Lausanne), and Dr Edmond Locard developed his famous 'Lockard's exchange principle' that the perpetrator of a crime will bring something into the crime scene and leave with something from it, and that both can be used as forensic evidence [2].

It was not until comparatively recently that DNA-based forensic evidence was introduced by Sir Alec Jeffreys in the 1980s [3–5]. Jeffreys found that by breaking apart the DNA at particular sequences (a process called restriction fragment length polymorphism, or RFLP) the difference in sequences between people would lead to different-sized DNA fragments. When stained, these fragment patterns (commonly called DNA fingerprints) could be used to distinguish between people or look for similarities between people who were thought to be related. The first instance of 'DNA profiling' used in a criminal investigation was in 1986 by Jeffreys in the rape and murder of a 15-year-old girl, Dawn Ashworth, in Leicestershire, England [6]. The suspect held by police at the time (who had confessed to the crime) was DNA profiled using Jeffrey's new technique, as was done for a sample for the 1986 offence and for a sample from a similar crime committed in 1983. While the crime scene samples matched, the reference of the suspect did not. This sets in motion a genetic dragnet by police, who typed around 4000 applicable male individuals from the area. No match was found. Later, a conversation by a man named Colin Pitchfork was overheard, where Colin revealed the details of how he had convinced a friend to provide a sample on his behalf. When Colin was arrested, and his profile generated, it matched the two crime scene profiles and the offender was identified.

As the end of the 20th century drew near, forensic science evolved, particularly in the area of DNA profiling. DNA profiling kits had moved on from the original type of analysis carried out by Jeffreys and used a process called polymerase chain reaction (PCR), developed in the mid-1980s by Kary Mullis [7]. PCR could make multiple copies of the regions of interest, so they could be visualised, even with very small starting amounts of genetic material. DNA profiling technology improved through the beginning of the 21st century to where it is today, the capability to generate highly discriminating DNA profiles from only a few cells. In conjunction with this ability to generate DNA profiles came the statistical methodologies to analyse the DNA profile data in sophisticated and complex ways, including mixtures of DNA from multiple individuals (currently up to six, although that number increases with technological advancements).

The power of this technology meant that very small amounts of genetic material (which can easily be transferred from object to object) were increasingly being used as the basis of evidence presented in court. Consequently, much of the focus of the justice system shifted from identifying whose DNA may be present on an item (a fact that through the developments of technology was practically taken for granted) to finding out how that DNA may have come to be on it in the first place.

This shift in focus has led the forensic biology field to start considering the meaning of their results within the context of actions or activities in the case. From the pioneering work on this topic in the late 20th century by members of the Forensic Science Service in the UK [8], the field has grown in popularity and sophistication. In recent years, many forensic laboratories are considering how they can carry out these types of evaluations within their own legal systems.

As is common for anyone hearing about a new concept or practice, those initial learnings of activity level evaluations have shown some feelings of trepidation. This often stems from thoughts along the lines of 'we can't say the person of interest (POI) and complainant had sexual intercourse, all we can say is the profiles match'. A related concern sometimes expressed by the legal community is that the scientist cannot make comments on the activities of the defendant as that is the duty of the court. Both thoughts are based on a misunderstanding of the role of a scientist in court.

The role of the scientist is to evaluate the scientific observations within a framework of circumstances and give some competing versions of events. Provided that the scientist comments on the evidence and not on the events, there is no breach of the role of the scientist in their report or during their testimony. The recipients of this evaluation (judge or jury) will incorporate the scientific opinion with the other testimonies they have heard and will decide whether sexual intercourse has taken place and whether it is consensual. In fact, it is commonplace during scientific testimony on DNA profiling results that the scientist will be asked whether they would expect to have obtained these profiles given different potential events occurring. Answering these questions, ad hoc on the stand, will force the scientist to provide vague answers that are less informative than they could have otherwise been. Common examples of this are answers such as 'it is possible' or 'that could explain the results'. It is in effect no different to providing an evaluation given activity level propositions from the start. The difference is that such an ad hoc question has the disadvantage of catching the scientist without the ability to have properly reviewed literature or prepared a logical construction of how the various elements of the framework of circumstances interact. In addition, these vague responses can still be suggestive of an activity, to those hearing them.

At the time of writing this book, most countries in the world have forensic laboratories, many doing DNA profiling, toxicology, examination of chemical traces, pathology, document examination, ballistics, digital evidence examination, fingerprints and examinations in many other disciplines.

This book seeks to explore evaluating observations given activity level propositions. Many of the examples are from the field of forensic biology, but the methods described are applicable to evaluations in other fields of forensic science, or indeed any science.

The book starts with some introductory material on the basic building block of evaluation, that is, dealing with uncertainty, the principles of evaluation and Bayesian inference. Moving through the chapters of the book, the application of these principles to different elements of evaluations is explored and examples are given, motivated by real casework. Explanation is provided on methods of evaluation, both manual formulaic derivation and utilising graphical software tools to assist in this task. It is shown how different aspects of circumstances surrounding the crime and the items need to be considered within the evaluation in order for it to be robust and logical. Once constructed, the methods are explored for testing the robustness of the evaluation, and the interpretation of these tests is explained. Importantly, information and examples are provided on how these types of evaluation can be implemented in a forensic laboratory, including aspects of reporting and court testimony. The book finishes with evaluations examples in disciplines outside of forensic biology and demonstrates how multi-disciplinary evaluations can be carried out.

The first two chapters of the book are introductory and provide information about the concepts of evidence evaluation that will be used and built upon throughout the book. The laws of probability are explained, and it is shown how they can be used to derive, from first principles, the likelihood ratio (used throughout the book to express the strength of evidence for any evaluation). Concepts such as the hierarchy of propositions, the difference between an investigation and evaluation, and the practice of case assessment and interpretation [9] are explained in order to provide the reader with a broad grounding in the topics that are important to understanding evaluation. The second chapter focuses on activity level evaluation and the reason for its importance. It is a chapter that presents a number of issues that have been faced in court cases that could have been addressed by activity level evaluation and addresses common concerns with activity level evaluation. In Chapter 2, four cases are introduced and will be used throughout the book to apply the concepts of evaluation being discussed. These cases have been chosen as they represent key examples of different types of evaluation that forensic practitioners will face. These cases are a common thread and tie the chapters and concepts together, so it can be seen how each concept is important to the overall picture.

By the time Chapter 3 is reached, the reader will have an understanding of the concepts of activity level evaluation and the need for it to be applied in casework. In this chapter, key activity concepts that must be considered in any activity level evaluation are introduced. In forensic biology, these are the (prob)ability for biological material to be transferred from one object to another, the ability to subsequently persist on an item for a period of time or through an event, the expectations of the prevalence of biological material on objects in the environment and the ability to recover the biological material from the object when sampled for analysis. These concepts of transfer, persistence, prevalence and recovery are discussed, along with the factors that affect each of them.

With all the necessary background information provided in Chapters 1–3, Chapter 4 starts the process of evaluation. Explanation about the evaluation process, how to structure case information and importantly how to formulate propositions is provided. The way that different case circumstances lead to proposition formation is explained, and guidance is given on the type of information included in propositions. The chapter ends with the application of the four example cases so that propositions are formulated and explained.

Chapter 5 is the first of the mathematically technical chapters of the book. Using the first three case examples, Chapter 5 explains how a likelihood ratio formula can be derived in a step-wise manner to evaluate forensic observations. These steps are followed through for three of the cases, with in-depth explanations of the concepts as the process is worked through. At the end of this chapter, the reader will have gained the ability to take a case, map out the pathways of importance for the evaluation, convert the pathways to a mathematical formula and assign numerical values to the probabilities required in the evaluation.

Chapter 6 introduces a commonly used tool in activity level evaluations, the Bayesian network. This chapter presents an introduction to the topic of Bayesian networks, and the concepts are built on in the following two chapters. The three cases that were evaluated in Chapter 5 by deriving mathematical expressions for the likelihood ratio are re-evaluated in Chapter 6 using Bayesian networks. Much like how Chapter 5 moved through a series of steps to derive the formula, Chapter 6 works through steps on how to construct a Bayesian network for a forensic case. Again, as topics arise during the case evaluations, they are explained in depth throughout the chapter. The result of the chapter is that the three cases end up with the same final evaluation outcome as they did in Chapter 5. By showing how the two different methods of evaluation (manual derivation in Chapter 5 and Bayesian networks in Chapter 6) provide the same outcome, a deep understanding of the whole evaluation process is gained.

Having worked through Chapter 6, the reader will have a fundamental understanding of Bayesian networks, what they represent and how they can be used in evaluations. Chapter 7 builds on this knowledge by introducing a series of more advanced abilities of Bayesian networks, mainly the use of Object-Oriented Bayesian networks and the use of probability distributions and interval nodes rather than point values for node states. It is in this chapter that the fourth example case is evaluated, left until this point in the text due to the complexity of the evaluation requiring these advanced concepts. At the end of the chapter, guidance is given on data modelling and how that applies to advanced concepts of activity level evaluations using Bayesian networks.

Chapter 8 is all about how the evaluation can be tested for robustness. This is an important chapter as it guides the reader in the various tests they can perform to make sure they are providing appropriate information to stakeholders. The results of any test or evaluation provided to the court must be robust and sound. Sometimes in activity level evaluation, it can be difficult to show this is the case, without knowing the right analyses to run. Again, this chapter uses the example cases to demonstrate different aspects of testing an evaluation for robustness, from very simple tests to those that are advanced.

Chapter 9 demonstrates the way that the methods used in activity level evaluations can be applied to questions about body fluids. The evaluation may be carried out to form an opinion about the body fluid type in a case, or uncertainty about body fluid can be incorporated into an activity level evaluation. This chapter utilises the concepts built on in the previous chapters and examples of different cell type tests and their results' evaluation are explored.

Chapters 10 and 11 are aimed at forensic institutions that wish to implement activity level evaluation into their active casework. The aspects of reporting the results (Chapter 10) and how to implement activity level evaluation amongst the very regimented legal and forensic frameworks (Chapter 11) are discussed. These chapters provide examples and advice learned through the experiences of both authors. The chapters will answer the practical implementation questions that arise from many forensic institutions after they have read about the theory of activity level evaluations. Once again, the worked cases are used as examples. By the time the reader has completed these chapters, they will have seen the cases from the initial police reports in Chapter 2, through the evaluations in Chapter 5, the analysis of the case evaluations for robustness in Chapter 8 and finally the reports generated for the cases in Chapter 10.

Chapter 12 is a general chapter that takes the reader out of the world of forensic genetics and shows how the concepts of evidence evaluation, transfer, persistence, prevalence, recovery, Bayesian network construction and assessment, reporting and implementation can be applied in any forensic discipline. It is designed in a way that laboratories that may wish to start with activity level evaluation in forensic genetics can start to look beyond biology in order to apply the practice to other disciplines. Examples are given of how this has been done in practice demonstrating that activity level evaluation provides a holistic framework that applies to all areas of forensic science.

The book finishes with Chapter 13, which takes a look at where the field of forensic evaluation may progress in the future, followed by Chapter 14, which provides worked answers to questions provided throughout the other chapters of the book.

REFERENCES

1. S. Ci, 洗冤錄 *(Collected Cases of Injustice Rectified or the Washing Away of Wrongs)*, Hu Wenhuan, Hangzhou, Zhejiang China, 1247.
2. E. Locard, *L'enquête criminelle et les méthodes scientifiques*, Ernst Flammarion, Paris, 1920.
3. P. Gill, A. Jeffreys, D. Werrett, Forensic application of DNA 'fingerprints', *Nature* 318(12 December) (1985) 577–579.
4. A.J. Jeffreys, A.C. Wilson, S.L. Thein, Individual specific "fingerprints" of human DNA, *Nature* 316 (1985) 75–79.
5. A.J. Jeffreys, V. Wilson, S.L. Thein, D.J. Weatherall, B.A.J. Ponder, DNA 'fingerprints' and segregation analysis of multiple markers in human pedigrees, *American Journal of Human Genetics* 39 (1986) 11–24.
6. M.A. Jobling, Curiosity in the genes: The DNA fingerprinting story, *Investigative Genetics* 4(1) (2013) 20.
7. K.B. Mullis, F.A. Faloona, S. Scharf, R. Saiki, G. Horn, H. Erlich, Specific enzymatic amplification of DNA in vitro: The polymerase chain reaction, *Cold Spring Harbor Symposia on Quantitative Biology* 51 (1986) 263–273.
8. R. Cook, I.W. Evett, G. Jackson, P.J. Jones, J.A. Lambert, A hierarchy of propositions: Deciding which level to address in casework, *Science & Justice* 38(4) (1998) 231–240.
9. R. Cook, I.W. Evett, G. Jackson, P.P. Jones, J.A. Lambert, A model for case assessment and interpretation, *Science and Justice* 38(3) (1998) 151–156.

Acknowledgements

We are immensely grateful for the support and understanding of our partners Jasmin and Nicolette.

We also wish to thank several colleagues for reading earlier drafts of the chapters of this book and providing valuable feedback:

- Simone van Soest
- Jord Nagel
- Bart Aarts
- Yvonne van de Wal
- Charles Berger
- Jan de Koeijer
- Oliva Handt

About the Authors

Duncan Taylor is the Chief Scientist of Forensic Statistics at Forensic Science South Australia, Adelaide, Australia. He has studied and worked in the forensic field for over 20 years, supervising criminal casework in hundreds of cases for both prosecution and defence, and testifying in Courts around Australia. He holds PhDs in molecular biology and statistics, both obtained from Flinders University. Dr Taylor is one of the developers of STRmix™, a probabilistic DNA interpretation software being used in forensic DNA laboratories around the world, and a developer of the neural network component of FaSTR™, a DNA profile reading software. He is an Associate Professor of Biology and a Distinguished Alumnus at Flinders University, supervising honours, masters and PhD students in biology and statistics projects. Duncan collaborates with colleagues around the world and has over 120 peer-reviewed publications and a book on DNA evidence interpretation. In 2017, Duncan was awarded the SA Science Excellence STEM professional award and in 2018 he was part of the group that won the New Zealand Prime Minister's Science Award. In 2021, Duncan was awarded the Public Service Medal for contributions to forensic statistics.

Bas Kokshoorn is a Principal Scientist at the Netherlands Forensic Institute (NFI) in The Hague (The International City of Peace and Justice). Bas started his career at the NFI in 2008 and since then has submitted over 1000 reports on forensic biology examinations and interpretations to the Dutch criminal justice system as well as to international tribunals and other jurisdictions in continental Europe, the UK and Australia. Bas holds a PhD in evolutionary biology from Leiden University and is currently appointed as a Professor in *Forensic Trace Dynamics* at the Amsterdam University of Applied Sciences. In this role, he supervises bachelor, master's and PhD students working on research projects aimed at understanding the dynamics of transfer, persistence, prevalence and recovery of biological and other types of trace evidence and the interpretation of such findings in criminal cases.

1 Principles of Evaluation

Duncan Taylor

CONTENTS

1.1 PROBABILITY

While the complexity and sophistication of forensic analyses have grown along with the disciplines in which they now fall, one aspect has not changed over time. At some point, a question arises in the pursuit of truth, which cannot be satisfactorily answered by confessions or eyewitness accounts. To answer the question, objects(s) are examined, and observations are made (as the result of measurement or testing). These observations, when considered within the set of circumstances surrounding the question, lead the investigator to believe a particular version of events has occurred. The more directly the observations can be linked to the question of interest, the greater the use they are to the person asking. But what happens when there is more than one explanation for the observations? What would Song Ci have done (see Preface, page vii) if the owner of the sickle claimed that, instead of being attracted to the blood of the murder victim, the flies had simply grown weary of flying at that point and needed to rest on the nearest object, which just happened to be his sickle?

Disputes over the interpretation of the observations are at the heart of the intersection of forensic science and statistics and transcend all the forensic domains, making it a coherent scientific discipline. If there are two explanations for the observations, then which explanation is better? In which should a person judging the two propositions place their belief? Early persuasions would likely have relied on anecdotal reasoning or common-sense beliefs. But as the methods used to generate the observations became more sophisticated, and as the public's need for fair and balanced treatment of criminal matters grew, so too did the need to use more than just common sense or anecdote. Measured and numerical frequencies became prominent in reasoning and methods for dealing with uncertainty were developed.

Of note in the pursuit of evaluating observations and findings and dealing with uncertainty is the work of the Presbyterian minister Thomas Bayes in the 1700s. Bayes did not publish his

DOI: 10.4324/9781003273189-1

1

theories on probability himself. They were communicated after his death by Richard Price in 'An Essay towards Solving a Problem in the Doctrine of Chances' [1]. In this work, Bayes outlined theorems on conditional probability, which are the basis of Bayes' theorem, that is, a method to calculate the probability of a proposition based on its prior probability and the addition of some relevant evidence (or observations). The concept of probability was different to that of frequency because while frequency is something that can be measured through a series of experiments (as explained below), a probability is assigned to an event – even in the absence of observations.

Frequency reports on how common an event has been measured, and it is reported in units that align with that measurement, for example, 'adults drink wine three times per week', or 'a heartbeat is heard at 78 beats per minute'. Relative frequency divides the count of the event being measured by the total number of observations and so the sum of the relative frequencies of all the different values being measured will be one. For example, '1 in 1000 Swedish men between 25 and 35 years old are over 200 cm tall', which when combined with the relative frequency of Swedish men between 25 and 35 years who are shorter than 100 cm, 100 to 150 cm, and 150 to 200 cm will sum to one. A context that would be familiar within forensic genetics is allele frequency, that is, the relative frequency of observing an allele in a random survey of alleles in the population. The sum of all relative allele frequencies must equal one.

When talking of probability, it is important to consider that probabilities are a personal belief in an event occurring. They are personal, as they rely on information, and each person has different knowledge and experience and so will be basing their probability on different information. In this way, probabilities are considered to be a statement about our knowledge of the world, rather than a statement about the world itself. This idea is the fundamental concept behind the premise that all probabilities are personal and subjective, which was very elegantly explained by Denis Lindley in *Understanding Uncertainty* [2]. Because probabilities are considered personal and subjective there is no true value for a probability to take. Therefore, when providing a numerical value, it is common to talk of assigning a value to a probability rather than to talk of 'the probability's value'. Probability has no units, as it is a prediction of occurrence based on knowledge and information. Also, unlike frequency, it does not need to be based on a long run of experiments in order to be given a numerical value. Indeed, the probabilities of many events within an evaluation being carried out have not been directly observed and could not ethically be observed. For example, it would not be ethical to carry out a study on the probability that a person cuts themselves when stabbing another person.

Describing exactly what probability *is* requires careful consideration so that the explanation does not use other terminology such as likelihood, chance, rate and frequency, which merely substitutes one term of uncertainty for another and risks the misuse of terminology. The definition must also encapsulate the idea of probability being personal. One method to describe probability as used by Lindley is to compare it to a gamble [2]. Accept that all events in nature are uncertain. Some events may seem unwaveringly true, such as that the sun will rise tomorrow (but even then, it may be possible that some astronomical event will prevent this from occurring). Other events may seem unwaveringly false, but belief in most events occurring will fall somewhere in between these two extremes. Consider now just one event and that a friend offers you a wager on that event occurring, say that it will rain tomorrow. The wager they provide you is that you bet $100 on rain occurring, and if it rains then you will win $1. If it rains, you keep your original bet of $100 and 'win' an additional $1. If it does not rain, then you lose the money of your $100 bet and end up with nothing. This does not appear to be a very smart wager to take, as there is so much risk and so little rewards for a relatively volatile event, based on the whims of meteorology. The level of your belief that it will rain is not so great that the risk of this gamble is worth taking.

Having refused his wager, your friend now offers you another wager, you bet $1 and if it rains then you win $100 (i.e. you keep your original bet of $1 and win $100) and if it doesn't rain then you lose the money you bet and end up with nothing. This now appears to be a very attractive prospect as the potential rewards are so much greater than the risk of losing a single dollar. In this instance the small risk you are making far exceeds what you would be willing to gamble on the weather being rainy. Even if your belief in rain occurring was only very slight, this would likely make an attractive wager.

Your friend realises that this is not a very good wager for him and withdraws and makes another counteroffer. You can imagine this process of making various wagers of differing extremities continuing and at some point of wager, you will become indifferent as to whether the wager is a good one to take or not. In other words, your belief that it will rain is matched with your inclination to take the wager. At this point, you have found your personal level of belief, or probability, in rain occurring tomorrow. During this process, the balance of potential gain and loss will align with your personal belief that it will rain tomorrow. Note that in this analogy consideration of the amount must be taken out of the equation as people will tend to become more cautious as amounts of money increase (and some people are simply morally against gambling).

In the first wager, the amount being requested by your friend was very high, and the return comparatively low. One might only take this wager on an event that they believe was virtually certain to occur (such as the sun rising the next day). Imagine the size of the initial asking amount increasing with no additional payout (e.g. a $1000 wager that would return $1 if won or a $1 000 000 wager that would return $1 if won). To take these increasing wagers would require ever more belief that the event on which was being bet was going to occur. The ratio of the amount of money you start with to the amount of money you would end up with if you won gets closer to 1 as the initial wager amount increases. For example, if you started with $1000 and won $1 then the ratio of starting to ending money is 1000 to 1001, which is just below 1. If you started with $1 000 000 and won $1 then the ratio is 1 000 000 to 1 000 001, which again is just below 1, but close to 1 than when the bet was $1000. This leads to the first part of the first law of probability, all probabilities are less than or equal to 1.

In a similar thought experiment, but at the other end of the wager spectrum, the asking amount in the wager decreases and the winnings increase. If the wager was not to be taken, then your belief in an event occurring must be very low. At the extreme, the ratio of the amount of money you start with compared to the amount of money you would end up with if you won approaches zero. For example, if you started with $1 and won $1000 then the ratio of starting to ending money is 1 to 1001, which is just above 0. If you started with $1 and won $1 000 000 then the ratio of starting to ending money is 1 to 1 000 001, which again is just above 0, but closer to 0 than when the potential winnings were $1000. This provides the other bound of the first law of probability, which is that all probabilities are greater than or equal to zero. Together these two points can be summarised as:

The first law of probability states that all probabilities lie between 0 and 1.

A probability of 0 represents an impossible event, for which no further information can ever change our mind that the event simply will not occur. A probability of 1 represents an event of absolute certainty, one which will definitely happen regardless of anything else in the world.

At this point, some nomenclature is introduced that will be used throughout this book. When talking of probability, the abbreviation 'Pr' is used and followed by a bracketed term, which is a placeholder used to represent the event to which the probability is assigned. For example, the term $Pr(A)$ represents the probability of A being true (or occurring). As with standard mathematical nomenclature, an equality between two terms is represented by '=', when one term is greater than the other it is represented by '>' and if it can be greater than or take the same value '≥' is used. Similarly, less than is '<' and less than or equal to is '≤'. With this terminology the first law of probability for event 'A' can be written as:

First law of probability: $0 \leq \text{Pr}(A) \leq 1$

In many instances, the probability of more than one event occurring may be of interest. These can be considered in various ways, but one way would be to consider any one of a set of events occurring. In probability nomenclature, this would be the probability of event A or event B occurring and can be represented by the symbol '$\cup$', or the word 'or', for example:

$$\text{Pr}(A \text{ or } B) \text{ or } \text{Pr}(A \cup B).$$

Two further concepts must be defined before continuing to the second law of probability, and these are 'mutually exclusive' and 'exhaustive'. When considering a set of potential events, sometimes two of these events can occur at the same time. For example, with the simple activity of rolling a die, consider the two events: 'an even number is rolled' and 'a number less than 4 is rolled'. The possible outcomes would be the die rolling a:

1 – in which the second event (a number less than 4 is rolled) is true
2 – in which the first event (an even number is rolled) and the second event (a number less than 4 is rolled) is true
3 – in which the second event (a number less than 4 is rolled) is true
4 – in which the first event (an even number is rolled) is true
5 – in which case neither event is true
6 – in which the first event (an even number is rolled) is true

In these two events, it is possible that a 2 is rolled which would see that both events have occurred. For other sets of events, any outcome can at most only cause one event to be true. For example, if in the die example the second event was changed to 'a five will be rolled' then the possible outcomes would be the die rolling a:

1 – in which case neither event is true
2 – in which the first event (an even number is rolled) is true
3 – in which case neither event is true
4 – in which the first event (an even number is rolled) is true
5 – in which the second event (a five is rolled) is true
6 – in which the first event (an even number is rolled) is true

Now there are no outcomes that lead to more than one of the events being true and so they are said to be mutually exclusive. Graphically this can be shown by considering a Venn diagram as seen in Figure 1.1, in which the outer rectangle represents all possible outcomes that can occur (i.e. 1, 2, 3, 4, 5 or 6). In statistical parlance, this is known as the 'sample space'. The space that is represented by $A \cup B$ in Figure 1.1 is the area within the box that is coloured, note that in the upper box the coloured areas overlap, whereas in the lower box they do not.

Mutually exclusive events are those that cannot occur at the same time. Therefore, within a set of mutually exclusive events, at most one of them can be true. In Figure 1.1, the upper box, representing the first pair of events

• An even number is rolled
• A number less than 4 is rolled

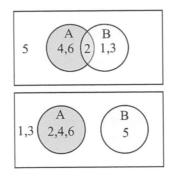

FIGURE 1.1 Venn diagram showing the outcome of rolling a dice. The top panel shows the two events; *A* – 'an even number is rolled' and *B* – 'a number less than 4 is rolled'. In the lower panel the two events being represented are *A* – 'an even number is rolled' and *B* – 'a five will be rolled'.

are not mutually exclusive as they can both be true when 2 is rolled, whereas the lower box, representing the second pair of events

- An even number is rolled
- A five is rolled

are mutually exclusive as only one can be true given any of the possible outcomes in the sample space. Note that in both these examples there are outcomes of the die roll that can lead to neither event being true, i.e. a 5 in the first event pair or a 1 or 3 in the second event pair. This means that the two events have not exhaustively covered all the possible outcomes of the die roll. As the description suggests then neither event pair is exhaustive. Technically, an exhaustive set of events are those where at least one (and possibly more than one) must be true. Using the die roll example, a set of exhaustive events could be achieved by adding an event to the first pair, so that they are now:

- An even number is rolled
- A number less than 4 is rolled
- A prime number is rolled

Now, no matter what the outcome of the die roll, at least one of these events will be true. They are an exhaustive set. Note that because rolling a 3 will lead to two events being true and rolling a 2 will mean that all three are true, they are not mutually exclusive. If a set of events are exhaustive and mutually exclusive, then always exactly one of the sets will be true. This concept will become important in later chapters when discussing proposition formation in the context of a criminal matter.

Having defined mutually exclusive events (and seeing it visually in the lower part of Figure 1.1) the second law of probability can be given, by stating:

> *If two events, A and B, are mutually exclusive then the probability of A or B occurring is the sum of the probability of A occurring and the probability of B occurring*

In statistical nomenclature this can be represented as:

> *Second law of probability:* $\Pr(A \cup B) = \Pr(A) + \Pr(B)$

However, if the events are not mutually exclusive, some account for the fact that both can occur at the same time must be made. This moves on to the third law of probability, which deals with multiple events occurring together. A term for 'and' is introduced, which is represented by the symbol $\cap$. Therefore, the probability of events A and B occurring can be written as:

$$\Pr(A \ and \ B) \ \text{or} \ \Pr(A \cap B) \ \text{or even sometimes the simplified} \ \Pr(AB).$$

In the top panel of Figure 1.1, the probability of A and B is represented by the overlapping area of the two, coloured circles. Before moving on to the third law of probability, some further terminologies must be introduced, and this is the concept of independence.

Imagine that you hold a certain belief in the occurrence of a proposition, that is, you have assigned a probability of its occurrence. As you receive more information you may revise your assigned probability, for example, if the proposition was that a particular football team would win their next match, then the information that their star player has been injured would change your belief in their ability to win. You may receive information about the predicted weather during the match, and with knowledge of their performance in wet or dry conditions, again you would update your probability that they will win. There will also be an abundance of information that you receive throughout the day that will not affect your opinion on the winning prospects of your team. You may be told that the lawn needs mowing, or that you have passed a recent test, or that your friend is having chicken for dinner, all of which will not affect your football team's performance (unless some information affects your mood, which could irrationally affect your assessment positively or negatively). Your *rational* belief in your team winning the football match is quite independent of this myriad of inconsequential pieces of information. Therefore, a formal definition of independence is that two events are independent if knowledge of one event does not affect your belief in the other. Within the third law of probability this plays a role when stating:

The probability of two independent events occurring is the product of the probability of each event occurring

Formulaically this is given by:

Third law of probability: $\Pr(A \cap B) = \Pr(A)\Pr(B)$

It is now possible to consider the second law of probability and the consequence of the events not being mutually exclusive. In the bottom panel of Figure 1.1, when the two events were mutually exclusive, the probability of either occurring could be calculated by summing the coloured areas of the two circles. In the top panel of Figure 1.1, if the same procedure were carried out, the area in which the two circles overlap would be counted twice (once as part of circle A and once as part of circle B). In order to calculate the probability of events A and B when they are not mutually exclusive, the probability that either has occurred is summed and the probability that both have occurred is subtracted to account for the double counting of the area of overlap. Therefore, a more general second law of probability can be reformulated as:

Generalised second law of probability: $\Pr(A \cup B) = \Pr(A) + \Pr(B) - \Pr(A \cap B)$

Now consider what occurs if the assumption of independence is false when considering the probability of two events occurring (as is the case in the stated third law of probability given above). To assign a probability to A and B occurring, if they are not independent, then knowledge of the probability that one event would occur if the other one has occurred is required. This concept is encapsulated within 'conditional probability'. A conditional probability is the probability of an event occurring, conditioning on the fact that some other event has occurred. The nomenclature representation of a conditional probability is by use of a vertical bar 'I', known as a conditioning bar. When the term:

$$\Pr(A|B)$$

is written, this means the probability of event 'A', conditional on event B, or given that event B has occurred. The effect of the third law of probability given above is that the product is between a probability and a conditional probability:

Generalised third law of probability: $\Pr(A \cap B) = \Pr(A \mid B)\Pr(B) = \Pr(B \mid A)\Pr(A)$

The three laws of probability provide the ability to describe uncertainty in all situations, i.e. they are all that is needed. They can be used to develop further insights, one of which is referred to is the 'law of total probability'. The law of total probability (sometimes called the law of extending the conversation) is commonly invoked in forensic evaluations. This makes use of the second and third laws of probability and can be used when the probability of an event cannot be assigned directly without consideration of some other dependant factor. Imagine a probability must be assigned for a house in Europe having solar panels; however, there is no organisation that assigned that probability directly. In fact, the only way that information is available is by considering the country as a factor, i.e. each country has information about the probability of a house within its borders having solar panels. In that case, the probability of a house in Europe could be calculated by considering the probability of a house in each country having solar panels, multiplied by the probability of a house in Europe being within that country (for which the information on house number or, more indirectly, population size, might be available). Here another term is introduced. When an event has a horizontal line above it, then this means 'that event has not occurred'. For example:

$$\Pr(\overline{A})$$

represents the probability that A did not occur. Formally:

$$\Pr(\overline{A}) = 1 - \Pr(A)$$

It follows from this then that A and $\overline{A}$ are mutually exclusive and exhaustive, i.e. either A has occurred or it has not occurred, and these represent all possible outcomes. Therefore, the probability of event 'A' occurring could be assigned by summing the probability of 'A' occurring if 'B' occurred, $\Pr(A \cap B)$, and the probability of 'A' occurring if 'B' did not occur, $\Pr(A \cap \overline{B})$. Using this terminology, the law of total probability can be represented as:

$$\Pr(A) = \Pr\left((A \cap B) \cup (A \cap \overline{B})\right) = \Pr(A \cap B) + \Pr(A \cap \overline{B}) = \Pr(A \mid B)\Pr(B) + \Pr(A \mid \overline{B})\Pr(\overline{B})$$

This law need not be confined to two possibilities for the dependent factor, it can be extended to any number of mutually exclusive and exhaustive demarcations of a factor. This law will become essential in Chapter 5 when deriving a formula for the probability of observations.

Before finishing this section on probability, one final concept is introduced. This is the difference between categorical variables (which is what has been dealt with in all examples so far) and continuous variables. Imagine a scientist wanted to measure the height of people, or more specifically wanted some record of the range of heights of people in the population and the probability that someone would be any particular height. The scientist could go around measuring people and would end up with a list of heights. Depending on how accurately the scientist measured those heights, there may be many instances of the same height occurring but imagine that the scientist wanted to be very precise and so measured the heights with extreme precision. So precise that in fact no two people on the list had exactly the same height at that level of resolution. The question now faced by the scientists is how best to represent the data that had just been accumulated. One way would be to group all the people who have similar heights, say into brackets of 50 cm, by counting all the people who fall within 0 and 50 cm, 50 and 100 cm, etc. all the way up the bracket that encapsulates the highest observation. The count of people that fall within each bracket, divided by the total number of people measured would give the proportion that each bracket makes up of the total. These proportions could then be assigned as the probability that anyone chosen from the population would fall within any of these particular height brackets. Now imagine that the scientist wanted their observations to have a greater resolution than 50 cm brackets and instead broke the data into brackets of 10 cm. The exact same process could be undertaken and now the scientist would have the ability to provide a probability that anyone from the population would fall within any particular 10 cm height bracket. In fact, the scientist could continue to decrease the bracket size to get ever more resolution in their dataset (1 cm, 1 mm, 1 nm, etc.), as long as the number of observations they had still allows sufficient coverage of the ever-smaller brackets. At some point, the width of the brackets could be decreased down to nothing, so that instead of a bracket of sizes, the scientist was considering a single size value. In doing so, at any particular size, the scientist would have either one individual observed at that height, or none (recall each height was different). It may be difficult to see how this data could then be used. Instead of reducing bracket sizes in this way, the scientist could draw a smooth line that represented the density of people at any particular point, i.e. the range in which the most people fell would have a high density and the range which contained very few observations would have low density. This change from considering height as a series of brackets to a line representing a distribution is the shift from considering a variable as discrete to considering it as continuous. This concept is discussed in more detail in Chapter 5.

When considering the variable as discrete, the proportion of the total that each bracket made up was considered. These proportions sum to one. When a variable is modelled as continuous distribution there is no longer talk of proportions within a bracket, but rather the 'density' of the distribution at any particular value. Now, the area under the distribution sums to one. Figure 1.2 shows the concept of probability and density by drawing out a normal distribution with a mean of 15 and a standard deviation of 4. At the mean of this distribution, the density is 0.1 (as seen on the graph by tracing across from the highest point of the distribution to the left axis). The

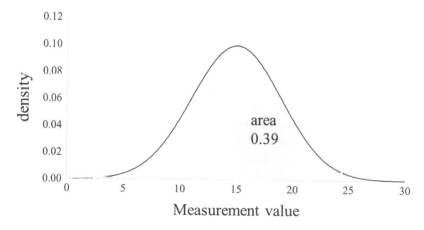

FIGURE 1.2 Graph showing normal distribution with a mean of 15 and a standard deviation of 4. The shaded area represents 0.39 of the area under the distribution.

total area under the curve is 1, and the area under a part of the curve (i.e. between two measurement values) can be considered as the probability of observing a measurement value within this range. For example, the area under the normal distribution between 15 and 20 is 0.39, meaning that there is a 39% chance of seeing a measurement value between these bounds for a random value that has been drawn from this distribution. If this distribution was to be broken up into discrete brackets of 5 units then for the bracket [15–20] the density of the distribution the mid-measurement value of 17.5 is 0.082. This represents the height of the bar that would be used for the discrete distribution (and is akin to the density) with the probability that it represented being obtained by multiplying the height by the length of the bracket, in this case, 5. Doing so yields a value of 0.3625, which is close to the area under the continuous distribution of 0.39 (different only because the curve between 15 and 20 would have to be a straight line in order for the areas to be exactly the same).

There may be times throughout this book when probability density is being referred to, and this is denoted by a lowercase 'p', for example, $p(A)$.

Having laid out the basic foundations of probability, these fundamentals can be used to build a system of evidence evaluation as will be shown when moving through the book.

1.2 BAYES' THEOREM

Recall the third law of probability sought the probability of two events A and B, both occurring. Note that it does not matter in which order these events are referred, i.e. they could just as well be thought of as the probability of B and A. The same two events are considered, and it is expected that the same probability would be assigned. The third law of probability, for events A and B, gave:

$$\Pr(A \cap B) = \Pr(A \mid B)\Pr(B)$$

But using the concept that the order that these probabilities does not matter, the third law of probability could equally be stated as:

$$\Pr(A \cap B) = \Pr(B \cap A) = \Pr(B \mid A)\Pr(A)$$

By equating the two right-hand sides of the equations just given, it can be seen the equality holds that:

$$\Pr(B \mid A)\Pr(A) = \Pr(A \mid B)\Pr(B)$$

and by dividing both sides of the equation by $\Pr(A)$ the classic formulation of Bayes' 'inverse probability' (commonly called today as Bayes' theorem) is obtained:

$$\Pr(B \mid A) = \frac{\Pr(A \mid B)\Pr(B)}{\Pr(A)}$$

In words, this equation provides the level of support that A provides for B (sometimes referred to as the likelihood of B on A) when information about A is available (assuming B is true) and given prior knowledge of A and B. Bayes' theorem can thus be used to calculate a posterior probability on a proposition, given prior probabilities updated by some evidence.

A common example to demonstrate Bayes' theorem is to consider a medical test for a disease and the probability that a positive result to this test means that a person has the disease. Note that this example is often used as people will confuse the accuracy or true positive rates with the posterior probability of having the disease. Imagine that a disease within a community is very rare and has an occurrence of approximately 1 in every 1 million people. This rate of occurrence is the prior probability of having the disease (assuming everybody, including you, is equally likely to carry the disease). If '*D*' represents having the disease and $\overline{D}$ as not having the disease, then Pr(*D*) = 0.000001. A patient gives a positive result to a test for this disease, which is known to have a 99.9% true positive rate (i.e. the rate of positive results for persons that have the disease) and a 0.001% false positive rate (i.e. the rate of positive results for persons that do not have the disease). Note here these values are given as relative frequencies, and not probabilities. However, in this special case, the probability is informed by the relative frequency (so that a 99.9% rate is used to assign a probability of 0.999). The information that the person wants to know is, having had a positive result, what is the probability that they actually have the disease. Define a positive test result as '*P*' and a negative test result as $\overline{P}$. By using Bayes' theorem with these defined terms:

$$\Pr(D \mid P) = \frac{\Pr(P \mid D)\Pr(D)}{\Pr(P)}$$

The prior probability of having the disease, Pr(*D*), is already given as 0.000001. It is also known that the probability of giving a positive result if you have the disease is 0.999 (i.e. the true positive rate) so that Pr(*P*|*D*) = 0.999. This leaves only the prior probability of giving a positive test result, Pr(*P*), to calculate. To do this the law of total probability can be used:

$$\Pr(P) = \Pr(P \mid D)\Pr(D) + \Pr(P \mid \overline{D})\Pr(\overline{D})$$

The probability of obtaining a positive test result if you don't have the disease (i.e. the false positive rate) is Pr(*P* | $\overline{D}$) = 0.00001, and so the posterior probability of having the disease given the prior probability of having the disease and updated with the information about obtaining a positive test result can be calculated by:

$$\Pr(D \mid P) = \frac{\Pr(P \mid D)\Pr(D)}{\Pr(P \mid D)\Pr(D) + \Pr(P \mid \overline{D})\Pr(\overline{D})}$$

$$= \frac{0.999 \times 0.000001}{0.999 \times 0.000001 + 0.00001 \times 0.999999}$$

$$\approx 0.091$$

or approximately 9.1%. Quite low considering what most people would think when hearing that they had just tested positive for a disease when the test they used had a 99.9% true positive rate and a 0.001% false positive rate. Equating the true positive rate with the posterior probability of having the disease is known as the 'base rate fallacy', that is, failing to take into account the base rate of the disease in the population. This demonstrated base rate fallacy has direct equivalence to issues of evidence interpretation by a jury when they fail to properly take into account the prior probability of guilt into their deliberations.

Probabilities can be described in what is called 'odds form', which is a ratio of two numbers rather than as a single probability value between 0 and 1. Using odds is common in gambling, and those who have been to horse races will be used to see the odds on a race given in the form X:1. This is in fact the form in which the initial description of probability was given in this chapter. Odds and probability can be interchanged by the formula:

$$O(X) = \frac{\Pr(X)}{1 - \Pr(X)} = \frac{\Pr(X)}{\Pr(\overline{X})}$$

The term $O(X)$ is used to signify the odds of event X occurring. Converting from odds back to probability involves a manipulation of the above formula to:

$$\Pr(X) = \frac{O(X)}{1 + O(X)}$$

As an example, consider that the probability of rolling an even number on a 6-sided die is 0.5. Substituting this probability into the conversion above yields $O(X) = 1$. In other words, the odds of rolling an even number on a 6-sided die are 50:50 (or 1:1), or equal to rolling an odd number.

1.3 THE LIKELIHOOD RATIO

In everyday life, it is common to use descriptors for the property of an object. As humans, it is common to provide descriptors that are tied to ourselves, e.g. an object is described as heavy if we would require a large exertion of energy to lift it, something is moving quickly if we could not easily meet its speed, an object is bright if we find it uncomfortable to gaze at. When one person uses a descriptive word for an object, our knowledge of the world gives us an impression of the object. This idea can be extended so that the descriptor is relative to something other than a human, but that would be familiar to most people. For example, rather than stating 'that car is fast', which may imply fast compared to a human, the statement could be made 'that is a fast car' implying that the car is fast, even compared to other cars.

There are different ways to describe an event to someone such that they can use the information in forming an opinion. The most basic form of reporting is to provide descriptive statements of fact. In the example of a fast car suppose the statement was that the car was travelling at 175 km/h. This statement gives the recipient the same base information and leaves any interpretation of that information to them. In a forensic sense, this is akin to making a statement that the components of two DNA profiles are the same or the physical properties of two traces (such as fibres, or paint flecks) are similar. For those that have knowledge of the field in which the information is being given, cars/DNA/fibres/paint, the significance of such a statement will be known. For those who are unfamiliar with these fields, the statements of a match may mean very little – maybe all fibres have the same properties, and so the information is providing nothing more than if it were just stated that fibres were present. On the other hand, perhaps there is so much variation in fibre construction that no two fibres from different garments are alike. There is in fact a real risk that the significance of matching traces is overstated in the minds of non-scientists if not properly explained and for this reason, it is generally agreed that more than a simple statement of the observations should accompany any forensic analysis.

To address a lack of contextual knowledge, additional information can be given about some properties of the world that adds context. A frequency-based statement is a step up from a descriptive result. In the car example, in order to impress upon the receiver of information that the car being described (that was seen travelling at 175 km/h) is fast, a frequency-based statement like 'only 1 in 1000 cars can travel at 175 km/h' can be provided. In a forensic context, it may be stated that the frequency of seeing a fibre at random in the population is 1 in 200 or that 1 in 1 billion people would possess the same DNA profile. These frequency-based statements can assist people, unfamiliar with the significance of the results being discussed, form a belief about their evidential value. However, even providing a frequency-based statement can be easily misunderstood, when the significance of that frequency is not placed into the context of the question being asked.

This idea of considering the probability of one event in comparison to another is encapsulated by the odds form of Bayesian inference. Take the formula for Bayes' theorem from the previous section:

$$\Pr(B \mid A) = \frac{\Pr(A \mid B)\Pr(B)}{\Pr(A)}$$

It would be equally valid to state the posterior probability that B did not occur given A is given by:

$$\Pr(\overline{B} \mid A) = \frac{\Pr(A \mid \overline{B})\Pr(\overline{B})}{\Pr(A)}$$

and noting that $\Pr(\overline{B} \mid A) = 1 - \Pr(B \mid A)$. By taking the ratio of the terms of these equations and noting that $\Pr(A)$ cancels out on the right-hand side, the result is:

$$\frac{\Pr(B \mid A)}{\Pr(\overline{B} \mid A)} = \frac{\Pr(A \mid B)\Pr(B)}{\Pr(A \mid \overline{B})\Pr(\overline{B})}$$

This is the classic odds form of Bayes' theorem and is the cornerstone of evaluations used throughout this book. Now some terms can be defined that apply the formula to a forensic context. Let 'B' be a proposition or hypothesis put forward by a prosecutor, and named H_p. The negation of this is the proposition being put forward by the defence, which is denoted as H_d. There are different forms that these propositions typically take in forensic literature, and often the pair H_1 and H_2 are used to denote a more impartial stance (i.e. one that does not presume to know the propositions of the parties). Particularly when the matter is not criminal (such as civil paternity) the use of H_1 and H_2 may be more appropriate. In other instances, H_p is used for the prosecution proposition as it is common that the forensic work will have been commissioned by the police or prosecution on behalf of the state and their allegations are known. However, the defence are under no obligation to provide a proposition, and so as not to imply that the proposition chosen by the scientist is that of the defendant, the alternative to the prosecution proposition is referred to as H_a ('a' for alternative). This book uses H_1 and H_2 simply by convention when speaking of propositions in an abstract sense, and H_p and H_d if specifically assigning the propositions to the positions of prosecution and defence in an evaluation.

Next, define 'A' as some observations that have been obtained through forensic testing and call these 'E' (for evidence). The odds form of the Bayes formula is then:

$$\frac{\Pr(H_1 \mid E)}{\Pr(H_2 \mid E)} = \frac{\Pr(E \mid H_1)}{\Pr(E \mid H_2)} \times \frac{\Pr(H_1)}{\Pr(H_2)}$$

This formula takes on the general form of:

Posterior odds = likelihood ratio x prior odds

In the context of a criminal trial, the jury (or Judge) starts the trial, before hearing any testimony, with their prior probability of the person (before they became a defendant) being guilty. This is represented by the $\dfrac{\Pr(H_1)}{\Pr(H_2)}$ term in the equation above. At the beginning of the trial, to be consistent with the presumption of innocence, this probability should rationally reflect their belief that the person standing accused is no more likely guilty than any other person in the population. Note that there is much discussion about what should be considered as 'the population', but that discussion is deferred until the next section. As each piece of information is provided to the Judge and/or jurors (possibly in the form of testimony of lay witnesses, or evaluation or opinions regarding scientific observations from experts) they will update their beliefs in guilt or innocence. In a strictly Bayesian framework, this information is being provided to the jury, who consider how likely it is in the context of guilt or innocence, which is the likelihood ratio (LR), equating to the term $\dfrac{\Pr(E \mid H_1)}{\Pr(E \mid H_2)}$. The jurors then use that information, along with their prior belief to form a posterior belief in guilt or innocence, represented by the term $\dfrac{\Pr(H_1 \mid E)}{\Pr(H_2 \mid E)}$. These posterior odds then become the juror's new prior odds to be used and updated with the next round of information. Ultimately at the end of

the trial, when all information has been passed to the jury, they form their ultimate posterior odds on guilt and if this exceeds some internal amount that equates to 'beyond reasonable doubt' then a guilty verdict will be returned. The description here is a very analytical view of the workings of the court and a jury. In reality, it is unlikely that a jury will be consciously adjusting their beliefs in guilt or innocence with each piece of information, and almost certainly this will not be done in any sort of a numerical sense. There are also likely to be many dependencies within the separate pieces of information that mean that even if assessed numerically, their combined effect could not simply be obtained by multiplication of the parts (an error that was famously made in the case of R v Collins [5]). There is some evidence to suggest that jurors more typically take in all information and consider it all together at the end, during deliberations.

In writing the LR as $\dfrac{\Pr(E \mid H_1)}{\Pr(E \mid H_2)}$, the hypothesis of the prosecution is placed in the numerator. This is by convention only and it is equally acceptable, and equally an LR, if this were inverted so

that the defence proposition was in the numerator and the prosecution proposition in the denominator. In fact, the inversion of the LR with a 'conventional' proposition assignment is often reported when the value of the LR assigned supports the defence proposition over the prosecution proposition (as will be discussed shortly). Throughout this book, the LR is formulated with the propositions assigned in this conventional way, unless otherwise stated.

If the probability of the observations is higher when considering the prosecution proposition than when considering the defence proposition the LR will be greater than one. This can be described as an LR that supports the prosecution proposition over the defence proposition. It will occur when the observations are better explained by the prosecution proposition than the defence proposition. The extreme of this is when the observations cannot be explained by the defence proposition (but can be explained by the prosecution proposition), in which case the LR is infinite in size, and it can be said that the prosecution proposition must be true (at least within the context of considering only these two propositions).

If the observations are explained equally well given either proposition (i.e. the probability of the observations is the same in either case) then the LR will be one. This can be thought of as neutral or uninformative, although care must be given when using these words, so as they are understood to apply to the immediate evaluation and not mistaken as applying to the case. For example, $LR = 1$ means that the observations may be uninformative in the quest to discriminate between the propositions, but at a case level, this result may be extremely useful to the judge or jury in deciding on the case.

If the observations are more probable given the defence proposition than given the prosecution proposition, then the LR will be less than one. This can be described as an LR that supports the defence proposition over the prosecution proposition and will occur when the observations are better explained by the defence proposition than the prosecution proposition. The extreme in this regard is when the observations cannot be explained by the prosecution proposition in which case the LR is 0 and it can be said that the defence proposition must be true (again, at least within the context of considering only these two propositions).

From this explanation, it can be seen that the LR can take any value from 0 to infinity (as is true of a ratio of any two non-negative numbers). Sometimes the phrase 'negative LR' will be used to describe an *LR* that supports the defence proposition over the prosecution proposition. This term comes from considering the *LR* on a logarithmic scale (see Box 1.1).

BOX 1.1   EXPRESSING *LR*S ON A LOGARITHMIC SCALE

A logarithmic scale is one which scales numbers by a power series. The most common logarithmic scale for transforming *LR*s is a base-10 scale; that is the logarithmic number represents the power to which 10 is raised in order to achieve the desired value. This is represented as $\log_{10}(x)$, meaning the log base-10 of x. In a base-10 scale, whole positive numbers represent the number of zeros behind a one for its equivalent on a natural scale. Whole negative numbers represent the one more than the number of zeros between a one and the decimal place for its equivalent on a natural scale. For example:

$\log_{10}(1000) = 3$

$\log_{10}(10) = 1$

$\log_{10}(1) = 0$

$\log_{10}(0.1) = -1$

$\log_{10}(0.001) = -3$

Log base-10 scales are used for a couple of reasons. First, a log scale represents a symmetrical scale with respect to any number and neutrality. For example, 3 is as far away from 0 (0 being the equivalent value for neutral on a $\log_{10}(LR)$ scale) as is −3, and 3 provides as much support for the prosecution proposition over the defence proposition as −3 provides support for the defence over the prosecution proposition. Contrast this to the natural scale equivalent noting that 1000 is much further way from 1 (1 being neutral on a natural scale) than 0.001.

The second reason that *LR*s are commonly considered on a logarithmic scale is to do with the representation of the *LR* as a 'weight' of evidence. Think of the classic image of Lady Justice holding her scales and that evidence is added to either side so that the scales eventually tip to innocence or guilt. The *LR* on a natural scale acts as an updater of belief in a multiplicative manner, i.e. if the results of an evaluation are given for which *LR* = 2 in favour of the prosecution proposition over the defence proposition, then prior beliefs are updated (multiplied) by a factor of 2. A multiplication on a natural scale, however, is an addition on a logarithmic scale, and so instead of multiplying by a factor of 2 on a log scale, this could be considered as adding $\log_{10}(2)$. Hence, the information is truly then adding some weight to one of the pans of the scales of justice.

A derivation of the *LR* has been provided, starting from the laws of probability, and it has been explained how the *LR* can be used to evaluate observations for presentation in court. But it has not yet been explained why the *LR* should be used as a measure of the strength of evidence in place of any other kind of value. It was shown by Good in the 1980s [6] and 1990s [7] that the evaluation of observations (or more precisely the evaluation of uncertainty regarding the propositions) will naturally lead to the use of a likelihood ratio if reasonable assumptions are made, and further that the *LR* is the only reasonable means by which to do so. Much of the foundational thinking for this work can be credited to Alan Turing's work in the 1940s. The terminology and general argument construction of Aitken [3] will be used in this proof. So far, the situation has been considered where there are observations (defined by the letter '*E*') and two explanations for these observations, which are called propositions, H_1 and H_2. For simplicity, the formulaic proof will not include any term for background information but note that it is ever present in any evaluation. Define the value of the evidence as 'v' and note that it must be some function of E, H_1 and H_2, i.e. the value of the evidence must be derived by some consideration of these elements, which is represented formulaically as $v = f(E, H_1, H_2)$. Making the assumption that the uncertainty in the observations can be expressed probabilistically and using the third law of probability 'v' can be expressed as $v = f(E \mid H_1, H_2) f(H_1, H_2)$. Note that no definition of the function has been given, other than the elements which it must consider. As the propositions are the province of the court and the second element in the equation for v only contains propositions, the proof can continue by concentrating on the component that involves the observations $f(E \mid H_1, H_2)$. The propositions are assumed to be mutually exclusive and exhaustive (i.e. exactly one of them has occurred). This means the value of the observations can be considered as a combination of two separate probabilities: one that involved H_1, defined as x, where $x = \Pr(E \mid H_1)$, and one that involves H_2, defined as y, where $y = \Pr(E \mid H_2)$. The value of the observations can be expressed as a function of these two terms, $v = f(x, y)$.

Now consider an irrelevant piece of information, T. Consider a simple example where a DNA profile obtained at a crime scene matches the reference DNA profile of a person of interest (POI). In this scenario then E is the matching profiles, H_1 is that the POI is the source of the crime scene DNA and H_2 is that someone unrelated to the POI is the source of the crime scene DNA. The irrelevant information could be any one of an infinite number of considerations but as an example define the term T as the fact that the crime was committed on Tuesday. To consider how the irrelevant information affects the probability of the observations, note that:

$$\Pr(E, T \mid H_1) = \Pr(E \mid T, H_1) \Pr(T \mid H_1)$$

By definition, an irrelevant piece of information does not have any dependency on the factor that it is irrelevant for, nor does that factor have any dependency on the irrelevant information itself. If T is truly irrelevant then it means it must be independent of E and H_1 so that:

$$\Pr(E \mid T, H_1) \Pr(T \mid H_1) = \Pr(E \mid H_1) \Pr(T) = x \Pr(T)$$

Going back to the simple matching DNA profile example from earlier, consider that the probability of obtaining a matching profile if the POI is the source of the crime scene DNA is 1, i.e. $x = 1$ and the chance of it being Tuesday is $\Pr(T) = 1/7$ and note now that the probability of the observations given the prosecution proposition and the irrelevant information, $x \Pr(T)$, has reduced by a factor of 7.

Carrying out the same task of introducing the irrelevant information into the probability of the observations considering the defence proposition, then the following is obtained:

$$\Pr(E, T \mid H_2) = \Pr(E \mid T, H_2) \Pr(T \mid H_2) = \Pr(E \mid H_2) \Pr(T) = y \Pr(T)$$

And so, the value of the observations given the defence proposition will also have been reduced by a factor of 7 by the introduction of the irrelevant information. Initially:

$$v = f(x, y)$$

and introducing the irrelevant information updates this to:

$$v = f(x \Pr(T), y \Pr(T))$$

But surely when considering the value of the observations, they should not be affected by irrelevant information. Therefore, whatever the function that is represented by 'f', it would be expected:

$$v = f(x,y) = f(x\Pr(T), y\Pr(T))$$

and that this must be the case whatever the value of Pr(T) takes within the range of 0 to 1. The only class of functions for which this can be the case is x/y, or in other words:

$$\frac{\Pr(E \mid H_1)}{\Pr(E \mid H_2)}$$

i.e. the likelihood ratio. Therefore, the value of the observations must be a function of the LR. Continuing the example whereby T is the fact that it is Tuesday, by expressing the results in LR form the change in probabilities from introducing Pr(T) is matched in the numerator and denominator and so cancel each other.

The LR has been spoken about very generally in this section, specifically its derivation and use as a measure for the value (on a natural scale) or weight (on a log scale) of evidence. The LR will be used exclusively as the tool for evaluating observations throughout this book, hence the amount of time spent introducing it.

There are many aspects of understanding to the LR, and these are particularly important in the logical formation of propositions and the evaluation of observations given those propositions. There are also logical extensions of the LR that flow into the principles of evidence evaluation, which are discussed in the following section.

1.4   PRINCIPLES OF EVALUATION

This section is started by again providing the odds form of the Bayes formula, including the background information:

$$\frac{\Pr(H_1 \mid E,I)}{\Pr(H_2 \mid E,I)} = \frac{\Pr(E \mid H_1,I)}{\Pr(E \mid H_2,I)} \times \frac{\Pr(H_1 \mid I)}{\Pr(H_2 \mid I)}$$

Note that the difference between this formula and the formula from the previous section is the inclusion of an 'I' term, which represents background information. Throughout the examples given for which an evaluation of observations is required, or would be useful, there is always a set of circumstances surrounding the scenario. Even the simplest of probability assignments, say for example the probability of rolling a 6 on a die, brings with it a series of contextual information. In the case of very common-knowledge examples (such as rolling a die), it is assumed that this contextual information aligns with the normative set of background information. However, it is always possible to explicitly state the circumstances or assumptions that are required in order to assign this probability. For example, the assumption is (unless told otherwise) that it is a standard 6-sided die, with the numbers 1 to 6 present on the faces, that the die is evenly balanced, that the surfaces have not been coated with any adhesive substance and that it has not been 'rolled' in such a way that would dictate the outcome (for example, dropping the die with the six facing upwards from 1 mm above a surface). Having (or assuming) this information then allows an assignment of a probability of 1/6 for rolling a six. It is making these assumptions that allows a probability assignment that would appropriately describe the uncertainty of rolling a six.

Another example, which demonstrates the importance of the framework of circumstances surrounding an event, uses information provided sequentially. If I were to ask 'How tall is Clara?', you may think of the average height of a woman and provide an answer in the vicinity of 160 cm. I then provide you with the information (representing a small part of the framework of circumstances surrounding Clara) that Clara is 1 year old. Now having this information, you realise that your assumption that Clara was a grown woman was incorrect, you would likely revise your estimate to something in the vicinity of 65 cm. I now provide the second and final piece of information, which is that Clara is a pet mouse. With the full set of background information, it can be seen that the first estimate of Clara's height was not at all representative of the true height of Clara, which is likely to be only a few centimetres. It is worth noting that none of the estimates, made at any stage of the information being revealed were incorrect, given the assumptions about the framework of circumstances. The estimates acted on the available information and made assumptions about the most likely set of circumstances that would apply. The assumptions may have been incorrect, and as more information was provided, the question when from very general (the height of anyone or anything that could be named Clara) to very specific (the height of a one-year-old mouse names Clara). It is the process of updating beliefs when more information is obtained that sits at the heart of Bayesian evaluation. This small demonstration also shows the importance of stating assumptions whenever providing the results of an evaluation.

The larger message relates to one of the three principles of evidence evaluation as outlined by Evett and Weir [8].

First principle of evidence evaluation: All observations should be evaluated within a framework of circumstances

This concept is commonly represented in the formula for Bayesian inference by the background information term, '*I*' and also includes the knowledge, experience, beliefs and reasoning of the individual carrying out the evaluation. Evaluations make use of observations that must consider multiple elements of uncertainty surrounding the events. It is therefore very important to the recipient of the evaluation's results that the framework of circumstances surrounding the event (as they are understood when carrying out the evaluation) and any assumptions that are being made are clearly stated. The full framework of circumstances may be broken into multiple smaller frameworks, that apply to different aspects of the evaluation. For example, consider the case of a robbery where a shoe-mark has been left at the scene, and a suspect who possesses shoes that may be the source of the shoe-mark is arrested. There will be a framework of circumstances surrounding the shoe-mark at the scene which will have information about where the shoe-mark was found and under what conditions. There will also be a framework of circumstances surrounding the suspect, and where they were or what shoes they own. Finally, there will be a framework of circumstances surrounding the shoes of the suspect themselves, and who has worn them, when and in what manner.

A commonly used example that demonstrates the ease at which frequency-based statements can be misunderstood is to consider a physician who is tasked with uncovering whether a child has been abused. The physician tells the investigator that the child is rocking back and forth, which is a descriptive statement of behaviour. This by itself does not assist the investigator as to whether the child has been abused, as they are not familiar with child psychology, and so the physician tells the investigator that 7 out of 10 abused children rock back and forth like this. The investigator, having been told this frequency-based information, decides that this behaviour means that this child has probably been abused. However, the investigator is still missing a key piece of information that will make the observation of rocking back and forth useful to answer their question – the rate at which non-abused children rock back and forth like this. If the answer to this question is that 1 in 100 non-abused children behave like this then there is some support for the child having been abused. If, however, the physician states that 7 out of 10 non-abused children behave like this then the finding means nothing in the context of determining if the child has been abused (one would then perhaps ask why it was even mentioned in the first place). In fact, the physician could state that 9 out of 10 non-abused children rock back and forth, in which case observing this behaviour actually supports the child not having been abused. The important point is that the single frequency could not give the support that the observation provides for one or the other proposition.

The second principle of evidence evaluation is demonstrated nicely with this example of the rocking child. In this example, while a probability could be assigned to a child rocking if they had been abused, it was not a useful piece of information by itself. In fact, worse than that, the probability when provided alone, had the potential to mislead. It was not until this probability was then contrasted to the probability of a non-abused child rocking that its use to the authority could be seen in their enquiry as to whether the child had been abused. It tells us that the value of any observation in support of a proposition can only be known when seen in the light of at least one alternative. An example of this effect in forensic science is to consider the value of two matching DNA profiles: one obtained from a swab of the handle of a murder weapon and the other obtained from a reference sample of a suspect. Given the rarity of DNA profiles, this finding may seem to provide quite strong support to the guilt of the suspect in the minds of the jury. However, the alternative must be considered. If the suspect claims that they have never seen the murder weapon, never had contact with the murder victim and were not responsible for the murder, then the finding of a profile matching the suspect on the murder weapon seems improbable. However, if the suspect claims that the weapon belonged to them but had recently been stolen and must have been used by another individual in the commission of the crime, then the finding of the suspect's DNA on the weapon is perhaps not so improbable. In this latter scenario, the finding of a DNA profile matching the suspect on the weapon would not provide as much support towards guilt when compared to the first circumstances. A scenario could also be considered where the suspect claimed that his evil identical twin brother was the one who committed the crime. As standard STR DNA profiling cannot distinguish between identical twins (evil or otherwise), the probability of the DNA observations would be equal if either the suspect or his identical twin was the DNA donor and so would provide no support for guilt. These examples lead to the second principle of evidence evaluation.

Second principle of evidence evaluation: Observations should be evaluated with respect to at least two competing propositions

Again, this principle is reflected in the formula for Bayesian inference given at the beginning of the section, by the fact that the *LR* is a ratio of the probabilities of the observations given competing propositions. Details on how propositions should be set and the properties of good propositions are explored in Chapter 4.

This leads to the third principle of evidence evaluation, which regards the ability of the person carrying out the evaluation to comment on the various parts that make up the *LR*. A well-known example that introduces this topic is given in the book *Interpreting DNA Evidence* by Evett and Weir [8]. Imagine a person is interested in whether a particular animal is an elephant. The only information available to the person is that the animal has four legs, and so they wish to consider the probability of an animal having four legs if it is an elephant. Define the observation of four legs as '*E*' and the proposition that the animal is an elephant as '*H*', then formulaically the person seeks $\Pr(E|H)$. In formal words this is the probability of *E*, conditioned on the proposition *H*. The person, having basic knowledge of elephants, considers that the probability of seeing four legs if an elephant has been observed is close to 1 (barring any injured or mutant specimens). Consider what would occur if the person transposed the *E* and the *H* in their consideration, giving $\Pr(H|E)$, an activity known as transposing the conditional. When being done unwittingly, it is referred to as the fallacy of the transposed conditional). Consider the probabilistic statement, the probability of an animal being an elephant if all that is known is that the animal has four legs. It can be seen now that this probability is much less than 1, as there are many four-legged animals other than elephants. In fact, in order to assign a probability to whether or not the person was viewing an elephant given the animal has four legs would require much additional information, such as where in the world the person was, what the terrain is like and what else they have observed (note that if this information was absent, one could still fall back on general worldwide values). At the very least it can be seen that $\Pr(H|E) \neq \Pr(E|H)$. This is also true for forensic evaluations, albeit perhaps more difficult to intuitively 'see' the issue. For example, an evaluation may be provided for matching suspect reference and evidence profiles which states that the probability of observing this match is 1 million times higher if the suspect was the source of the DNA rather than someone else in the population. To fall prey to transposing the conditional in this instance would be to make the statement that the probability the suspect, rather than someone else from the population, was the source of the DNA was 1 million times higher given the profiles match. This of course is incorrect, as the probability that the suspect is the source of the DNA relies on much additional information about the framework of circumstances surrounding the suspect and the crime (including information such as motive, opportunity, character, eye witnesses, etc.) as covered in the third principle.

> ***Third principle of evidence evaluation: The role of the expert should be to consider the probability of the observations given the propositions and not the probability of the propositions themselves***

And again, this is encapsulated in the *LR* formula by conditional probabilities involved.

It can be seen that the three principles of evidence evaluation are all encapsulated in the formulaic representation of the *LR* as given at the beginning of this section. All thoughts on evaluation and the many refinements that are discussed throughout this book flow from these three principles and their application to real-world scenarios.

1.5   HIERARCHY OF PROPOSITIONS

The ultimate end to many forensic science evaluations is to present the observations to a court of law and to explain their meaning and significance within the framework of circumstances. The jury (or judge) will take on board the evaluation (in the form of an *LR*) that is provided by the scientist. In doing so the judge or jury update what they thought prior to hearing the scientific evaluation, to produce their posterior belief in guilt or innocence. There is of course no requirement for the judiciary to take the full weight of any *LR* provided by the scientist as their own [9]. They will take the *LR*, along with many other aspects of the evaluation not specifically incorporated by the scientist into the *LR* (such as the trustworthiness of the individual or organisation in which the work was done, the apparent quality and thoroughness of the work, etc.) to produce their own *LR*.

In most trials, the forensic evaluation will make up a small part of the overall case being heard by the judge or jury. The trial can be thought of as a pyramid, with innocence or guilt sitting at the pinnacle of the structure, being supported by a few blocks that represent the essential components that make up the elements of the alleged crime. Below those are rows of even more blocks that support the elements of the alleged crime. Below those, more and more blocks that represent pieces of information that are even more removed from the ultimate issue, but each contributing some small element to the overall structure of the case. Much like building a pyramid, these foundational pieces must be laid first, and be solid and reliable for any further layers to be built upon them.

When points are being addressed, or questions being asked, they can be done so at any point in the structure. Issues that are asked at a base level may be far removed from the ultimate issue but provide an important element of information that is required before anything more directly relevant can be considered. On the other hand, there may be issues that sit very close to the main issue (metaphorically close to the pinnacle of the pyramid). This pyramid view of a case reveals a hierarchy of issues that can be addressed, which has particular importance to the manner in which scientific observations are evaluated [10, 11]. In forensic science, this hierarchy has been defined into discrete levels:

Level 3: Offence
Level 2: Activity
Level 1: Source
Sub-level 1: Sub-source
Sub-level 2: Sub-sub-source

The higher the propositions considered in the evaluation are in the hierarchy, the more directly useful to the court it will be. This fact will become evident when moving through an example of an alleged rape case and considering how at different levels in the hierarchy the evaluations can proceed.

1.5.1   SUB-LEVEL 2: SUB-SUB-SOURCE

In explanation, it is easiest to start from the lowest level in the hierarchy and ascend. Consider a case of alleged rape and a body of biological forensic work being carried out. In this case, the police have taken a swab from the vagina of the complainant a day after the alleged rape. This was submitted to the forensic laboratory, along with a reference DNA sample from the complainant and a person of interest (POI), suspected of the alleged crime. A microscope slide was prepared from the intimate swab and revealed the presence of sperm. The swab was processed in the laboratory using a method that separates sperm from other cell types. The 'sperm fraction' produced a mixed DNA profile from the POI and the complainant (note that a mixture in these circumstances is not unusual as the cell type separation techniques rarely achieve 100% separation). The peaks in the DNA profile appeared to fall into two distinct height groupings (representing different DNA donation amounts by the two contributors). The larger peaks were the same as those in the reference of the POI, and the smaller peaks were the same as those in the reference of the complainant. Given this scenario, it is now possible to demonstrate the issues and evaluations that can occur at the different levels in the hierarchy of issues.

Initially, an analyst may consider just the major component of the mixed evidence profile and in doing so obtain the observation that the major component matches the reference of the POI. The word 'match' is used here due to its common usage to mean that the components (called alleles) at each region examined (called loci, or locus for singular) are the same between the interpreted major component and the reference DNA profile. Note that the word 'match' has become somewhat of an awkward term. The advent of systems that can analyse mixed DNA profiles, whereby uncertainty in the peak pairing means that the DNA donor may possess any of a number of possible alleles, and so do not really 'match'. Additionally, a pre-evaluation step of determining if something matches is generally not required when dealing with observations probabilistically (as is more often the case as time goes on). Regardless, the word 'match' is used here for simplicity, but it will not feature much throughout this book. The analyst may consider the issue of who is the donor of this major component of DNA. They could assist the court in addressing this issue by evaluating the probability of obtaining the matching profiles given two propositions:

- The POI is the source of the major component of the mixture
- A [*person*] is the source of the major component of the mixture

It has not been specified who the 'person' is, and it is not required to do so for the purposes of this explanation. Typically, the alternate source of DNA is considered to be a person unrelated to the POI or a relative of the POI of a specific type or someone else in the population who could be related or unrelated to the POI. Given the propositions above, the analyst will be able to assign a numerical value to the *LR*, which will be based on the rarity of the component that makes up the profile, and some properties of the population from which the POI and potential alternate offender originate (such as the level of inbreeding within the population). Note that the issue being addressed is quite far removed from the issue of whether the POI raped the victim. The set of propositions given above address a sub-sub-source level issue and hence are considered to be sub-sub-source level propositions (see [12] for a more comprehensive explanation of this level in the hierarchy). As DNA profiles become more complex, and there are no clearly distinguishable components, the concept of sub-sub-source evaluation becomes more abstract as the propositions must indicate a POI being the source of a specific (but unresolved component) of the profile. A jury being given the *LR* and having had the propositions outlined to them would have to make a number of leaps of inference in order to use the evaluation in their ultimate deliberations on guilt or innocence. For example, even if they accepted that the DNA was that of the POI, they would need to know how complex the mixture was and what sort of contribution the POI had made to it, which cellular source the DNA came from, what the presence of the DNA or cellular source meant with regard to the potential for the POI to have sexual intercourse with the complainant and whether any sexual intercourse was consensual.

1.5.2 Sub-Level 1: Sub-Source

Both sub-levels 2 and 1 tend to only apply to DNA, and there is often no analogous subset of information in other trace types. In the sub-sub-source level propositions described above, the analyst carrying out the evaluation had to make a pre-assessment of the DNA profile in order to choose which component to compare to the reference profile of the POI. In this example as it is described, the task of choosing the component for evaluation is relatively obvious; however, there are many situations when such an assignment is not so obvious, or not appropriate. Thinking of this situation in terms of prior and posterior probabilities, there was no reason, a priori that the major component would be any more applicable to compare to the POI than the minor. Indeed, if the minor component had been the one that matched the POI then this is the component on which the *LR* would be based. This perhaps hints at a subtle bias, i.e. that an *LR* can be calculated for the comparison of the POI to each component of a mixed DNA profile and the maximum of those *LR*s given in the evaluation. Note that the maximum of a set of *LR*s is now no longer an *LR* representing the full set of available information (in the same way that if a set of values was broken into subsets and the average of the subsets calculated then the maximum of the subset averages is no longer an average with regard to the whole set). To overcome this issue, a step up the hierarchy could be taken to consider the issue of whether the POI was a donor of DNA to the sample as a whole (i.e. rather than just a component of it). This issue is now at the sub-source level and sub-source propositions can be set in the form:

- The POI is a contributor of DNA to the mixture
- The POI is not a contributor of DNA to the mixture

The above examples of propositions are a very general sub-source level pair, and there are a couple of points to be made on their construction. First, the second proposition is simply a negation of the first and does not provide much information about what is being suggested as having occurred. Providing an alternate proposition that is simply a negation of the first runs the risk of being vague, causing difficulty in the evaluation [13]. Both propositions are also vague in the sense that they do not specify how many contributors are being assigned to the mixture, or whether anyone can be assumed to have contributed. In this example, it would be common to assume the complainant is a contributor of DNA. In practice, propositions would be formed along the lines of:

- The complainant and POI are the sources of the DNA in the mixture
- The complainant and an unknown person are the sources of the DNA in the mixture

All discussions on appropriate proposition settings are left to Chapter 4.

An evaluation given sub-source level propositions will still require information on the same aspects as the evaluation given sub-sub-source level propositions (i.e. the rarity of the profile and properties of the population of potential contributors). It will also require information on the additional aspects of the number of loci in the DNA profile, the number of contributors (questioned or assumed) and the level of resolution that exists between them, i.e. how well the contributors in the mixture can be distinguished.

Note that the propositions at the sub-source level do not specify a cellular source of DNA, e.g. the nature of the cellular material from which the DNA was obtained. In this example, it may be that one of the components of the mixture has originated from sperm. In order to state this, either an assumption must be made or the possibility treated probabilistically. For an example of the latter see [13], but note that the work involves aspects of Bayesian networks which we will only get to later on in this book. It may be that the DNA has come from a contamination event. If this is the case the propositions given above are still as valid as when the DNA had come to be in the sample from another transfer event (contamination and transfer are explored in Chapter 3). It is possible to take small steps within the sub-source level propositions that still only speak to the source of DNA. If an assumption is made that contamination has not occurred, then the propositions can speak to the source of DNA in or on objects, rather than the DNA obtained from laboratory testing [14]. For example, if contamination is assumed not to have occurred during the time the sample was processed in the forensic laboratory, then propositions can be formed that question the source of DNA on a sample submitted to the forensic institution (rather than giving propositions that talk about the sample at the time it was analysed within the laboratory). Further, if contamination by police is also assumed not to have occurred then the sub-source level propositions can refer to the source of DNA on the item at the scene. Note though that it is generally not advisable to assume contamination has not occurred. A better treatment is to consider the probability that it may have occurred, but this takes us to activity level consideration, which is explored soon.

1.5.3   LEVEL 1: SOURCE

Most cells in the human body contain DNA. As mentioned in the section on the sub-source level, the issue did not extend to determining the cellular source of the DNA. Level 1 considers the source of the DNA which could originate from a biological fluid such as blood, semen or saliva, or from non-body-fluid sources (such as DNA deposited by touch, which is discussed more in Chapter 3). This may be of importance to a case, particularly if the presence of DNA is not in dispute. If the issue being addressed considers the cellular source of the DNA then it is a source level issue and the propositions that address this issue are source level propositions. There are different ways in which a source level issue can be thought about. It may be that the cellular source of the biological material is not in question, but the donor is, in which case the source level propositions may take the form:

- The POI is the source of sperm on the intimate swab from the complainant
- Someone else is the source of the sperm on the intimate swab from the complainant

In the situation where we would use source level propositions such as these, it is implicit in their construction that the fact that the DNA has come from sperm is not being disputed. Whatever testing had been conducted to identify the cellular source (such as microscopy or a test for acid phosphatase or prostate-specific antigen, which are both common in forensic laboratories when screening for semen) will have to be accepted as such, meaning that sperm is indeed present and that the male DNA detected has come from that sperm. The evaluation of observations given these source level propositions will depend on the same factors as the evaluation given sub-source level propositions, as all aspects of uncertainty about the source are removed by the construction of the propositions.

Alternatively, it could be that the source of the DNA was not in dispute, but the cellular source was. For example, it may be that the laboratory screening results were not strongly indicative of semen being present, or that the microscopy showed relatively few sperm. It may be that in an attempt to separate the sperm from other cellular material in the laboratory, there was additional male DNA detected in the non-sperm fraction and there is some question as to what the male DNA component was being detected within the sperm fraction. If these results are coupled with the fact that the complainant and POI had previous contact with each other and there was the possibility (or even an expectation) of the POI's DNA on the victim then propositions may be formed such as:

- The POI is the source of sperm on the intimate swab from the complainant
- The POI is the source of only non-sperm cellular material on the intimate swab from the complainant

Again, these propositions make the implicit assumption that the DNA of the POI is present and not in dispute.

The evaluation of observations given these source level propositions will depend not on intrinsic aspects of the DNA profile (as in the evaluation given sub-source level propositions) but rather on the extrinsic aspects such as the amount of DNA present and the level of degradation. It will also require consideration of additional screening tests, the result they gave (see [15] for an example of a source level evaluation on potential blood staining), the ability of the analysts to interpret those results (see [16] for examples of human ability to score sperm levels on microscope slides), the false positive and false negative rates of testing and the conditions under which they occur [17].

It may be that both the nature of the cellular source and the individual who contributed the material are at issue. Addressing this would require at least three propositions that would be evaluated in two sets [18]:

- H_1. The POI contributed semen in the sample
- H_2. The POI contributed cellular material other than semen in the sample
- H_3. The POI did not contribute any biological material in the sample

Given these propositions, H_1 vs H_2 would address the issue of the nature of the cellular material (implicitly assuming that the POI has contributed DNA), while H_1 vs H_3 (or H_2 vs H_3) would effectively address the issue of DNA contribution (regardless of the type of cellular material). If LRs were calculated using proposition pairs H_1 vs H_3 and H_2 vs H_3 then it would be seen that they were the same (both based on the rarity of the POI's profile in the population) and an inference could be made that cell type did not play a role in the LR, i.e. the question is whether or not the POI donated any cellular material, it does not matter which cell type.

It is not common to see evaluations of observations given source level propositions alone. Often the uncertainty surrounding the cellular source of DNA and the propositions which flow from the framework of circumstances are intertwined with the activity level considerations. It can be difficult (and often inappropriate) to untangle the aspects of activity level issues from source level issues and when there is uncertainty surrounding cellular source they will be incorporated into a larger evaluation of the observations given activity level propositions. Cell type testing will be explored in more detail in Chapter 9.

1.5.4 LEVEL 2: ACTIVITY

As just demonstrated in the source level section when forming the source level propositions either the cellular source or the actor is in dispute, which leads to two different categories of propositions. These flow naturally when either the activity or the actor is in dispute when considering activity level issues [19]. Activity level issues where the actor is disputed, using the running example would result in activity level propositions such as:

- The POI had sexual intercourse with the complaint
- Someone else had sexual intercourse with the complaint

Again, implicit in this type of a proposition set-up is the fact that someone has had sexual intercourse with the victim. The decision to concede this activity (which can only be made by the members of the court) does not depend on the results of forensic laboratory testing, i.e. the results of laboratory testing can be evaluated using these two propositions regardless of whether sperm was observed, or DNA obtained, or DNA matching the POI obtained. The activity level in the hierarchy of issues is the first level at which such a statement can be made. In previous levels, it would make no sense to carry out the evaluation with the propositions stated if the results had not aligned themselves with that particular construction. For example, if no sperm had been observed on the intimate swab from the complainant, then it makes no sense to formulate propositions that queried the source of sperm. Or if DNA had not been obtained in the sample, again it would make no sense to question the source of DNA. However, when considering activity level issues, the absence of biological material (which itself is still a finding) is of use in the evaluation (see [20] for an example of evaluating such results given activity level propositions). The fact that evaluation can still occur regardless of the laboratory observations, allows a pre-assessment of a case considering the observations that might be obtained, in order to decide how best to proceed (a process called Case Assessment and Interpretation, which is elaborated on later in this chapter).

An alternate construction of propositions will occur when the activity is in dispute. In the example of an alleged rape, it may be that the POI states that they lent the complainant their unwashed underwear, which had a fresh semen stain on it. In that case, the activity level propositions could take the form:

- The POI and complainant had sexual intercourse
- The complainant wore underwear from the POI, they did not have sexual intercourse

Note now that both propositions concede the presence of sperm from the POI but dispute the mechanism by which it came to be on the intimate swab from the complainant. In order to evaluate the observations given, these propositions require a knowledge of the rates of transfer of sperm from underwear into the vagina, the persistence of sperm in the vagina, the level of transfer and persistence of other seminal constituents targeted by forensic laboratory testing in underwear and the vagina, and the ability of the laboratory to recover these biological materials. There may also be observations from several items combined in order to carry out the assessment (i.e. in this case the underwear from the POI as well as the intimate swab from the complainant).

A further example of an activity level issue is provided, which at first appears to have conceded the activity. Imagine in the running example that the POI and complainant were a sexually active couple and that the complainant states she was raped by the POI. Meanwhile, the POI states that he did not rape the complainant but engaged in consensual sexual intercourse with her two nights before the alleged rape. It would appear that the activity of sexual intercourse is now being conceded by both parties. However, in this case, the timing of the activities is different between the two versions of events, and so technically there arc two different activities: sex on the night of the alleged offence and sex two nights before the alleged offence. In this instance, it becomes important to include a timeframe for the activities (in fact it is important to consider timeframes whenever issues of DNA persistence must be considered). The resulting activity level propositions would be:

- The POI and complainant had sexual intercourse on the night alleged by the complainant, and not two nights before
- The POI and complainant had sexual intercourse two nights before the night alleged by the complainant

In the evaluation of observations using these propositions, the persistence of seminal components in the vagina will be the most informative factor.

1.5.5   LEVEL 3: OFFENCE

In the metaphor of the pyramid used earlier, activity level issues sit very close to the tip of the structure, just below the peak. There can sometimes be a fine line distinguishing between the activities and the offence. For example, in a case of alleged assault swabs from the hands of the POI may reveal a stain that has properties of blood when tested and yields a DNA profile matching the complainant. The activity level proposition put forward by the prosecution could be that the POI punched the complainant. This would seem to address the ultimate issue in the case, but it must be remembered that an activity is not an offence. For any activity, there is always a possibility for an innocent explanation. For example, punching someone may be legal if the situation is a professionally organised fight, or if the punch came as a result of self-defence. The aspect which turns the activity issue into an offence issue is the context; i.e. in the case of the alleged assault if the POI carried out the activity unprovoked and with the intent to harm the complainant then it is an offence. In the case of the alleged rape, the offence level issue will require that both the activity of sexual intercourse has taken place and that consent was not given by the complainant to the POI. For the running example of an alleged rape, there are again multiple different types of offence level issues depending on whether it is the actor that is disputed, or the activity, or another variant considering the mental state of the actors. Dealing first with the offence level issue of a disputed actor, an example of offence level propositions is:

- The POI raped the complainant
- Someone else raped the complainant

Note again the implicit assumption that someone has raped the complainant. As with an evaluation that considers activity level propositions, an evaluation that considers offence level propositions makes sense regardless of the results that scientific examination yields.

Offence level propositions are the remit of the court and deal directly with the ultimate issue. However, at this point, it is worth asking whether an evaluation of the scientific observations could be done considering offence level propositions. It might surprise the reader that it would still be possible to provide an evaluation of scientific observations given offence level propositions without impinging on the role of the court. This is because the scientist would still only be commenting on the probability of the observations, and not on the propositions themselves. Evett [21] gives an example of evaluations given offence level propositions by incorporating aspects into the evaluation such as the number of offenders, the relevance of items and the potential for an innocent explanation (all aspects which are for the court to provide prior belief to). The issue with using offence level propositions to evaluate the observations is that the main difference between offence and activity level issues is a state of mind, which cannot be incorporated into an evaluation of scientific observations. This means that the evaluation given the offence level propositions cannot provide any different result from the evaluation given the activity level propositions. An example of this would be that the probability of finding sperm in the vagina after sexual intercourse where ejaculation has occurred is the same as the probability of finding sperm in the vagina after sexual intercourse where ejaculation has occurred when the sexual intercourse was not consensual. The risk lies in the fact that by stating the propositions at the offence level, it gives the impression to the recipients of the information that the issue of consent has somehow been incorporated into the evaluation in a meaningful way. In forensic genetics, it is unlikely that such issues could be addressed. There is an argument that in other disciplines evaluations may be able to consider issues such as consent, for example in forensic medicine the presence of physical harm or defensive wounds may be evaluated in light of offence level issues. By definition then the issue of consent is irrelevant (independent of the probability of obtaining the observations) to the evaluation of the forensic observations and so should be omitted from the proposition, hence taking it back to just the activity. This risk of misunderstanding what has been evaluated is unnecessary and the general convention when providing propositions is to state only the relevant information (that is, not to say that consent is not relevant to the offence but rather not relevant to the activity level evaluation). Hence scientists should be cautious to foray into evaluations that consider offence level propositions. Importantly, communication of results should indicate the limitations of the evaluation and make clear when a component of proposition that is being put to them is not meaningfully being contributed to in the evaluation that has been undertaken.

The second form of offence level issue is when the activity is in dispute which will lead to propositions that take the form:

- The POI raped the complainant
- The complainant was not raped (and did not engage in any sexual activity with the POI)

In this form, the offence itself is being disputed and naturally leads to the disputed activity when considering activity level issues. The third potential type of offence level issue relates to the state of mind aspect and would lead to propositions such as:

- The POI raped the complainant
- The POI and complainant had consensual sexual intercourse

Notice that there is no activity level issue in forensic genetics that this offence level issue leads to as it uniquely applies to an offence that does not imply a materially different activity.

A diagrammatic representation of the hierarchy of issues is shown in Figure 1.3.

Note that the hierarchy exists for the interpretation of all forms of forensic observations. Its use has been demonstrated in a forensic biology example, and the reason for this is that the lower levels in the hierarchy, i.e. the sub-source and the sub-sub-source level issues, only really apply to DNA observations. However, the source, activity and offence level issues can be useful when interpreting other forms of trace observations such as fibres, glass fragments or paint. Similarly, but subtly different, terminologies such as 'cause of death' and 'manner of death' (which are being used in forensic pathology) can be considered analogous to source level and activity level issues, respectively. For some disciplines, the source and activity level may appear virtually indistinguishable. For example, if a fingermark is found on an item, then the source level issue of whose fingerprint that is and the activity issue of who touched the item cannot be separated (except perhaps if considering the rather far-fetched possibility of fingerprints somehow being manufactured and planted at a scene). However, the manner or timing of contact may still be in dispute, allowing us to evaluate observations from fingermark comparisons given activity level propositions (given factors like direction and location of marks, as well as the part of friction ridge skin that left the mark) [22]. This is discussed in more detail in Chapter 12. Examples of the apparent hierarchy jump exist even within forensic genetics, for example finding DNA, associated with semen, on an internal vaginal swab taken from a minor. In this instance, the observations at the sub-source level are highly informative on issues right up to the offence.

As can hopefully be seen from the explanation of the hierarchy of issues and propositions, the higher up the hierarchy that the propositions can be formed for the evaluation of the observations, the more directly useful to the court the evaluation will be. The scientists, as a general rule, best limit themselves to considerations of activity level, not because an evaluation cannot be performed considering offence level propositions, but because of the risk of the court misunderstanding what has been used in the evaluation. Note that, as with every evaluation, all assumptions and limitations need to be clearly communicated to the court. So too with an exceptional evaluation at the offence level when this is considered the best approach in a case. As the scientists consider propositions moving up the hierarchy they must consider many additional elements of uncertainty, such as the ability of biological material to transfer to items, the ability for it to persist on an item for a length of time or through various events, the level of DNA expected to be found on items and the ability to

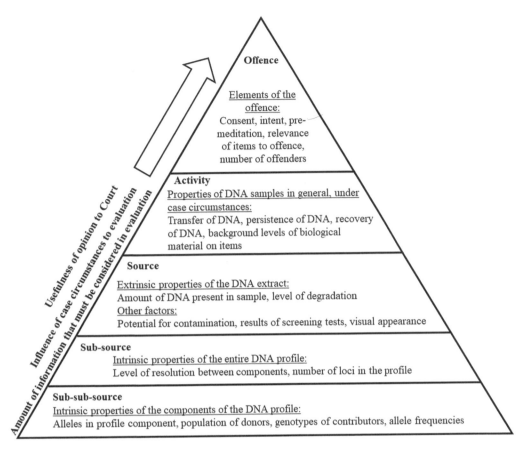

FIGURE 1.3 Pyramid showing the levels in the hierarchy of propositions and the components requiring consideration when evaluating observations given propositions at those levels.

recover DNA from an item. These factors will be discussed in more detail in Chapter 3. Considering these factors in the evaluation takes that burden from the court and places it with the scientist. This is where this responsibility should lie, as it is the scientist, an expert in their field, who will have the knowledge and experience (or access to the knowledge through published literature and validation studies) that is required to address these factors. This is the use and the power of evaluating observations given activity level propositions and the reason for the expansion of the field.

1.6 SCIENTIST WORKING AS INVESTIGATOR VERSUS EVALUATOR

Up until this point the evaluation process has been considered as occurring once the observations have been obtained and a case is destined for court. A court-going case is the very end of a long process and will only occur in a small percentage of criminal cases. There are many instances where the techniques and principles of evaluation can (and should) be used at a much earlier stage in the process. Typically, in forensic evaluation literature, the distinction between the early stages of the criminal investigation and the latter, court going, stage of the case is defined by the terms 'investigative' and 'evaluative' stages. These terms will be defined more formally soon, but first, it is worth spending some time on the different forms of inference: induction, deduction and abduction [23].

Deductive reasoning (sometimes called top-down logic) requires full knowledge and understanding of the world in which observations exist and utilises rules that can be applied to observations to make inferences. In an abstract example, picture a world where it is known that if *A* occurs then it will lead to observation *B*. If tests are carried out and observations *B* are seen then it can be deduced that *A* has occurred. The classic literary figure Sherlock Holmes was claimed to be a master of deduction for his ability to deduce what must have occurred based on a keen set of observations and his seemingly boundless knowledge and understanding of the world. Although in truth much of the reasoning undertaken by Holmes was in fact inductive, which is covered next. This type of deductive reasoning is limited in its use in forensic science and may be most successfully applied to closed-set problems (or classification problems). An example is given that is similar to that of Jackson *et al.* [24] who give a deductive statement 'if *A*, *B* and *C* are true then *X* must be true' and apply it to the closed-set problem of a murder committed. Consider that the murder must have been committed by one of five people (statement A), that DNA at the crime scene has come from the offender (statement B) and that all five people have different DNA profiles (statement C). Then it can be deduced that if one of the five people has a reference DNA profile generated and it matches the crime scene profile, that person is the murderer.

Of more common use in forensic science evaluations is inductive reasoning. In this type of reasoning, observations are used to extrapolate back and consider what might have given rise to them. Inductive reasoning is able to work within a framework in which there exists incomplete knowledge and uncertainty (unlike deductive reasoning which is exact). As might be expected from this terminology used to describe inductive reasoning, it works in a framework of probability. Bayesian inference is a method for the application of inductive reasoning, by which prior beliefs in some propositions are updated by observations in order to obtain posterior beliefs (or uncertainty) on what might have occurred. Take the example from earlier and remove condition C (that all individuals have different DNA profiles). It may be the situation that only one member of the group is profiled and matches the crime scene profile. This person may be the offender, or it may be one of the other individuals, as they may have a matching profile. Note the two possible inductive outcomes, which would be assigned probabilities based on the priors and updated with the DNA profiling results.

Abductive reasoning leaves some doubt as to the cause of an observation. Deductive reasoning says that because only A leads to B therefore observing B means that A must have occurred. Abductive reasoning states that A leads to B, but so to might other events, therefore observing B might mean that A has occurred. In forensic science, abductive reasoning can be thought of as listing out a number of possible explanations for a set of observations and working through these to find the most likely.

These types of inference can be used at different points throughout a case. Abduction is used to generate the propositions from the case circumstances, deduction is excluding some of those propositions given the observations and induction is used when finding evidence for or against the propositions [25]. The method of inference that has mostly been spoken about so far employs inductive reasoning and would be considered in the evaluative stage of a case. When the scientist is acting as an evaluator, they have two propositions in mind and are evaluating the probability of a set of observations given those two competing propositions. This is known as the scientists being in 'evaluation mode'. As much relevant information as possible is obtained in order to carry out the evaluation and any assumptions that are made in order for the evaluation to progress must be thought-out, justifiable and fully explained. The hierarchy of issues when in evaluation mode is proposition-centred, i.e.:

Sub-source: Was the POI's DNA in the sample?
Source: Did the POI leave the biological material?
Activity: Did the POI carry out the activity (or which of the two activities were carried out by the POI)?
Offence: Did the POI commit the offence?

The propositions reflect these issues in their construction. Note that it may not be that two clear versions of events exist for evaluation. The defendant may not have provided any information on what they are stating as having occurred, or there may be information that is simply not available, or not known. In these instances, it will be up to the scientist to assign justifiable propositions (which will be formed by trying to predict the most relevant issues to the legal parties), and this is discussed in Chapter 4.

In the early stages of a case, there may be little information available and there may not be a POI. This stage of a case falls into the investigative stage, and the role of the scientist carrying out an evaluation is different to that at the evaluative stage. This is known as the scientists being in 'investigation mode'. In the investigative stage, the focus is observations-centred rather than POI-centred. In contrast to the issues that were faced at the evaluative stage, the issues now will be about identifying what has occurred:

Sub-source: Whose DNA is in the sample?
Source: What is the biological material present on the item/scene?
Activity: What were the activities that could have occurred?
Offence: What offence has occurred (or has any offence occurred)?

From these issues, it can be seen that the use of abductive reasoning or deductive reasoning may be useful at this stage of the case. The analyst or the police may list out a number of possible explanations that could account for what has been observed. This list is likely to be open-ended and not every explanation possible will be considered, many will be so unlikely (in statistical terms the prior probability of their occurrence is very low) that they are discounted and dropped from the plausible list. The list of remaining potential explanations then may direct further scientific testing, which leads to further explanation list refinement until some useful investigative information can be provided.

An example of an investigative phase of a forensic DNA investigation is in a case where no POI is identified and a DNA profile (that has a reasonable prior probability of being from the offender and hence is relevant to the case) has been obtained. Technically one could consider the list of possible sub-source explanations as each person on the database being the source of DNA. During the investigative stage, the profile will then be searched against the DNA profile database to (hopefully) identify a matching DNA profile. At this stage, the information is passed to the police and it may be, after they have investigated the potential offender, that the POI is arrested. After obtaining a statement from them and further information is obtained from the investigation about the framework of circumstances surrounding the POI, a more complete picture can lead to the scientist moving to an evaluative stage. Now the DNA profiles in the case are evaluated with much more defined propositions being considered and potentially at a higher level in the hierarchy of propositions than the original sub-source propositions being used in the investigative phase.

1.7 CASE ASSESSMENT AND INTERPRETATION

As has been mentioned in previous sections, when propositions reach the level of activities in the hierarchy of issues, they are able to be meaningfully and sensibly used to evaluate observations, regardless of what those observations are. From the previous section it was explained how the role of the forensics scientist as an investigator in the early phase of a case can lead to the listing of possible propositions for a set of observations. This idea can be extended further, to consider a range of possible observations that could be obtained if an item (or multiple items) is examined. For example, if a swab of a weapon handle was being considered, it may be that:

- No DNA is obtained from the swab (call this O_1, for Observation 1)
- DNA matching the POI is found (call this O_2)
- DNA from only an unknown person is found (call this O_3)
- DNA matching the POI and DNA from an unknown person is found (call this O_4)

Consider that the POI used the gun in an armed hold-up (call this H_1). Using this information, it is possible to assign probabilities to each of the potential DNA results listed above. This may be based on bespoke experiments, carried out with a view to match the elements of the propositions for the alleged crime as closely as possible. It may be by examining forensic literature and finding experiments carried out in other laboratories that are close in their experimental set-up to the features of the alleged crime. Alternatively, it may be based purely on personal experience. The spectrum of information sources described above demonstrates the range of data that can be used in assigning probabilities, ranging from purely empirical to purely intuitive. Later chapters of this book will go into detail on how probabilities can be assigned. For now, accept that probabilities can be assigned to obtaining each of the above observations given the scenario being considered. Suppose they are assigned the following values:

$$Pr(O_1 \mid H_1) = 0.2$$

$$Pr(O_2 \mid H_1) = 0.3$$

$$Pr(O_3 \mid H_1) = 0.1$$

$$Pr(O_4 \mid H_1) = 0.4$$

where the terms above are conditional probabilities, and for example, the equality $Pr(O_1 \mid H_1) = 0.2$ means the probability of observing no DNA if the POI has handled the gun during the alleged crime is 0.2. With the observations O_3 and O_4, consideration must be given to the fact that the unknown has come from background DNA. This consideration is encompassed within the probability assigned (the process of which will be explained in Chapter 7). To complete an evaluation, consider a competing proposition to create an exhaustive (at least within the context of the case) and mutually exclusive pair. Consider for this example the proposition, H_2, that the POI has had no contact with the firearm, directly or indirectly (for example, perhaps the POI claims to have been out of the city for some time leading up to and including the time of the offence). In this case an alternate offender (AO) must have handled the stolen firearm during the offence. Then it must be considered that the evidence profiles matching the POI must be because the AO happens to have the same DNA profile (note we are excluding any other forms of transfer of DNA from the POI to the gun). For the purpose of this example the probability of an AO having the same DNA profile as the POI is assigned as 1 in 1 million.

Again, it is possible to consider the probabilities of obtaining the four sets of observations given H_2. It can be seen that the probabilities assigned to observations including a profile that matches the POI have a very low probability, as they will have had to take into account the chance of matching profiles.

$$Pr(O_1 \mid H_2) = 0.2$$

$$Pr(O_2 \mid H_2) = 0.0000025$$

$$Pr(O_3 \mid H_2) = 0.79999$$

$$Pr(O_4 \mid H_2) = 0.0000075$$

Having now assigned probabilities for obtaining the four posited observations under two competing propositions, it is a simple matter to calculate *LR*s, by diving the two paired probabilities by each other. Doing this yields:

TABLE 1.1

Results of Considering the Probability of the Observations Given Each Proposition and the Resulting *LR*

Observation (O)	$\Pr(O \mid H_1)$	$\Pr(O \mid H_2)$	*LR*
1	0.2	0.2	1
2	0.3	0.0000025	120 000
3	0.1	0.79999	0.125
4	0.4	0.0000075	53 333

$$LR_1 = \frac{\Pr(O_1 \mid H_1)}{\Pr(O_1 \mid H_2)} = \frac{0.2}{0.2} = 1$$

$$LR_2 = \frac{\Pr(O_2 \mid H_1)}{\Pr(O_2 \mid H_2)} = \frac{0.3}{0.0000025} = 120000$$

$$LR_3 = \frac{\Pr(O_3 \mid H_1)}{\Pr(O_3 \mid H_2)} = \frac{0.1}{0.79999} = 0.125$$

$$LR_4 = \frac{\Pr(O_4 \mid H_1)}{\Pr(O_4 \mid H_2)} = \frac{0.4}{0.0000075} = 53333$$

Collectively the probabilities of the observations given each proposition and the resulting *LR* can be tabulated as seen in Table 1.1.

Having gone through the exercise of considering the different DNA profiling results from the firearm used in the offence, the probability of obtaining those observations given two competing propositions and the resulting *LR*, it is possible to provide some guidance as to whether forensic DNA work is justified in this case (justified in the sense that carrying out the work has a reasonable possibility of providing observations that can discriminate between the competing propositions). Imagine now that a police officer came to the forensic science laboratory and presented the framework of circumstances surrounding the case as the armed hold-up described above. They wanted to know whether it was worth submitting a swab of the firearm for analysis. Given the work carried out in this section it would be possible to provide the following advice:

If the POI did handle the firearm, then:

There is a 20% chance that the results will be neutral with respect to whether it was the POI or an AO who handled the firearm (obtained from the values of $\Pr(O_1 \mid H_1) = 0.2$ and $LR_1 = 1$).

- There is approximately a 70% chance that the result will very strongly support the POI having handled the firearm rather than an AO. The term here 'very strong' is used here to reflect a numerical *LR* value above 1000. Using a verbal descriptor for the numerical value of the *LR* is a common practice and aspects of reporting the outcome of the evaluation in this way are discussed in Chapter 10. For now, accept the descriptors used in this example as they are given.
- There is approximately a 10% chance that the results will slightly support an AO having handled the firearm rather than the POI.

If the AO did not handle the firearm, then:

- There is a 20% chance that the results will be neutral with respect to whether it was the POI or an AO who handled the firearm
- There is a 0.0001% chance that the result will very strongly support the POI having handled the firearm rather than an AO
- There is an approximately 80% chance that the results will slightly support an AO having handled the firearm rather than the POI

This exercise demonstrates that a defence for an innocent POI will, on average, benefit from the examination. A defence with a guilty client will, on average, not benefit from any effort to find the truth.

Given this assessment, a decision can be made as to whether the swabs will be submitted. For this, the information on the resources required (time, cost) will be important. Also, aspects such as the seriousness of the crime, other cases that are competing to be completed in a timeframe, may be important.

The process of carrying out the evaluation of an exhaustive set of possible outcomes given the framework of circumstances of the case is known as the case pre-assessment stage of 'Case Assessment and Interpretation' (CAI). CAI was developed by the Forensic Science Service (FSS) in the UK during the late 1990s [26, 27]. It was developed during a period of the FSS financially accounting for the work they carried out, even to the point where stakeholders such as police were considered customers. In CAI, case pre-assessment sits between 'customer requirements' and 'service delivery'. The customer requirements phase determines the needs of the customer, ideally settling on a pair of propositions that can be considered in an evaluation. In the case pre-assessment stage (as was just shown) the scientist considered the observations that may be obtained and generates a probability distribution for the expected strength of the observations given the context. In the service delivery stage, laboratory work is undertaken and the formal evaluation is conducted. Today the value of the process of CAI is seen in determining whether there is value in pursuing work with the view of carrying out an assessment of observations given activity level propositions. It also assists in formulating the correct propositions and can assist in determining an item examination strategy. If case assessment is carried out before work begins on a case, then another advantage is the evaluation cannot be affected by the potential for results-lead thinking or biases (although often activity level evaluations in practice are usually carried out after source level work has already been completed).

One final note on CAI is that it depends on the architecture of the policing, forensic and legal framework. The formal process of conducting CAI is most useful in a structure where the evaluation is destined to be carried out considering activity level propositions from the start. The effectiveness of carrying out CAI is diminished when the forensic work is carried out considering sub-source level propositions first and then a decision on whether to evaluate considering activity level propositions is made afterwards. In effect it is likely that the decision to carry out the higher-level evaluation is done with an intuitive feeling that the observations have some power to discriminate between the propositions. It is the same factors that would be considered within CAI that should be intuitively guiding the choice of sampling at the scene and item triage, even without CAI formally being conducted. In other words, if the item has no power to distinguish between the potential activities that could have occurred, then there seems little point in sampling it. By training the decision-makers in the process of CAI, they may understand better how to reason with evidence and make better decisions when submitting items (reducing uninformative work) and understanding the meaning of observations when results are obtained. Training is discussed more in Chapter 11.

1.8 PRACTICE QUESTIONS

Q1) Which of the following sets of events are exhaustive? For those that are not exhaustive, what additional states could the event take?

A. [rolling a 6-sided dice and getting a 1, 2, 3, 4, 5 or 6]
B. [Days of the week ending in 'y']
C. [Months of the year ending in 'y']
D. [even or prime numbers on a 6-sided die]
E. [even or prime numbers on a 10-sided die]

Note that a prime number is any number which has exactly two numbers that it is divisible by, 1 and itself.

Q2) Which of the following pairs of sets are mutually exclusive? For those that are not mutually exclusive, which states can apply to both sets?

A. [colours of the rainbow] & [prime colours]
B. [Mr Smith drinking a glass of water] & [someone other than Mr Smith drinking a glass of water]
C. [Mr Smith drinking a glass of water] & [Mr Smith not drinking a glass of water]
D. [Months of the year ending in 'y'] & [Months of the year with 31 days]
E. [even numbers on a 6-sided die] & [prime numbers on a 6-sided die]

Q3) Explain the difference between the generalised second and third laws of probability and the non-generalised second and third laws of probability as given in this chapter.

Q4) Mr Smith is walking along the road under some apple trees, on a windy day. Many birds are flying from branch to branch. Mr Smith's hat is knocked off and as he looks around he sees someone walking behind him. List the different ways that Mr Smith's hat could have been knocked off and what we would expect to see as evidence of these possibilities. What type of reasoning is being used here?

Q5) Following on from question 4, Mr Smith notices that there is a fresh bird-poo on his shoulder and that the person behind him is 4 metres away. How do these observations affect Mr Smith's belief in the possible theories from Question 4? What type of reasoning is being used here?

Q6) Convert the following probabilities to odds:

A. 0.9
B. 0.5
C. 0.01
D. 0.0005

And the following odds to probabilities:
E. 1:1
F. 55:1
G. 5:4
H. 1:20

Q7) A test for a disease has a false positive rate of 0.02% and a false negative rate of 2%. Approximately 1 in every 15 000 people in the population has this disease. If you are tested for the disease using this test and the result is negative, what is the probability that you do have the disease?

Q8) In the example given in Q7, if the test was positive, what is the probability that you do have the disease?

Q9) Why is it important to consider two different possibilities for the observations when forming an opinion?

Q10) What are the advantages of carrying out evaluations at the activity level? And why can scientists usually not evaluate observations at the offence level?

REFERENCES

1. Bayes. Mr, Price. Mr, An essay towards solving a problem in the doctrine of chances. By the Late Rev. Mr. Bayes, F. R. S. Communicated by Mr. Price, in a Letter to John Canton, A. M. F. R. S, *Philosophical Transactions (1683-1775)* 53 (1763) 370–418.
2. D.V. Lindley, *Understanding Uncertainty*, John Wiley and Sons, Hoboken, 2007.
3. C. Aitken, A. Nordgaard, F. Taroni, A. Biedermann, Comment: "Likelihood ratio as weight of forensic evidence: a closer look (S.P. Lund and H. Iyer; Journal of Research of National Institute of Science and Technology, 2017)", *Frontiers in Genetics (Provisionally accepted)* 9(224) (2018).
4. A. Biedermann, T. Hicks, F. Taroni, C. Champod, C. Aitken, On the use of the likelihood ratio for forensic evaluation: Response to Fenton et al, *Science & Justice: Journal of the Forensic Science Society* 54(4) (2014) 316–8.
5. W.B. Fairley, F. Mosteller, A conversation about *Collins, The University of Chicago Law Review* 41 (1974) 242–253.
6. I.J. Good, Weight of evidence and a compelling metaprinciple, *Journal of Statistical Computation and Simulation* 31 (1989) 121–123.
7. C.G.G. Aitken, D.A. Stoney, The use of statistics in forensic science, in: J. Robertson (Ed.) *Ellis Horwood Series in Forensic Science*, Ellis Horwood Ltd., Chichester, 1991, p. 242.
8. I. Evett, B. Weir, *Interpreting DNA Evidence: Statistical Genetics for Forensic Scientists*, Sinauer Associates, Sunderland, 1998.
9. S. Gittelson, C.E.H. Berger, G. Jackson, I.W. Evett, C. Champod, B. Robertson, J.M. Curran, D. Taylor, B.S. Weir, M.D. Coble, J.S. Buckleton, A response to "Likelihood ratio as weight of evidence: A closer look" by Lund and Iyer, *Forensic Science International* 288 (2018) e15–e19.
10. R. Cook, I.W. Evett, G. Jackson, P.J. Jones, J.A. Lambert, A hierarchy of propositions: Deciding which level to address in casework, *Science & Justice* 38(4) (1998) 231–240.
11. I.W. Evett, G. Jackson, J.A. Lambert, More on the hierarchy of propositions: Exploring the distinction between explanations and propositions, *Science & Justice* 40(1) (2000) 3–10.
12. D. Taylor, J.A. Bright, J. Buckleton, The 'factor of two' issue in mixed DNA profiles, *Journal of Theoretical Biology* 363 (2014) 300–306.
13. D. Taylor, Probabilistically determining the cellular source of DNA derived from differential extractions in sexual assault scenarios, *Forensic Science International: Genetics* 24 (2016) 124–135.
14. D. Taylor, D. Balding, How can courts take into account the uncertainty in a likelihood ratio?, *Forensic Science International: Genetics* 48 (2020) 102361.
15. D. Taylor, D. Abarno, C. Champod, T. Hicks, Evaluating forensic biology results given source level propositions, *Forensic Science International: Genetics* 21 (2016) 54–67.
16. S. Tobe, L. Dennany, M. Vennemann, An assessment of the subjectivity of sperm scoring, *Forensic Science International* 251 (2015) 83–86.
17. T.R.D. Wolff, A.J. Kal, C.E.H. Berger, B. Kokshoorn, A probabilistic approach to body fluid typing interpretation: An exploratory study on forensic saliva testing, *Law, Probability and Risk* 14(4) (2015) 323–339.
18. J. de Zoete, W. Oosterman, B. Kokshoorn, M. Sjerps, Cell type determination and association with the DNA donor, *Forensic Science International: Genetics* 25 (2016) 97–111.

19. B. Kokshoorn, B.J. Blankers, J. de Zoete, C.E.H. Berger, Activity level DNA evidence evaluation: On propositions addressing the actor or the activity, *Forensic Science International* 278 (2017) 115–124.
20. D. Taylor, The evaluation of exclusionary DNA results: A discussion of issues in R v. Drummond, *Law, Probability and Risk* 15(1) (2016) 175–197.
21. I.W. Evett, Establishing the evidential value of a small quantity of material found at a crime scene, *Journal of the Forensic Science Society* 33(2) (1993) 83–86.
22. A. de Ronde, B. Kokshoorn, C.J. de Poot, M. de Puit, The evaluation of fingermarks given activity level propositions, *Forensic Science International* 302 (2019) 109904.
23. J. Nordby, *Dead Reckoning: The Art of Forensic Detection*, CRC, Boca Raton, FL, 2000.
24. G. Jackson, S. Jones, G. Booth, C. Champod, I.W. Evett, The nature of forensic science opinion – A possible framework to guide thinking and practice in investigation and in court proceedings, *Science & Justice* 46(1) (2006) 33–44.
25. C. Berger, Criminalistics is reasoning backwards, *Nederlands Juristenblad* 85 (2010) 784–789.
26. R. Cook, I.W. Evett, G. Jackson, P.P. Jones, J.A. Lambert, A model for case assessment and interpretation, *Science and Justice* 38(3) (1998) 151–156.
27. R. Cook, I.W. Evett, G. Jackson, P.J. Jones, J.A. Lambert, Case pre-assessment and review in a two-way transfer case, *Science and Justice* 39(2) (1999) 103–111.

2 Evaluation of Observations Given Activity Level Propositions

Duncan Taylor

CONTENTS

DOI: 10.4324/9781003273189-2

2.1   WHAT ARE 'ACTIVITY LEVEL EVALUATIONS'?

In Chapter 1, it was explained how the scientist can act at an early, investigative stage of a criminal matter or later when that matter is in court as an evaluator of the observations. For the latter scenario the ultimate purpose of forensic analyses is to assist a judge or jury in making a decision on the guilt or innocence of an accused person. Unlike many other types of testimonies, such as lay witnesses, or eyewitnesses, the testimony of forensic analysts is one of opinion. It is expected that the scientist has analysed one or many traces that relate to a crime by applying analytical techniques that can provide complex and detailed data. This data must be interpreted, but more than that, it must be in a format so that it can be used to assist the court in its ultimate purpose. There is little benefit in simply recounting a set of observations to the court (and one could argue that this is not providing an opinion but rather acting as an information relay device for a laboratory instrument, much like a monitor or printer). This is particularly true when those observations require specialist knowledge to understand, not just at an observational level (i.e. as in understanding that two sets of numbers such as those that represent allele designations are the same) but at understanding in a wider case context (i.e. the issues around population frequency of those alleles). The role of the scientist is therefore to use specialised knowledge of the methods used to generate the observations and put them in a case context. The more directly applicable the opinion provided by the scientist to the ultimate issue the court is convened to answer, the more assistance this provides the court.

While the testimony provided is an opinion, it must occur within the bounds of the scientific method. The bounds are fundamentally captured in the principles of evidence interpretation given in Chapter 1. The scientists must consider the framework of circumstances in the case, they must consider competing possibilities for the observations and they must only comment on the observations and not on the propositions. In addition, scientists should not stray up the hierarchy of propositions to the offence level, as these propositions always contain a legal element to which the scientist cannot contribute.

Traditionally, forensic analyses of DNA samples have led to reporting the results of evaluations that consider different sources of DNA. Such reports have provided the courts with valuable information over the decades. However, as techniques for sampling and generating DNA profiles become more sensitive and methods for DNA profile analysis and interpretation become more sophisticated, often the presentation of DNA profile evaluation given sub-source level evaluations is followed up by questions like:

- Could this result be from laboratory contamination?
- Could the DNA be on the item from secondary transfer?
- If the defendant was at a party with the complainant could that explain the presence of their DNA on the complainant's clothes?
- Could DNA have been shed from the defendant onto an object they were near but didn't touch?
- Could these items from the defendant's home have the defendant's DNA on them even if the defendant did not touch them?

All these questions require more information than is used in a sub-source level evaluation. Of course, the scientist will not answer any of these questions directly, that is for the court, but they will be able to comment on the probability of the observations given that activities or events address these points. In order to do so, information is required about events such as:

- The frequency of contamination in a laboratory
- The probability of DNA transfer from item to item
- The probability of DNA transfer and persistence arising from social interactions
- The amount of DNA shed by individuals in their surroundings
- The prevalence of DNA on objects in people's homes

This type of information is not common knowledge, and so it is not reasonable to expect that lay members of a jury will be able to assess these aspects when they incorporate scientific observations into their deliberations. It is not even reasonable to expect that a lay jury will know what factors are important to consider (let alone their probabilities of occurrence), or how to combine these factors in a logical way. All of these aspects of evaluating observations require specialist training and knowledge. They therefore sit within the responsibilities of the scientist, who can use their training and knowledge to advise the court of the greater significance of the forensic observations within the framework of circumstances of the alleged crime. In essence this is the meaning and purpose of carrying out evaluations given activity level propositions.

Evaluating forensic observations in light of activity level propositions has a relatively long-standing history. In the late 1990s work was carried out in the UK (i.e., former Forensic Science Service) that defined the notion of activity level within a hierarchy of propositions [1] and how evaluations using this type of propositions can be used to inform customers and to pre-assess cases [2, 3]. Examples of interpreting small quantities of DNA using Bayesian Networks (BNs) and evaluating them given activity level propositions were published in 2002 by Evett *et al.* [4].

The prevalence of research in the area of evaluations given activity level propositions, the use of graphical models to assist in evaluations and studies into the transfer and persistence of biological material has increased considerably over the past 20 years. For example, the number of documents that are flagged per year on Google Scholar when the phrase 'Forensic DNA Bayesian Network' is searched increased from less than a hundred to over two thousand (Figure 2.1).

The increase in interest in this area is likely due to a combination of reasons. Some of the contributing factors are:

- DNA profiling technology has improved in two different aspects. Originally, DNA profiling required large quantities of genetic material. ABO blood grouping required nanograms of DNA and so was only applicable in a very limited number of cases where large amounts of DNA were expected to have been left by an offender (such as a sexual assault or a scenario in which the offender bled). Advancements shifted forensic typing from blood grouping to sequences that were elucidated through restriction fragment length polymorphisms (RFLP) leading to the first images that were classically known as 'DNA fingerprints' [5]. As our understanding of biology progressed, the use of polymerase chain reaction (PCR)

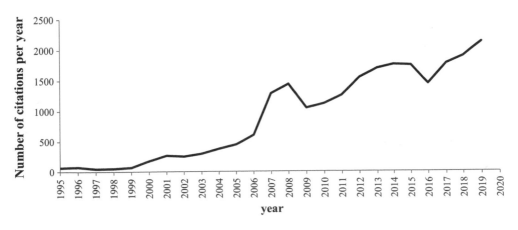

FIGURE 2.1   Number of documents identified (per year) on Google Scholar by searching the phrase 'Forensic DNA Bayesian Network'.

was developed and provided the ability to copy areas of DNA of interest, allowing much smaller amounts of starting material to be targeted [6]. With continual refinements in the reagents and methodologies used in PCR, the sensitivity has improved to the point where it is possible to generate DNA profiles from as little as a single human cell [7]. As less cellular material was required in order to detect someone's DNA, it did become more probable that their DNA had come to be on an item through being inadvertently transferred (i.e. not through the commission of a crime). As this increase in sensitivity was occurring, so too was the discrimination power of the information being produced. From the original blood grouping providing a chance match of around 1 in 1.66 [8], the latest DNA profiling kits provide chance match probabilities of around 1 in 10^{30}. Not only were forensic investigations more likely to detect DNA, but they also were more strongly able to associate it with an individual. These two factors working together caused questions in Court to largely shift from who the donor of recovered trace material is, to the mechanism by which the material was deposited on the examined item/surface. This is particularly relevant when the profile has been generated from very low levels of starting DNA that cannot be assigned to a particular body fluid. Such DNA is commonly called 'trace DNA', 'contact DNA', 'touch DNA', 'latent DNA' or 'low template (lt) DNA' and gained prominence after a landmark publication demonstrating the ability to obtain DNA from fingermarks in the late 1990s [9].

- With early forensic tests such as blood grouping, the need to deal with the complications of mixed DNA profiles was limited as typically large stains (most commonly from a single individual) were targeted. As the type of sample shifted to trace DNA and the sensitivity of DNA profiling technology increased, it became more common to encounter mixed DNA profiles. Initially, the statistical models used to evaluate mixed DNA profiles could only do so under restrictive assumptions that resulted in relatively few evaluations. Population genetic models and statistical models grew over the late 1990s and 2000s to allow a larger proportion of mixed profiles to be evaluated. The validity of DNA profiling, population genetics and statistical models initially drew heavy challenge in court during a period known as the 'DNA wars'. In the 2010s, more sophisticated methods of DNA profile evaluation, known as probabilistic genotyping methods [10], were developed that overcame many of the criticisms of subjectivity and limitations that were raised in earlier challenges. The introduction of probabilistic genotyping again was met with heavy courtroom challenge in the late 2010s in the period referred to by some as the 'DNA war II'. These challenges have subsided in recent years, and in many countries, the courts have largely come to accept that DNA profile evaluation has now reached a point where numerous software programs exist which can reliably evaluate complex and low-level DNA profile data [11–19]. These software programs provide evaluations, which will often lead to the presence or absence of DNA from nominated individuals being agreed upon by both prosecution and defence. These observations provide a strong foundation upon which evaluations, considering higher-level propositions in the hierarchy, can be carried out.

- Early evaluations of any kind (but specifically those considering activity level propositions for the purposes of our discussion) generally required manual derivation of *LR* formulae. This need tended to relegate the evaluations to those that were very comfortable with algebraic manipulations. The concept of graphically displaying the factors of importance to a probabilistic evaluation was introduced in the 1980s [20] and work began on developing software that could take advantage of this more user-friendly format. In the past 20 years, there has been an increase in the availability of software programs (and computing power to run them) that can be used to graphically display the factors, and their dependencies, that are important in the evaluation of DNA observations given activity level propositions. This has opened up the practice of carrying out evaluations to a much wider audience and for a more complex range of cases.

- There have been several high-profile cases that received international attention, which revolved around DNA transfer, persistence, prevalence and recovery (TPPR) issues (such as the Meredith Kercher case [21, 22] and the Fitzgerald case [23]). These high-profile cases are not exceptional but reflect the increasing emphasis that is given in many jurisdictions to DNA transfer and persistence rather than questions of source only. In particular, there has also been a rise of true crime docuseries on streaming platforms such as Netflix or Prime that popularise the concepts of activity level evaluation, through their in-depth questioning of the significance of the standard forensic evidence.

- In order to carry out an evaluation considering issues of transfer, persistence, prevalence and recovery it is preferable to have data that can be used to inform an opinion on the probability of events. While there are various survey-style studies that provided important information even back in the 1980s (for example see [24] for a survey of suspects of homicides for the presence of bloodstains) it wasn't until the popularisation of DNA profiling on trace samples that TPPR studies started to accumulate in large numbers. An ever-increasing body of scientific literature has furthered our understanding of DNA TPPR issues (see the Oorschot *et al.* review [25]). The data provided in the publications, or as supplementary material to the publications, allow for more detailed modelling of these factors. An example can be seen in Taylor *et al.* [26], which includes a high level of detail regarding factors such as contact surface area, amount of DNA etc.

- There are courses and training programs that teach the principles of evidence evaluation (some specifically when dealing with activity level propositions). The online course 'Challenging Forensic Science: How Science Should Speak to Court' is a freely available course on the Coursera platform. There are a number of courses that vary in length that are run by instructors through the University of Lausanne (https://www.formation-continue-unil-epfl.ch/formation/statistics-evaluation-forensic-evidence-cas/). It is also common for workshops to be presented as part of forensic conferences that teach aspects of evidence evaluation and evaluations given activity level propositions. All of these events make the information on such evaluations more accessible to the forensic (and wider) community.

2.2 WHY ARE ACTIVITY LEVEL EVALUATIONS NEEDED?

Some philosophical reasons for the provision of evaluations that consider activity level propositions have been provided. These reasons are more than simply hypothetical ideals. This may be best demonstrated by a number of high-profile court cases where activity level issues such as the method or timing of deposition of biological traces were at issue. The non-exhaustive list of cases below is those that received some international attention or scientific interest. In each of these cases, the court could have been assisted by an expert applying their knowledge of DNA TPPR, combined with the available forensic literature in order to place the forensic observations into context.

- The Queen v Joyce [2002] NTSC 70 (Australia)
- 'Deventer murder case (1999)' [2004] AO3222 (The Netherlands)
- R v David Butler (2005) (UK)
- Private J. Kovko, Kovko Report (2006) (Australia)
- The Queen v Hillier [2007] HCA 13 (Australia)
- R v Reed and Reed, R v Garmson [2009] EWCA Crim 2698 (UK)
- Knox and R. Sollecito; Massei Report [2009] (Italy)
- 'Phantom of Heilbronn' (2009) (Germany)
- Farah Jama; Vincent Report (2010) (Australia)
- R v Weller [2010] EWCA Crim 1085 (UK)
- 'Putten murder case (1994)' [2011] BU3933 (The Netherlands)

- Adam Scott (2011) (UK)
- Commonwealth v Dirk K. Greineder [2013] (USA)
- Lukis D. Anderson [2013] (USA)
- R v Drummond [2013] SASCFC 135 (Australia)
- R v Fitzgerald [2013] SASCFC 82, & [2014] HCA 28 (Australia)
- R v Pfennig [2016] SASC 170/171 (Australia)
- R v QUIST [2021] (Australia)

It is common that in cases which have forensic biology observations, where the innocence of the defendant is being argued, that some issues of DNA transfer, persistence, prevalence or recovery will be raised. Often these issues are raised cursorily with questions that start with the phrase 'Is it possible that …'. Rarely are these issues the lynchpin on which innocence or guilt will be decided by the jury, but in some circumstances, such as those in which are described here, they are. These aspects will be described and discussed below. They are illustrated using the circumstances of cases where the particular issue was a crucial element following from the prosecution and defence positions.

Often these issues were raised too late, and by this, it meant they have been raised after a person was convicted based on the misunderstanding of the meaning and significance of the forensic observations in the greater case context. Consequently, the issues raised below were generally not addressed through evaluation considering activities in these cases. Note that although the cases are grouped according to a specific issue, when assessing the observations under activity level propositions all of the issues are important to consider within a case evaluation.

2.2.1 Case Issue: Indirect Transfer

A primary transfer occurs when there is a transfer of biological material (i.e. a person) directly to an object. Often this is thought of as occurring through touching, but it could also extend to a transfer from airborne transfer (for example by sneezing, coughing or talking over), or even accumulation through mere proximity to an item or surface. Secondary transfer is the process whereby the DNA of an individual is transferred by an intermediate vector (such as another person or an object). It is through this intermediate vector that target DNA is transferred to an item of interest (i.e. target surface). If the DNA that has ultimately been deposited onto the target object has travelled through two intermediaries, then this is called tertiary transfer, three intermediaries are called quaternary transfer and so on. Collectively, any transfer that has occurred through one or more intermediaries can be referred to as an indirect transfer. It has been the authors' experiences that indirect transfer is one of the most commonly brought up events in court, particularly in cross-examination in adversarial systems. Four case examples are provided where issues of indirect transfer have played a pivotal role in the case.

- R v Fitzgerald

  Fitzgerald was accused of being involved in homicide. The DNA evidence consisted of a DNA profile corresponding to the profile of Fitzgerald on a didgeridoo located at the house of the victim of an assault. There was no evidence that the didgeridoo was used in the attack; however, it was found near the body of the deceased victim. This DNA sample was close to the only evidence tying Fitzgerald to the crime. Fitzgerald denied ever having been to the house but claimed that an individual he shook hands with visited the house of the victim. The defence claim was then that the DNA of Fitzgerald had transferred to this friend and then to the didgeridoo. This is therefore an example of a claim that a 'secondary transfer' may have occurred. Fitzgerald was convicted but subsequently released on appeal as secondary transfer could not be discounted. For a more detailed analysis of this case see Szkuta *et al.* [23].

- Lukis D. Anderson

   Lukis became a suspect in a Monte Sereno (California) murder case. Lukis was hospitalized during the time of the incident and so had a strong alibi. On investigation into the DNA result, it was concluded that Lukis' DNA was probably transferred to the victim by paramedics attending Mr. Anderson and then the victim that night. After five months in prison, Anderson was released [27] and the case against him dropped.

- R v David Butler

   This case investigated the homicide of a woman in the UK. Samples taken from the deceased were analysed in the local Forensic Laboratory. The DNA profile of the traces obtained supported Butler as a DNA donor. Butler was a taxi driver and had a skin condition, which meant that he may shed more than usual amounts of skin. The victim was wearing nail polish that attracted dirt, dust and likely skin flakes. The possibility was raised by the defence of secondary transfer. Specific examples posited include that Butler's DNA had been shed to money, or a passenger and then transferred to the victim. Before being acquitted, Butler spent eight months in jail on remand.

- R v Pfennig

   A young girl, Lousie Bell, went missing from her bedroom in 1983. Her pyjama top was found a few days after her disappearance on the lawn of a nearby house. A sample from the top was tested and a DNA profile that had a male component was obtained. The profile corresponded to that of the accused, Pfennig. He alleged that his DNA transferred to the top through his daughter, who attended the same school as Bell. The mode of transfer being suggested here was tertiary with DNA from Pfennig transferring primarily to his daughter, then secondarily to the top of the victim and then tertiarily to the pyjamas of the victim. There were also indications that the top had been submerged in a nearby river (so issues of DNA persistence could also be considered in this matter). Pfennig was convicted of the offence. He later appealed on the grounds that the tertiary transfer scenario had not been adequately discounted during the trial; however, the conviction was upheld. The appeal Judge stated in his ruling that the defence proposition was 'so unlikely that I consider the trial judge was correct to exclude the appellant's hypothesis as a reasonable possibility' [28]. For an example of an evaluation based on the Pfennig case see Taylor *et al.* [29].

2.2.2   CASE ISSUE: CONTAMINATION

It is important to make a distinction between background levels of human DNA (i.e. not relevant to the activities under consideration) that occur on items as a consequence of existing in an environment with humans and DNA transferred through a defined mechanism (such as a primary contact). DNA transfer can be from individuals (e.g. victims, witnesses, defendants), and the presence or absence (and potential amount) are the observations used in an evaluation. DNA transfer can also occur when the DNA has been added as a result of investigating the alleged offence (either at the scene, during transport, or within a laboratory). Only this latter type of transfer is often thought of as contamination, as defined by the ISFG commission in section 9 of [30].

   Contamination occurs in two general ways, person to exhibit or exhibit to exhibit. With regards to the former, protective measures are specifically designed to help avoid contamination. These include measures such as personal protective equipment, regular cleaning of work surfaces or items and the use of disposable items. These measures seek to minimise contamination from occurring. It is also common for laboratories to have practices that identify contamination when it has occurred. These include the use of screening DNA databases that comprise DNA profiles from staff members, police, doctors, or employees of consumable manufacturers, or the use of negative controls and blanks that are designed to identify contamination from a person to a reagent.

   Exhibit-to-exhibit contamination is a specific type of secondary transfer. This could occur when the presence of DNA of an individual on an item has resulted from inadvertent transfer from another item, for example through the use of an unchanged glove. Another mechanism by which exhibit-to-exhibit contamination can occur is by directly touching two separate items together (for example placing an item on a bench or floor may transfer DNA from that surface to the item). This second type of contamination is very difficult to identify as (unlike person-to-item contamination) it is not known that the DNA was not transferred to the item as a result of the alleged crime. The cases below represent examples of both person-to-sample and sample-to-sample contamination.

- Adam Scott case [31]

    Adam Scott was suspected of a sexual offence that occurred in Manchester. The evidence in the case was his reference DNA profile corresponding to that of an evidence DNA profile generated from an exhibit in the case. The original report issued by the forensic laboratory (LGC Forensics) only gave an evaluation of the DNA profile considering sub-source level propositions, for which the *LR* was approximately 1 billion. The sample in which semen was found was contaminated with Adam Scott's DNA from a reference sample. The semen in the sample was from the boyfriend of the victim, deposited from a consensual act. The contamination by Scott's DNA occurred through an inadvertently re-used disposable plastic container in the laboratory. It was incorrectly assumed that the DNA on the sample from Scott was from the semen (rather than contamination). As well as an example of contamination, it is also an example of an incorrect assumption that the DNA in a sample had come from a specific cellular source. Scott was held for several months before it was found that the semen found was not his and he was released.

- The Farah Jama case

    This is a case of suspected sexual assault that occurred in Victoria, Australia in 2006. A woman was found unconscious in a nightclub toilet, allegedly she had been sexually assaulted. She attended a medical facility, where a number of intimate swabs were taken. When processed in the forensic laboratory one of the swabs showed levels of sperm. The DNA profile developed from the sperm corresponded to the reference of Farah Jama, identified on the database. Despite no other evidence (and in fact some evidence that pointed towards innocence), Jama was convicted of the crime. It was eventually determined that contamination had occurred at the medical facility. A sample containing Jama's sperm had been present in the examination room immediately prior to the intimate samples being taken from the victim. The discovery of the contamination led to a detailed investigation into how such a miscarriage of justice had occurred, and importantly how it could be avoided in the future. Details can be found in the Vincent report [32]. One suggestion that has come in the years since is that the sperms were only identified on one of a number of the sexual assault swabs taken. Given the unusual occurrence of a single sperm-positive swab result, it may be that an activity level evaluation that took into account the possibility of contamination would have identified that the results did not support a sexual encounter.

- 'Phantom of Heilbronn' issue

    This is one of the most famous instances of contamination, due to the breadth and scale of occurrence. The DNA profile of an unknown female was found in a number of high-profile cases in Germany, Austria and France, for cases that included several murder cases, burglaries and robberies. Investigators initially suspect a serial killer was involved but in 2009 it became apparent that the DNA profile could be traced to an individual who worked packaging the cotton swabs at a factory. The Heilbronn issue received much publicity since its discovery [33–35]. While the 'case' never went to court, it is worth thinking about in relation to how activity level evaluation could have assisted (even at an investigative stage). Activity level propositions can still be helpful in this investigative setting since they encourage scientists to think not only about intrinsic features (i.e., corresponding DNA profiles), but also about extrinsic features relating to the quality and quantity of traces, the position in which they were found and, related to this, aspects of transfer and persistence.

2.2.3   CASE ISSUE: PREVALENCE OF DNA OF A KNOWN INDIVIDUAL

When a known individual has deposited DNA on to an item then questions may turn to how the amount of that person's DNA can assist in distinguishing the different proposed activities. The consideration of these issues may include the amount of DNA from an individual expected to be on an item due to their regular proximity (e.g. in their home) or due to the previous contact.

- R v Weller

    Weller was charged with the sexual assault of a woman who claimed that when she was drunk Weller put her to bed and then penetrated her vagina with his fingers. Weller claimed that he assisted her to bed after she had become intoxicated. This included assisting her while she vomited due to being highly intoxicated. His claim was that there was no sexual activity and therefore he did not assault her. The argument in court was then whether the victim's DNA found on the fingers of Weller was more probable given the digital penetration of the victim's vagina or given assisting the victim with going to bed. Ultimately the jury decided that the sexual assault was more probable and Weller was convicted in 2006 and sentenced to 3 years of imprisonment.

- Deventer murder case [36]

    An elderly lady was found murdered in her house in the city of Deventer in 1999. DNA of a male individual was recovered from multiple areas and multiple different trace types were on the blouse of the victim. The DNA profile corresponded to that of the financial consultant of the victim (Louwes) who claimed that the DNA from the traces of the blouse and from the bloodstain on the collar was the result of their meeting sometime prior to the incident. Louwes was known to have conditions that meant he was more prone to shed DNA when speaking than the average person. He was also said to have a habit whereby peeling skin on his fingers would lead to there often being his own blood present on his hands. Louwes was convicted of the crime in 2003 and sentenced to 12 years. In 2003 the Dutch Supreme Court reopened the case and Louwes was released due to the discovery of errors in the results of a knife that was examined (and later found not to even be the murder weapon). In 2004 DNA testing of the victim's blouse saw Louwes convicted again and served an additional five years until his release in 2009.

- Meredith Kercher case [21, 22]

    Amanda Knox was convicted and then later exonerated of the murder of her roommate, Meredith Kercher. A number of pieces of evidence that were used to build the prosecution case were from the apartment where the murder occurred and where both Knox and Kercher cohabited. There was evidence of Knox's DNA on a knife handle. The knife was claimed as a murder weapon by the prosecution, but this fact was disputed by the defence. Even if the knife were the murder weapon it was a knife that Knox would have had regular access to and so a level of her DNA would have been expected to be present, regardless of whether it was the murder weapon or Knox committed the murder.

2.2.4   CASE ISSUE: PREVALENCE OF DNA OF AN UNKNOWN
INDIVIDUAL AND BACKGROUND DNA

Prevalence relates to the presence of DNA of one or more unknown individuals (within the context of the case) on an item or person. Recall the distinction between prevalent DNA and background DNA. Prevalent DNA corresponds to the presence of DNA that cannot be accounted for by someone in the case, but which it is expected (through a defined transfer route) for their DNA to be present. Background DNA is unknown DNA from a non-specific source, which is expected to be on an item. The presence of unknown DNA can be important to consider during an evaluation when an alternate offender is being considered under the defence proposition (and there is an innocent explanation for the presence of the defendant's DNA). In these instances, the presence of unknown DNA must be considered as either background or the alternate offender.

- R v Quist

    This is a case that occurred in South Australia. In this case Quist was seen running from the scene of a fire in a public toilet in a shopping centre. In the toilet cubicle multiple bottles of accelerant, not yet burnt, were found. The bottles were carried out of the cubicle by police and into the main shopping centre (through the same area where Quist had passed 90 minutes earlier) where they were sampled for DNA analysis. Quist's DNA, and DNA from unknown individuals, were found on two of the bottles and no DNA was found on the remaining bottles. The prosecution alleged that Quist's DNA had come to be on the bottles because she placed them in the toilet. The defence argued that Quist's DNA had come to be on an item as a result of scene contamination. Importantly in the context of unknown DNA, whether the DNA was from background DNA or an alternate offender. In an evaluation of this case, Taylor *et al.* [37] investigated the importance of considering multiple sources of unknown DNA. Also investigated was the importance of considering whether the unknowns on the two bottles from which DNA profiles were obtained could have come from the same unknown individual.

2.2.5 Case Issue: Persistence Issues

In some cases, the presence of DNA from an individual is not in contention, only the mechanism by which it came to be on an item. For cases in which the two competing propositions have two different timeframes (or when they represent two different exhibit histories, such as washing and non-washing) aspects of persistence are important to consider. Persistence issues can be particularly important in sexual assaults, as there is a relatively quick degradation of foreign DNA in areas such as the vagina or mouth. It may be that the complainant states involuntary intercourse occurred at a given time, while the accused alleges that the intercourse took place voluntarily some hours or days prior. In this case the presence (or absence) of bodily fluids must be considered at the differing time frames.

- Putten murder case [38]

    This case related to the rape and murder of a young woman in the village of Putten, Netherlands, in 1994. Swabs were taken from the body of the women and semen was located on swabs taken from internal and external areas of the body. Initially, without a suspect, the profile obtained from the sperm was an unknown male. In 2008, a match to the unknown male profile was obtained through a search of the National DNA Database and the suspect was arrested. He claimed to have been in a relationship (unknown to anyone else) with the victim and that they had intercourse shortly before the murder. At the conclusion of the trial, he was found guilty and convicted in 2009. While persistence was not part of the evaluation of this matter, it could have been considered during the evaluation of the evidence. It may be that the sexual intercourse was considered as 'rape' by the prosecution and occurred at the same time as the homicide. Alternatively, the sexual intercourse could be considered consensual by the defence and occurring sometime prior to the homicide (i.e. that there was no rape at all). The evaluation would then need to consider the probability of obtaining the levels of semen observed, given these two competing timeframes. The persistence of semen internally and externally would be important in this evaluation.

2.2.6 Case Issue: Absence of DNA

The absence of DNA from an individual should not be considered an 'absence of evidence'. The finding that no DNA of a person of interest has been detected in fact does have evidential value. The absence of DNA will rely on probabilities of transfer, persistence and recovery of DNA, given the proposed activities put forward by the prosecution and defence. Although the non-detection of DNA is not necessarily evidence of the absence of an individual, it may provide support for the proposition under which an absence of DNA is expected.

- Drummond case

   In this case a schoolgirl, walking home along a road after school one day claims that someone pulled off the road and attempted to pull her into his van and kidnap her. The evidence provided to police was a licence plate, which led to the arrest of a suspect, Drummond. The alleged attempted kidnap involved the victim hitting Drummond's top in order to fend him off and Drummond grabbing the victim's top to drag her into his vehicle. Both tops (i.e. from the suspect and victim) were submitted for DNA profiling and neither showed signs of DNA of the other person. The lack of transfer of DNA was explained in court but a misunderstanding about its significance lead to the conviction of Drummond for the offence. An appeal was lodged, and Drummond was later exonerated of the crime. For an example of evaluation of the observations in the Drummond matter, see [39]; however, this case is used as an example in this book to show the application of methods of evaluation. More information to follow later in this chapter.

2.3 WHO IS DOING IT?

A number of advisory bodies and leading thinkers in the field of forensic science advocate the evaluation of given propositions regarding activities, rather than given propositions at lower hierarchical levels (i.e., source, sub-source or even sub-sub-source levels), specifically:

- The European Network of Forensic Science Institutes (ENFSI) [40]
- The Association of Forensic Science Providers (AFSP) [41]
- The Royal Statistical Society (RSS) [3, 42–44]
- Forensic scientists across the world [31, 45–48]
- The International Society of Forensic Genetics [49]

The cases where forensic results ought to be reported considering activity level propositions have been delineated in the ENFSI guideline for evaluative reporting [40]. It is needed, in particular, when the amount of collected trace material is low and when considerations of transfer, persistence and recovery require specialised forensic knowledge. There is a widespread recognition that there is danger in leaving such assessments to non-forensic scientists and that is the duty of the scientists to guide the court appropriately in these matters [50]. Other typical cases where the observations lead themselves naturally to an interpretation considering activity level propositions is when the source of the trace material is not disputed in the case but only the mechanisms whereby the trace material was transferred are debated.

The ENFSI guideline said on the matter:

'Source level propositions are adequate in cases where there is no risk that the court will misinterpret them in the context of the alleged activities in the case.'

Biedermann *et al.* [50] state:

Not pursuing this topic bears the risk of leaving recipients of expert information without guidance. Reliance on recipients' own devices is prone to conclusions that are based on (sub-) source level propositions being wrongly carried over to conclusions about activity level propositions.

It has also been recognised that if the Court's question relates to activities, then a scientist should carry out an evaluation using propositions that relate to activities and not to source only (and not attempt an ad-hoc consideration of the observations given posited activities during oral testimony) [47].

There have been numerous publications demonstrating the importance of considering activity level propositions when evaluating forensic results [39, 46, 50–58]. From these references, it is clear there is an increasing awareness in the forensic community of the role of the scientist in the assessment of DNA analysis results given activity level propositions. The expert plays a crucial role in the evaluation of the evidence by imparting knowledge on DNA transfer, persistence, prevalence and recovery (TPPR) to the court. This recognition of the need for activity level assessment extends to the legal community. Margot [59] states that:

> The real problems of interpreting poor quality traces and mixtures have only come to the fore in recent years. These problems have illuminated the important challenge that forensic science is facing: interpreting results in view of conflicting versions of events and activities.

More recently the National Institute of Standards and Technology (NIST) in the USA has conducted a foundational review of DNA mixture interpretation, on which they published a draft report [60]. This review, which had the main aim of identifying the capabilities and limitations of DNA profile interpretation, recognised the importance of placing those observations into a framework that considers activity level issues. The authors dedicate an entire chapter to activity level considerations and provide key takeaway points, four of which are provided below:

- **KEY TAKEAWAY #5.1**: DNA can be transferred from one surface or person to another, and this can potentially happen multiple times. Therefore, the DNA present on an evidence item may be unrelated (irrelevant) to the crime being investigated.
- **KEY TAKEAWAY #5.2**: Highly sensitive DNA methods increase the likelihood of detecting irrelevant DNA. When assessing evidence that involves very small quantities of DNA, it is especially important to consider relevance.
- **KEY TAKEAWAY #5.4**: DNA statistical results such as a sub-source likelihood ratio do not provide information about how or when DNA was transferred, or whether it is relevant to a case. Therefore, using the likelihood ratio as a standalone number without context can be misleading.
- **KEY TAKEAWAY #5.5**: The fact that DNA transfers easily between objects does not negate the value of DNA evidence. However, the value of DNA evidence depends on the circumstances of the case.

Within the Netherlands exists a group called the Netherlands Register for Court Experts (NRGD; https://english.nrgd.nl/). It is an independent organisation that sets the standards for different forensic fields and provides competency tests for individuals who wish to be recognised as experts in a chosen field. The first field for which standards were set by the NRGD was in 2010 in the field of 'DNA analysis and interpretation – source level'. Since then, the fields of 'DNA analysis and interpretation – sub-source level' and 'DNA analysis and interpretation – kinship' have been developed. In 2021 the NRGD has added the field 'DNA analysis and interpretation – activity level'. This field recognised not only the importance of the work conducted in this area but also the specialised skills required to carry out evaluations of this type compared to those required for the analysis of DNA profile data considering sub-source level propositions alone. More on the demarcation of fields of expertise by the NRGD will be discussed in Chapter 11. Other accreditation bodies have also recognised the existence of activity level evaluations within their accreditation schemes, such as the National Association of Testing Authorities although have not designated the practise as requiring a separate sub-class of accreditation.

There is also a rise in the availability of online courses in evidence evaluation as evidence of increasing global interest in the forensic community. One example of such a course is the 'Challenging Forensic Science: How Science Should Speak to Court' massive open online course (MOOC) available through the Coursera platform (https://www.coursera.org/learn/challenging-forensic-science). Another example is the online course on evidence evaluation 'Statistics and the Evaluation of Forensic Evidence' available through the University of Lausanne (https://www.formation-continue-unil-epfl.ch/formation/statistics-evaluation-forensic-evidence-cas/).

2.4 INTERPRETATION FALLACIES AND CONCERNS WITH ACTIVITY LEVEL EVALUATION

2.4.1 COMMON EVIDENCE INTERPRETATION FALLACIES

There are numerous fallacies, errors, arguments, misconceptions, biases and faults of logic that have been described and published with regard to forensic science. A selection of this extensive list is provided below. These are the most commonly encountered when providing probabilistic evaluation evidence to those not familiar with such areas.

2.4.1.1 The Transposed Conditional (a.k.a. the prosecutor's fallacy)

In Chapter 1 the most referred to fallacy was presented; the fallacy of the transposed conditional. This occurs when one of the principles of evidence evaluation is not adhered to, i.e. that the scientist must comment on the observations and not on the propositions. This fallacy is also commonly referred to as the 'prosecutor's fallacy' [61], although this name seems to unfairly target prosecutors for a fallacy that is quite widely committed by any group of non-scientific experts. In formulaic terms the fallacy occurs when the individual who is meant to be commenting on the probability of the evidence (E), given the proposition (H), $\Pr(E \mid H)$, instead comments on the proposition itself, i.e. they are incorrectly commenting on $\Pr(H \mid E)$. In forensic biology the fallacy is committed when making statements along the lines of 'These profiles match, therefore the DNA is 100 times more probable to be from the suspect that someone else' rather than correctly commenting on the evidence, which would yield a statement along the lines of 'The probability of obtaining these matching profiles is 100 times higher if the DNA is from the suspect that someone else'.

2.4.1.2 The Defence Fallacy

Another fallacy commonly committed is commonly called the 'defence fallacy' [61] (again somewhat unfairly, but so named as it is during cross-examination that the fallacy is most commonly stated). A typical statement of the fallacy would be as follows for an example of a DNA profile from a POI to an evidence profile:

> The probability of obtaining these matching profiles is 1 million times higher if the source of the DNA is the defendant, rather than an unrelated individual. There are approximately 20 million people residing in the country in which this offence has been committed. This means that there are 20 people that might be expected to have this same DNA profile. Therefore, there is only a 1 in 20 chance that this DNA has come from the defendant.

Most of the logic in this statement is sound. The error comes in the last statement; i.e. because there is a large group of individuals with the same characteristics as the tested defendant and crime scene, the observations of matching profiles actually amount to very little evidence in support of the defendant's DNA contribution. This is fallacious however, because not all the individuals with the same matching profiles are equally suspected of the crime. In a Bayesian sense the matching profile has narrowed down the pool of suspects from everyone in the population to a select few and so has very strongly updated the belief that the suspect is the source of the DNA.

2.4.1.3   The Base Rate Fallacy

This fallacy (also called base rate neglect) is based on people's failure to account for prior odds (or the base rate of an event) when calculating posterior odds [62]. Formulaically the leap made with this type of fallacy is:

$$Pr(A \mid B) = Pr(B \mid A)$$

In a court setting, this can occur when those hearing the evidence do not appropriately take into account the prior probability for their beliefs when incorporating forensic evidence into the posterior. A famous example of people presenting this behaviour was given by Kahneman *et al.* [62] where participants were given prior probability for graduates being from specific fields (in their example, lawyers or engineers). Participants were then provided with a personality sketch that they use to assign a probability of a person being from that field. Finally, participants were asked to assign a probability to a recently graduated student who had either the lawyer or engineer personality sketch. The researchers found that the probability assigned to the students being from specific fields was assigned the same as the probability of the fields given a personality sketch, but the base rate of the fields did not seem to be considered at all.

2.4.1.4   The Association Fallacy

Also known as the source probability error [63] this is the chain of reasoning that a reported set of matching DNA profiles is indicative that the profiles are a 'true match' (i.e. there has been no error in the generation or reporting of result) and then that the suspect with the matching profile is the source of the trace and then that this suspect is the perpetrator of the crime. Gill [31] presented a number of examples where, through ambiguous or misstated testimony, the results of an evaluation given source level propositions were equated with an evaluation given activity level propositions. When such a carrying over occurs between sub-source and source, Gill terms this an association fallacy, though the same misinterpretation can occur between source, sub-source and activity. These latter two variants were named by Meakin *et al.* [64] who termed them 'source level fallacy' and 'activity level fallacy'. The chain of reasoning that leads to these fallacious leaps fails to consider all the additional data required to evaluate the evidence at higher levels of the hierarchy of issues. In the original source probability error example, there is also a failure to distinguish commenting on the evidence from commenting on the propositions.

2.4.1.5   The Probability of Another Match Error (or the birthday problem)

This error of logic comes from a subtle misrepresentation of the difference between the probability of finding another set of specific traits and the probability of two people having the same traits [63]. A common example of this misunderstanding came with the advent of searchable DNA databases. The apparent 'problem' would be stated as:

> You say that the probability that someone else in the population has the same DNA profile as the defendant is 1 in 1 million, and yet in your database of 10000 people you already have multiple instances of unrelated people with the same DNA profile. Clearly the frequency of your DNA profiles are much higher than what you say they are.

The logic problem here comes from not seeing the difference between the probability of observing a nominated DNA profile in the population and the probability of two people having the same DNA profile. It is commonly referred to as the 'birthday problem', so called due to the example used to explain the logic. Consider how many people you need in a room before you expect to see two people with the same birthday. It may be an initial reaction to have the line of thought that because each person has a birthday on 1 out of 365 days that you will need 365 people to be in the room before you expect to see a matching birthday. In fact, the number is much lower than this, around 23.

The reason is that the interest is not in any specific birthday, and so rather than needing to consider the number of people to find a matching birthday, the number of pairwise comparisons between people must be considered. The number of pairwise comparisons for N people can be calculated using the formula $N(N-1)/2$. Therefore with 24 people, there are $23 \times 22/2 = 253$ pairwise comparisons, which means that there is a $1-(1-1/365)^{253} = 0.5005$ probability of finding two people with the same birthday, i.e. it is more likely to find two people with the same birthday than not. If, however, the interest lay in finding someone with a specific birthday (for example, 23 July) then 253 people would be required in the room in order to expect to see someone with that birthday.

2.4.1.6 The Numerical Conversion Error

This error comes from incorrectly considering how many people are required in order to see matching traits [63]. The issue has already been partially demonstrated in the 'probability of another match error' section. Consider again the birthday problem and the situation where the interest lay in determining the probability of finding someone in the room with that birthday. The erroneous thinking is that when one person is in the room then there is a 1/365 chance of seeing someone with a birthday of 23rd July and when there are 2 people in the room there is a 2/365 chance of seeing someone with that birthday and when there are 100 people then there is a 100/365 chance of seeing someone with that birthday, etc. This logic very demonstrably fails though when you consider, for example, 365 people in the room as by that logic the expectation would be a 100% chance of seeing someone with that birthday, but of course, this cannot be the case (there must be a chance that no one with that birthday was selected). The correct way to calculate the probability of someone in the room containing N people having a specific birthday is:

$$\Pr(specifc\ birthday \mid N\ people) = 1-(1-1/365)^{N}$$

2.4.1.7 Selection Bias

Selection bias is closely related to reinforcement bias. It relates to the action of choosing evidence seemingly in support of a proposition amongst a large group of 'multi-opportunity' searches when in fact it was supportive by chance when viewed amongst the many other searches performed [65]. This type of issue occurs in many guises across many disciplines. In essence, the issue can be boiled down to the thought that the more tests you conduct, the more likely it is you will find something in support of your proposition. Selection bias is why it is important to present all evidence and consider all observations in an evaluation and not fall into the trap of thinking that the failed, insufficient, or negative test results do not add any power to the evaluation.

2.4.1.8 The Individualisation Fallacy

This is the belief that once an LR becomes large, or a believed frequency of characteristics becomes small, the observations can be considered as effectively an individualisation [66]. This type of thinking is classically presented in traditional forensic sciences such as fingermarks, or shoeprints (although the myriad of good work must be noted that has been carried out to evaluate both these evidence types probabilistically from various groups). Individualisation opinions can be recognised from statements such as 'it is my opinion that this fingerprint comes from Mr X to the exclusion of all other individuals'. As put by Stoney [67] there are no mathematical calculations that can take the step from providing a very small probability (or a very large LR) to a finding of individualisation. The last step must be taken as a 'leap of faith' on the part of the evaluator. In forensic genetics, the claim of individualisation can sometimes, incorrectly, be made when the discrimination power of the DNA observations is large enough that the LR exceeds the population of the Earth many times over. In effect the claim of individualisation is a statement about the proposition itself (i.e. that the individual is the source of the DNA) and not the observations (the DNA profile) made by the belief that the LR is so large that the prior probabilities could no longer shift the enormity of support for the proposition.

2.4.1.9   The Conjunction Fallacy

The conjunction fallacy [68] is another fallacy based on the failure to recognise probabilistic logic. In essence, the conjunction fallacy tells us that if B is a subset of A, then the probability of A must be greater than the probability of B. Formulaically:

If $A \supset B$ then $\Pr(A) \geq \Pr(B)$

Because the set of possibilities associated with A and B is included in the set of possibilities associated with B another way to think about this is that the probability of multiple events occurring cannot be greater than the probability of any one of the constituents. Formulaically this can be written as:

$$\Pr(A \,\&\, B) \leq \Pr(A)$$

There are many examples of instances where an individual's intuition does not follow the logic of the conjunction statement above.

One example given by Tversky *et al.* [68] was to ask people:

'In a 2000-word document how many words do they estimate would take the form _ _ _ _ ing'

(seven letter word ending in 'ing') and:

'how many words do they estimate would take the form _ _ _ _ _ n _'

(a seven-letter word with 'n' as the 6th letter).

As the first condition (_ _ _ _ ing) is a subset of the second condition (_ _ _ _ _ n _) it should have a lower probability; however, people in the study guessed an average of 13.4 for the first condition and 4.7 for the second.

2.4.1.10   Pseudodiagnosticity

This fallacy represents the situation where individuals tend to seek out the wrong information in order to develop a posterior belief regarding a proposition [69]. Take the standard form of Bayesian inference, where two propositions (H_1 and H_2) are considered and two different sets of data (D_1 and D_2, but which are referred to in generality as D_i).

$$\Pr(H_1 \mid D_i) = \frac{\Pr(H_1)\Pr(D_i \mid H_1)}{\Pr(H_1)\Pr(D_i \mid H_1) + \Pr(H_2)\Pr(D_i \mid H_2)}$$

In order to be able to assign a probability to the posterior probability of proposition 1, as well as know the base rate for each proposition, information is required regarding the probability of one dataset given each of the propositions. For example, to calculate $Pr(H_1 \mid D_1)$ both $Pr(D_1 \mid H_1)$ and $Pr(D_1 \mid H_2)$ are needed. Those falling prey to the fallacy of pseudodiagnosticity seek out $Pr(D_1 \mid H_1)$ and $Pr(D_2 \mid H_1)$ and then incorrectly draw a conclusion regarding $Pr(H_1 \mid D_1)$.

2.4.2   RECURRENT CONCERNS WITH ACTIVITY LEVEL EVALUATIONS

It has been the authors' experiences that, when a scientific discipline or legal system first encounters evaluations given activity level propositions, particularly if they have only been used to source or sub-source level evaluations, a number of concerns or objections are initially raised. This was also found in a recent collection of opinions of forensic scientists in the United States on this topic [70].

These objections and concerns are often due to misunderstandings about what forensic evaluation is, how it is conducted and the role of the scientist in court. Biedermann *et al.* [50] outline a number of concerns they have heard raised regarding higher-level evaluations. These can be grouped into the main categories:

- That the information about the alleged crime is not known or specific
- That the scientific data available is not adequate
- That the evaluations infringe on the duties of the court

The following is added to this list from our own experiences:

- That the *LR*, or Bayesian inference, is incompatible with the legal process (this is not specific to propositions set at the activity level, but has been used in this context)

2.4.2.1   The Information about the Alleged Crime Is Not Known or Not Specific

There are several variations of the way this concern is raised with regard to the specifics of a case. One concern is that it is rare to receive detailed information about an alleged crime. Even if considering a single DNA transfer event arising from contact between a person and an object it is virtually impossible to receive the information that would allow us to address each of the factors that are known to affect DNA transfer. Information that may not be known includes the length of contact, the pressure of the contact, the amount of movement or friction that occurred, the amount of time since washing, the shedder status of the individual, the myriad of small subconscious activities undertaken prior to the contact (such as coughing, scratching, adjusting hair), the level of sweatiness when contact was made, etc.

Further to this type of uncertainty around the specifics of an event is when specific events are not known. For example, it might be that a DNA transfer from a defendant to a complainant's hands is being claimed as possibly occurring due to the complainant spending time at a party occurring in the defendant's home (i.e. DNA transfer through social interaction). Such a scenario distinguishes itself from the previous example of uncertainty as it is not known what activities had taken place during the social interaction (each of which will have all the uncertainties mentioned above). It may be that the defendant and complainant spoke, or danced, or drank together; they may have shared a glass; the complainant may have sat on the defendant's couch or used their bathroom. The possible activities that could have taken place (and will not have been noted to the point of being able to be recalled by the complainant) are virtually endless.

An extension to the concept of the previous paragraph is when there is no information available to inform the defence proposition. Such a situation is commonly referred to as a 'no comment' defence stance. It may be that the complainant is claiming some offence has occurred (for example, that they were assaulted in some way), and the only information known from the defence side is that they claim to have had no involvement in the alleged crime, and no knowledge of any events. In this situation, even general activity (such as social interaction) is not available. It is not clear whether the alternative proposition to the prosecution should be to dispute that the offence occurred at all, or whether it should dispute the person who carried it out.

Initially, this lack of knowledge on the framework of circumstances surrounding the alleged crime may seem insurmountable and evaluating observations given activity level propositions cannot proceed. However, the appropriate treatment of different factors that need to be considered is beyond the knowledge and abilities of an average juror. What is more, the members of the jury do not have the luxury of being able to decline to make a decision about the innocence or guilt of a defendant at the conclusion of a trial. Even with the lack of clear direction on what should be considered as an alternative to the prosecution case, what activities that may entail and the specifics of those activities, the jury must still ultimately deliberate. It is timely to remember the role of the expert witness in court. They are testifying to the observations in order to assist the court in their deliberations. As discussed earlier in this chapter, the different factors that are important to consider in the evaluation of forensic observations, the way in which they should be considered within the framework of circumstances, and in relation to each other, best sits with the scientist. Declining to carry out an evaluation due to an incomplete set of information unfairly places the burden of uncertainty on those that are less equipped to handle it appropriately. Imagine going to a doctor with intense chest pain and being told that you might be suffering from Magenblase syndrome. You are told by the doctor that surgery was possible, but due to the lack of information, they were declining to provide any opinion or further information and leaving it to you to decide how to proceed. This may seem quite serious and as you are not educated in the medical field you decide to proceed to surgery. However, if the doctor had been more forthcoming, they could have told you that Magenblase syndrome was simply a pain or feeling of fullness that is relieved by belching.

By its very nature, science deals with uncertainty. It does this best through the use of probability, which is designed to convey the level of uncertainty possessed about different aspects of the world around us. In later chapters of the book, it will be shown how uncertainty in the events that occurred can be accounted for in an evaluation and how tests can be performed for the impact that the lack of knowledge has on the evaluation. Often the experimental data being relied upon will cover many of the aspects of uncertainty. For example, if the aspect of pressure in a grab is unknown then experiments that mock a grabbing scenario, will likely present a distribution of pressure amounts and so the uncertainty is accounted for in the dataset (this is spoken of in more detail in Chapter 5). Also, it may be that the aspects which are uncertain are not those that will have the greatest impact on evaluation (something that is not necessarily known until the evaluation is carried out and tested).

The important point here is that incomplete knowledge of all aspects of the framework of circumstances is not a reason that evaluation given activity level cannot, or should not, proceed.

2.4.2.2 That the Scientific Data Available Is Not Adequate

The concern expressed here is that even if all the aspects of the framework of circumstances about the case were known there is insufficient data available to assign probabilities required for evaluation. This may be because scientific literature do not address the required transfer or persistence event, or the aspects of the study do not exactly align with the case circumstances, or even if bespoke experiments could be done, they can never mimic exactly the circumstances of the crime (possible because they are not known, as in the first concern previously mentioned). A variant of this concern is that there is too much variation in the results obtained to be reliable. For example, DNA transfer experiments for different types of contact have a notoriously large spread of possible DNA amounts.

The response to this point is very similar to the response to the previous point in that it is up to the scientist to make the best use of all available forensic literature in order to provide the best evaluation that forensics can provide at this point in time. Remember that the scientist is best equipped to handle these aspects of evaluation and can take into account the size of experimental data available, and any misalignments with the case through probabilistic adjustments or investigations into how impactful those probabilities are on the overall evaluation.

Even when 'hard' data is not available to the scientist (i.e. data based on controlled scientific studies) then 'soft' probability assignments can be used. These assignments are when the scientist assigns a probability based on their understanding of the aspects of the event being considered and their experience in the dynamics of transfer and persistence in the discipline in which they work (for example, see the use of expert elicitation in [71]). This may make the reader uncomfortable, as it seems highly subjective; however, subjectivity itself should not be shied away from (all evaluations are subjective in some way, in the data chosen, the models used and the assumptions made, or the prior probabilities set). A simple example can demonstrate this point. Imagine that a stabbing murder of a homeless person occurred in a burnt-out vehicle. The handle of the knife is tested and 100 ng of DNA is obtained, which produces a profile matching the defendant in the case. The prosecution's proposition is that the defendant stabbed the homeless person.

The defence claims that the knife belonged to the defendant, which had been lent to a friend. It was the friend's car that had been burnt, and they must have left the knife in the car, and it must have still been present after the car was burnt, picked up by the true offender and used to carry out the stabbing. Even without examining the literature, it is known that 100 ng of DNA is high when talking about contact DNA transfers. It is also known that the probability of that amount of DNA persisting through the intense heat of a fire is very small. A very small, but still conservative, probability could be assigned to such an event just based on the knowledge of DNA transfer and persistence, and the evaluation would provide strong support for the prosecution proposition over the defence proposition.

It is also worth noting that it is often the case that scientists who provide evaluations of DNA profile evidence given sub-source level propositions are asked questions about transfer and persistence on the stand. In their answers, the scientists are providing ad-hoc opinions, or evaluations, considering activity level considerations. They are simply doing so with only the knowledge they can recall on the stand, and without being able to place it in a logical context of all observations. Vuille *et al.* [47] state:

> If the question of interest to the fact-finder pertains to activities (such as possible contaminations, alternative transfers, innocent explanations for the presence of the material on the crime scene), the expert's report must contain a detailed description of her evaluation of the evidence under hypotheses mentioning activities. Indeed, reporting results under source level hypotheses in this context is unacceptable, because it renders the review by the defence of the expert's conclusions almost impossible: complex questions need to be assessed thoroughly, not on the stand with no preparation.

The impression here should not be that scientists can 'pluck' any number from the air and use it in an evaluation, claiming experience alone has guided them to that value. All probability assignments used in evaluations should be completely transparent to anyone who receives the evaluation. If there is a limitation to the data then this needs to be highlighted and its impact on the evaluation tested. These tests will be discussed later.

2.4.2.3 That the Evaluations Infringe on the Duties of the Court

This point is raised often in conjunction with the thought that it is best for the scientist to only provide sub-source level evaluations and just flag activity level considerations to the court. Typical objections come in the form 'you cannot state that the defendant had sex with the victim, that is up to the jury to decide!'. This line of thinking confuses the role of the scientist with respect to the propositions. The scientist certainly does not state that the two people have had sex. The scientist only comments on the probability of obtaining the observations if the complainant and defendant had sex compared to if some alternate activity took place (or a person other than the defendant had sex with the complainant). Evaluations carried out using activity level propositions may receive their greatest objection when the activity itself seems to speak to the ultimate issue (such as if the complainant was under the legal age at which sexual intercourse can be considered consensual); however, the same reasoning still applies.

Recall that if the scientist only provides evaluation about DNA observations using sub-source level propositions it is left up to the jury to decide, unaided, whether this means that sexual intercourse has taken place. Even providing information about the various factors that would need to be considered during an activity level evaluation (but not actually carrying out the evaluation) leaves the court in the precarious situation that they could simply transfer the sub-source level *LR* in their minds to apply it to the activity level issue.

2.4.2.4 That the *LR*, or Bayesian Inference, Is Incompatible with the Legal Process

Note that this criticism is not specific to propositions at the activity level but has been used in this context. These same criticisms have also been used against providing *LR*s for DNA evidence using sub-source level propositions (see [72] and response [73]). There are several arguments that have been put forward:

1) That the use of Bayesian inference requires the jury to consider prior probabilities for guilt or innocence and a presumption of innocence should equate to a prior probability for the guilt of 0.
2) That in the *LR* the activity being described by the prosecution is assumed to have occurred and this violates the presumption of innocence of the defendant
3) That the *LR* forces jurors to balance two sides of competing scenarios against each other, and that a fundamental tenet of the court process is that the defence need not provide any scenario, it is solely up to the prosecution to prove their case beyond reasonable doubt
4) That the *LR* forces jurors to consider that proof beyond reasonable doubt as a numerical value, which is inherently against the tenants of how reasonable doubt should be considered

The first of these issues relates to the prior probability of guilt (often called the base rate of guilt). Note that presumption of innocence does not mean a prior probability of 0. If it were zero then the implication is that no further evidence, regardless of how compelling, could even change the mind of the jurors about the innocence of the defendant (and then why bother having a trial). Nor is the prior probability of guilt or innocence 0.5; this is known as the base rate fallacy [74]. A presumption of innocence refers to the fact that at the start of the trial, before any testimony has been heard, the defendant should be considered no more probably guilty than anyone else in the population. This is clearly not zero but is likely to be a small number (particularly if the population being considered is the entire city, state or even country).

The second point is again falling prey to the fallacy of the transposed conditional. As stated in the principles of evaluation given in Chapter 1, the role of the scientist is to comment on the observations and not the propositions. The propositions are not assumed to be occurring, other than to consider the probability of obtaining the observations if one proposition was true compared to if a competing proposition was true. Consider the following thought experiments:

> The ratio of men to women in a country is called the sex ratio. Qatar has one of the highest sex ratios with approximately 3.4 males per each female across all age brackets. In other words, the probability of being male if you are from Qatar is approximately 0.77. This does not assume you are from Qatar, it simply makes a statement about being male if you were from Qatar.

The next point is nuanced. If there are two mutually exclusive and exhaustive outcomes for a trial (the guilt or innocence of a person on trial), the belief in guilt beyond some level (whether that be reasonable doubt or some number) must then be inversely proportional to disbelief in the alternative. Hence there must be two options being weighed against each other, whether that is how they are being thought of or not. There are risks of not considering both sides and how well they compare. Imagine that a prosecution case is very strong and a jury decides to convict on that basis; of course, if the defence case was made and was equally strong the jury could no longer convict, but this sort of reasoning is only possible with a comparison of the two sides. Of course, the jury are under no obligation to think in any particular way when carrying out deliberations and it is not being suggested they should think in the way described above. Nor do they need to take the *LR* as provided by the expert and either accept it without questions or accept it fully or try to incorporate it numerically into their deliberations. The jury can use the information provided to them throughout the trial in any manner they see fit when carrying out deliberations.

The argument for the final point comes from the thinking that assigning a probability beyond a reasonable doubt is to accept that a percentage of innocent individuals will be convicted and that this cannot be an aim of the legal system. Perhaps best described by an analogy given by Nessan [75]:

In an enclosed yard are 25 identically dressed prisoners and a prison guard. The sole witness is too far away to distinguish individual features. He sees the guard, recognizable by his uniform, trip and fall, apparently knocking himself out. The prisoners huddle and argue. One breaks away from the others and goes to a shed in the corner of the yard to hide. The other twenty-four set upon the fallen guard and kill him. After the killing, the hidden prisoner emerges from the shed and mixes with the other prisoners. When the authorities later enter the yard, they find the dead guard and the twenty-five prisoners. Given these facts, twenty-four of the twenty-five are guilty of murder. Suppose that a murder indictment is brought against one of the prisoners – call him Prisoner 1. If the only evidence at trial is the testimony of our distant witness, it would seem that a verdict of acquittal must be directed for the defendant. The prosecution's best case is purely statistical.

The argument being made by Nessan here is that if a probabilistic threshold of say 95% was set for beyond reasonable doubt, then all 25 prisoners would be convicted of the crime, knowing full well that this would result in one innocent man being convicted. This, according to Nessan, is fundamentally opposed to the idea of beyond reasonable doubt. There are a few different points to think about in this example. Accept for now that a threshold for 'reasonable doubt' should not exist. The first issue with Nessan's argument is that the jury are still entitled to hear all the numerical evidence. In Nessan's example this is the number of individuals in the yard and the number who were known to be involved in the crime. In DNA evidence evaluation the numerical evidence is an *LR*. The jury can incorporate the numerical information into their own deliberations in whatever way they wish. This argument (that the *LR* given by the scientists does not automatically have to be the *LR* of the juror) was made by Gittelson *et al.* [76] in response to Lund *et al.* [77]. The second argument against this example lies in the rights of jurors to define their own reasonable doubt. If there is no way to put a threshold on reasonable doubt, and as suggested, this is purely a construction of the beliefs of the court then who is to say that there is only a 1 in 25 chance that the person on trial wasn't involved in the homicide is not enough to convict all 25 prisoners to a level of beyond reasonable doubt? An even more extreme scenario could be considered where there were only two prisoners in the yard and one was involved in the homicide. Perhaps 12 jurors would still find this good enough to be beyond reasonable doubt (this of course would require a very broad definition of what is reasonable) to convict either or both the prisoners of the crime. The point here is that in such a situation, the judge is likely to warn against a verdict of guilt, cautioning the burden of proof required for reasonable doubt against the very real probability that an innocent man could be convicted (in reality the trial would likely never even be able to start if this was the only evidence available). This caution cannot be done without some inherent belief that a probability of 0.5 is not enough to be beyond a reasonable doubt.

Note that this is not an attempt to make the argument that a numerical threshold should be applied to the concept of beyond reasonable doubt. The only point being made is that information provided in numerical form to a court (including in the form of an *LR*) does not act in opposition to the process of deliberation. As shown in the prisoner example, there is even likely to be some inherent consideration of probability (although not in a conscious or strictly numerical sense) that is indeed present during deliberation; otherwise, how could one know what was reasonable? After all, probability simply reflects our individual beliefs.

2.5 INTRODUCING THE CASES

Throughout the chapters of this book, four case examples will be used to demonstrate the concepts that are being discussed. These cases all have a foundation in real casework that has been evaluated by the authors and are chosen as they represent a number of key points, important to an evaluation in general. These cases will be referred back to throughout almost every chapter of the book. Using these example cases will show how a case progresses from first being considered, all the way through to reporting.

Each of the four cases possesses multiple elements that will need to be considered during evaluation. As you read through the cases, think about how the different elements in this chapter apply to each case. The elements will be expanded on as the cases are explored in depth throughout the book.

The first case was mentioned in Section 2.2.6 when describing the case issue of an absence of DNA recovery. This case has been chosen as it represents a simple case scenario but reveals the many factors about the framework of circumstances of the case, the many assumptions required when developing propositions and the considerations that go into assigning probabilities. The second case introduces more complex issues of transfer and persistence, and, in particular, demonstrates the importance of dependencies within the evaluation. The third case introduces the use of body fluid tests. The evaluation of this case will show the importance of distinguishing DNA from cellular sources to avoid making the source level fallacy that has been mentioned earlier in this chapter. The final case is used as an example of a more advanced evaluation that models DNA amounts. The same considerations mentioned throughout this chapter need to be taken into account but within the more advanced construction.

2.5.1 CASE 1: THE ATTEMPTED KIDNAP

This case is based on the Australian case of R v Drummond. In 2010 the complainant, a high school student, was walking home from school along Prospect Road in Adelaide, South Australia. According to the complainant, she saw a white station wagon drive towards her, and she became worried and entered the vehicle's licence plate number on her mobile phone. The vehicle stopped next to her; a male exited the vehicle, grabbed her by the arm and attempted to pull her into the car. During this altercation, the complainant was able to break free from the defendant by pushing him away and hitting him around the collar bone area. The defendant then drove off in his car. The complainant called the police stating that she had been attacked and provided the licence plate. Shortly after, police attended the home of the defendant and seized the clothing he had been wearing at the time in question. The following day police seized as evidence the clothes worn by the complainant.

The man arrested for the crime was Adrian Drummond, a 39-year-old Glazier, who, according to his version of events, had been driving home along Prospect Road at the time of the alleged offence but had neither stopped the car nor seen the complainant. Drummond also claimed not knowing the complainant and never to have had any previous contact with her to his knowledge.

The forensic laboratory in South Australia received the top of Drummond and the top of the complainant for testing. In each case a piece of tape (called a tapelift) was used to sample the surface of the garments, to look for potential DNA. In the case of the top from the complainant, the purpose of the tapelift was to identify DNA from an attacker and compare this to Drummond. In the case of Drummond, the purpose of the tapelift was to see whether there was any indication of the presence of DNA from the complainant. The forensic observations were as follows:

- On the sample from the top of the complainant: A mixed DNA profile was obtained, explainable by a contribution of DNA from three people. The complainant accounted for the major component of the mixed DNA profile. The second component was male and was later identified as a school friend whom the complainant had hugged shortly before leaving school that day. Drummond was excluded as being the final DNA donor to the mixture. A further test was carried out to target male DNA (targeting regions of DNA on the Y chromosome), and this identified a single male contributor, who could be accounted for as the friend of the complainant. In summary, the DNA results from the top of the complainant showed DNA from the complainant, the friend of the complainant and a low-level female who is assumed to be unrelated to the alleged offence (as the attacker in the scenario was definitely male)
- On the sample from the top of Drummond: A mixed DNA profile was obtained, explainable by a contribution of DNA from two people. Drummond accounted for the major

component of the DNA detected. The complainant was excluded as a contributor of DNA to the sample. The second contributor of DNA to this sample was from an unidentified source, assumed to be unrelated to the alleged offence

In court, testimony was given to the absence of DNA from the complainant on the top of Drummond and the absence of DNA from Drummond on the top of the complainant. Focus turned to factors of transfer and persistence, which indicate that the court was interested in the significance of the absence of DNA transfer between the complainant and Drummond, given the different versions of events being put forward. An evaluation of the observations given such activity level considerations had not been carried out, as this was not a service offered by the laboratory at the time. In summary, the scientist testified that:

1. The likelihood of DNA being left on a surface is dependent in part on the nature of the surface, the nature of the contact with that surface and a person's propensity to shed DNA.
2. DNA is more likely to be left on a surface such as wood or fabric than on a surface such as glass.
3. DNA is more likely to be left on a surface where there has been prolonged or vigorous contact.
4. A failure to obtain DNA from an item does not preclude the possibility that contact with that item occurred.
5. The DNA testing conducted on the clothing of the complainant and the defendant did not preclude contact having taken place between the complainant and the defendant.
6. A small study into the success of sampling at Forensic Science SA disclosed that DNA that may be uploaded onto the database is recovered in about 10% of cases. This study related to samples where it was unknown whether DNA had in fact been left on each sample and, as a result, was only a 'sort of indication' of how useful DNA samples are for uploading onto the database.

In 2012 Drummond was convicted of the attempted kidnapping of the complainant and sentenced to 5 years imprisonment. Drummond appealed to the Court of Criminal Appeals and the High Court, but both failed. In 2013 South Australia introduced new legislation that allowed new rights of appeal when 'fresh and compelling' evidence was available. In 2015 Drummond appealed under this new legislation and the conviction was overturned based on 'flawed DNA evidence'.

This case is interesting and worthy of an example for a few reasons. One of the reasons is that it demonstrates the ability to carry out an evaluation of the observations when they are exclusionary. A sometimes-used catchphrase for those wishing to explain that exclusionary DNA results do not preclude the offence from having occurred is 'absence of evidence is not evidence of absence'. This adage, however, is not the case (at least not without further clarification). There needs to be a distinction between the absence of evidence meaning that a particular source of DNA was not present in a sample tested and that no testing (and hence no evidence at all) is available. The former is an informative result and does indeed provide some support for the absence of activity, whereas the latter is simply a lack of any information. In the appeals, two defence experts were engaged, and one said in their report:

no conclusions about contact or the lack of it may be drawn from the absence of a DNA result, i.e. a 'negative' DNA result. The 'negative' results obtained from the garments in this case provide absolutely no evidence to prove that contact between victim and accused had or had not taken place.

This opinion aligns with the absence of evidence adage, and it will be shown that the absence of finding any cross transfer of DNA does indeed provide some support that no attempted kidnapping took place.

There are a number of issues identified from this case scenario and the trials that ensued. For greater detail, the interested reader can read the original ruling [78, 79], or a paper written on the evaluation of the DNA observations in the Drummond case, which is the basis of the example provided in the following chapters [39].

2.5.2   CASE 2: THE THREE BURGLARS

The circumstances surrounding this case are that a woman (the complainant, C) arrived home after having dinner at a restaurant. As she entered her home, she was confronted by three men, all wearing balaclavas running down the stairs from the upper story of her house. One of the men grabbed her by the hair and forced her to the ground by pulling her hair and threatening her with a knife. C was then dragged by the hair into her bathroom. C stated that during the ordeal the offender who grabbed her hair was not wearing gloves. Once C was in the bathroom the offenders told her to stay there and they left her home. Immediately after the offence C called the police who arrived at her home within 30 minutes.

Inside C's home the police found a knife on the floor that C recognised as being the knife she was threatened with. She also identified it as a knife that she owned and that it had come from the knife block in her kitchen. The following samples were taken for DNA profiling approximately one hour after the incident occurred:

- A swab of the handle of the knife and
- A swab of C's hair
- A reference sample from C
- A reference sample from C's husband, H, who lives in the house with C

C stated that she had not washed her hair since the offence.

Later that night (approximately 3 hours after the offence) a group of men in nearby streets were stopped and questioned about the offence. One man (the defendant, D) was arrested and questioned. He claimed that he had nothing to do with the burglary, did not know the victim and had never been to her house or interacted with her socially. However, D did admit to being involved with a group of individuals that he suspected committed burglaries similar to this one. D recently lent a pair of gloves to one of these individuals but refused to provide that individual's name and so police were unable to identify him or take his reference sample. Police take the following samples from D:

- A reference from D

All samples that were taken were sent to the forensic laboratory. No tests were carried out for cell type determination at the forensic laboratory, but all samples were submitted for DNA profiling.

From the handle of the knife, a mixed DNA profile was obtained that appeared to originate from three individuals. The profile could be completely accounted for by the presence of DNA from C, H and D and an *LR* was calculated with the sub-source propositions:

- H_p: The DNA originates from C, H and D
- H_d: The DNA originates from C, H and an unknown individual

The *LR* produced from this sub-source evaluation is 100 million. In other words, the probability of obtaining the DNA profile from the handle of the knife is 100 million times higher if the sources of DNA are C, H and D rather than C, H and an unknown person. This provides extreme support for D having contributed DNA to the knife handle sample, and it is conceded by both parties that this is the case.

From the hair sample, a mixture was obtained that appeared to originate from two individuals. The profile could be completely accounted for by the presence of DNA from C and D and an LR was calculated with the sub-source propositions:

- H_p: The DNA originates from C and D
- H_d: The DNA originates from C and an unknown individual

The *LR* produced from this sub-source evaluation is 1 trillion. In other words, the probability of obtaining the DNA profile from the hair of C is 1 trillion times higher if the sources of DNA are C and D rather than C and an unknown person. This provides extreme support for D having contributed DNA to the hair sample, and it is conceded by both parties that this is the case.

2.5.3 CASE 3: THE FAMILY ASSAULT

The circumstances of this case are that A 24-year-old girl (C), who normally lives with her biological mother (M) and father (F) has stayed for a week at her older brother's (D) house. A friend of the girl receives a phone call from the girl stating that her brother has bitten her on the vagina, over her underwear. The friend picks up the girl and they go immediately to the police, where the underwear is seized and a reference from the girl is taken. The police then arrest the brother and take a reference DNA sample from him. The defence states that she has been staying at his house, but that no biting occurred.

The underwear is examined at the local forensic science centre and the following was found:

1. Faecal staining was present on the inner and outer crotch of C's underwear.
2. An RSID test for saliva on the crotch of C's underwear gave a positive reaction.
3. A tapelift of the outer crotch of C's underwear yielded a single source autosomal DNA profile that matched C. The quantification result revealed the presence of a low level of male DNA. The presence of the complainant's DNA in such high amounts meant that male DNA was not able to be profiled using autosomal profiling systems.
4. Y-STR profiling of the DNA extract from the outer crotch tapelift of C's underwear yielded a single source profile that matched D's Y-STR reference.

2.5.4 CASE 4: THE GUN IN THE LAUNDRY

The circumstances surrounding this case are that a defendant (D) is accused of robbing a service station armed with a handgun. Two days after the robbery police attend D's home and carry out a search for the firearm. The firearm is found in a laundry basket covered with D's unwashed laundry. D claims that he has never seen the firearm and that it must have been put in his laundry basket by someone else (an Alternate Offender, AO). D states that he last did his laundry yesterday and so whoever put the firearm in his laundry must have done so within the last 24 hours. The defendant states that he has not been home much in the last 24 hours and couldn't tell if anyone had been in the house.

Police swab the firearm for DNA and submit it to the forensic science laboratory for testing. At the request of the forensic science laboratory, the Police also tapelift several items of clothing from the laundry and submit those samples for DNA testing. Also submitted is a reference from D for comparison to the samples taken from the firearm and laundry.

All samples that have been taken are sent to the forensic laboratory. There are no tests carried out for cell type determination at the forensic laboratory, but all samples are submitted for DNA profiling.

From a swab of the entire outer surface of the firearm, a profile was obtained that appeared to originate from four individuals. An *LR* was calculated with the sub-source propositions:

- H_p: The DNA originates from D and three unknown individuals
- H_d: The DNA originates from four unknown individuals

The *LR* produced from this sub-source evaluation is 100 million. In other words, the probability of obtaining the DNA profile from a firearm is 100 million times higher if the sources of DNA are D and three unknown individuals rather than four unknown individuals. This provides extremely strong support for D having contributed DNA to the firearm, and it is conceded by both parties that this is the case.

2.6   FINAL WORDS

The information in this chapter provides some foundation for what evaluations given activity level propositions are, why they should be used and counters some common misunderstandings about their use. From this point onwards, in the book, the information becomes more focussed on setting propositions, finding or using data, carrying out evaluations and testing the robustness of those evaluations. If there are concepts that have not made sense in these first chapters do not be discouraged. Continue through the book and come back to re-read these chapters at a later stage, and you will find that the information provided has a more practical context.

2.7   PRACTICE QUESTIONS

Q1) What is the issue with the statement 'the absence of evidence is evidence of absence'?

Q2) When might it be important to consider contamination? When might it not be important to consider contamination?

Q3) What are the different sources of unknown DNA that might need to be considered in a case?

Q4) Which of the following statement(s) are incorrect? How could they be corrected? Could the clarity of any of the correct statement(s) be improved?

   A.   The DNA profile from the evidence matches the defendant. The probability of this match is 1 million times higher if the defendant is the source of DNA than if someone else is the source of the DNA.
   B.   The DNA profile from the evidence matches the defendant. The probability of another person in the population having this profile is 1 in 1 million.
   C.   The DNA profile from the evidence matches the defendant. The probability of the defendant being the source of DNA then is 1 million times higher than someone else being the source of the DNA.
   D.   The DNA profile from the evidence matches the defendant. It is 1 million times higher if the defendant is the source of DNA than if someone else is the source of the DNA.

Q5) If a case cannot be evaluated given activity level propositions (because the experimental data required to do the evaluation is not available), is it ok to simply report the sub-source level propositions? Why/why not?

Q6) You have a single source evidence DNA profile that matches a defendant. You provide an *LR* using the propositions:

- The defendant is the source of the DNA
- Someone other than the defendant is the source of the DNA

A defence lawyer puts you on the stand that you should not be allowed to use this *LR* in court because you have assumed the defendant is present in the DNA. How would you respond to this claim? Does the lawyer have a point?

Q7) Under what circumstances might it be important to consider persistence? Are there any circumstances when it is not important to do so?

Q8) What are some different types of persistence events that could arise in casework?

Q9) What types of contamination could be considered within an evaluation? Do the types of contamination change when considering sub-source level propositions compared to activity level propositions?

Q10) How many people would you require to be in a room in order to have at least a 75% chance of finding two with the same birthday?

REFERENCES

1. R. Cook, I.W. Evett, G. Jackson, P.J. Jones, J.A. Lambert, A hierarchy of propositions: Deciding which level to address in casework, *Science & Justice* 38(4) (1998) 231–240.
2. R. Cook, I.W. Evett, G. Jackson, P.P. Jones, J.A. Lambert, A model for case assessment and interpretation, *Science and Justice* 38(3) (1998) 151–156.
3. G. Jackson, C. Aitken, P. Roberts, *Case Assessment and Interpretation of Expert Evidence: Guidance for Judges, Lawyers, Forensic Scientists and Expert Witnesses*, Royal Statistical Society, 2015.
4. I.W. Evett, P.D. Gill, G. Jackson, J. Whitaker, C. Champod, Interpreting small quantities of DNA: The hierarchy of propositions and the use of Bayesian networks, *Journal of Forensic Sciences* 47(3) (2002) 520–530.
5. P. Gill, A. Jeffreys, D. Werrett, Forensic application of DNA 'fingerprints', *Nature* 318(12 December) (1985) 577–579.
6. K.B. Mullis, F.A. Faloona, S. Scharf, R. Saiki, G. Horn, H. Erlich, Specific enzymatic amplification of DNA in vitro: The polymerase chain reaction, *Cold Spring Harbor Symposia on Quantitative Biology* 51 (1986) 263–273.
7. T. Geng, R. Novak, R.A. Mathies, Single-cell forensic short tandem repeat typing within microfluidic droplets, *Analytical Chemistry* 86(1) (2014) 703–712.
8. I.W. Evett, J.S. Buckleton, Some aspects of the Bayesian approach to evidence evaluation, *Journal of the Forensic Science Society* 29(5) (1989) 317–324.
9. R.A. Van Oorschot, M. Jones, DNA fingerprints from fingerprints, *Nature* 387(6635) (1997) 767.
10. M. Coble, J.-A. Bright, Probabilistic genotyping software: An overview, *Forensic Science International: Genetics* 38 (2019) 219–224.
11. P. Gill, J.P. Whitaker, C. Flaxman, N. Brown, J.S. Buckleton, An investigation of the rigor of interpretation rules for STR's derived from less than 100 pg of DNA, *Forensic Science International* 112(1) (2000) 17–40.
12. D.J. Balding, J. Buckleton, Interpreting low template DNA profiles, *Forensic Science International: Genetics* 4(1) (2009) 1–10.
13. D. Balding, Evaluation of mixed-source, low-template DNA profiles in forensic science, *Proceedings of the National Academy of Sciences of USA* 110(30) (2013) 12241–12246.
14. H. Haned, Forensim: An open-source initiative for the evaluation of statistical methods in forensic genetics, *Forensic Science International: Genetics* 5(4) (2011) 265–268.
15. K. Lohmueller, N. Rudin, Calculating the weight of evidence in low-template forensic DNA casework, *Journal of Forensic Sciences* 58(1) Supplement 1 (2013) s234–259.

16. D. Taylor, J.-A. Bright, J. Buckleton, The interpretation of single source and mixed DNA profiles, *Forensic Science International: Genetics* 7(5) (2013) 516–528.

17. M.W. Perlin, A. Sinelnikov, An information gap in DNA evidence interpretation, *PLoS ONE* 4(12) (2009) e8327.

18. M.W. Perlin, M.M. Legler, C.E. Spencer, J.L. Smith, W.P. Allan, J.L. Belrose, B.W. Duceman, Validating TrueAllele® DNA mixture interpretation, *Journal of Forensic Sciences* 56 (2011) 1430–1447.

19. Ø. Bleka, EuroForMix: An open source software based on a continuous model to evaluate STR DNA profiles from a mixture of contributors with artefacts, *Forensic Science International: Genetics* 21 (2016) 35–44.

20. J. Pearl, Bayesian networks: A model of self-activated memory for evidential reasoning, in: Proceedings of the 7th Conference of the Cognitive Science Society, University of California (1985) 15–17.

21. P. Gill, Analysis and implications of the miscarriages of justice of Amanda Knox and Raffaele Sollecito, *Forensic Science International: Genetics* 23 (2016) 9–18.

22. J. Vuille, A. Biedermann, F. Taroni, The importance of having a logical framework for expert conclusions in forensic DNA profiling: Illustrations from the Amanda Knox case, in: C.R. Huff and M. Killias, (Eds.) *Wrongful Convictions and Miscarriages of Justice: Causes and Remedies in North American and European Criminal Justice Systems*, Routledge, New York, 2013, 137–159.

23. B. Szkuta, K.N. Ballantyne, B. Kokshoorn, R.A.H. van Oorschot, Transfer and persistence of non-self DNA on hands over time: Using empirical data to evaluate DNA evidence given activity level propositions, *Forensic Science International: Genetics* 33 (2018) 84–97.

24. T.J. Briggs, The probative value of bloodstains on clothing, *Medicine, Science and the Law* 18(2) (1978) 79–83.

25. R.A.H. van Oorschot, B. Szkuta, G.E. Meakin, B. Kokshoorn, M. Goray, DNA transfer in forensic science: A review, *Forensic Science International: Genetics* 38 (2019) 140–166.

26. D. Taylor, A. Biedermann, L. Samie, K.-M. Pun, T. Hicks, C. Champod, Helping to distinguish primary from secondary transfer events for trace DNA, *Forensic Science International: Genetics* 28 (2017) 155–177.

27. T. Kaplan, Monte Sereno murder case casts doubt on DNA evidence, *San Jose Mercury News*, June 28 (2014).

28. R v Pfennig, *SASCFC* 27, 2018.

29. D. Taylor, L. Samie, C. Champod, Using Bayesian networks to track DNA movement through complex transfer scenarios, *Forensic Science International: Genetics* 42 (2019) 69–80.

30. P. Gill, C.H. Brenner, J.S. Buckleton, A. Carracedo, M. Krawczak, W.R. Mayr, N. Morling, M. Prinz, P.M. Schneider, B.S. Weir, DNA commission of the international society of forensic genetics: Recommendations on the interpretation of mixtures, *Forensic Science International* 160 (2006) 90–101.

31. P. Gill, *Misleading DNA Evidence: Reasons for Miscarriages of Justice*, Elsevier, 2014.

32. F.H.R. Vincent, *Inquiry into the Circumstances that Led to the Conviction of Mr Farah Abdulkadir Jama 2010*.

33. E. Gasiorowski, *The Mystery of the Phantom of Heilbronn*, https://www.iso.org/news/2016/07/Ref2094. html, 2016 (accessed 14/02/2018).

34. C. Himmelreich, *Germany's Phantom Serial Killer: A DNA Blunder.* content.time.com/time/world/article/0,08599,1888126,00.html, 2009 (accessed 14/02/2018).

35. T. Paterson, *DNA Blunder Creates Phantom Serial Killer.* www.independent.co.uk/news/world/europe/dna-blunder-creates-phantom-serial-killer-1655375.html, 2009 (accessed 14/02/2018).

36. G.J.A. Knoops, *Redressing Miscarriages of Justice. Practice and Procedures in (International) Criminal Cases* (2nd revised ed.), Martinus Nijhoff Publishers, Leiden, 2013.

37. D. Taylor, L. Volgin, B. Kokshoorn, C. Champod, The importance of considering common sources of unknown DNA when evaluating findings given activity level propositions, *Forensic Science International: Genetics* (2021) PMID 102518.

38. W. Wagenaar, False confessions after repeated interrogation: The Putten murder case, *European Review* 10(4) (2002) 519–537.

39. D. Taylor, The evaluation of exclusionary DNA results: A discussion of issues in R v. Drummond, *Law, Probability and Risk* 15(1) (2016) 175–197.

40. S.M. Willis, L. McKenna, S. McDermott, G. O'Donell, A. Barrett, B. Rasmusson, A. Nordgaard, C.E.H. Berger, M.J. Sjerps, J.-J. Lucena-Molina, G. Zadora, C. Aitken, T. Lovelock, L. Lunt, C. Champod, A. Biedermann, T.N. Hicks, F. Taroni, *ENFSI Guideline for Evaluative Reporting in Forensic Science*, European Network of Forensic Science Institutes. http://enfsi.eu/sites/default/files/documents/external_publications/m1_guideline.pdf, 2015.

41. Association of Forensic Science Providers, Standards for the formulation of evaluative forensic science expert opinion, *Science & Justice* 49(3) (2009) 161–164.
42. P. Roberts, C. Aitken, *The Logic of Forensic Proof: Inferential Reasoning in Criminal Evidence and Forensic Science: Guidance for Judges, Lawyers, Forensic Scientists and Expert Witnesses*, Royal Statistical Society, 2015.
43. R. Puch-Solis, P. Roberts, S. Pope, C. Aitken, *Assessing the Probative Value of DNA: Guidance for Judges, Lawyers, Forensic Scientists and Expert Witnesses*, Royal Statistical Society, 2015.
44. C. Aitken, P. Roberts, G. Jackson, Fundamentals of probability and statistical evidence in criminal *Proceedings: Guidance for Judges, Lawyers, Forensic Scientists and Expert Witnesses*, Royal Statistical Society, 2015.
45. C. Berger, J. Buckleton, C. Champod, I. Evett, G. Jackson, Expressing evaluative opinions: A position statement, *Science & Justice* 51(1) (2011) 1–2.
46. F. Taroni, A, Biedermann J. Vuille, N. Morling, Whose DNA is this? How relevant a question? (a note for forensic scientists), *Forensic Science International: Genetics* 7 (2013) 467–470.
47. J. Vuille, L. Luparia, F. Taroni, Scientific evidence and the right to a fair trial under Article 6 ECHR, *Law, Probability and Risk* 16 (2017) 55–68.
48. T. Hicks, J. Buckleton, V. Castella, I. Evett, G. Jackson, A logical framework for forensic DNA interpretation, *Genes* 13(6) (2022) 957.
49. P. Gill, T. Hicks, J.M. Butler, E. Connolly, L. Gusmão, B. Kokshoorn, N. Morling, R.A.H. van Oorschot, W. Parson, M. Prinz, P.M. Schneider, T. Sijen, D. Taylor, DNA commission of the International society for forensic genetics: Assessing the value of forensic biological evidence: Guidelines highlighting the importance of propositions. Part II: Evaluation of biological traces considering activity level propositions, *Forensic Science International: Genetics* 44 (2020) 102186.
50. A. Biedermann, C. Champod, G. Jackson, P. Gill, D. Taylor, J. Butler, N. Morling, T.H. Champod, J. Vuille, F. Taroni, Evaluation of forensic DNA traces when propositions of interest relate to activities: Analysis and discussion of recurrent concerns, *Frontiers in Genetics* 7 (2016) 215–220.
51. S. Gittelson, A. Biedermann, S. Bozza, F. Taroni, Bayesian networks and the value of the evidence for the forensic two-trace transfer problem, *Journal of Forensic Sciences* 57(5) (2012) 1199–1216.
52. A. Biedermann, S. Bozza, F. Taroni, Probabilistic evidential assessment of gunshot residue particle evidence (Part I): Likelihood ratio calculation and case pre-assessment using Bayesian networks, *Forensic Science International* 191 (2009) 24–35.
53. R. Wieten, J de Zoete, B. Blankers, B. Kokshoorn, The interpretation of traces found on adhesive tapes, *Law, Probability and Risk* 14(4) (2015) 305–322.
54. M. Breathnach, E. Moore, Oral intercourse or secondary transfer? A Bayesian approach of salivary amylase and foreign DNA findings, *Forensic Science International* 229 (2013) 52–59.
55. M. Breathnach, E. Moore, Background levels of salivary-α-amylase plus foreign DNA in cases of oral intercourse: A female perspective, *Journal of Forensic Sciences* 60(6) (2015) 1563–1570.
56. M. Breathnach, L. Williams, L. McKenna, E. Moore, Probability of detection of DNA deposited by habitual wearer and/or the second individual who touched the garment, *Forensic Science International: Genetics* 20 (2016) 53–60.
57. J.E. Allard, The collection of data from findings in cases of sexual assault and the significance of spermatozoa on vaginal, anal and oral swabs, *Science & Justice* 37 (1997) 99–108.
58. L. McKenna, Understanding DNA results within the case context: Importance of the alternative proposition, *Frontiers in Genetics* 4(242) (2013) 1–4.
59. P. Margot, Commentary on: The need for a research culture in the forensic sciences, *UCLA Law Review* 58 (2011) 795–801.
60. J.M. Butler, H. Iyer, R. Press, M.K. Taylor, P.M. Vallone, S. Willis, *DNA Mixture Interpretation: A NIST Scientific Foundation Review (NISTIR 8351-DRAFT)*, National Institute of Standards and Technology: U.S. Department of Commerce, Gaithersburg, 2021.
61. W.C. Thompson, E.L. Schumann, Interpretation of statistical evidence in criminal trials: The prosecutors fallacy and the defence attorneys fallacy, *Law and Human Behavior* 11 (1987) 167–187.
62. D. Kahneman, A. Tversky, On the psychology of prediction, *Psychological Review* 80(4) (1973) 237–251.
63. J.J. Koehler, Error and exaggeration in the presentation of DNA evidence at trial, *Jurimetrics Journal* 34(1) (1993) 21–39.
64. G.E. Meakin, B. Kokshoorn, R. van Oorschot, B. Szkuta, Evaluating forensic DNA evidence: Connecting the dots, *Wiley Interdisciplinary Reviews: Forensic Science* 3 (2020).
65. J.J. Koehler, W.C. Thompson, Mock Jurors' reactions to selective presentation of evidence from multiple-opportunity searches, *Law and Human Behavior* 30(4) (2006) 455–468.

66. M.J. Saks, J.J. Koehler, The individualisation fallacy in forensic science evidence, *Vanderbilt Law Review* 61(1) (2008) 199–219.

67. D.A. Stoney, What made us ever think we could individualize using statistics, *Journal of the Forensic Science Society* 31(2) (1991) 197–199.

68. A. Tversky, D. Kahneman, Extensional versus intuitive reasoning: The conjunction fallacy in probability judgment, *Psychological Review* 90(4) (1983) 293–315.

69. M.E. Doherty, C.R. Mynatt, R.D. Tweney, M.D. Schiavo, Pseudodiagnosticity, *Acta Psychologica* 43(2) (1979) 111–121.

70. Y.J. Yang, M. Prinz, H. McKiernan, F. Oldoni, American forensic DNA practitioners' opinion on activity level evaluative reporting, *Journal of Forensic Sciences* 67(4) (2022) 1357–1369.

71. T.R.D. Wolff, A.J. Kal, C.E.H. Berger, B. Kokshoorn, A probabilistic approach to body fluid typing interpretation: An exploratory study on forensic saliva testing, *Law, Probability and Risk* 14(4) (2015) 323–339.

72. B. Stiffelman, No longer the gold standard: Probabilistic genotyping is changing the nature of DNA evidence in criminal trials, *Berkeley Journal of Criminal Law* 24 (2019) 110–146.

73. J. Buckleton, B. Robertson, J. Curran, C. Berger, D. Taylor, J.-A. Bright, T. Hicks, S. Gittelson, I. Evett, S. Pugh, G. Jackson, H. Kelly, T. Kalafut, F.R. Bieber, A review of likelihood ratios in forensic science based on a critique of Stiffelman "No longer the Gold standard: Probabilistic genotyping is changing the nature of DNA evidence in criminal trials", *Forensic Science International* 310 (2020) 110251.

74. M. Bar-Hillel, The base-rate fallacy in probability judgments, *Acta Psychologica* 44(3) (1980) 211–233.

75. C. Nesson, Reasonable doubt and permissive inferences: The value of complexity, *Harvard Law Review* 1225 (92) (1979) 1187.

76. S. Gittelson, C.E.H. Berger, G. Jackson, I.W. Evett, C. Champod, B. Robertson, J.M. Curran, D. Taylor, B.S. Weir, M.D. Coble, J.S. Buckleton, A response to "Likelihood ratio as weight of evidence: A closer look" by Lund and Iyer, *Forensic Science International* 288 (2018) e15–e19.

77. H. Iyer, S. Lund, Likelihood ratio as weight of forensic evidence: A closer look, *Journal of Research, National Institute of Standards and Technology, Gaithersburg, MD* 122(27) (2017).

78. R v Drummond, *Supreme Court of South Australia - SASCFC* 135, 2013.

79. R v Drummond (No. 2), *Supreme Court of South Australia - SASCFC* 82, 2015.

3 Transfer, Persistence, Prevalence and Recovery of Biological Traces

Bas Kokshoorn

CONTENTS

3.1   INTRODUCTION

When addressing issues at the activity level, we are assessing the probability of observations given specific activities. To do this, we have to consider factors that are different from those considered relevant when evaluating findings given sub-source or source level propositions. With propositions at the sub-source level we would consider, for instance, the probability of observing a specific allele in the relevant population, the number of contributors to a mixed DNA profile or the occurrence of certain stochastic effects like allele drop-out or heightened stutter in STR DNA profiles generated from a low template DNA extract. With source level propositions we would consider type I and type

II errors in body fluid testing, and potentially the association between the body fluid and a specific donor of DNA in a mixture.

The primary factors that we would consider relevant when evaluating observations (and we will elaborate on what these observations may be later in this chapter) given activity level propositions relate to the dynamics of traces. How do biological traces 'behave' in time and physical space? The dynamics of traces given these activities will be dictated by their transfer to a surface, their persistence on that surface from the moment of deposition to them being sampled and their recovery during the forensic process by the methods applied to collection and further analysis. A fourth factor that plays an important role in itself when evaluating our findings, but also potentially affects the first three, is the prevalence of human biological traces in an environment.

In Section 2 of this chapter we will define the four factors Transfer, Persistence, Prevalence and Recovery (TPPR) in more detail, as well as define other relevant terminology. In Section 3 we will provide an overview of our current understanding of TPPR of biological traces, and particularly from that, what factors we need to consider when evaluating our findings given activity level propositions. In Section 4 we will discuss the experimental design of studies looking at DNA TPPR, as well as ways in which data from such studies are shared.

3.2   WHAT IS TPPR (AND ASSOCIATED TERMINOLOGY)?

There is terminology in use that is associated with trace dynamics in the broadest sense. Examples are terminology related to mechanisms (like primary or secondary transfer), terminology used to define a type of trace (like touch DNA or wearer DNA) and terminology associated with the legal perspectives on these issues (like 'innocent DNA' or 'incriminating trace'). While there is evidently a need for a terminology to refer to concepts in our communication about them, there is an inherent risk when we use the same terminology while different users apply a different meaning to them. This risk may be minimal when we communicate with our peers (be it scientist to scientist or judge to judge) but could lead to misconceptions when we communicate between communities and laypersons like a jury. Take, for instance, the term 'touch DNA'. This term is quite regularly used in scientific literature to refer to DNA traces collected after handling an item. This makes sense because the DNA was deposited through touch, which was observed in controlled experimental conditions. However, we see that the term also finds its way into casework and courtroom discussion. Given that the questioned activities in criminal cases are by definition not controlled and observed, the use of the term 'touch DNA', when we cannot reasonably assume the mechanism by which the trace was deposited, may be misleading. It is important that we realise that such terminology is context dependent and, as such, may convey a different meaning in a different context.

In discussion with police such information may be considered 'intelligence', misdirecting their efforts towards identifying the individual that *must* have touched the questioned item (implicitly excluding the possibility of indirect transfer of the biological material). The use of this same term in court may lead to misconceptions with the jury or judges about the activity that leads to deposition of the trace ('He must have fired the gun, because we found his touch DNA!'). It is therefore crucial to define the terminology that is being used, as well as to use the appropriate terminology in appropriate situations. In this paragraph we define the terminology that is commonly encountered in scientific literature and the legal domain when dealing with trace dynamics. This is not an attempt at making an exhaustive list of all terms that are being used when dealing with the dynamics of biological traces. Rather we discuss a number of terms that are commonly used. They serve to illustrate the different uses of the same term and highlight the potential pitfalls if they are not used correctly.

Transfer of biological materials is generally referred to as 'DNA transfer', even though it commonly means the transfer of a biological material that contains the DNA. While this refers to the transfer event of DNA specifically, it is used as a common term to mean transfer, persistence and recovery of any biological material. Although we often refer to the probability of transfer for the combined probability of transfer, persistence, and recovery (TPR), the better term would be the probability of recovery [1].

Indirect transfer is used to refer to any mechanism by which biological material of an individual transfers to a surface by one or more intermediate vectors. A commonly used example is a handshake between persons A and B, after which DNA of person A is deposited on a surface via the hands of person B. Related terms that are being used are 'secondary transfer', 'tertiary transfer', etc. by which two-step or three-step indirect transfer events are specified. Nonetheless, 'secondary transfer' has been used incorrectly as a blanket term to mean indirect transfer.

Direct transfer is used as opposed to indirect transfer, meaning the one-step transfer of biological material from an individual to a surface. It is commonly referred to as primary transfer. An example is the transfer of material through the handling of an item.

Aerial transfer refers to the transfer of biological material through the air. Examples of this are the transfer of saliva from an individual to a nearby surface through speaking, or the transfer of flakes of dried blood from an object with dried bloodstains to another surface. An aerial transfer could be considered a specific subset of direct or primary transfer if used to refer to transfer from an individual to a surface (speaking, coughing, sneezing, etc. or an arterial squirt from a severe wound resulting in blood being projected to a surface). The agitation of an object to release traces that thereby transfer to another surface, like dried body fluids that flake when the object is handled or even aerosols that are created when opening and closing tubes in the laboratory, would be examples of indirect transfer.

TPR is an acronym (for transfer, persistence and recovery) that relates to the entire route from the transfer of the biological material of an individual to a surface, its persistence over time (and possible further indirect transfer to other surfaces and persistence on those surfaces) and its recovery during the forensic examination (either at the scene of a crime or in a laboratory setting when the item carrying the trace has been secured).

TPPR is an acronym (for transfer, persistence, prevalence and recovery) that is used to refer to the four main factors to be considered when evaluating findings given activity level propositions.

'[…]' DNA is generally used as a descriptor of the activity that has led to transfer, persistence and recovery of the biological material. Examples are 'touch' DNA, 'wearer' DNA, 'handler' DNA, 'contact' DNA, etc. These terms can be appropriately used in a research environment, where under controlled conditions activities and the resulting traces can be observed. We also see that such terms can be used in casework when setting an examination strategy and formulating working hypotheses. One could, for example, decide to sample an item for DNA of the wearer, or 'wearer DNA', by recovering material from the inside collar of a sweater. However, it would be improper to refer to the resulting findings as 'wearer DNA', since the mechanism by which the DNA was deposited on the sweater cannot be known with certainty (as the history of the item will generally not have been recorded or observed). Using the term in the latter context may therefore be misleading. Neutral terminology like 'trace DNA' or 'latent DNA' (analogous to latent fingermarks) is best used to refer to traces containing DNA that are not linked to a specific cell type.

Passive or active transfer has been used by [2] to refer to the mechanism by which biological materials have transferred. Specifically, 'active transfer' was meant to relate to transfer during the commission of a crime, while 'passive transfer' relates to events unrelated to the crime. As such the terms are somewhat confusing as they are suggestive of the mechanism by which traces were deposited, relating to the intensity of contact and the mechanism of transfer. Transfer during the incident may, for instance, also occur 'passively' (a gun placed on a table and later removed, transferring DNA from the gun to the table and vice versa), while transfer unrelated to the incident may occur 'actively' (by an individual actively cleaning the table prior to the incident). Possibly because of the fuzzy definition of the terms they have not caught on and have not been widely used.

Persistence refers to the retention of biological material on a surface after deposition until its recovery during the forensic process. The term can refer to the persistence of material on an object through a *specific* event (like washing or handling of the item of interest by another person), or a *non-specific* event (like passive decrease over time due to environmental conditions).

Unknown DNA is any source of DNA found that cannot be attributed to any of the known reference samples (or database samples) that are available in a case. Unknown DNA might be attributed to either an unknown person who performed alleged activities related to an incident (an 'unknown offender') or more generic to 'background DNA'.

Background DNA constitutes any source of unknown DNA that was recovered from a surface and that was deposited through an unknown mechanism (and which is not modelled specifically. This is discussed in more detail in Chapters 5 to 8). As the presence of background DNA in a case may be informative in the context of evaluations given activity level propositions, the number and relevance of reference samples included in comparisons needs to be considered, as does the fact whether there is a single source or multiple sources of unknown DNA.

Prevalence refers to the presence of biological materials on a surface that are unrelated to the questioned activities in a case. Here we need to make a distinction between 'prevalent DNA' and 'background DNA'. Prevalent DNA relates to the presence of DNA of a specified individual through an assumed mechanism. This could, for instance, be the prevalence of DNA of the wearer of an item of clothing under examination. Prevalent DNA may as such relate to either a known individual within the context of a case (like the prevalence of DNA of the victim on their underwear in a sexual assault case) or an unknown individual (like the prevalence of DNA of the unknown perpetrator on a balaclava recovered from the scene of a robbery). Background DNA relates to the presence of DNA on a surface of one or more unknown individuals through mechanisms that are not explicitly considered or modelled in an evaluation.

Recovery refers to obtaining biological materials as part of the forensic process. This includes the sampling strategy and method, and further analysis and interpretation of the resulting data. An inter-laboratory study by Steensma et al. [3] has shown how the process of recovery will impact on the probability of transfer, persistence *and* recovery of biological traces. This illustrates again why we should refer to the probability of TPR or the probability of recovery of biological traces rather than the probability of transfer alone.

Evidence, findings and observations are commonly used to refer to the results of forensic examinations. When we discuss the 'weight of evidence' we commonly talk about, for instance, 'evaluation of our findings given activity level propositions', 'assessment of the evidence given the relevant propositions' or 'our observations are this much more probable if one proposition is true than if the other is'. In this context we see that terms like 'findings', 'evidence' and 'observations' are used interchangeably (also remember the 'E' for 'evidence' in $\Pr(E|H,I)$). While these generic terms can be used to convey a plethora of results from forensic examinations, it must always be clear to all parties what we are referring to. In that sense it is also important to realise that 'evidence' has a specific meaning in the legal context. One could say that 'observations' only become 'findings' after interpretation by the scientist, and that this becomes 'evidence' when these findings are able to provide more support for one of two relevant propositions in a case. Following this line of thinking, we suggest that the scientist refrains from using the term evidence, but rather refers to their findings (i.e. when discussing the results of comparative DNA examinations) or observations (for instance, when talking about a positive RSID saliva test result).

3.3   OUR CURRENT KNOWLEDGE BASE ON TPPR

The concept that traces are left behind after an event forms the fundamental basis for most, if not all, of the forensic sciences. This comes back to Locard's exchange principle (which was mentioned in the Preface):

> It is impossible for a criminal to act, especially considering the intensity of the crime, without leaving traces of this presence.

While it is believed that the principle holds true for all types of particulates, it has been demonstrated through experiments in many disciplines. Particles may be left after a contact, be it fibres,

DNA, gunshot residue, fingermarks or any other materials. Activities on electronic devices will leave traces of such activities in databases and files. Impacts on the human body will leave their mark in the form of wounds or abrasions. As Locard already mentions in his famous quote, the intensity of the interaction may impact on the probability that such traces are left behind. Over time a body of scientific work has been developed that addresses the factors impacting on the transfer, persistence and recovery of traces, as well as the prevalence of trace evidence types in the environment. Here we will focus on human biological traces, while some general comments on this topic in other forensic disciplines are given in Chapter 12. This paragraph will not be an extensive review for two reasons; first, several reviews on this topic have been published recently [4–10] that we don't intend to repeat or duplicate. We refer the reader to these papers as a valuable resource. Second, because a high number of papers are continuously being published on DNA TPPR, this section would be outdated the moment this book is published.

3.3.1 Transfer of Biological Materials

In a landmark study published in 1997, van Oorschot and Jones [11] described a study into the recovery of DNA from a range of objects handled by volunteers. On gloves they reported:

> We found alleles in addition to those of the wearer in samples from two of the gloves, which could be due to secondary transfer.

Hence from the early days of DNA profiling applications in the forensic domain it was recognised that traces may originate from a direct (touch) or an indirect transfer to a surface. When assigning probabilities to these TPR events in casework we need to consider the potential routes of transfer, persistence and recovery in a timeline. Figure 3.1 graphically represents five examples of such TPR routes. A transfer may occur at any time prior to, during or after the incident. The transfer may also occur during the forensic process, which is considered a form of contamination.

Example route A is a situation where the transfer of interest occurred early on during the incident. The recovery of the trace was done early during the forensic process. As a consequence, we need to consider the persistence of the trace for the duration of the incident and post-incident. This example could be a burglar breaking a window to enter a house and cutting themselves on the broken glass.

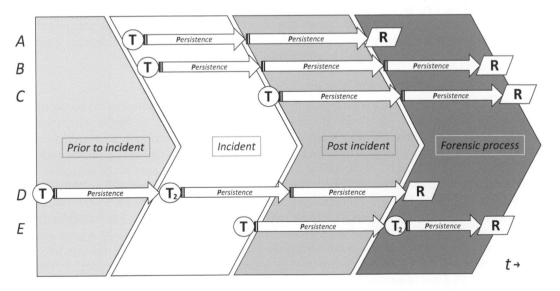

FIGURE 3.1 Five examples of TPR routes of traces. T = primary transfer, T_2 = secondary transfer, R = moment of recovery.

Blood transferred to the windowsill early during the incident. We then need to consider the persistence of the stain during the incident (was the stain, for instance, wiped by the clothing of the burglar when climbing in and later out through the window?) and post-incident (did the owner of the house clean the windowsill prior to calling the police?). The recovery of the trace was possibly done quite early during the crime scene examination, which would mean that persistence during that time is not at issue.

Example route B could be a similar situation as described under example A. The difference could be that the blood was not transferred to the windowsill but rather to a crowbar which was subsequently left at the scene. We now need to consider the probability that the bloodstain persisted on the crowbar while it was used during the incident, while it was thrown on the carpeted floor, while it was handled by the children of the owner when they got home post-incident and during the packaging and transport of the crowbar to the forensic laboratory during the forensic process.

Example route C describes a situation where the trace of interest was transferred at the very end of the incident, for instance on a balaclava that was left on the escape route of the robber of a supermarket. We still need to consider the persistence of the trace on the balaclava post-incident (to what environmental conditions was it exposed?) and during the process of packaging, transporting and storing the balaclava at the forensic laboratory.

Example routes D and E represent situations where the trace of interest was deposited through a secondary transfer event.

Example route D could be a situation where person A and B shake hands at which time the DNA of person A transfers to the hands of person B. We need to consider the persistence of A's DNA on B's hands prior to the incident and subsequently the secondary transfer of A's DNA to another surface, for instance a door handle. We then need to consider the persistence of A's DNA on the door handle through the incident and post-incident, before it is recovered early on in the forensic process, by sampling the door handle at the scene of the crime.

Example route E represents a situation where the secondary transfer of DNA occurs during the forensic process. This could be an example where the perpetrator leaves a room and touches the door handle while doing so. The crime scene officer touches the door handle when entering the scene and subsequently handles a knife while securing it for further examination in the lab. We need to consider the transfer of DNA of the perpetrator to the door handle, its persistence post-incident (did anybody touch the door handle during this time? Is it the inside or outside handle being exposed to what environmental conditions?), the secondary transfer from the door handle to the knife, and the persistence of DNA of the perpetrator on the knife during the packaging, transport and storage of the knife at the forensic laboratory.

Numerous studies have explored factors that may affect the transfer, persistence, prevalence and recovery of biological materials. These studies lead us to ask ourselves the following questions whenever we are tasked with assigning probabilities to TPPR events.

3.3.1.1 Who Transferred? (and how much?)

The individual who deposited the biological material may be a factor influencing the probability of transfer. There is variation reported in the amount of DNA that individuals deposit after the same contact event [12–17]. This leads to the conclusion that there may be variation between individuals in their ability to deposit biological materials. This inter-individual variability is commonly referred to as 'shedder status'. Shedder or shedder status is a term that has been used to define the nature of the release of trace particulates by a person or object (be it fibres from a garment or DNA from an individual). In forensic biology, traditionally, it has been used to refer to the ease with which individuals 'shed' skin cells [18]. A 'good' shedder may release more cells than a 'poor' shedder, resulting in greater amounts of DNA being recovered from a surface after contact with a good shedder relative to a poor shedder. Recent research suggests that other factors than 'shedding' skin cells play a role in the amount of DNA being transferred by individuals [19]. The term 'shedder' or 'shedder status' is still used, but encompasses all factors that relate to the specific individual which may influence the amount of biological material that is consistently being transferred through skin-to-surface contact, relative to other individuals from the same population given the same type of contact. This includes, but is not limited to, personal hygiene, skin disease or hormonal fluctuations. Aspects of personal habits and hygiene may play a role in determining shedder status, as do health [20], biological gender [21], age [22, 23] and likely a combination of any of these factors.

As well as high shedders being thought of in the classical sense of someone depositing large amounts of DNA when they touch an item, it has been determined that some people emit more DNA (mostly through saliva) than others when talking. Port et al. [24] defined an 'oral shedder' term to describe the difference in DNA amount people emit when speaking. This phenomenon has also been recognised by Asadi et al. [25] who recognised 'super-emitters', individuals who emitted 10 times the amount of saliva when speaking compared to the general population. The difference in the amount of DNA expelled when breathing, and the concept of a 'breathing high producer' was investigated by Edwards et al. [26] and putting these different aerosolised DNA emission mechanisms, along with the possibility of shedding skin cells to the surroundings, Puliatti et al. [27] referred to an 'environmental shedder' status.

The aspect of shedder status may need to be taken into account if there is an indication or suggestion in a case that a person-of-interest deviates from the average population in their shedding ability.

3.3.1.2 What Was Transferred?

Lacerenza et al. [21] studied the prevalence of biological materials on hands. They found that samples from hands contain a diversity of body fluids and cell types. Most samples do contain skin cells, as is expected, and saliva and blood are most often found in addition to this. Other cell types that were recorded in samples from the hands of sixty volunteers were semen, vaginal mucosa and menstrual blood. Zoppis et al. [28] looked at the shedding ability of different skin areas. They found that sebaceous fluid represents another important vector for the transfer of DNA.

We know that different bodily fluids, and hence body fluids expelled from a body, contain different amounts of nucleated cells. This leads to substantially different DNA amounts present in different body fluid traces. Table 3.1 gives an overview.

Studies have also shown that 'cell-free DNA' can constitute a significant proportion of the DNA in a 'touch' deposit [30, 31]. In samples from traces obtained in a range of studies on the transfer of DNA through direct contact with hands, the amount of DNA varies quite strongly. Meakin et al. [9] suggested that the source of the material may be a factor to explain this variation, as cell type tests are not commonly performed with such studies.

We therefore need to be mindful of the nature of the material that may have transferred, dictated by the circumstances of the case.

3.3.1.3 How Was It Transferred?

The nature and duration of contact may impact on the amount of DNA transferred. It has been reported that the amount of force [32–35] as well as the amount of friction [32, 33] between hand

TABLE 3.1

Amount of DNA in Selected Body Fluids, after Lee and Ladd [29]

Type of material	Amount of DNA
Liquid semen	150,000–300,000 ng/ml
• Postcoital vaginal swab	10–3000 ng/swab
Liquid blood	20,000–40,000 ng/ml
• Bloodstain	250–500 ng/cm^3
Liquid saliva	1,000–10,000 ng/ml
• Oral swab	100–1500 ng/swab
Liquid urine	1–20 ng/ml

and surface may impact on the amount of DNA being transferred. Duration in itself does not appear to be a factor that impacts on the probability of DNA transfer (as time in itself is not a factor), but it may add to variation given any of the other relevant factors impacting the on transfer (like shedder status, nature of the contact, etc.). Repeated contacts or uses of an item may result in more DNA being recovered [36]. The interplay between these factors is not yet well understood and needs additional study.

The aerial transfer is another factor to consider, as the transfer of saliva (through speaking or shouting [24]), semen (through ejaculation) or blood (for instance, by swinging a wounded hand) may result in biological material being transferred without physical contact. Material may also transfer inadvertently through flaking of dried body fluid stains and subsequent aerial transfer, an indirect route of transfer that is more probable to occur some moment after the event (e.g. after stains have dried) and is thus a potential route for indirect transfer that may also occur during the forensic process (defined as contamination).

3.3.1.4   Where Is the Material Transferred From?

With indirect transfers of biological material, one needs to consider the nature of the intermediate surface. The structure of a surface and the nature of the material it is composed of will impact on the ability of a surface to both receive and release biological materials. Goray et al. [32, 33] have shown, for instance, that porous, textile materials are less prone to release any biological materials present on these surfaces than are non-porous, plastic surfaces.

Another aspect to consider is the size of the areas in contact. It intuitively makes sense that larger contact areas will result in larger amounts of biological materials transferring compared to smaller contact areas, and this has, for instance, been modelled as such in [37, 38]. However, this assumption of a linear relationship between the surface area contacted and the amount of biological material transferred does not necessarily hold. An example comes from studies into the direct transfer of biological material from the hands to the surface. Studies by Oleiwi et al. [39] and McCol et al. [40] have shown that the amount of DNA transferred from different parts of the hand does not necessarily correlate with the surface area.

3.3.1.5   Where Is the Material Transferred To?

As with the shedding surface, the nature of the receiving surface will impact the probability of biological materials transferring to that surface [32, 33]. The nature of that surface may also impact on the persistence of traces over time, which is discussed next.

3.3.2   Persistence of Biological Materials

3.3.2.1   Was the Item Exposed to Negative Circumstances?

Biological materials will degrade over time. While time itself does not affect biological materials, it is a relevant factor in combination with any negative conditions the biological materials are exposed to. There is quite an extensive list of factors that will impact on the degradation of biological materials, which includes (but is not limited to) radiation (UV, nuclear), moisture (through either a washing effect or enhancement of microbial activity), heat and chemical exposure. This prompts us to consider, for instance, the weather conditions an item has been exposed to before it was secured for forensic examination.

3.3.2.2   What Happened to the Item After the Activity of Interest?

Other factors that affect the persistence of traces are any subsequent activities that are performed that affect the surface of an item bearing a trace. Studies have shown that subsequent use of an item by another individual will affect the persistence of DNA of a prior user [41, 42]. We therefore need to consider, in our case, any activities that may have been performed after the activity of interest. This point was convincingly made by Van Oorschot et al. [43] who studied the number of contacts of hands with surfaces in everyday activities. They found that individuals touched a substrate an average of 15 times per minute with their dominant hand. If we consider a scenario where DNA may have transferred indirectly after a handshake, it is crucial that we also consider the time between the handshake and the moment of indirect deposition on the sampled surface. With an extended period of time between these two activities of focus, many activities may have led to the loss and redistribution of traces on the hand of the vector. This was confirmed by Szkuta et al. [44] who studied the persistence of non-self DNA on hands over time. An increase in the time between handshake and deposition reduced the frequency in which the person of interest was detected in the deposit.

3.3.2.3   How Was the Item Handled During the Forensic Process?

Crucially, we also need to consider any activities that occurred during the forensic process which may affect the integrity of traces. Consider the handling of an item at the scene, the packing of the item, transport, storage and subsequent examination. Any contact with or exposure to the trace-bearing surface may again remove or redistribute latent DNA traces or body fluid stains. Many examples have been described, both in experimental studies [7, 45–47] and in case studies, the most famous of which is probably 'The phantom of Heilbronn' (http://content.time.com/time/world/article/0,8599,1888126,00.html [last accessed February 2023]).

It has, for instance, been found that up to half of a stain may be lost from knives to their packaging [48], which was confirmed by Steensma et al. [3] using cable ties.

When asking ourselves questions about the handling of the item during the forensic process, we also need answers to other examinations that were performed prior to DNA trace recovery. Such examinations may also affect the persistence of biological traces (not excluding the impact of unpacking and repacking an item in between examinations). An example can be taken from the treatment of items to visualise latent fingermarks. A range of methods may be applied to a surface (like cyano acrylate, methyl violet, vacuum metal deposition, physical developer, etc.) all containing chemicals or other treatments that may affect the persistence of DNA to different degrees [49].

3.3.3   Prevalence of Biological Materials

3.3.3.1   What Was Already There?

When we consider the prevalence of biological materials in an environment or on an item, we need to distinguish between two broad categories: the background presence of biological materials and prevalent materials. Background materials or 'background DNA' relates to material that cannot be traced back to any of the reference materials in a case, nor to any specific mechanisms for deposition. It is DNA or other biological substance from unknown individuals in the context of a case. Prevalent materials or 'prevalent DNA' relates to DNA from individuals that are present due to activities that are specifically considered (like wearing an item of clothing, prior use of knives, etc.) An example of this could be the presence of DNA of a suspect in a house that he used to frequent regularly prior to the incident occurring in that house. DNA of that suspect could be prevalent in the house through activities unrelated to the incident. Prevalent DNA may also relate to unknown individuals within the context of a case (like the prevalence of DNA of an unknown wearer of an item of clothing).

When we define these two categories it becomes immediately clear that the probability of finding background DNA is directly related to the effort that has been made to collect reference samples from all those that may have left their DNA on the sampled location. While in cases where the collection of reference samples is often limited to those involved in the incident or closely related to such individuals, a broader collection of reference samples of associates of the persons of interest, as well as of those frequenting the crime scene location, may assist in lowering the probability of observing background DNA. This is illustrated by studies by Szkuta et al. [50, 51] on the prevalence of DNA on worn upper garments. In these studies, an effort has been made to collect reference samples of associates to the wearers of the garments, and in a high number of samples from nearly all garments, DNA of associates was found. If such studies were used in a case where reference samples of associates are not made available, the probability of finding background DNA (which was also found in both studies on garments) or prevalent DNA (if we are considering an associate who has performed a specific action we are modelling) would need to be adjusted for all those associates from whom no reference samples were obtained.

A good source for data to inform the probability of finding background DNA on a specific item is files of past cases worked on by the laboratory. While the use of casefile data to inform probabilities of transfer and persistence is generally hampered by the lack of knowledge of the 'ground truth' of the history of the item, this is less of an issue when informing the probability of events that are not modelled specifically, like finding background DNA. By critically selecting cases with similar items, the results will tell how frequently DNA of unknown individuals from background DNA in the context of a case is found on similar items. As the methodologies used for sampling and analysis of samples are generally the same between cases in the same laboratory, frequencies obtained from such past cases may quite reasonably be converted to probabilities in the case under consideration.

The probability of finding background DNA on items can be assigned from quite general literature as there (by definition) is no transfer mechanism being explicitly modelled. However, when assigning a probability to the presence of DNA from a specific person (whether known or unknown by virtue of a reference sample being available) from a defined mechanism then it becomes more difficult to find literature or casework data that aligns with case circumstances. This means that various assumptions need to be made about the data or the case or both. The probability of observing prevalent DNA will be highly case specific and is therefore best informed by collecting 'control samples' from other locations or items. We can specify four categories of items or samples to be taken as controls for this purpose (based on Ton et al. [52]):

1. *Samples from the same item*
   An example could be an item of clothing of a victim that was sampled at locations contacted by an offender. If the accused states that there exists some prior relation between them and the victim or their clothing (suggesting that their DNA is prevalent on the clothing), other locations may be sampled that do not directly relate to the activities under consideration. The sampling strategy could be guided by specifics of the prevalent DNA scenario (e.g. folding of clothes vs the victim sitting in a chair owned by the accused, etc.)

2. *Samples from items that were in contact with the item under examination*
   An example could be a firearm found under a pillow. The accused states that he had regular contact with the pillow and his DNA is prevalent on the pillow, and indirectly, on the firearm. Samples from the pillow may allow us to assess the probability of recovering prevalent DNA from the firearm.

3. *Samples from similar items*
   If an accused states that their DNA is on a knife used in a stabbing because they regularly use knives from the kitchen that the knife under examination was taken from, samples from other knives from that kitchen may assist in assigning a probability for the prevalence of DNA of the accused on the knife in question.

4. *Samples from the broader crime scene*
   This is a broad category, which we encounter when an accused states that they visited a location any time prior to the incident. Their presence there could be argued to have resulted in their DNA being prevalent at the scene and thus also at the locations sampled. Samples from multiple locations at the scene that may be contacted by visitors could assist in assigning the probability of their DNA being prevalent in that environment.

We note that this approach would require an effort early on during the crime scene investigation. A suspect may not have been identified at that point, and alternative scenarios may therefore not have been postulated. However, considering these four broad categories will allow crime scene examiners to anticipate these alternative scenarios as we encounter these alternatives quite routinely in our casework practice.

3.3.3.2 What Was Prevalent on Surfaces Interacting During or After the Activities of Interest?

We have of course discussed factors impacting DNA transfer given activities of interest earlier in this chapter. There are, however, two routes for the transfer of biological materials that we need to consider when assigning a probability to encountering either background or prevalent DNA.

The first is any interaction between the handler of an item and its environment. If we have information that the handler of the item under investigation has interacted with other (unknown) individuals just prior to or during the handling of the item, we might consider that this interaction increases the probability of finding background DNA on the hands of the handler and subsequently on the item. The concept that we need to consider here is the co-transfer of DNA of the handler and the non-self DNA prevalent on their hands.

The second is any handling of the item after the incident. Any contact with individuals or surfaces that contain unknown DNA may increase the probability of finding DNA of unknowns (and thus increase the probability of finding background DNA) as well as increase the probability of finding prevalent DNA of known individuals in the context of that case.

In relation to the latter, the concept of 'DNA parking' [53] should also be considered. The concept refers to DNA that was transferred at some point which was temporarily deposited on a surface and re-collected at a later point in time to subsequently be transferred to the surface of interest. An example of this could be DNA of person X being transferred to the coat of person Y during a hug. Person Y stores the coat for some time before wearing it again. In the process of putting on the coat and wearing it, DNA of person X is transferred to the clothing of person Y which she bought some time after the last encounter with person X. In this example the DNA from person X was 'parked' on the coat for some time before being transferred to the clothing of person Y.

3.3.3.3 What Has Been Added During the Forensic Process?

Any additions of DNA, or its displacement on an item or scene, in the course of the forensic process may collectively be called 'contamination'. This aspect, together with the uncontrolled removal through wiping, packaging or transport, is directly related to the integrity of the traces recovered and as such is an important consideration that should not be overlooked in evaluations given activity level propositions. We should distinguish contamination of samples or items containing traces of DNA from persons involved in the forensic examination (like the scene of crime officers or laboratory analysts) whose reference DNA profiles are in an elimination database from other sources of contamination. The first source of contamination will generally be identified early in the DNA analysis process as the comparison of traces to elimination databases is a routine operating procedure. The latter type of contamination is much harder to identify and may affect an evaluation given activity level propositions more directly as it may, for instance, be considered a source of background DNA.

When dealing with contamination there are a number of potential events (or routes of transfer) that need consideration:

1. The probability of contamination of a sample with DNA of a known person-of-interest (for instance, a suspect or victim)
2. The probability of contamination of a sample with DNA of a random unknown individual (unknown in the context of the case)
3. The probability of removal or displacement of traces by handling, packaging or transportation of an item

The first probability is generally quite small as we would require a source for the contamination with DNA of that specific individual. For instance, DNA of a victim on the clothing of a suspect would require (for a contamination event to be the source) the victim or an item which has been in contact with the victim to be present in the course of the forensic examination. This may happen when an officer interviews the victim before going over to the suspect and securing their clothing. Or when items of clothing of both victim and suspect are examined simultaneously in the same laboratory space. It is a mode of indirect transfer that may need to be considered in cases where no more probable alternative routes for TPR of that person's DNA are available (like a social interaction between suspect and victim before the alleged incident). It is, however, a factor that could affect our assignment of probability to finding the prevalent DNA of an individual in a sample.

The probability of contamination of a sample with DNA of a random unknown individual is likely higher than the previous one we discussed (and less easy to detect) as it requires any source of DNA that cannot be traced back to any of the individuals in the case or any of the forensic examiners or scientists (through an elimination database). This type of contamination may occur at any stage in the forensic process and may therefore (depending on case procedures, item handling and examinations performed) be a factor that could significantly impact on the probability of observing background DNA.

The probability of removal or relocation of traces has been studied for some types of items and some methods of packaging [3, 48]. These studies have shown that significant amounts of DNA may be lost in this part of the forensic process. Traces may also relocate to other areas of an item. This could result in different scenarios of handling being considered or in a different weight being assigned to the findings given the case-relevant propositions. These aspects are important to consider when assigning probabilities to DNA TPR, as most published studies recover traces from items that have been handled in an experimental setting *without* going through the process of packaging the item and transporting it. When using data from published studies it is therefore crucial to consider whether or not the packaging and transportation of an item in the case may have influenced the persistence of the traces. The number of traces and their location may be observations that could assist in distinguishing between direct transfer or indirect transfer (contamination) scenarios.

3.3.4 Recovery of Biological Materials

3.3.4.1 What Happened to the Item at Any Point in the Forensic Process?

Recovery relates to the entire forensic process, from securing an item at the scene of a crime (were – given the scenarios being considered – all relevant samples taken and items secured?), the sampling strategy of the item in the laboratory (were the right areas targeted for sampling?), the sampling method, the DNA analysis process, as well as the interpretation methods applied. Each step in this process may impact on the probability of recovery of traces and when tasked with assigning a probability to a TPR event, we need to be mindful of what the impact of the procedures would be on the potential for recovery of a trace.

This becomes explicit when we use published studies to inform our assignment of the probability of a TPR event. With regard to recovery, we need to consider the methodology used in the experimental study and compare it to the methodologies applied in the case at hand.

As an example, we compare two studies on the recovery of DNA from knife handles. Both studies addressed the transfer of DNA of handlers to knives, but we will not discuss the differences in the experimental design here. In Table 3.2 we look at the differences in the factors that may affect recovery of traces between these two studies, which could be representative of differences we encounter between a case and a potentially relevant study to inform our probability for DNA TPR.

From Table 3.2 we can take two important points:

1. Aspects of the methods that affect recovery may differ quite substantially
2. Published studies do not always specify in detail the methods applied

TABLE 3.2

Comparison of Aspects of Recovery of Traces in Two DNA Transfer Studies

	Samie et al. (2016) [54]	Meakin et al. (2017) [55]
Sampling strategy	Not specified	Entire surface of the handle
Sampling method	Double swab (cotton, first wet – liquid not specified – then a dry swab)	Single mini-tape (type of tape specified by the manufacturer who produces a single-type only)
DNA extraction of traces	QIAshredder and QIAAmp kit, elution in 50 µL (water of buffer not reported) with Microcon 30 spin Columns	QIAamp DNA Investigator Kit, as per the manufacturer's instructions for extracting DNA from cotton swabs. The mini-tapes were cut up into thin strips using DNA-free scissors prior to the addition of lysis buffers, and 1 µg carrier RNA was added to improve the DNA extraction process. Thirty-five µl of Buffer ATE was used in the final elution
DNA quantification	7500 Real Time PCR System, using Investigator Quantiplex HYres (Qiagen) kit following standard protocols	Quantifier Human DNA Quantification Kit with the ABI PRISM 7900HT Sequence Detection System, as per the manufacturer's instructions
DNA amplification	Template amount not reported, using NGM SElect Kit, 30 cycle PCR protocol on a PCR system 9700. Triplicate analysis for each sample	10 µl template, using the AmpFlSTR NGM SElectTM PCR Amplification Kit, 30-cycle PCR protocol (Single replicate? Not reported)
Fragment separation	3130xl Genetic Analyser (injection voltage and duration not reported)	DNA Analyzer 3730xl (injection voltage and duration not reported)
Profile analysis	GeneMapper IDX Software, for all these protocols the manufacturer's instructions were followed	GeneMapper 4.0 software, all as per the manufacturer's instructions or the internal validation study of the laboratory
Profile interpretation	All replicates (with all loci) from the same DNA trace was analysed together with STRmixTM v2.3.05 software [with extensive documentation on proposition setting and parameter values assigned]	The RMP was calculated manually and the *LR*s were determined using LRmix Studio 2.0 as per the developer's instructions. These calculations used UK Caucasian allele frequencies and an FST value of 0.03, as per directions from the UK's Forensic Science Regulator's guidance

As the two studies differ in their research question and thus in their experimental design, we cannot directly compare the outcomes of these studies to see if and how the methods affected the probability of recovery of DNA. Such differences between methodologies need to be addressed when using published data to assign probabilities. To this end, validation studies of the methods and systems may be used to assess the impact of such differences on the frequencies of TPR events occurring, and thus to inform the assignment of a probability that extends beyond the use of frequencies from experiments only.

Another approach is to conduct collaborative studies with multiple laboratories participating in the same experimental design. An example of such a study is that by Steensma et al. [3], where DNA TPR to cable ties was studied by four independent laboratories. The observed differences can be linked to differences in methodologies applied by the laboratories and may assist in assigning probabilities that best reflect the methodologies applied in the case at hand. A specific aspect to consider is the differences in sensitivity between older and newer generation STR typing systems. Newer systems are more sensitive in detection, as well as less sensitive to inhibition from contaminants. Both could lead to a difference in the probability of obtaining a result between older and newer generation kits.

A summary of factors impacting on the TPPR of biological materials is shown in Figure 3.2.

We need to be aware of the factors that may affect the probability of recovery and of those discussed previously that relate to transfer, persistence or prevalence of traces, which may affect the probability of observing our findings given the propositions and case context. This we will either do by making these factors explicit in our statistical model (on which we will elaborate in the following chapters) or they are part of our considerations when assigning probabilities to TPPR events. We could, for instance, model the impact of time on the probability of persistence if the time interval is not known. The different time intervals between deposition of a trace and its recovery will result in different probabilities assigned to the persistence of the biological material. If the time interval is known, we might decide not to model this parameter. Many of the factors that we currently know that may affect the transfer, persistence or recovery of biological materials (like the nature of the depositing and receiving substrates, the type of biological material and the state of it (e.g. wet or dry), etc.) will generally not be modelled explicitly. We nevertheless do need to take them into account when assigning our probability to the TPR event under consideration. We do this by selecting the most relevant experimental studies to support our probability assignments (those that most closely align with the conditions of the case) or by designing studies where we take these factors into account in the experimental design. The factors will also dictate our probability assignment in that they may be used to adjust from the 'raw' frequencies obtained from a study. Say we have a case where we need to assign a probability to the transfer of DNA to the handle of a knife by grabbing it. We may find a study describing such an activity with plastic knife handles. This fictitious study describes 10 iterations in which 10 volunteers grabbed the knife handle. Say that in 7 out of 10 studies DNA of the grabber was recovered from the knife handle, resulting in a frequency of 0.7. The knife in our case, however, has a rough wooden handle. Can we translate the frequency of

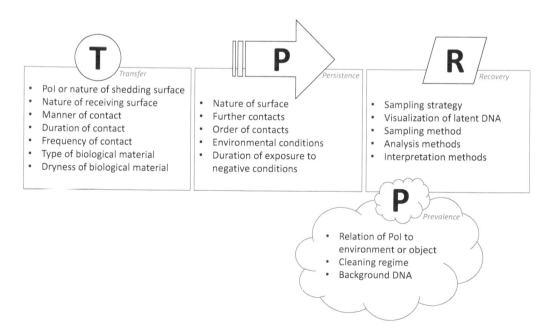

FIGURE 3.2 Some factors impacting the probability of biological materials transfer, persistence and recovery, as well as the prevalence of traces potentially unrelated to the incident. Note that the prevalence of DNA in the environment and on a target surface might also be a factor affecting the recovery of a deposit of interest.

occurrence of recovery from the study directly to the probability of transfer given our case circumstances? We will need to consider if we expect a difference in the probability of transfer to plastic or a rough wooden substrate. Studies like that of Daly et al. [54] and Goray et al. [33] have shown that biological materials are more likely to transfer to rough surfaces than to smooth surfaces. Based on that knowledge we might decide to assign a (personal!) probability that is somewhat higher than the frequency observed in the fictitious knife study, say 0.8. In the same way we will need to factor in all other relevant aspects that might differ between the datasets available to us and the known conditions of a case. It is unlikely that we have data available to us that align perfectly with the case circumstances. We are therefore always in the position that there is some level of subjectivity in the assignment of probabilities. Even if we feel that we should simply use frequencies obtained from available data to assign probabilities to TPPR events, this is a subjective decision made by the expert. Either way, the impact of variations in the assigned probabilities needs to be explored using sensitivity analyses, on which we will elaborate in Chapter 7. A final note is that it is generally difficult to separate out probabilities for transfer, persistence and recovery. As experiments would always require methods to obtain data, recovery will be part of any experiment. One could design an experiment on the transfer of DNA with no significant factors impacting persistence (for instance, by immediately sampling the item after the activity and extracting the DNA), but it will always include factors impacting on the recovery of the traces. The same holds for experiments that focus on persistence (for instance, with a controlled starting amount of biological material being deposited on a surface to rule out the factor transfer). However, again we will include factors related to recovery in such an experiment. So, while we are able to separate factors of transfer and persistence in our experimental design, recovery will always be part of it. One needs to be mindful, if modelling transfer and persistence separately, to include the factor recovery in only one of those so as not to have a recovery impact twice in the probability assignment. Because this is rather difficult, transfer, persistence and recovery are frequently combined in a probability assignment. More on this in Chapter 6.

The second point taken from the comparison of the two studies in Table 3.2 is that it is not always possible to gauge from the published article what methods were applied. Both studies in Table 3.2 are quite detailed in their report on the methods that were used, but information is still lacking from one or the other to properly compare the two sets of methodologies. This may hamper the expert in their ability to select papers that properly align with methods applied in their case, thereby increasing uncertainty and reducing their usefulness when assigning probabilities. It is therefore crucial that such methods are recorded in great detail. This leads us on to the design and sharing of data from studies on TPPR of biological traces, which will be discussed in Section 3.4.

3.3.5   The Transfer, Persistence, Prevalence and Recovery of Human Hairs

One type of human biological trace that is encountered frequently in casework is human hairs. Up to this point we have not discussed this type of trace with regard to their transfer, persistence, prevalence and recovery. Evidently hairs may be a very relevant trace to consider when addressing an issue at the activity level. Examples of situations where hairs may be very informative are sexual assault cases where pubic hairs may be transferred between victim and perpetrator, or violent crimes where hairs may be shed or pulled. Hairs have been a useful object for study in forensics for a long time. Attribution of hairs to donors has been done, and is still being performed, by morphological comparison [55, 56] as well as by autosomal and mitochondrial DNA analysis [57, 58] and protein sequencing [59]. As such they are strongly embedded in forensic biology casework.

Few studies have been published on the TPPR of human hairs. Some examples are:

- Transfer of pubic and scalp hairs in sexual and other violent assaults [60]. In this retrospective study of six years of case file data, hairs were attributed to either victim or (convicted) offender based on the morphological comparison. They found amongst other things that in

3 out of 68 pairs of underwear of the victim's head hair of the perpetrator was found (4%). In two pairs of underwear pubic hairs of the perpetrator were found (3%). On the outerwear of a victim they found head hair of the perpetrator in 8 out of 62 cases (13%).

- Transfer of pubic hairs during sexual intercourse [61]. This study looked at the transfer of pubic hairs between consenting partners. They found overall transfer rates to be low (17% of combings) but suggestive of higher rates of transfer from female to male (24%) than vice versa (11%).
- The persistence of human hairs on clothing [62]. This study looked at characteristics of hair (dyed or not, root or not) and different types of fabric to study the persistence of hairs. They found that the hair characteristics showed no statistically significant impact on persistence but that the type of fabric does. Rougher fabrics tend to have higher retention of fibres and hairs than smoother fabrics [63–65].

The recovery rates in most of these studies are based on morphological comparisons to attribute the hairs to specific donors. We therefore need to be careful in translating them to cases where the attribution of the hairs was done by DNA analysis. Studies have shown that the attribution of hairs to individuals based on morphology or DNA analysis may differ substantially [66]. Cultural differences in personal grooming are also likely to influence rates of transfer and persistence of hair [67].

Hairs behave differently from other human biological traces, in that their 'behaviour' is more akin to fibres or animal hairs than to body fluids or latent DNA traces. This leads us to the question whether the study of human hair's given activity level propositions is something that is within the domain of the forensic biologist, or rather with a forensic expert in fibres. The partial overlap or 'grey' areas between forensic disciplines will be further discussed in Chapter 12.

3.4 DESIGNING EXPERIMENTS ON DNA TPPR AND SHARING THEIR RESULTS

We use published scientific studies as one of the main sources to obtain frequencies of occurrence of certain TPPR events. Such frequencies are what we use to inform the probabilities that we need to assign to certain outcomes of these events occurring. This could, for instance, be the probability of recovering DNA of the accused from the handle of a knife that he allegedly stabbed the victim with. Such a probability could be informed based on the frequencies reported for volunteers in the studies by (for instance) Samie et al. [68] or Meakin et al. [69] which were discussed in the previous paragraph.

Many such studies are not designed specifically to inform probability assignments. The purpose of the studies could be quite different, for instance, to study specific mechanistic aspects of DNA transfer or persistence. The purpose of the study will dictate the experimental design, as well as the way in which the findings are reported in a journal article. This may cause some issues when attempting to use the data from published studies in casework.

3.4.1 NOTES ON EXPERIMENTAL DESIGN AND PURPOSE OF A STUDY

As the purpose of a study and the underlying research question will dictate the experimental design, it is crucial that we recognise that studies into TPPR can have two main purposes:

1. Understanding mechanisms that impact on any aspect of TPPR
2. Generating data to assign probabilities to TPPR events occurring given specific scenarios

Neither does the first purpose exclude the use of the data for the second nor does the second prevent insights into mechanisms. However, both broad purposes will introduce limitations that will reduce their ability to inform the other. There are many studies currently published that try to encompass both purposes, and as a consequence are not reaching either goal [4]. Meakin et al. [9] further elaborated on this topic by defining the type of repeat experiments that are needed to address the two purposes. For understanding mechanisms that impact TPPR, one should conduct *repetitions* of strictly controlled experiments. For example, have a single volunteer conduct repeated activities or contacts and vary one factor only. Strictly controlling the experimental conditions will necessarily affect the extent to which the resulting data can be aligned with casework conditions. An example would be that it is common to use pre-cleaned surfaces to detect the transfer of DNA of a volunteer. While this makes sense when attempting to control the conditions of experiments, it is unlikely to encounter pre-cleaned items in casework. Another common difference between experiments and casework is the effect of packaging and transport of items, which is not accounted for in most controlled experiments. Both these examples may impact on the probability of recovery of DNA and are thus important to consider when using experimental data when tasked with assigning probabilities to DNA TPPR events. A detailed discussion of such aspects of a study by Cale et al. [70] can be found in responses to that study by Goray et al. [71] and Kokshoorn et al. [72].

In contrast, to gain data on the frequency of occurrence of an event (which is of interest when assigning probabilities to such events) we would need *iterations* of events. These could be repeated activities by a group of volunteers that is representative of the persons-of-interest in a case (or general case conditions). The latter are generally less controlled by their nature. As many aspects of such studies are commonly not controlled (items not pre-cleaned, history of handwashing and – contacts self-reported, etc.) this strongly limits the ability of such studies to assess the impact of specific factors on the TPPR events.

Having said this, any source of data may be found useful to assign probabilities to TPPR events. It is up to the professional judgment of the expert to decide whether data sufficiently align with the case circumstances to inform aspects of TPPR that are dictated by the scenarios in that case.

3.4.2 REPORTING FINDINGS FROM A STUDY ON TPPR

The greater the difference between the circumstances of the case and that of the published study, the less the resulting data will be useful to directly assign (unadjusted) probabilities to DNA TPPR events. Details on the design of a study, the volunteers, the methods applied, etc. are therefore crucial when determining if the resulting data from such a DNA TPPR study align with the case circumstances. Any information omitted from the paper will leave the expert in doubt about the applicability of the study for the task at hand. It could also mean that different experts will have a different take on the same data, resulting in quite different probability assignments.

An example of this was given by Kokshoorn et al. [73] who described a mock case evaluation given activity level propositions. Two participating laboratories were compared in their use of data from Daly et al. [54] to assign probabilities to the transfer of DNA from hands to an airbag. It was pointed out that the Daly et al. study did not specify their criteria for determining whether a DNA profile was reportable or when a person-of-interest was considered a contributor. This led the two laboratories to set their own criteria on this to categorise the data. Both laboratories used the same four categories to assign probabilities of observation to: 'Single source DNA profile of driver', 'Mixed DNA profile with driver major contributor', 'Other result (interpretable profile excluding the driver)' and 'No (reportable) DNA profile'. However, based on different assumptions and different definitions of these categories, the probabilities that were assigned, *based on the same study*, were quite different, as shown in Table 3.3. This was primarily caused by the definition of what a single source DNA profile is and the extent to which a DNA profile was considered reportable.

This example illustrates (as does the comparison of the two 'knife' studies in Table 3.2.) that it is crucial that the methods used in a study, from sampling strategy and method all the way through to the DNA profile interpretation, need to be documented in great detail. If this is not done, it will introduce an additional factor of uncertainty in the data, that will reduce its potential for use to discriminate between proposed activities in a case.

TABLE 3.3

Probability Assignments to Result Categories by Two Laboratories Based on the Study by Daly et al. [54] (based on Kokshoorn et al. [73])

	Laboratory 1	Laboratory 3
Single source DNA profile of the driver	0.32	0.06
Mixed DNA profile with driver major contributor	0.28	0.29
Other results (interpretable profile excluding the driver)	0.09	0.03
No (reportable) DNA profile	0.31	0.64

When carrying out experimentation with the intention to provide data that might be useful in probability assignments, whether it be for in-house use only or intended for publication, some pointers and suggestions can be found in Box 3.1.

We conclude with the remark that the body of published scientific data will never be complete, as in that it will never be possible to find experimental data that perfectly align every set of case circumstances. It is up to the expert to explore all available sources to find those studies and data that are considered to be useful in the case at hand. Any limitations associated with study design or reporting should not dissuade experts from using the data to inform their personal probabilities. As with any forensic science endeavour, limitations to the study or reported opinion will always need to be made transparent to the user of the report. We will expand on that in Chapter 10.

We will explore the use of data further in the following chapters, where in Chapter 7 we go into 'sensitivity analyses' as a method to deal with uncertainties introduced by the use of data to inform probabilities.

BOX 3.1   POINTERS AND SUGGESTIONS FOR EXPERIMENTATION AND GENERATION OF DATA TO INFORM PROBABILITY ASSIGNMENTS FOR TPPR OF BIOLOGICAL MATERIALS

Provide experimental data in the supplementary material for any published paper

A publication will necessarily summarise the findings with a focus on the study's purpose. However, the data are likely to be of broader value to those wishing to mine them for their own purposes.

RECORD AS MUCH INFORMATION AS POSSIBLE, EVEN IF DOWNSTREAM ANALYSIS FOUND THE PARAMETER TO BE UNIMPORTANT

In studies there will be multiple factors of potential significance that are recorded with each observation. In many studies (particularly those that involve student projects) the exact analysis that will be performed is not fully set until halfway through the project and so there is a tendency to over-record information. Additionally, when the data is analysed for publication, it is common that there will be a number of factors or combinations of factors that ultimately are found not to be significant to describing the response variable and so will not make up part of the final model. Even when the factors are not found to be significant, or they are not even analysed, it is still preferable to include them in the raw data that is provided with a publication. It may be the case that some factors that were found insignificant in the study are important, but there was simply insufficient sample size in the study to identify the significance. A later study that pooled the results of multiple studies together may have the power to identify a dependency that none of the individual studies could observe. This is only possible when each of the individual studies makes the data available, even for the factors that were not found to be significant. Additionally, it may be that relationships between factors other than those examined in the study are investigated. For example, a study may look at the amount of DNA transferred from grabbing an object, with instrumentation that could measure both the pressure and time of the grab. It may be that neither pressure nor time is found to significantly contribute to the amount of DNA transfer. However, another study may be interested in whether the pressure exerted during a grab of an object depends on the amount of time the object was held for. This was not even considered in the initial experiment, and again, by including the information in the raw data, it makes the findings more broadly useful to other studies.

RECORD THE RAW OBSERVATIONS, AS WELL AS INTERPRETATIONS

There are different ways that data can be interpreted, presented in publications and used in probability assignments. It may be that the DNA amount, number of alleles, profile interpretation or likelihood ratio are being considered as the response variable (or observation in the case). Also mentioned previously is that the further downstream the response variable is in the laboratory process, the more the laboratory process used to generate the data will have to align with the process of the laboratory using it for evaluations. For example, if DNA amount is measured, then only the DNA extraction methods need to align (if we assume that the method of quantitation is accurate – but see [74] – and does not need to align between publication and laboratory). If the number of alleles is measured then the extraction, PCR and capillary electrophoresis methods need to align and then if the profile interpretation is measured then in addition to all laboratory processes aligning, then so too does the method of interpretation. If only the profile interpretation is given as the response variable in the provided dataset then a laboratory has no option but to use profile interpretation on their own data in order to use the published study in an evaluation (even if this is not the usual way in which the data would be interpreted in their laboratory). However, if the study also provides DNA amounts, and mixture properties (i.e. the number of contributors, mixture proportions, number of POI alleles present, the position the POI aligns with in the mixture and the profile interpretation) then any laboratory can use the raw data to carry out regressions using the response variable that best aligns with their processes.

CLEARLY LABEL AND EXPLAIN THE MEANING OF THE DATA

This is mainly a housekeeping matter and provides as much benefit for the laboratory generating the data as it does for another laboratory using it. Whenever a dataset is provided then the data should be presented in a clearly identifiable way. This includes:

- Providing the data in a spreadsheet or text file with columns that represent variables and rows that represent observations
- Not using the same heading for multiple columns/variables and making the headings helpful. Ideally, the column heading themselves should have some intuitive name for what they represent. For example, the column heading such as 'DNA' can be ambiguous. Is this DNA concentration, the total DNA amount or the DNA donated by the POI? A better heading may be 'total_DNA_ng'
- Presenting the data so that each row only represents a single response variable. This is not strictly required, as the data can be reformatted as needed, but most statistical programs will need the data presented in this format before it can be analysed. Note that this style of data presentation can be quite different to how we think about presenting the data in a publication table. For example, if an experiment measured the amount of DNA that transferred

from object 1 to object 2 to object 3 then we might present the data in a publication table with headings 'Experiment', 'DNA on object 1', 'DNA on object 2', 'DNA on object 3'. However, this would present three response variable measurements per row. For presentation as supplementary material to be used in a statistical package a more useful presentation would be to have columns 'Experiment', 'DNA amount' and 'object', so that each row in the previously presented format would become three rows in the latter presented format

- Providing a document that clearly explains what each column in the dataset represents and what units in which it is measured. If the column represents a categorical variable, then the list of all possible states that the category can contain should be provided

USE CONSISTENT VARIABLES BETWEEN SUB-DATASETS IN THE EXPERIMENT WHERE POSSIBLE

Again, this is a point of housekeeping but is often not adhered to. If a study breaks the total experiments into subsets of similar sub-experiments then as much as possible the same measurements should be taken and recorded in the same way, and presented in the dataset with the same variable headings. For example, imagine a study that looks at the amount of DNA transferred to glass or plastic by grabbing, and the decision is made to split the experiment into two components (one for each substrate). When providing the raw data, if it is also split into two files, then headings such as 'DNA transferred to plastic' and 'DNA transferred to glass' should be avoided. Instead, the heading could be 'DNA transferred' and another column with 'substrate' could be added. Then if the data is ever to be analysed together, the two separate data files can simply be read and combined. These types of considerations are particularly important if the study occurs across multiple laboratories or is carried out by multiple people all semi-independently. Measurement and data recording should be settled on before any experiments are performed.

PROVIDE FULL DETAILS OF ANALYTICAL METHODS USED

This should be already recorded in the publication itself; however, there are often small aspects of the data recording that are ambiguous, and it is important for these to be clarified for proper data interpretation. For example, if the variable being measured is POI alleles, then specify whether at a locus where the POI is homozygous and the evidence possesses that allele and does this count for 1 or 2 of the POI alleles. Also, if the dataset is not part of a publication and rather comes from an in-house validation or is made available through an online database then it will be important to fully specify the method (as there will be no published paper to refer to).

PROVIDE DETAILS OF SIMULATION

It is becoming more common as time advances to use simulation as part of data interpretation/ collection. This comes about as more data is available, more computer power is available and people in the forensic field advance in their mathematical, statistical and programming ability. There are many fields that have groups working on ways in which simulation and machine learning can be used to answer questions that previously were out of reach, or that we did not have enough power to address. If a simulation is used as part of a study, for example modelling DNA amount from two objects and simulating DNA from those distributions to obtain a transfer distribution, then the exact details of the simulation should be provided, i.e. what distributions were modelled, how were they modelled, what language base was used, whether any libraries were used in the simulation, the number of simulations and the way in which the simulation results have been summarised and even the seed that was used for random number generation if it is available. Ideally the code used in the simulation would also be provided as supplementary material to allow other users to exactly recreate the experiment that has been carried out.

CONSIDER ADDING TO EXISTING KNOWLEDGE DATABASES

Colleagues from the University of Kiel (Germany) have launched a database with TPPR studies in forensic biology called 'DNA-TrAC', which is accessible through https://bit.ly/2R4bFgL [4] (last accessed February 2023). This database hosts references to published and peer-reviewed studies, including details on relevant parameters of the study design, methods used and the results. The database was created and filled with data on publications up to 16 December 2019 by the authors/custodians, which is (at the time of writing) still the current version. The authors have called for the forensic genetics community to support the database by providing data on new publications, or by assisting in updating the database with previously published studies. The database includes a list of studies that the custodians are aware of, but have not yet been included in the database. This list (again at the time of writing) was last updated on 25 October 2022.

Another database for published studies on TPR of traces is the TTADB (Transfer Traces Activity DataBase) hosted by the University of Québec at Trois-Rivières (UQTR), Québec, Canada [75]. The database contains data on transfer studies in a broad range of fields in forensic science, including DNA and body fluids. The database is actively maintained and updated. This database is discussed in more detail in Chapter 12.

Adding to central and openly accessible databases such as 'DNA-TrAC' and 'TTADB' makes the study more accessible to a wider audience.

3.5   PRACTICE QUESTIONS

Q1. Consider the following series of activities:

- You open your front door
- You walk down the hallway
- You open the door to the kitchen
- You open a drawer
- You take out a knife
- You walk back out the front door
- You use the knife to remove chewing gum that someone stuck to your front door
- You go inside and close the front door
- You walk back through the open door to the kitchen
- You place the knife on the countertop

What routes for transfer of 'unknown DNA' (assume you have a reference DNA profile of yourself) to the knife handle would you consider?

Q2. Considering the previous question, what if you washed your hands prior to opening the drawer and taking out the knife?

Q3. Consider example case 2 ('The three burglars', see Chapter 2, Section 2.5.2). What routes of transfer of DNA of the defendant (D) to the hair of the victim would you consider?

Q4. Consider example case 3 ('The family assault', see Chapter 2, Section 2.5.3). What routes of transfer of the biological material of the brother (D) to the underwear of the sister (C) would you consider?

Q5. Consider the following series of activities.

- You don a pair of laboratory gloves
- You open the door to the lab and close it behind you
- You grab an evidence bag from a cart containing an item of clothing and place it on a side table
- You don the second pair of gloves and clean the examination table
- You open the evidence bag and let the clothing slide out on the examination table
- You place the empty bag on the side table
- You fold open the clothing
- You grab a camera and take a photo of the item

Which of these activities may impact the probability of contamination of the item of clothing? And how?

Q6. Consider example case 1 ('The attempted kidnap', see Chapter 2, Section 2.5.1). From the top of the complainant a sample was taken. Assuming DNA from the perpetrator transferred to the top during the attempted kidnap; which factors impacting its persistence would you consider? And why?

Q7. Consider example case 4 ('The gun in the laundry', see Chapter 2, Section 2.5.4). A gun was found in a laundry basket covered with laundry from the defendant (D). What factors would you consider when assigning a probability to transfer, persistence and recovery of DNA of D from the laundry to the outside surfaces of the gun?

Q8. Consider example case 2 ('The three burglars', see Chapter 2, Section 2.5.2). In this case under the prosecution proposition a DNA TPR event is postulated from the hands of the defendant to the hair of the victim (through grabbing the hair and pulling it).

You may consider the publication by Daly et al. [54] as a potential source of data to inform your probability assignment for the TPR event. What arguments would you consider to decide whether or not to use the study for this purpose?

Q9. How would you define (in your own words) the term 'shedder status'?

Q10. Please explain how 'shedder status' may impact your assignment of a probability to a DNA TPR event.

REFERENCES

1. P. Gill, T. Hicks, J.M. Butler, E. Connolly, L. Gusmão, B. Kokshoorn, N. Morling, R.A.H. van Oorschot, W. Parson, M. Prinz, P.M. Schneider, T. Sijen, D. Taylor, DNA commission of the International society for forensic genetics: Assessing the value of forensic biological evidence: Guidelines highlighting the importance of propositions. Part II: Evaluation of biological traces considering activity level propositions, *Forensic Science International: Genetics* 44 (2020) 102186.
2. P. Gill, *Misleading DNA Evidence: Reasons for Miscarriages of Justice*, Elsevier, 2014. pp. 1–100.
3. K. Steensma, R. Ansell, L. Clarisse, E. Connolly, A.D. Kloosterman, L.G. McKenna, R.A.H. van Oorschot, B. Szkuta, B. Kokshoorn, An inter-laboratory comparison study on transfer, persistence and recovery of DNA from cable ties, *Forensic Science International: Genetics* 31 (2017) 95–104.
4. A. Gosch, C. Courts, On DNA transfer: The lack and difficulty of systematic research and how to do it better, *Forensic Science International: Genetics* 40 (2019) 24–36.
5. G. Meakin, A. Jamieson, DNA transfer: Review and implications for casework, *Forensic Science International: Genetics* 7 (2013) 434–443.
6. R.A.H. van Oorschot, K.N. Ballantyne, R.J. Mitchell, Forensic trace DNA: A review, *Investigative Genetics* 1(1) (2010) 14.

7. R.A.H. van Oorschot, B. Szkuta, G.E. Meakin, B. Kokshoorn, M. Goray, DNA transfer in forensic science: A review, *Forensic Science International: Genetics* 38 (2019) 140–166.
8. R.A. Wickenheiser, Trace DNA: A review, discussion of theory, and application of the transfer of trace quantities of DNA through skin contact, *Journal of Forensic Sciences* 47(3) (2002) 442–450.
9. G.E. Meakin, B. Kokshoorn, R.A.H. van Oorschot, B. Szkuta, Evaluating forensic DNA evidence: Connecting the dots, *WIREs Forensic Science* 3(4) (2021) e1404.
10. R.A.H. van Oorschot, G.E. Meakin, B. Kokshoorn, M. Goray, B. Szkuta, DNA transfer in forensic science: Recent progress towards meeting challenges, *Genes* 12(11) (2021) 1766.
11. R.A. Van Oorschot, M. Jones, DNA fingerprints from fingerprints, *Nature* 387(6635) (1997) 767.
12. A. Lowe, C. Murray, J. Whitaker, G. Tully, P. Gill, The propensity of individuals to deposit DNA and secondary transfer of low level DNA from individuals to inert surfaces., *Forensic Science International* 129 (2002) 25–34.
13. M. Phipps, S. Petricevic, The tendency of individuals to transfer DNA to handled items, *Forensic Science International* 168 (2007) 162–168.
14. R.K. Farmen, R. Jaghø, P. Cortez, E.S. Frøyland, Assessment of individual shedder status and implication for secondary transfer, *Forensic Science International: Genetics Supplementary Series* 1 (2008) 415–417.
15. A.E. Fonneløp, M. Ramse, T. Egeland, P. Gill, The implications of shedder status and background DNA on direct and secondary transfer in an attack scenario, *Forensic Science International: Genetics* 29 (2017) 48–60.
16. P. Kanokwongnuwut, B. Martin, K.P. Kirkbride, A. Linacre, Shedding light on shedders, *Forensic Science International: Genetics* 36 (2018) 20–25.
17. M. Goray, S. Fowler, B. Szkuta, R.A.H. van Oorschot, An analysis of self and non-self DNA in multiple handprints deposited by the same individuals over time, *Forensic Science International: Genetics* 23 (2016) 190–196.
18. P. Wiegand, M. Kleiber, DNA typing of epithelial cells after strangulation, *International Journal of Legal Medicine* 110(4) (1997) 181–3.
19. J. Burrill, B. Daniel, N. Frascione, A review of trace "Touch DNA" deposits: Variability factors and an exploration of cellular composition, *Forensic Science International: Genetics* 39 (2019) 8–18.
20. T. Kamphausen, S.B. Fandel, J.S. Gutmann, T. Bajanowski, M. Poetsch, Everything clean? Transfer of DNA traces between textiles in the washtub, *International Journal of Legal Medicine* 129(4) (2015) 709–14.
21. D. Lacerenza, S. Aneli, M. Omedei, S. Gino, S. Pasino, P. Berchialla, C. Robino, A molecular exploration of human DNA/RNA co-extracted from the palmar surface of the hands and fingers, *Forensic Science International: Genetics* 22 (2016) 44–53.
22. M. Poetsch, T. Bajanowski, T. Kamphausen, Influence of an individual's age on the amount and interpretability of DNA left on touched items, *International Journal of Legal Medicine* 127 (2013) 1093–1096.
23. P. Manoli, A. Antoniou, E. Bashiardes, S. Xenophontos, M. Photiades, V. Stribley, M. Mylona, C. Demetriou, M.A. Cariolou, Sex-specific age association with primary DNA transfer, *International Journal of Legal Medicine* 130(1) (2016) 103–112.
24. N.J. Port, V.L. Bowyer, E.A.M. Graham, M.S. Batuwangala, G.N. Rutty, How long does it take a static speaking individual to contaminate the immediate environment?, *Forensic Science, Medicine, and Pathology* 2(3) (2006) 157–163.
25. S. Asadi, A.S. Wexler, C.D. Cappa, S. Barreda, N.M. Bouvier, W.D. Ristenpart, Aerosol emission and superemission during human speech increase with voice loudness, *Scientific Reports* 9(1) (2019) 2348.
26. D.A. Edwards, J.C. Man, P. Brand, J.P. Katstra, K. Sommerer, H.A. Stone, E. Nardell, G. Scheuch, H. Brenner, Inhaling to mitigate exhaled bioaerosols, *Proceedings of the National Academy of Sciences of the United States of America* 101(50) (2004) 17383–17388.
27. L. Puliatti, O. Handt, D. Taylor, The level of DNA an individual transfers to untouched items in their immediate surroundings, *Forensic Science International: Genetics* 54 (2021).
28. S. Zoppis, B. Muciaccia, A. D'Alessio, E. Ziparo, C. Vecchiotti, A. Filippini, DNA fingerprinting secondary transfer from different skin areas: Morphological and genetic studies, *Forensic Science International: Genetics* 11 (2014) 137–143.
29. H.C. Lee, C. Ladd, Preservation and collection of biological evidence, *Croatian Medical Journal* 42(3) (2001) 225–228.
30. I. Quinones, B. Daniel, Cell free DNA as a component of forensic evidence recovered from touched surfaces, *Forensic Science International: Genetics* 6(1) (2012) 26–30.

31. M. Vandewoestyne, D. Van Hoofstat, A. Franssen, F. Van Nieuwerburgh, D. Deforce, Presence and potential of cell free DNA in different types of forensic samples, *Forensic Science International: Genetics* 7(2) (2013) 316–20.

32. M. Goray, E. Eken, R.J. Mitchell, R.A.H. van Oorschot, Secondary DNA transfer of biological substances under varying test conditions, *Forensic Science International: Genetics* 4(2) (2010) 62–67.

33. M. Goray, R.J. Mitchell, R.A.H. van Oorschot, Investigation of secondary transfer of skin cells under controlled conditions, *Legal Medicine* 12 (2010) 117–120.

34. S.H.A. Tobias, G.S. Jacques, R.M. Morgan, G.E. Meakin, The effect of pressure on DNA deposition by touch, *Forensic Science International: Genetics Supplement Series* 6 (2017) e12–e14.

35. I. Hefetz, N. Einot, M. Faerman, M. Horowitz, J. Almog, Touch DNA: The effect of the deposition pressure on the quality of latent fingermarks and STR profiles, *Forensic Science International: Genetics* 38 (2019) 105–112.

36. C.M. Pfeifer, P. Wiegand, Persistence of touch DNA on burglary-related tools, *International Journal of Legal Medicine* 131(4) (2017) 941–953.

37. D. Taylor, A. Biedermann, L. Samie, K.-M. Pun, T. Hicks, C. Champod, Helping to distinguish primary from secondary transfer events for trace DNA, *Forensic Science International: Genetics* 28 (2017) 155–177.

38. D. Taylor, L. Samie, C. Champod, Using Bayesian networks to track DNA movement through complex transfer scenarios, *Forensic Science International: Genetics* 42 (2019) 69–80.

39. A.A. Oleiwi, M.R. Morris, W.M. Schmerer, R. Sutton, The relative DNA-shedding propensity of the palm and finger surfaces, *Science & Justice* 55(5) (2015) 329–334.

40. D.L. McColl, M.L. Harvey, R.A.H. van Oorschot, DNA transfer by different parts of a hand, *Forensic Science International: Genetics Supplement Series* 6 (2017) e29–e31.

41. F. Oldoni, V. Castella, D. Hall, Shedding light on the relative DNA contribution of two persons handling the same object, *Forensic Science International: Genetics* 24 (2016) 148–157.

42. R.A.H. van Oorschot, G. Glavich, R.J. Mitchell, Persistence of DNA deposited by the original user on objects after subsequent use by a second person, *Forensic Science International: Genetics* 8(1) (2014) 219–225.

43. R.A.H. van Oorschot, D.L. McColl, J.E. Alderton, M.L. Harvey, R.J. Mitchell, B. Szkuta, Activities between activities of focus—Relevant when assessing DNA transfer probabilities, *Forensic Science International: Genetics Supplement Series* 5 (2015) e75–e77.

44. B. Szkuta, K.N. Ballantyne, B. Kokshoorn, R.A.H. van Oorschot, Transfer and persistence of non-self DNA on hands over time: Using empirical data to evaluate DNA evidence given activity level propositions, *Forensic Science International: Genetics* 33 (2018) 84–97.

45. G.N. Rutty, A. Hopwood, A. Tucker, The effectiveness of protective clothing in the reduction of potential DNA contamination of the scene of crime, *International Journal of Legal Medicine* 117 (2003) 170–174.

46. R. Daniel, R.A.H. van Oorschot, An investigation of the presence of DNA on unused laboratory gloves, *Forensic Science International: Genetics Supplement Series* 3(1) (2011) e45–e46.

47. B. Szkuta, M.L. Harvey, K.N. Ballantyne, R.A.H. van Oorschot, DNA transfer by examination tools: A risk for forensic casework?, *Forensic Science International: Genetics* 16 (2015) 246–254.

48. M. Goray, R. van Oorschot, J. Mitchell, DNA transfer within forensic exhibit packaging: Potential for DNA loss and relocation, *Forensic Science International: Genetics* 6 (2012) 158–166.

49. P. Kumar, R. Gupta, R. Singh, O.P. Jasuja, Effects of latent fingerprint development reagents on subsequent forensic DNA typing: A review, *Journal of Forensic and Legal Medicine* 32 (2015) 64–9.

50. B. Szkuta, R. Ansell, L. Boiso, E. Connolly, A.D. Kloosterman, B. Kokshoorn, L.G. McKenna, K. Steensma, R.A.H. van Oorschot, Assessment of the transfer, persistence, prevalence and recovery of DNA traces from clothing: An inter-laboratory study on worn upper garments, *Forensic Science International: Genetics* 42 (2019) 56–68.

51. B. Szkuta, R. Ansell, L. Boiso, E. Connolly, A.D. Kloosterman, B. Kokshoorn, L.G. McKenna, K. Steensma, R.A.H. van Oorschot, DNA transfer to worn upper garments during different activities and contacts: An inter-laboratory study, *Forensic Science International: Genetics* 46 (2020) 102268.

52. E.M. Ton, J. Limborgh, L.H.J. Aarts, B. Kokshoorn, J. de Koeijer, M.C. Zuidberg, Plaats delict-onderzoek met vooruitziende blik: Anticiperen op alternatieve scenario's tijdens het sporenonderzoek op de plaats delict, *Expertise en Recht* 4 (2018) 144–149.

53. B. Szkuta, K.N. Ballantyne, R.A.H. van Oorschot, Transfer and persistence of DNA on the hands and the influence of activities performed, *Forensic Science International: Genetics* 28 (2017) 10–20.

54. D.J. Daly, C. Murphy, S.D. McDermott, The transfer of touch DNA from hands to glass, fabric and wood, *Forensic Science International: Genetics* 6 (2013) 41–46.

55. L. Wilkinson, C. Gwinnett, An international survey into the analysis and interpretation of microscopic hair evidence by forensic hair examiners, *Forensic Science International: Genetics* 308 (2020) 110158.

56. M. Airlie, J. Robertson, E. Brooks, Forensic hair analysis: Worldwide survey results, *Forensic Science International: Genetics* 327 (2021) 110966.

57. K.S. Grisedale, G.M. Murphy, H. Brown, M.R. Wilson, S.K. Sinha, Successful nuclear DNA profiling of rootless hair shafts: A novel approach, *International Journal of Legal Medicine* 132(1) (2018) 107–115.

58. K.J. van der Gaag, S. Desmyter, S. Smit, L. Prieto, T. Sijen, Reducing the number of mismatches between hairs and buccal references when analysing mtDNA heteroplasmic variation by massively parallel sequencing, *Genes* 11(11) (2020). 1355.

59. K.F. Jones, T.L. Carlson, B.A. Eckenrode, J. Donfack, Assessing protein sequencing in human single hair shafts of decreasing lengths, *Forensic Science International: Genetics* 44 (2020) 102145.

60. M.J. Mann, Hair transfers in sexual assault: A six-year case study, *Journal of Forensic Sciences* 35(4) (1990) 951–5.

61. D.L. Exline, F.P. Smith, S.G. Drexler, Frequency of pubic hair transfer during sexual intercourse, *Journal of Forensic Sciences* 43(3) (1998) 505–8.

62. J. Dachs, I.J. McNaught, J. Robertson, The persistence of human scalp hair on clothing fabrics, *Forensic Science International* 138(1) (2003) 27–36.

63. C.A. Pounds, K.W. Smalldon, The transfer of fibres between clothing materials during simulated contacts and their persistence during wear. Part I--Fibre transference, *Journal: Forensic Science Society* 15(1) (1975) 17–27.

64. C.A. Pounds, K.W. Smalldon, The transfer of fibres between clothing materials during simulated contacts and their persistence during wear: Part II—Fibre persistence, *Journal of the Forensic Science Society* 15(1) (1975) 29–37.

65. C.A. Pounds, K.W. Smalldon, The transfer of fibres between clothing materials during simulated contacts and their persistence during wear. Part III--Fibre persistence, *Journal: Forensic Science Society* 15(1) (1975) 29–37.

66. M.M. Houck, B. Budowle, Correlation of microscopic and mitochondrial DNA hair comparisons, *Journal of Forensic Sciences* 47(5) (2002) 964–967.

67. L.K. Craig, P.B. Gray, Pubic hair removal practices in cross-cultural perspective, *Cross-Cultural Research* 53(2) (2019) 215–237.

68. L. Samie, T. Hicks, V. Castella, F. Taroni, Stabbing simulations and DNA transfer, *Forensic Science International: Genetics* 22 (2016) 73–80.

69. G.E. Meakin, E.V. Butcher, R.A.H. van Oorschot, R.M. Morgan, Trace DNA evidence dynamics: An investigation into the deposition and persistence of directly- and indirectly-transferred DNA on regularly-used knives, *Forensic Science International: Genetics* 29 (2017) 38–47.

70. C.M. Cale, M.E. Earll, K.E. Latham, G.L. Bush, Could secondary DNA transfer falsely place someone at the scene of a crime?, *Journal of Forensic Sciences* 61(1) (2016) 196–203.

71. M. Goray, K.N. Ballantyne, B. Szkuta, R.A.H. van Oorschot, Cale CM, Earll ME, Latham KE, Bush GL. Could secondary DNA transfer falsely place someone at the scene of a crime?, *Journal of Forensic Sciences* 61(1) (2016) 196–203, *Journal of Forensic Sciences* 61(5) (2016) 1396–1398.

72. B. Kokshoorn, B. Aarts, R. Ansell, L. McKenna, E. Connolly, W. Drotz, A.D. Kloosterman, Cale CM, Earll ME, Latham KE, Bush GL. Could secondary DNA transfer falsely place someone at the scene of a crime? *Journal of Forensic Sciences* 61(1) (2016) 196–203, *Journal of Forensic Sciences* 61(5) (2016) 1401–1402.

73. B. Kokshoorn, L.H.J. Aarts, R. Ansell, E. Connolly, W. Drotz, A.D. Kloosterman, L.G. McKenna, B. Szkuta, R.A.H. van Oorschot, Sharing data on DNA transfer, persistence, prevalence and recovery: Arguments for harmonization and standardization, *Forensic science International: Genetics* 37 (2018) 260–269.

74. K. Nielsen, H.S. Mogensen, J. Hedman, H. Niederstätter, W. Parson, N. Morling, Comparison of five DNA quantification methods, *Forensic Science International: Genetics* 2(3) (2008) 226–230.

75. L. Cadola, M. Charest, C. Lavallée, F. Crispino, The occurrence and genesis of transfer traces in forensic science: A structured knowledge database, *Canadian Society of Forensic Science Journal* 54(2) (2021) 86–100.

4 Structuring Case Information

Bas Kokshoorn

CONTENTS

DOI: 10.4324/9781003273189-4

4.1 INTRODUCTION

This chapter is about the 'H' and the 'I' in $Pr(E|H,I)$. As we have seen in Chapter 1, we are considering our findings given propositions and all other relevant information. In this chapter we discuss the concepts of propositions, their formulation and the structuring of other case and task-relevant information.

4.1.1 Propositions or Hypotheses

These terms are commonly used interchangeably and to a certain extent, they are interchangeable. However, they do have different connotations which may lead to miscommunication between users if those connotations do not align. Hypothesis testing, as it is used in frequentist statistics, means that based on a test for statistical significance, a hypothesis will be rejected or not. In the context of interpreting findings in a forensic context, as scientists we are not interested in rejecting the premise (hypothesis) that is put to us. As explained in Chapter 2, we are to address our findings 'given' the hypothesis or proposition, and not the hypothesis itself. Using the term 'hypothesis' may suggest that the interpretation of the findings by the scientist given the hypothesis results in a conclusion about the hypothesis. One of a number of small aspects of phrasing (like 'activity level reporting', which suggests the scientist reports on the activity rather than their findings) might induce the reader to transpose the conditional. We therefore follow Aitken and Taroni [1] in their preference for the term 'proposition' when assessing the weight of our findings in a Bayesian statistical framework. In addition, they argue in favour of 'proposition' over 'hypothesis':

> Propositions may be complementary in the same way as events are said to be complementary. One and only one can be true and together they exhaust all possibilities.

4.1.2 What Are Propositions

In a criminal case the police and prosecution, as well as the defence counsel of a defendant will generally present scenarios to the court (but note that the defence are in their right to present nothing at all). These scenarios are chronological descriptions of the events that have taken place according to these parties. Certain elements in these scenarios will be conceded, in that there is no dispute between the parties that certain events have taken place. An example of this could be legitimate activities between a complainant and defendant like them socialising at a particular time prior to an incident. Other aspects of the scenarios will be disputed. Some of these disputed aspects may not be within the domain of the forensic scientist (such as the motive for a criminal act or whether pre-meditation was part of the crime). Disputed events that can be addressed by the application of forensic testing form the basis for one or more sets of propositions. Such events could include the activity of stabbing a complainant, where either the activity itself may be in dispute (was the complainant stabbed? – the wounds sustained by the complainant may be examined by a forensic medic) or the actor who stabbed the complainant (was it the defendant or somebody else? – which may be addressed by examining traces on the knife, like DNA or fingermarks). A number of disputed aspects taken from the scenarios will form the 'core propositions'. From these core propositions, sub-propositions will be formulated that can be addressed by different types of forensic analysis. A core proposition for the prosecution in a case with a complainant who was stabbed could, for instance, be:

- The complainant was stabbed three times by the defendant with knife X at 3 am on Friday, 1 October 2021.

Sub-propositions can be derived from this core proposition:

1. The complainant was stabbed three times
2. The complainant was stabbed with knife X

3. The complainant was stabbed by the defendant

4. The complainant was stabbed at 3 am on Friday, 1 October 2021

The first sub-proposition may be addressed by forensic medicine by examining the wounds and lacerations sustained by the complainant. The second sub-proposition could be part of the forensic medical examination (do the wounds correspond to the dimensions of the blade of the knife?) as well as by examination of biological traces from the complainant on the blade (is there blood or tissue of the complainant on the blade?). The third sub-proposition covers an aspect that is commonly addressed by forensic biology and fingermarks examinations. The fourth sub-proposition may be addressed by digital forensic examinations. For instance, by determining when the defendant and complainant were at the scene of the incident from data on their cell phones. Combining evidence from different forensic disciplines will be further covered in Chapter 12.

Note that all these propositions are at the activity level. Further sub-propositions may be formulated at lower levels in the hierarchy of propositions, e.g. at the source, sub-source or sub-sub-source levels. At the sub-source level, we could, for instance, formulate propositions like 'The sample from the handle of the knife contains DNA of the defendant and two unknown individuals' versus 'The sample from the handle of the knife contains DNA of three unknown individuals'. With those propositions we would also need to formulate assumptions like 'we assume that the individuals who contributed DNA to the sample are not related to each other or to the defendant'. Based on our findings *and information provided to us*, we could also assume the presence of DNA of a complainant in the sample, changing the propositions to 'The sample contains DNA of the complainant, the defendant and one unknown individual' versus 'The sample contains DNA of the complainant and two unknown individuals'.

In this example of formulating a set of propositions at the sub-source level, we illustrate three aspects of structuring of case information that we discuss in more detail later in this chapter. First is the formulation of the propositions, second, the setting of assumptions about issues that are uncertain or unknown to us and third is the impact of other information on our evaluation. These issues are illustrated here with an example of an issue at the sub-source level. However, the same concept holds for the evaluation of findings at all levels in the hierarchy of propositions. Regardless of where we are operating within the hierarchy of issues, we will always need to set propositions, make certain assumptions and collect all other information that is relevant to our assessment of the findings. What will change is the nature of the information that we need at each level in the hierarchy, as well as the amount of information needed. At the sub-source level we generally need information on who the persons-of-interest are in a case, from whose house or personal item a trace was collected (to assess whether we can reasonably expect DNA of an individual on an item; think about expecting the complainants DNA on her own clothing) and potential relatives of the persons-of-interest to assess whether our assumption of non-relatedness is sensible. This is about the extent of the information that we would need at the sub-source level. At the activity level we would need information on, for instance, the user history of items, the nature of the questioned activities and prior interactions between persons-of-interest and with their environments, but also details on collecting items at the scene, packaging and transport, and analysis methods.

We will discuss the setting of propositions in Section 4.2. Structuring other case information will be the topic of Section 4.3. However, before we do so, we will need to discuss the concept of 'task-relevant information'.

4.1.3   TASK-RELEVANT INFORMATION

As forensic scientists we are tasked with the evaluation of our findings given propositions and other contextual information from a case. One could argue that, for that reason, the scientist should have access to all case information as this would make sure that the evaluation will be relevant to the considerations of the court. The case information, however, will likely also contain information that may bias the scientist (un)consciously. That such cognitive biases affect human decision-making has been shown empirically [2, 3]. Specific studies in different domains of forensic science have shown that cognitive biases also affect assessments of their findings by forensic scientists [4], including DNA profile interpretation [5]. Hence there is a need to minimise the risk of bias in forensic assessments. Ideally the forensic scientist receives only those pieces of information that are relevant to the task at hand. This 'task-relevant information' will be defined by the task that is performed. With DNA profile interpretation, for instance, a first task will be to assess whether an evidence DNA profile contains sufficient reliable information for it to be compared to reference DNA profiles. For this task the forensic scientist needs information on the quantity of DNA in a sample and the types of analyses performed. At this point, no other information is needed for this specific task. Information on the type of substrate that was sampled, the activities performed, persons involved, etc. might only bias the scientist in this task. Whether or not the DNA profile is suitable for comparison is a technical issue that is independent of the context of the case. When the profile is deemed suitable for comparison, reference DNA profiles of potentially uncontested contributors (for example the wearer of a piece of clothing from which the sample was taken) may be used to decide whether or not such individuals may be assumed as contributors to a sample. The next step may be to reveal the DNA profiles of questioned individuals. This stepwise approach to context information management is called 'linear sequential unmasking' [6].

Another example which illustrates the concept of task-relevant information could be an interview you were invited to because you saw a red-necked wallaby (*Macropus rufogriseus* (Desmarest, 1817)) in the wild (all is relative …) in the Netherlands. The reporter asks you how probable you consider your observation of this single foraging animal to be, given that either one wallaby recently escaped from captivity in the area where you observed the animal or given that they have extended their natural range to the Netherlands. In other words, the reporter asks you to assign a probability to your observation given two competing propositions.

Task-relevant information that you may consider here would be, for instance, the size of the natural area in which the observation was made, the amount of time spend in the area by you, the percentage of the area you covered, as well as whether wallabies are solitary or live in social groups, the size of such groups and the number of groups you might expect in an area of the relevant size.

Given that:

• Red-necked wallabies are mainly solitary animals
• You consider the area that you surveyed large enough to harbour a large number of these animals
• You spend only a brief amount of time there

you might conclude that your observation is more probable if the animals extended their natural range rather than if one escaped from captivity (since with a natural range extension you would expect more animals in the area and thus a higher probability of observing one in the limited time spent there).

Now this may initially seem counter-intuitive, since you wouldn't reasonably expect wallabies to migrate and extend their range from Eastern Australia, across Southeast and central Asia, the Middle East and into Europe, and you know that red-necked wallabies are being kept as petting zoo animals in the Netherlands. However, the latter two pieces of information are not relevant to the task assigned to you. You are considering the probability of your observations given the propositions and task-relevant context information ($\Pr(E|H,I)$). Information on wallaby migration and husbandry is information about the prior probability $\Pr(H)$ and so is only relevant when you are looking to assign probabilities to the propositions themselves ($\Pr(H|E,I)$). Such information is thus task-irrelevant to the scientist (in Chapter 1 we have explained why forensic scientists need to refrain from addressing the probability of the propositions themselves) and additionally has the potential to consciously or unconsciously bias you in your assignment of probabilities to the evidence given the propositions.

A counter-argument against providing the scientist with task-relevant information only is that crucial information to the assignment of the weight of the evidence is lost. While this is true if the scientist is the only person considering this information, this is not so when the fact finder (be it a judge or a jury) also considers their task-relevant information in their deliberations. Aitken and Nordgaard [7] discuss splitting the case information into (task-relevant) background information available to the scientist (I_b), task-relevant information for the decision-maker (judge or jury) (I_a), with the complete set of background information as $I_a \cup I_b$. They demonstrate by a derivation that:

$$\Pr(H_p \mid E, I_a \cup I_b) = \Pr(E \mid H_p, I_b) \Pr(H_p \mid I_a)$$

$$\Pr(H_d \mid E, I_a \cup I_b) = \Pr(E \mid H_d, I_b) \Pr(H_d \mid I_a)$$

which demonstrates that separating sets of information for different contributors to the decision (in this example the judge/jury and the scientist) should not impact the overall decision on the posterior probability of the propositions as all information is included in the assessment, either in the prior probability (based on information available to the decision-maker, I_a) or in the likelihood ratio (based on information available to the scientist, I_b).

It is now generally accepted that exposure of forensic scientists to task-irrelevant information should be limited. However, the solution should not be to keep all case information away from the forensic scientist. Task-relevant information is crucial to make proper decisions in a case and to make sure that the evaluation of the findings performed is relevant to the considerations of the court. Proper mitigation of risks of cognitive biases therefore requires the implementation of a form of case information management in the casework processes [8]. We will discuss practical aspects of case information management to mitigate the risks of cognitive bias in Chapter 11.

4.2 FORMULATING PROPOSITIONS ABOUT ACTIVITIES

Given that we have received our task-relevant information about a case, we will need to structure this information in such a way that we can use it to evaluate our findings given that information. The first step is to determine the issue that is under dispute. In other words, which activities or actors are being debated by prosecution and defence counsel? These disputes will form the basis for the evaluation and will be set in propositions. Formulating propositions thus is a key exercise whenever we endeavour case assessment or interpretation of our observations and findings.

4.2.1 GENERAL GUIDANCE ON FORMULATING PROPOSITIONS

From a theoretical and logical perspective, there are several 'rules' to which propositions must adhere. We will discuss the points to consider when formulating propositions starting with one of the most prominent 'rules' set by the purpose of forensic science in general; propositions must be relevant to the court.

4.2.1.1 Propositions Must Be Relevant to the Court

Propositions are used to evaluate our findings in the context of a case. The propositions must therefore align with scenarios that are considered to be relevant by one or more of the parties in the criminal justice system. Propositions will therefore outline what issues are being debated in court, usually between prosecution and defence. It is not the role of the scientist to formulate propositions based on their findings or based on the questions they feel they can answer (but which may not be relevant to the court).

This does not mean that the scientist has no role in the setting of propositions. Based on the context of a case, the scientist must consider all relevant routes for TPR of biological traces. It is therefore crucial that, if such routes are overlooked by the parties involved, the scientist makes these explicit.

4.2.1.2   Propositions Should Not Be Uncertain

We need propositions to evaluate our findings within a probabilistic framework. We are, as scientists, interested in the probability (uncertainty) of our findings given the propositions. The propositions themselves can therefore not contain uncertain elements. Examples of such uncertain statements are propositions formulated as questions (i.e. 'Has the defendant stabbed the complainant?') or uncertainty in the statement (i.e. 'The defendant may have stabbed the complainant'). We cannot assess the probability of transfer, persistence and recovery of DNA of the defendant to the handle of the knife based on such questions or uncertain statements. We can only do this if the propositions are formulated as a statement of a given activity (i.e. 'The defendant has stabbed the complainant'). The uncertainty around whether the defendant actually stabbed the victim or not is addressed by using at least two competing propositions.

4.2.1.3   Propositions Come in Pairs

The evaluation of our findings should always be balanced [9, 10]. Consider if we would simply assign a probability to our findings given only one proposition (i.e. H_1: 'The sample contains DNA of the defendant and two other individuals'). The probability of our findings (which consists, for instance, of the specific peak heights of the observed alleles, the alleles corresponding to those in the DNA profile of the defendant, as well as all the other alleles in the DNA profile which need to be accounted for by random people in the population) given this proposition is extremely small (often in the order of 1 in billions). Presenting this as a result to the court, without relating it to a probability for our findings given another relevant proposition (i.e. H_2: 'The sample contains DNA of three unknown individuals, unrelated to the defendant') would be misleading. Particularly since the probability of our findings (a full 'matching' profile) is orders of magnitude smaller given H_2 compared to H_1.

The same holds for propositions at every level in the hierarchy of propositions. If we are asked to assign a probability to recovering DNA of the defendant from the handle of a knife that is used in a stabbing event, we may conclude (given further task-relevant information that is provided to us) that this probability is high given the proposition that the defendant stabbed the complainant. This conclusion may provide the decision-makers with a high level of confidence that the findings provide strong support for the prosecution position. However, we fail to assess our findings given the defence proposition in which it is stated that the defendant handled the knife after the incident by wrestling it from the hands of the perpetrator and throwing it away. Under this scenario the probability of recovering DNA of the defendant from the handle of the knife may also be high, resulting in our findings not discriminating between the two scenarios. Without considering the probability of the findings given an alternative proposition, the weight of the findings cannot be determined and presenting such a one-sided assessment may be misleading to the court. The evaluation will thus be balanced by assessing our findings given at least two competing propositions.

More than two propositions may be considered relevant in a case. If this is so, multiple pairs of propositions can be constructed. An example could be where the prosecution states that the defendant wore a sweater when stabbing the complainant. The defendant through her defence counsel states that she has nothing to do with the stabbing, but that she has seen such a sweater in the past. She doesn't remember if she owned it and lent it to somebody else, or if she borrowed it once from a friend during a party. Here the following propositions could be constructed:

H_1: The defendant wore her own sweater when stabbing the complainant

H_2: Somebody other than the defendant wore the sweater, which was borrowed from the defendant, when stabbing the complainant

H_3: Somebody other than the defendant wore the sweater, which was worn once by the defendant prior to the incident, when stabbing the complainant

Here we have, based on the two proposed series of events by the defendant, three propositions. If all three propositions are considered relevant by the prosecution, defence and/or the court, two formal evaluations of our findings can be performed: once given proposition H_1 versus H_2 and once given proposition H_1 versus H_3. It may or may not be considered relevant to evaluate our findings given propositions H_2 versus H_3. It generally won't be relevant to assess our findings given two defence propositions only, since both scenarios consider 'legitimate' activities by the defendant and nobody is likely to care much whether the findings provide more support for the proposition that the defendant borrowed the sweater once herself or for the proposition that she lent her sweater to a friend.

Another way this situation could be addressed would be to consider a general defence proposition that covers both H_2 and H_3 simultaneously. For instance, 'Someone other than the defendant stabbed the complainant'. In such an evaluation two possibilities (two sub-propositions) would be considered, namely that the sweater belonged to the defendant and was leant to the offender and that the sweater belonged to someone else and was recently used by the defendant. This would require prior probabilities to be set for both of the two sub-propositions occurring (e.g. Are both proposed sub-propositions considered equally probable or should different prior probabilities be assigned?). The value to assign to these priors is likely to be considered the realm of the court, but the sensitivity of the LR to the probability value could be explored. For instance, by providing a table or graph that shows the LR given different prior probabilities assigned to either sub-proposition. Evidently, prior probabilities of 1 or 0 will result in the same LRs as the two separate evaluations of the findings given H_1 vs H_2 and H_1 vs H_3, respectively).

Different considerations may be relevant if there are multiple prosecution propositions. We may encounter this if there is a complex of criminal activities and the prosecution is charging the defendant with either one or more of these. An example is a case where a complainant (C) is beaten and subsequently sexually assaulted by two perpetrators. The prosecution may consider the following propositions to be relevant (note that from the generic offence level descriptions 'beating' and 'assault', we need to formulate specific activities):

H_1: The defendant has hit C on the head and subsequently penetrated her vagina with his fingers
H_2: The defendant has hit C on the head but has not penetrated her vagina with his fingers
H_3: The defendant has penetrated the vagina of C with his fingers but has not hit her on the head prior to this

Defence counsel may propose:

H_4: The defendant was not involved in either of the two alleged activities. Defendant and complainant co-inhabit in an apartment.

Under such conditions it will be relevant to evaluate our findings given H_1 versus H_4, H_2 versus H_4 and H_3 versus H_4. However, the court may also be interested to hear our opinion on the findings given H_1 versus H_2 or H_1 versus H_3.

This is because there are two questioned sets of activities here (the beating and the sexual assault). Since both may carry different legal consequences, the court may want to know to what extent our findings support either one or both of these sets of activities. Proposition H_1 is what is considered a 'package deal'.

4.2.1.4 Propositions as a 'Package Deal'
We do need to be mindful of combining multiple activities in propositions. We can, for instance, combine multiple unrelated activities in proposition H_1: 'the defendant cheated on his tax return papers, stole a bicycle, sexually assaulted his neighbour, and stabbed a driver in an act of road-rage'.

The defence may claim H_2: 'The defendant did none of these things'. Let us consider that our findings constitute (and only constitute) a forged signature on the defendant's tax return form. This might provide strong support for H_1 over H_2, suggesting there is strong support for him committing all the acts stated in proposition H_1, while in fact there are only findings supporting his cheating on his tax return. This quite extreme example of the package deal in propositions serves to illustrate that such a construct may be misleading to the court. We should therefore avoid combining multiple unrelated activities in a single proposition.

Consider a situation in which a defendant (D) is charged with two criminal activities, activity X and activity Y. Defence counsel may propose that an alternate offender (AO) did one or both activities. In a case (pre-)assessment we may formulate propositions in five ways:

A) Consider both activities given a single set of core propositions, where it is assumed that the same person committed both activities, the activities are not in dispute and that D does not concede to either:

   H_{1a}: D did X and Y
   H_{2a}: AO did X and Y

B) Consider both activities given multiple sets of core propositions, where the two activities may have been committed by different people, the activities are not in dispute and D may concede to carrying out one of the activities. The propositions that can be considered under H_2 (in addition to H_{2a}) are:

   H_{2b1}: D did X and AO did Y
   H_{2b2}: AO did X and D did Y
   H_{2b3}: AO1 did X and AO2 did Y (i.e. there are two AOs)

   In this situation findings may be evaluated given H_{1a} vs H_{2b1}, H_{2b2}, and/or H_{2b3}.

C) Consider both activities given separate sets of sub-propositions, where the activities are not in dispute. This leads to two pairs of propositions:

   H_{1c1}: D did X
   H_{2c1}: AO did X
   H_{1c2}: D did Y
   H_{2c2}: AO did Y

   In this situation separate evaluations will be performed given H_{1c1} vs H_{2c1} and H_{1c2} vs H_{2c2}.

D) Consider both activities given separate sets of sub-propositions, where the activities are in dispute, then additional H_2 propositions can be considered:

   H_{2d1}: X did not occur
   H_{2d2}: Y did not occur

   In this situation separate evaluations will be performed given H_{1c1} vs H_{2d1} and H_{1c2} vs H_{2d2}.

E) Consider both activities given multiple sets of core propositions, where the two activities may have been committed by different people, one or both the activities are in dispute and D may concede to carrying out one of the activities. The propositions that can be considered under H_2 are:

   H_{2e1}: AO did X and Y did not occur
   H_{2e2}: AO did Y and X did not occur
   H_{2e3}: D did X and Y did not occur
   H_{2e4}: D did Y and X did not occur
   H_{2e5}: X and Y did not occur

   In this situation separate evaluations will be performed given H_{1a} vs H_{2e1-5}.

Which set of propositions is most relevant to the decision-maker depends on the case circumstances, the scenario's being proposed and the findings. If activities X and Y are widely separated in time and physical space, and different actors are involved, it will generally not make sense to combine them in the propositions. We would intuitively consider separating the activities into multiple sets of sub-propositions (e.g. option C or D) in such a situation. An example may be a charge of sexual assault (a case of digital penetration) and a charge of petty theft two months earlier. This example also serves to illustrate the risk involved when such unrelated activities are combined in the same set of propositions. Consider a glove that was left at the shop by the perpetrator after the theft. DNA findings from the glove support D wearing it. If we would combine the two activities, the DNA evidence on the glove will provide support for D being the wearer of the glove, but also support him having digitally penetrated the complainant, as both activities are linked in the propositions. Reporting this as 'the findings from the glove are X times more probable if D wore the glove and digitally penetrated the complainant than if AO wore the glove and digitally penetrated the complainant' may be misleading the court in the diagnostic power of the findings, as support for one activity implies support for the other activity while no such support exists. Evidently consideration must be given to the prior probability that the same person carried out both activities.

However, separating evaluations and providing the court with the weight of evidence for findings given separate sets of sub-propositions carries a risk as well. If events are related, and a subsequent activity may impact the probability of the findings given the preceding activity, this dependence must be taken into account. Thus, making separate evaluations impractical.

If the same set of findings provides diagnostic info to both sets of sub-propositions, this dependence must be taken into account. If not, this forces the court to combine the evidence, carrying the risk of over- or understating the evidence since conditional dependencies are not taken into account. An example of this could be a case of sexual assault followed by the strangulation of the victim. As findings from the analysis of latent DNA traces on the clothing of the victim may have been impacted by both the sexual assault and the strangulation (during both activities latent DNA of the offender may have transferred, the persistence of traces from the assault may have been affected by the subsequent strangulation, etc.), these findings need to be considered given both sets of activities.

We are thus confronted with a trade-off; we try to be maximally transparent on the diagnostic value of the findings to each separate activity, while making sure that the combination of the evidence is properly addressed. To achieve this, we propose that there are five aspects to consider when deciding which of options A to E for the formulation of propositions is most appropriate to the case at hand. Depending on the outcome of these considerations, either a single set of core propositions, multiple sets of core propositions or separate sets of sub-propositions are the most appropriate way to evaluate the findings and to communicate the weight of the evidence. The considerations and their outcome are shown in Figure 4.1.

4.2.1.5 Propositions Must Be Mutually Exclusive

The propositions must be mutually exclusive. This means that if the statement under propositions 1 is true, the statement under proposition 2 cannot be true. Say that somebody is accused of stabbing a complainant with a knife found at some location that is not directly related to the stabbing incident. The defendant states that the knife is his but that he peeled an apple with it. The defence counsel also questions whether the stabbing really happened.

Propositions could be formulated as:

H_1. The defendant stabbed the complainant with his knife
H_2. The defendant peeled an apple with his knife

Without further information or assumptions these propositions are not mutually exclusive, since the defendant may have done both activities. The propositions as they are formulated do not preclude this. It is therefore crucial that we are explicit when formulating propositions about which activities

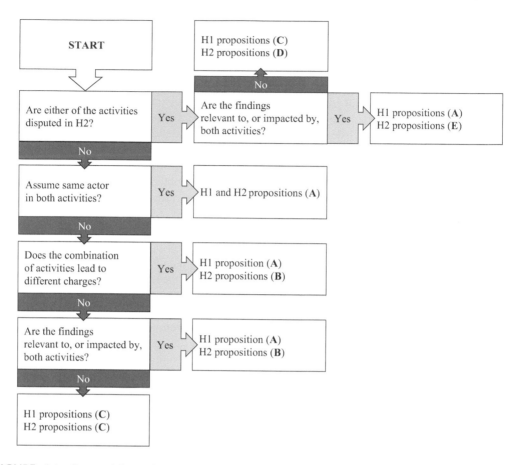

FIGURE 4.1 Sequential considerations leading to the formulation of appropriate proposition set(s). Propositions are referred to as H_1 or H_2, A to E: H_{1a}: D did X and Y; H_{2a}: AO did X and Y H_{2b1}: D did X and AO did Y; H_{2b2}: AO did X and D did Y; H_{2b3}: AO1 did X and AO2 did Y (i.e. there are two AOs); H_{1c1}: D did X; H_{2c1}: AO did X; H_{1c2}: D did Y; H_{2c2}: AO did Y; H_{2d1}: X did not occur; H_{2d2}: Y did not occur; H_{2e1}: AO did X and Y did not occur; H_{2e2}: AO did Y and X did not occur; H_{2e3}: D did X and Y did not occur; H_{2e4}: D did Y and X did not occur; H_{2e5}: X and Y did not occur.

are to be considered under each proposition. With this example we could formulate mutually exclusive propositions as:

H_{1b}: The defendant stabbed the complainant with his knife. He did not peel an apple with that knife.

H_{2b}: The defendant did not stab the complainant with his knife. He did peel an apple with that knife.

In these propositions it is made very clear which activities are being considered disputed (which as we explain later is the appropriate information to be provided in a proposition). Given these propositions, the presence or absence of blood of the complainant on the blade of the knife may be very informative.

If the example case was one where the stabbing incident was not questioned by the defence counsel, we must consider that propositions H_{1b} and H_{2b} do not provide an exhaustive list of possible and case-relevant events. In such a situation we need to also consider the involvement of an alternative offender, as well as another knife used in the stabbing. This leads us to the importance of considering an exhaustive set of propositions.

4.2.1.6 Propositions Are Exhaustive

Propositions do not need to cover all possible activities that may result in the findings. Assumptions can be made that greatly simplify the evaluation, but will still approximate an evaluation with exhaustive propositions, i.e. we do not need to consider activities that are not reasonable within the context of a case. With the previous example where the knife of the defendant has been examined, we do not need to consider the use of the knife by the complainant by which he may have cut himself resulting in his blood on the blade of the knife. Not if it is a given that the complainant and defendant did not socialise, and neither claims that the defendant lent his knife to the complainant. Technically speaking, the prior probability for such a proposition is close to zero. The evaluation of our findings will only be in the context of the case at hand. Hence propositions need only be considered if reasonable or explicit in the context of this case. Propositions are therefore not 'generally' exhaustive but do need to be exhaustive 'in the context of a case'.

In fact, the issue is not just with the propositions that need to be exhaustive within the context of a case. Crucially, the routes of TPR of biological material within the context set by the case circumstances need to be exhaustive [11] (see Box 4.1). Efforts therefore need to be made by all parties involved (including the scientist) to identify all relevant routes for TPR of biological materials.

BOX 4.1   AN EXHAUSTIVE SET OF ROUTES FOR TPR OF BIOLOGICAL TRACES – A CASE EXAMPLE

A suspect was arrested for drug trafficking. In the boot of their car, under the coat of the suspect, a suitcase was found. Upon opening by the police officer at the scene of the arrest, the suitcase was found to contain plastic-wrapped packages of powder. Swabs from these wraps were taken by police and submitted for DNA testing. Several samples resulted in DNA of the defendant being found. The prosecution stated that the defendant had handled the packages and was thus aware of the content of the suitcase. The defendant claimed that he was not involved, that he has handled and transported the suitcase for a (former) friend, and that his DNA would have gotten on the packages through the hands of this friend with whom he shook hands some weeks prior to the incident. The propositions are therefore:

H_1. The defendant handled the packaged drugs
H_2. The defendant's former friend handled the packaged drugs

Evaluating the findings given this direct versus indirect transfer scenario is likely to result in an *LR* that supports the prosecution proposition over that of the defendant.

The scientist asks for all documentation of the police procedures followed at the scene and subsequent handling of the evidence. Upon receipt it shows that the suitcase was handled by a police officer wearing disposable gloves. That this officer wearing the same gloves then took the packages out of the suitcase and placed them against the sides and on top of the suitcase to photograph them. The packages were subsequently sampled for DNA.

As it is reasonable to assume that the outside of the suitcase contained DNA of the defendant (he handled the suitcase, and it was under his coat in the booth of his car), DNA may have transferred from the outside of the suitcase to the packages through contact with the gloves of the police officer and through contact with the outside of the suitcase. This constitutes a reasonable risk of contamination which will need to be considered in the evaluation (under both propositions as it is not disputed that these activities were performed by the police officer). This additional route of transfer may impact strongly on the outcome of the evaluation, potentially reducing the *LR* to close to 1.

4.2.1.7   Propositions Must Be Specific

We have already seen that we need to be specific in our propositions on which activities we do consider. In this sense it is generally insufficient to formulate a proposition simply as the negation of the other. For example:

H_1. The defendant stabbed the complainant.
H_2. The defendant *did not* stab the complainant.

Under H_2 it is unclear what is being proposed. Was the complainant not stabbed at all? Or did somebody other than the defendant stab the complainant? Resulting in two quite different propositions:

H_{2a}. The complainant was not stabbed
H_{2b}. Somebody other than the defendant stabbed the complainant

This distinction is vitally important because it will strongly impact how we go about the forensic examination and evaluation of the findings. If it is disputed that the complainant was stabbed (proposition H_{2a}), the forensic investigation may focus on the forensic medical examination of the complainant. Does he have any wounds? Can they be related to a stabbing? Can they be dated so as to relate the time of the interaction between defendant and complainant? Under this proposition forensic genetics may not be the first field of forensic science to try and address this issue.

With propositions H_1 vs H_{2b} it is not disputed that the complainant was stabbed. Here the forensic examination of the knife that was used (whose DNA or fingermarks are on the handle?) may be considered relevant. This example serves to illustrate that we need to be very specific on what activities we consider to have, or not have, taken place.

4.2.1.8   Propositions Should Not Include Results or Mechanisms

Another aspect that is commonly encountered is that the 'mechanism' by which TPR occurred is included in the propositions rather than the 'activities'. An example of this is the use of propositions like:

- H_1. The DNA of the defendant transferred 'directly' onto the item
- H_2. The DNA of the defendant transferred 'indirectly' onto the item

Asides from being non-specific (what activity would have led to the direct or indirect transfer?), we see that the results (DNA from the defendant on the item) are included in the propositions. Results should in general not be included in the propositions. In this instance the *LR* would again be 1, as both propositions describe situations where DNA has been transferred. It is equivalent to stating sub-source level propositions such as:

H_1: The source of DNA is the defendant
H_2: The source of the DNA is someone other than the suspect who has the same DNA profile

Propositions should be distinguished from explanations for findings [12]. Those findings should also not be part of the propositions [13]. Case examples from several disciplines of forensic science where these distinctions are not made or are unclear are provided by Hicks et al [13]. We re-iterate two of their examples here as they illustrate crucial points to consider.

The first is their example 2 (a shoe-mark case). At the scene of a burglary, a shoe-mark is recovered. The mark is examined, and the findings are evaluated given source level propositions:

H_p: The mark was made by Mr Smith's Nike Air Jordan left shoe-sole
H_d: The mark was made by an unknown Nike Air Jordan left shoe-sole

While categorical features of the mark (brand and type of shoe) are included in the propositions, other features of the mark are not. Those other features are being evaluated given these propositions. The choice to include the brand and type of shoe in the propositions also defines the alternative population (all other Nike Air Jordan shoes), which will reduce the weight of the evidence compared to the selection of an alternative population that considers more types of shoes. Here we see that part of the information that is obtained from the trace is moved to a consideration of the prior probability (e.g. what was, without knowledge of the mark, the a priori probability that the perpetrator wore Nike Air Jordan shoes?) The question that needs to be considered is who is in the best position to assign this probability. Is it the decision-maker (e.g. can the relevant population of shoes worn by burglars be considered common knowledge? Can we assume that any member of the jury or a judge will be able to assign an informed probability?) or should this be considered part of the expertise of the shoe-mark expert who examines shoe prints left at crime scenes on a daily basis? If we consider the latter more reasonable, the brand and type of shoe should be left out of the propositions, broadening the alternative population of shoes, and making this piece of information part of the evaluation of the findings by the expert. Either way the decision to use or exclude part of the information obtained from the shoe-mark should clearly be communicated to the decision-maker. This is to avoid the information on the brand and type of shoe being overlooked or counted double in the assessment of the evidence.

The same issue is demonstrated in their example 7. This DNA example is about the following propositions at the activity level:

H_p: Mr A threw the package containing drugs away during the chase.
H_d: An unknown person threw the package away, but Mr A's DNA was transferred via Officer B.

As with the source level example we see here that part of the findings as well as the *mechanism* of transfer rather than the activity are included in proposition H_d. The transfer of DNA of Mr A is part of this proposition. This is an explanation based on results, as it would not be sensible to formulate this proposition before the findings are known (we wouldn't know if DNA from Mr A was recovered from the package).

In practice we often see that an evaluation of findings given propositions at the activity level is considered when the initial examinations and resulting findings are already known. This brings an inherent risk as explanations for the findings are often considered relevant propositions. Consider the following example propositions:

H_1. The DNA of the defendant got on the knife because he stabbed the victim
H_2. The DNA of the defendant got on the knife through secondary transfer

Such statements are considered explanations. Such explanations will generally be recognised because they contain the findings in the propositions ('the DNA of the defendant...'). Remember we are considering our findings given the propositions and other task-relevant case information ($Pr(E|H,I)$). With such explanations we move information from 'E' to 'H'. Hence, we cannot evaluate these specific findings (the presence of DNA of the defendant on the knife) given these propositions. We would be able to consider other findings (fingermarks on the handle, for instance), but we lose any power that the presence of DNA of the defendant on the knife would bring.

If, contrary to the framework for evaluation of findings, we would evaluate the presence of DNA of the defendant given these propositions, we still see that we have lost all power of discrimination. Consider what the probability of finding DNA of the defendant on the handle is, given 'that it is present' (which both propositions state). This probability is of course 1 under both propositions, resulting in a likelihood ratio of 1.

Thus, propositions should not contain findings (as this makes them explanations). Propositions should only contain statements of activities performed (or not) by specific individuals.

4.2.1.9   Propositions Should Be Concise

The findings in a case are to be evaluated given case-relevant propositions as well as other conditioning information, $\Pr(E|H,I)$. This begs the question which aspects of the case information should constitute part of the propositions (H) and which aspects are considered other task-relevant information (I). While it is theoretically possible to include all information in the propositions, it is best practice to keep the propositions concise. This is emphasised by the
guideline on DNA mixture interpretation from the forensic science regulator for England and Wales [14] which states that propositions '... should be as simple and concise as possible and must be clearly stated'.

Propositions thus act as 'placeholders' for the entire set of conditioning information, specifying which are the crucially disputed aspects of the activities or actors that performed them. Taylor et al. [11] have provided some guidance on formulating a concise set of propositions. They distinguish:

- Disputed aspects of the scenarios presented by the parties (which goes in propositions (H)),
- Undisputed aspects of the scenarios (which goes in the undisputed set of case information (I)),
- Uncertain or unknown aspects of the scenarios which are relevant to the evaluation (about which assumptions (A) need to be made).

Depending on the status of pieces of information they can thus be part of the propositions, assumptions or undisputed case information.

Let us assume a case in which a defendant (D) is charged with an activity (X). The defence postulates that an alternate offender (AO) was involved. Propositions could be formulated as:

- H_1. D did X
- H_2. AO did X

During the pre-assessment the scientist considers whether the proposed actors wore gloves during the activity. This could lead to three different outcomes.

1. The prosecution states that the defendant did not wear gloves because no mention of gloves was made by witnesses. Defence disputes this and states that it is entirely reasonable that the perpetrator (not being their client) wore gloves.
   In this situation the wearing of gloves by the actor is in dispute. This aspect therefore should be part of the propositions:
   H_1. D did X with their bare hands
   H_2. AO did X while wearing gloves
2. The prosecution and defence counsel agree that no information is available on the wearing of gloves. About this uncertain aspect of the case circumstances, an assumption needs to be made. This assumption could either be that the offender did not wear gloves or if the circumstances make this unlikely, the assumption could be that the offender did wear gloves. The evaluation may be performed under both assumptions, thereby making the impact of the assumption on the outcome of the evaluation transparent.
   H_1. D did X
   H_2. AO did X
   A: No gloves were worn by the person performing X.
3. Both parties agree that a witness statement that the perpetrator wore working gloves is true. In such a situation the fact that the offender wore working gloves is not disputed. As such this aspect becomes part of the undisputed case information.
   H_1. D did X
   H_2. AO did X
   I: The person performing X wore working gloves.

4.2.2 A Practical Approach to Formulating Propositions

In the previous section a number of recommendations for the formulation of propositions at the activity level were discussed. A summary of these can be found in Box 4.2.

**BOX 4.2   GUIDANCE ON FORMULATING
PROPOSITIONS AT THE ACTIVITY LEVEL**

From Section 4.2.1, the following general recommendations can be taken for the formulation of propositions at the activity level:

1. Use propositions that are relevant to the deliberations of the court. Avoid formulating questions to fit answers that you may be able to provide.
2. Propositions should be statements of fact. Avoid the use of terminology that is uncertain.
3. Always consider at least two competing propositions. This comes back to the general principles for the evaluation of evidence (Chapter 1, Section 1.4).
4. Only bring related activities together in propositions. Avoid unnecessary package deals.
5. Propositions are mutually exclusive. Activities in the competing propositions cannot both be true.
6. All case-relevant activities or events should be considered; propositions need to be exhaustive in the context of a case.
7. Activities are as specific as possible. Avoid vague descriptions like 'was in contact with'.
8. Mechanisms leading to results and results should not be part of the propositions. Avoid 'pseudo' activity level propositions like 'The DNA from the defendant was indirectly transferred to the clothing of the victim'.
9. Propositions should be concise. Only include disputed activities in the propositions. Other task-relevant information should either be in assumptions or the set of other, uncontested case information.

There is no uniform way for formulating propositions. There are nevertheless some conventions [15]. In this section we will go through a practical stepwise approach to formulating propositions that adhere to these conventions. Formulating propositions is effectively a process in which three questions need to be answered:

1. Is it disputed that an activity took place?
2. If the activity itself is not disputed, who performed the activity?
3. What alternative routes for TPR or the prevalence of materials are being disputed?

4.2.2.1 Is It Disputed That an Activity Took Place?

Depending on the case at hand, it may be disputed that an activity took place. Examples that may be encountered are:

- A victim whose cause of death is unclear: it may be disputed that a crime occurred
- An alleged sexual assault: a defendant may claim that nothing happened
- An alleged burglary: a defendant may state that no such thing occurred and that the complainant is trying to make a false claim to his insurance

When an activity is disputed, this will be reflected in the propositions. For the sexual assault example, the propositions could, for instance, be:

H_1. The defendant pulled down C's skirt and panties and penetrated her vagina
H_2. The alleged activities did not take place

In this first step in the process of formulating propositions we make sure that we are specific about whether the activity is in dispute or rather the 'actor' who performed the activity.

Note that 'sexual assault' (aside from being a legal term, which could be considered an offence level issue) in itself is not yet specific about the activities that actually took place. When we use such 'vague' terminology in the propositions, it is crucial that the actual activities that we consider to be part of the 'sexual assault' (i.e. oral sex, digital penetration, penetration with penis, etc.) 'as well as the order in which we consider them to have taken place' will be specified in the report. Note that the specific activities and the order in which they took place will affect the probability of transfer, persistence as well as recovery of body fluids and other biological traces.

4.2.2.2   If the Activity Itself Is Not Disputed, Who Performed the Activity?

More often than not the criminal activity itself is not in dispute. Take for example:

- A victim being found with multiple stab wounds; the stabbing will not be in dispute.
- A victim who has sustained wounds during a sexual assault; the assault itself may not be in dispute.
- A robbery which was recorded on CCTV; the robbery itself will not be in dispute.

If this is the case, then the propositions should reflect this. Again, for the sexual assault example:

- H_1. The defendant pulled down C's skirt and panties and penetrated her vagina. H_3. Somebody other than the defendant pulled down C's skirt and panties and penetrated her vagina.

Given these propositions it is not disputed that the complainant was sexually assaulted. In this second step it is crucial to be clear if we consider everybody other than the defendant a potential 'actor' or if this is restricted to a smaller subset of the population (or even one other nominated person). Depending on the case it may, for instance, be clear that we only need to consider the housemates of the complainant, the visitors to a party or a group of friends of the defendant. Incidentally, the defendant may point to one specific individual that may have committed the act.

For example:

H_1. The defendant pulled down C's skirt and panties and penetrated her vagina.
H_4. One of three housemates of the defendant pulled down C's skirt and panties and penetrated her vagina.

It is important to specify this in the proposition for two reasons:

1. The size of the population of 'alternate actors' will impact the prior probability of the propositions being true. The more extensive this population is, the lower the prior probability of the defendant being the person who assaulted the complainant (if a priori we consider all individuals to be equally likely to be the actor). While this aspect will neither impact the way the scientist sets up the evaluation nor impact the outcome of the evaluation, it should flag to the recipients of the report to consider an appropriate prior probability.
2. A limited set of 'alternate actors' should result in an endeavour to obtain reference samples from those individuals for comparison against trace DNA profiles found. Such additional examinations will provide additional findings that may provide additional power to discriminate between the two competing activity level propositions.

4.2.2.3   What Alternative Routes for TPR or Prevalence of Materials Are Being Disputed?

We do need to consider all reasonable routes for direct or indirect transfer, persistence and recovery of biological materials in the context of the case. These may, for instance, include:

- Prevalence of DNA of the defendant in his house where the assault occurred
- Indirect transfer of DNA of the defendant to the complainant by their use of items in the defendant's household
- Indirect transfer of DNA of the defendant through the hands of the unknown housemate who assaulted the defendant (under H_4)
- but also contamination of samples or items during the investigative process.

In a case the defence counsel may provide specific routes (or explanations) for the findings. The defendant may, for instance, state that he and the complainant went for drinks in a bar that afternoon and that during this interaction they hugged and held hands.

In this third step of formulating propositions, we need to consider if each potential route of transfer is disputed by the opposing party. The prosecution, for instance, may agree that the defendant and complainant had drinks together, but the complainant may state that they did not hug or hold hands. Part of the 'legitimate' activity is now disputed, which should be reflected in the propositions:

H_5. The defendant pulled down C's skirt and panties and penetrated her vagina, they did not hug or hold hands prior to the assault.

H_6. One of three housemates of the defendant pulled down C's skirt and panties and penetrated her vagina. The complainant and defendant hugged and held hands prior to the assault.

We see that them having drinks is not disputed, and such 'non-disputed case information' is best left out of the propositions. The same holds for potential contamination or the prevalence of DNA of the defendant in his house, etc. These are factual bits of information that are not in dispute. These aspects, although crucial to the evaluation of our findings, are part of the 'I', not the 'H' in our assessment of the $\Pr(E|H,I)$.

4.3   ASSUMPTIONS AND OTHER TASK-RELEVANT CASE INFORMATION

In the previous paragraph we have discussed the formulation of propositions based on task-relevant information that is provided to the scientist. But, as we have seen earlier, we will be evaluating our findings given propositions, assumptions and other undisputed case information (sometimes referred to as background information). In this paragraph we will elaborate on the differences between the three categories in which task-relevant information can be divided. We will go through a case example to illustrate the concepts.

At 2 am police break up a fight between a man and a woman. Both are placed under arrest and interrogated at the police station. The male states that the woman is his ex-girlfriend with whom he lived together for several months until two weeks ago. She has shown a history of abusing him and two weeks ago she again beat him, after which he was seen by a medic. He left their shared apartment and broke off the relationship. Tonight, he ran into her in a club where they chatted a bit, after which she followed him when he left. In the street she beat him again and they got into a fight.

The woman claims that she has been in a relationship with the man until about two weeks ago. He broke off the relationship two weeks ago and left their apartment. Tonight, in the club, he saw her with another man. They chatted a bit, after which he followed her outside where he attacked her, and they got in a fight. When asked she denies ever hitting him prior to defending herself tonight.

The woman is being prosecuted for repeatedly assaulting the man.

4.3.1 PROPOSITIONS

Propositions are statements that contain the disputed activities in a case. As propositions are mutually exclusive and exhaustive, any aspect of debated activities or interactions between persons-of-interest will be documented in the propositions.

In our case example we find that there is effectively only one activity that is in dispute. This is the abuse of the man by the woman that took place during their relationship up to and including the night he broke off their relationship. If the prosecution and the defence are interested in a forensic examination and subsequent evaluation of findings, the propositions to be addressed would be:

H_1. The defendant beat the complainant two weeks ago
H_2. The defendant 'did not' beat the complainant two weeks ago

We notice that the second proposition is simply a negation of the first. While H_2 may not be specific enough in itself, further contextual information provides the other potential routes for TPR of DNA.

4.3.2 ASSUMPTIONS

It is common that information is lacking on aspects of the case that may impact our assignment of probabilities to TPR events. An example could be whether or not a perpetrator wore gloves during an incident. This information is frequently unknown or not reported. Since the wearing of gloves is known to impact the probability of DNA of the perpetrator to transfer from their hands to a surface that is touched, we do need to make an assumption on this. We could, for instance, assume in our case example:

Assumption 1: The defendant did not wear gloves when beating the complainant.

Justification for assumption 1: No information is reported on this aspect, but we assume this because in domestic abuse situations we commonly see that no gloves are worn. The weather conditions on the day of the incident also did not necessitate wearing of gloves.

Note that we will need to make transparent what assumptions we make, but also motivate why we do so.

Alternatively, we could assume that the woman did wear gloves when beating the man. This would also require us to make an assumption on the type of glove, whose gloves they are, and the frequency and duration of wear by the defendant (since all these factors will impact the probability of transfer of DNA from the defendant to the complainant).

Other assumptions that we may make in a case like this (when an examination has been performed and we have findings to consider):

Assumption 2. We assume that any DNA found which can be related to either complainant or defendant is truly theirs.

Justification for assumption 2: This assumption is made because of the high likelihood ratios resulting from the findings given the sub-source level propositions and the fact that the source of the DNA is not disputed.

Note that the uncertainty at the sub-source level can be accommodated in the evaluation given activity level propositions. This is recommended when *LR*s at the sub-source level are low. However, making this assumption is pragmatic as it is unlikely to impact the *LR* at the activity level when sufficiently high, and because it tends to separate the discussion in court on sub-source vs activity level issues. Note that this is an example of using an assumption to eliminate an unreasonable explanation (i.e. that the alternate offender had a matching DNA profile) from the exhaustive propositions (as was discussed in Section 4.2.1.6).

With crucial assumptions it may be best to perform the evaluation under multiple different assumptions. For instance, evaluate the findings given that no gloves were worn, as well as evaluate the findings given that particular gloves were worn. This effectively amounts to evaluating the findings under a prior probability of 0 and 1 for wearing gloves. If (and this would generally be in the domain of the court) a specific prior probability can be assigned to the person wearing gloves, then this is another avenue to proceed with the evaluation (as was discussed in Section 4.2.1.3).

4.3.3 Undisputed Contextual Information

In any case there will be aspects that are factual and not a matter of dispute between the prosecution and the defence counsel. In our case example there are a number of potential routes for direct and indirect transfer, persistence and recovery of biological materials:

1. The woman lives in an apartment where the man lived up until two weeks ago. His DNA may be prevalent in the apartment, and on her clothing, simply from cohabitation (and regardless of whether the assault took place)
2. They met in the club and talked for a bit. Transfer of saliva (expelled during speaking) may be considered
3. They fought on the street on the night of the arrest. Two-way transfer of biological materials may be considered between complainant and defendant

4.3.4 Shifting between Propositions, Assumptions and Undisputed Case Information

Our case example also illustrates that information may shift between these categories if the nature or the status of the information changes. Say that the complainant states that the defendant wore gloves during the assault, this may be taken (if not disputed) as undisputed case information rather than having to make an assumption about this aspect of the case. Alternatively, if defence counsel disputes the statement of the complainant on this aspect, stating that his statement is inherently unreliable, and that the defendant doesn't own gloves, it should be part of the propositions:

H_1. The defendant hit the complainant two weeks ago and 'wore gloves while doing so'
H_2. The defendant 'did not hit' the complainant two weeks ago

Why shouldn't we simply put all relevant information in the propositions? This can be done, and it is not wrong. We could formulate very extensive propositions that detail all disputed and undisputed activities. This would make them quite complete but may also make it difficult to comprehend as it is not immediately obvious what the disputed issue is. Because we tend to repeat our propositions regularly in a report, it would also make reading the report tedious.

We therefore recommend that propositions are kept brief and to the point. They effectively act as 'place holders' for all task-relevant information that is used in the evaluation of the findings given activity level propositions. We must realise that our conclusion from any such evaluation is crucially dependent on the propositions, 'as well as the assumptions and other, undisputed task-relevant information'.

In this fictitious case example, we may perform a case assessment and conclude that the examination of biological traces on the body and clothing of the complainant or defendant will not assist in this matter. Because of the recent fight, we can expect to find their DNA on hands, fingernails and clothing, while any traces from an alleged assault two weeks prior are unlikely to have persisted.

We may therefore recommend that this issue is referred to a forensic medical examiner, who may check the body of the complainant for wounds or bruising that may be dated to the alleged incident two weeks ago.

4.4   CASE EXAMPLES

In Chapter 2 (Section 2.5) four case examples were introduced. These four examples will be briefly discussed here regarding the task-relevant information needed for an evaluation given propositions at the activity level. We will also formulate the propositions that are used with these case examples in the next chapters.

4.4.1   CASE 1: THE ATTEMPTED KIDNAP

We take the case circumstances from Taylor [16] (note that we change the terminology used by Taylor from T and Drummond to C and D, respectively, to align with the terminology used in this book):

> The prosecution case stated that the victim in this matter (C) was walking along Prospect Road on 24 November 2010 wearing a singlet top. It is alleged that a car pulled over, the defendant (D) exited, grabbed C by the arm and attempted to pull her into the vehicle. C struggled with D, which included her hitting him in the chest and he abandoned his attempt to kidnap C and drove off. C noted the licence plate of the vehicle.

> C called police stating that she had been attacked and provided the licence plate. Shortly after police attended the home of D and seized his clothing that he had been wearing at the time in question. The following day police seized as evidence the clothes worn by C.

> D stated that he was driving home along Prospect Road, but never stopped his vehicle. There was an implication that C had not been attacked at all, but this was not explored in any detail. The main issue in court appeared to be the identity of C's attacker.

If we follow the three-step approach to formulating propositions (Section 4.2.2), we will need to answer the following questions:

1. *Is it disputed that an activity took place?*
   In this instance, the defence appears to have suggested that the attack on the complainant did not take place. However, as this was not further explored in the case, we will leave this for what it is at this point.
2. *If the activity itself is not disputed, who performed the activity?*
   From the contextual information it is clear that the prosecution position is that the defendant (D) struggled with the complainant (C) to drag her into the car.

The prosecution proposition may therefore be formulated as:

- C and D struggled

The defence claims that it was someone other than the defendant who struggled with the victim. The proposition based on their scenario may therefore be formulated as:

- A male other than D struggled with C

As propositions need to be specific, we need to consider if the term 'struggle' is sufficient for an evaluation of findings. More information is available in the contextual information that was provided: '[the perpetrator] grabbed T by the arm and attempted to pull her into the vehicle. T struggled with [the perpetrator], which included her hitting him in the chest'. We may decide to include this information in the propositions as:

- C and D struggled which included D grabbing C's arm and C hitting D's chest
- A male other than D struggled with C which included him grabbing C's arm and C hitting him on the chest

Since these aspects of the struggle are the same under both propositions, we may also decide to leave them out and include them as undisputed contextual information:

H_1: C and D struggled
H_2: A male other than D struggled with C

I: The struggle included the perpetrator grabbing C's arm and C hitting the chest of the perpetrator.

This will keep the propositions short and 'snappy' and puts the focus on the disputed issue. The report will make clear that the conclusion of the evaluation is based not only on the propositions but is crucially dependent also on any assumptions made as well as the undisputed case information.

3. *What alternative routes for TPR or prevalence of materials are being disputed?*

No interaction has taken place between D and C prior to the incident (at which time the prosecution proposed they struggled). No alternative routes for TPR of their traces have been considered.

Note that the locations of contact between complainant and perpetrator are specified in the contextual information. This specification is done based on the statement made by the complainant, as the defendant (given their stance) has no knowledge of the attack. The sampling strategy of the clothing of defendant and complainant was also based on this information. This makes sense as the evaluation of the findings should be done given contextual information in the case.

A situation may occur in a case where it is unclear what the interaction and locations of contact were. In such a situation the sampling strategy may be broader, in that multiple locations of the items of clothing are sampled. If this results in some samples providing an incriminating result and others not, one may be tempted to specify those 'positive' locations in the propositions. However, this introduces the results of the analyses in the propositions, effectively making them explanations for the findings. As we have seen in Section 4.2.1 (on propositions vs explanations) including part of the findings in the propositions shifts the assessment of those findings to the prior probability. This is to be avoided if the scientist is in a better position to assess the distribution of traces on an item of clothing after an attack than a layperson.

We structure the task-relevant information for the evaluation of the findings as follows:

Propositions (H)

- H_1. C and D struggled
- H_2. A male other than D struggled with C

Undisputed case information (I)

- A person attempted to kidnap C.
- The subsequent struggle included the perpetrator grabbing C's arm and C hitting the chest of the perpetrator.
- D and C had no contact prior to the alleged attempted kidnapping.
- C hugged her friend within an hour prior to the attempted kidnapping.
- The offender was male.

- The area of D's top that was tapelifted corresponds to the area of contact between the hands of C and D's top.
- The area of C's top that was tapelifted corresponds to the area of contact between the hands of D and C's top.

Assumptions (A)

The assumptions on uncertain or unknown aspects of the case that can be made are:

- That the DNA of C's friend was present on C's top.
- That C's DNA was present on her own top.
- That there was no DNA from an offender found on C's top.
- That neither C, D and/or the offender were wearing gloves or had washed their hands within a few minutes of the offence.
- That neither C, D nor the offender had any skin conditions that meant there would be particularly high shedders of DNA.
- That neither C nor D had washed the clothing that was seized by police between the time of the alleged offence and the seizure.
- That the clothing supplied by C and D were the items worn at the time of the alleged offence.
- That no items of clothing were worn *over* the items that were sampled at the time of the incident and subsequently.
- That the items of clothing were appropriately stored and packaged after seizure and prior to submission to the forensic laboratory.

4.4.2 CASE 2: THE THREE BURGLARS

The circumstances of this case are taken from Chapter 2:

a woman (the complainant, C) arrives home after having dinner at a restaurant. As she enters her home, she is confronted by three men, all wearing balaclavas running down the stairs from the upper story of her house. One of the men grabs her by the hair and forces her to the ground by pulling her hair and threatening her with a knife. C was then dragged by the hair into her bathroom. C states that during the ordeal the offender who grabbed her hair was not wearing gloves. Once C was in the bathroom the offenders told her to stay there and they left her home. Immediately after the offence C called the police who arrived at her home within 30 minutes. The following samples were taken for DNA profiling approximately one hour after the incident occurred:

- A swab of the handle of the knife
- A swab of C's hair
- A reference sample from C
- A reference sample from C's husband (H) who lives in the house with C

C stated that she had not washed her hair since the offence.
Inside C's home the police found a knife on the floor that C recognised as being the knife she was threatened with. She also identified it as a knife that she owned and that it had come from the knife block in her kitchen.
[The defendant] claims that he has nothing to do with the burglary, does not know the victim and has never been to her house or interacted with her socially. However, D does admit to being involved with a group of individuals that he suspects commit burglaries similar to this one. D recently lent a pair of gloves to one of these individuals but refuses to provide that individual's name and so police are unable to identify him or take his reference sample.

1. *Is it disputed that an activity took place?*

    No, the burglary and attack of the victim are not disputed to have taken place.

2. *If the activity itself is not disputed, who performed the activity?*

    The prosecution's position is that the defendant (D) was one of the three burglars. He was the one who grabbed the hair of the complainant (C) and threatened her with a knife.

    Here we have a situation with two activities (grabbing of the hair and holding of the knife). As both activities have been performed by the same individual in the same time span, we can combine these activities in the propositions. The prosecution proposition can be formulated as:

    - D held the knife and grabbed the hair of the victim

    The defence position is that an alternative offender (AO) held the knife and grabbed the hair of the victim. The defence proposition could be formulated as:

    - AO held the knife and grabbed the hair of the victim

3. *What alternative routes for TPR or prevalence of materials are being disputed?*

    The defendant suggested that a pair of his gloves (which he wore regularly) were worn by the AO prior to the incident. This would suggest a route for the indirect transfer of DNA from the defendant through the gloves to the hands of the AO. As no gloves were worn by the perpetrator, these gloves would have been taken off prior to the incident. In the prosecution scenario there is no reason to assume that D wore gloves prior to the incident. It is also not likely to impact the probability of the transfer of his DNA during the incident. Hence prosecution disputes the alternative route of indirect transfer.

This results in the following structured set of case information:

Propositions (H)

- H_1. D held the knife and grabbed the hair of the victim
- H_2. AO held the knife and grabbed the hair of the victim, after wearing the gloves of D

Undisputed case information (I)

- The knife submitted for examination was used by the perpetrator.
- The perpetrator did not wear gloves while handling the knife or while interacting with the victim.

Assumptions (A)

In this case the following assumptions may be considered relevant:

- The suspect's fleece winter gloves were frequently worn and not recently washed prior to them being lent to AO.
- AO kept the gloves indoors and did not wear or use them prior to them being worn for a brief period of time on the day of the incident.
- Under H_2 the AO took off the suspect's gloves immediately prior to handling the knife and grabbing the hair of the victim.
- Samples were taken, secured, packaged and transported according to standard operating procedures.
- None of C, D, H or AO has conditions that would make them unusually high or low DNA shedder (i.e. they are representative of the typical spread of shedding ability within the population).

- The sample from the knife contains DNA of C, H and D, no DNA of other individuals.
- The sample from the hair contains DNA of C and D, no DNA of other individuals.

4.4.3 Case 3: The Family Assault

A brief description of the circumstances of this case is provided in Chapter 2:

> The circumstances of this case are that A 24-year-old girl (C), who normally lives with her biological mother (M) and father (F) has stayed for a week at her older brother's (D) house. A friend of the girl receives a phone call from the girl stating that her brother has bitten her on the vagina, over her underwear. The friend picks up the girl and they go immediately to the police, where the underwear is seized and a reference from the girl is taken. The police then arrest the brother and take a reference DNA sample from him. The defence states that she has been staying at his house, but that no biting occurred.

1. *Is it disputed that an activity took place?*
   Yes. While the prosecution's position is that D bit C over her underwear, the defence claims that no such biting has occurred. This is a clear example of a situation where the activity of interest is disputed. Propositions could be formulated as:
   - H_1: D bit C on the vagina over the underwear
   - H_2: C was not bitten
2. *If the activity itself is not disputed, who performed the activity?*
   This situation is not applicable as we are dealing with a disputed activity rather than a disputed actor.
3. *What alternative routes for TPR or prevalence of materials are being disputed?*
   C stayed with D for a week in his house. During this time direct and indirect transfer of biological material between C and D may have taken place. Indirect transfer of biological material from D to the underwear of C may reasonably be considered. Since it is not disputed that C stayed with D, the potential routes for indirect transfer will need to be considered under both propositions. This information is therefore part of the undisputed case information rather than the propositions.

Which leads to the following set of case information:

Propositions (H)

- H_1: D bit C on the vagina over the underwear
- H_2: C was not bitten

Undisputed case information (I)

- C stayed in the house of D with D for a week. Evidently more information is needed on the interaction between C and D during the week, the duration of wear of the underwear and the activities C performed while wearing the underwear, etc. If such information is not available, assumptions need to be made on these aspects.

Assumptions (A)

Assumptions need to be made on uncertain or unknown information. As the results of this case are based on Y-chromosomal DNA testing, any aspects of the relation between C, her underwear and her father (F) also need to be considered:

- No mutations have occurred between D and D's father that would affect the Y-STR profile (note that if this assumption is challenged, it could be tested by profiling F and comparing the profile to that of D).

- C's underwear has not been washed since the alleged offence (given the timeframe and activities stated by C and the witness, this is highly probable).
- C's underwear was not bought less than a week prior to the alleged offence (and so has been worn/washed at her normal residence).

4.4.4 CASE 4: THE GUN IN THE LAUNDRY

This fourth case is presented as Practice Question 10 at the end of this chapter. The answer to this question can be found in Chapter 14.

4.5 NO SCENARIO PRESENTED BY EITHER PROSECUTION OR DEFENCE

To conclude this chapter, we discuss a situation that may occur in casework. A situation where one of the parties involved is unable or unwilling to provide a scenario which contains elements that can be used to formulate propositions. Two examples illustrate how this may occur.

A victim is found dead in the bathroom of a pub. The victim is stabbed multiple times in the chest and back. Based on witness statements and CCTV imagery a suspect is identified and apprehended. The suspect is questioned and states that he was in the pub at a time prior to the incident and that he met the victim there whereby they shook hands and hugged each other. After a few drinks, he left the pub while the victim went to the bathroom. He hasn't seen the victim since.

The prosecution scenario follows that of the suspect, in that the legitimate social interaction is not in dispute. However, the prosecution claims that it was the suspect who went to the bathroom with the victim. Something must have happened there which resulted in the victim being stabbed to death by the suspect. However, given that no witnesses or CCTV were present there, the prosecution does not know, nor it wishes to make assumptions, about what transpired in the bathroom. Possible aspects of sexual assault, a fight, self-defence by the victim, etc. are unknown. Hence the prosecution scenario is that of legitimate social interaction between suspect and victim, plus some form of violent interaction. The defence denies involvement of the suspect in the assault on the victim.

In another example it may be the defence that is unable or unwilling to present a scenario. This could be because the defendant does not wish to make any statement or because they simply do not know how or when biological traces may have transferred. A truly innocent defendant may, for instance, have some idea about their last interaction with a victim or visit to the later scene of a crime. However, details may be lost, particularly if they are required sometimes years after the events occurred. Either way, significant aspects of the defence scenario may not be known in such situations.

Forensic scientists who are required to evaluate their findings given propositions at the activity level may be confronted with these types of situations. There are different routes the scientist may take to inform the decision-maker about the relevance of the findings.

1. *Do not perform an evaluation of the findings given propositions at the activity level*
   While this may seem like the easy and correct way, it may lead to misunderstanding by the court, particularly if sub-source level evaluations have been provided. Without an evaluation of the findings by the scientist, the court may consider the findings and assign weight to them while this is not possible due to a lack of information. It is therefore prudent that, if the scientist decides that sufficient information is lacking to properly assess the weight of the findings, they write a statement about this to the court, explaining why such an evaluation is not possible.
2. *Provide generic information on case-relevant TPPR studies*
   Instead of performing an evaluation of the findings given propositions at the activity level, the scientist may be asked to provide (orally or in writing) information about

scientific studies that may be relevant to inform probability assignments of TPPR of traces. This information is likely to lead the legal professionals involved to make their own assessment of the weight of the findings, based on insufficient information available. This approach may exacerbate the risk of misinterpretation of the findings as described under 1. It should thus be clearly communicated that a case-specific evaluation of the findings is not possible due to limited contextual information.

3. *Make assumptions on unknown or uncertain aspects of scenarios*

Whether this option is viable will depend on the number of aspects that are uncertain or unknown and on information available to the court to assess the relevance of the assumptions made. An evaluation of findings may be performed under assumptions (for instance, the assumption that no gloves were worn by the perpetrator or that they were). By performing the evaluation under both assumptions, it will be transparent to the court what the impact of the assumption is on the weight of the evidence. With large numbers of assumptions, the evaluation will become very complex, as all relevant combinations of assumptions would need to be assessed and reported. As such there may be a limit to the number of assumptions that can reasonably be tested.

The second aspect here is that the court will need to make a decision on the relevance of each assumption based on contextual information that is available in the case file or based on common knowledge. If such information is not available, it may not be sensible to proceed with an evaluation of the findings as the result will be useless to the recipient of the report. It may therefore be good practice to share the propositions and any assumptions that the scientist needs to make with the mandating authority. They should be able to assess the relevance of the assumptions that need to be made and decide whether or not an evaluation of the findings by the scientist is opportune.

4. *Formulate a scenario*

The scientist could, in the absence of a scenario from prosecution or defence, formulate a scenario in their stead. This may be reasonable if the scientist could do so based on their expertise or common knowledge. An example is a situation where a latent DNA trace of a defendant is found in a single sample from a relevant substrate. If no defence scenario is postulated, the scientist may decide to consider all potential routes of contamination (e.g. DNA TPR during the forensic process at the scene or the lab) as the relevant defence scenario. However, this may not be sensible if the probability of contamination is extremely low (for instance, in situations with a large number of traces on different items and locations with identifiable body fluids). In a situation where the findings 'scream for an explanation', the initiative for formulating case-relevant scenarios should really be with the prosecution and defence (or the court themselves).

4.6 PRACTICE QUESTIONS

Q1. Consider the following information:

- Fibres were found clustered on the sticky side of the tapes, suggesting gloves were worn by the perpetrator
- The fibres 'match' those from the fabric of gloves that were seized at the home of the suspect
- The tape was wrapped around the wrists of the victim
- The victim has bitten in the tape to remove it from their wrists
- A witness claims to have seen the defendant near the scene at the time of the incident

You are asked to evaluate the findings from DNA analyses of samples that were taken from the tape given propositions at the activity level. What information would you consider task-relevant and which information task-irrelevant?

Q2. Consider the following situation. A complainant (C) states that her ex-boyfriend sexually assaulted her three months after she ended their relation. The defendant (D) states that no such thing happened but that she kept his biological material in a freezer during that time and smeared his body fluids over her body to incriminate him in a false rape allegation.

The prosecution asks you to evaluate your findings from samples taken from the body of the complainant given propositions addressing the issue in this case. How would you formulate the propositions?

Q3. H_1: Saliva from D was deposited on the balaclava while wearing it during the robbery
H_2: Saliva from D was deposited on the balaclava when it was dropped on the street in a puddle of spit from D.

Would you consider these propositions for an evaluation of your findings? Why (not)?

Q4. H_1: D assaulted C
H_2: AO assaulted C

Would you consider these propositions for an evaluation of your findings? Why (not)?

Q5. Consider the following situation. The defendant (D) met the complainant (C) in a pub. They started chatting and when the pub closed at 11 pm, D forcefully tried to kiss C on the mouth. She ran away. Two weeks later D and C met in the street. C confronted D about his actions and he hit her on the head. C filed a complaint and D was arrested. Samples were taken from the head and face of the complainant for DNA analysis.

The following propositions are formulated to evaluate the findings from the DNA analyses.

H_1: D kissed C on the mouth and hit C over the head
H_2: D did not kiss C on the mouth, nor did he hit C over the head

Would you consider these propositions for an evaluation of your findings? Why (not)?

Q6. Consider the following situation. Four men get into an argument. One of them (offender 1; O1) grabs a knife from his pocket and stabs complainant A (CA). A second man (offender 2; O2) then takes the knife from O1 and stabs complainant B (CB). The knife is left at the scene when both O1 and O2 flee. A suspect is arrested and considered by the prosecution to be one of the two offenders. The defendant claims that he was not involved in the incident but that he recognises the knife as one he lost about a week prior to the incident. The knife is submitted for DNA analysis and the handle of the knife is sampled.

Formulate propositions at the activity level to evaluate the findings from the handle of the knife.

Q7. Consider the following situation. A woman makes a complaint about a sexual assault. The complainant states that: A man approached her in a park. He grabbed her shirt, forced her to the ground, removed her skirt and underwear, penetrated her with his finger, forced her to perform fellatio, penetrated her vaginally with his penis and ejaculated over her shirt. After this he ran off. She went home, showered and changed after which she made her complaint to the police. A sexual assault kit (SAK) was taken from the victim and her clothing was seized for examination. A suspect is arrested after some time. He claims that he was in the park that day and had spoken to the victim. He had however not touched her. Defence counsel does not dispute that the assault occurred but does claim that the defendant was not involved.

Formulate propositions at the activity level to evaluate the findings from the SAK and her clothing.

Q8. Consider the following case example.

A robbery takes place in a supermarket. The perpetrator grabs the shoulder of the owner (the complainant, C) who resists. The perpetrator then flees the scene before he gets any money. He leaves a backpack which is searched by a witness (W). The witness subsequently consoles the owner by putting her hand on his shoulder. A defendant (D) is questioned who claims not to be involved in the robbery, which must have been an alternate offender (AO). He states that the backpack was his but was stolen three months prior to the incident.

The shirt of the owner is submitted for DNA analysis. A sample from the shoulder results in a mixture of DNA of C and D. No DNA of other individuals is recovered from this sample.

The following propositions are formulated to evaluate the findings from the shirt:

H_1: D grabbed the shoulder of C during the incident
H_2: AO grabbed the shoulder of C during the incident
I: W searched for a backpack from D, after which she held C by the shoulder.

Would you assume in the evaluation that the backpack contained DNA of D under both propositions?

Q9. Consider an incident where it is uncertain whether the perpetrator wore gloves. How would you deal with this situation when evaluating findings given propositions at the activity level?

Q10. The case circumstances of example case 4 (The Gun in the Laundry) as described in Chapter 2 are as follows:

The circumstances surrounding this case are that a defendant (D) is accused of robbing a service station armed with a handgun. Two days after the robbery police attend D's home and carry out a search for the firearm. The firearm is found in a laundry basket covered with D's unwashed laundry. D claims that he has never seen the firearm and that it must have been put in his laundry basket by someone else (an Alternate Offender, AO). D states that he last did his laundry yesterday, and so whoever put the firearm in his laundry must have done so within the past 24 hours. The defendant states that he has not been home much in the past 24 hours and couldn't tell if anyone had been in the house.

For this example case;
   a. Formulate relevant propositions at the activity level
   b. List task-relevant, undisputed case information
   c. Formulate assumptions that are relevant to the evaluation of the findings

4.7 REFERENCES

1. C.G.G. Aitken, F. Taroni, *Statistics and the Evaluation of Evidence for Forensic Scientists* (2nd ed.), John Wiley & Sons, Ltd., Hoboken, 2004.
2. A. Tversky, D. Kahneman, Judgment under uncertainty: Heuristics and biases, *Science* 185(September 27) (1974) 1124–1131.
3. R.E. Nisbett, T.D. Wilson, Telling more than we can know: Verbal reports on mental processes, *Psychological Review* 84 (1977) 231–259.
4. G.S. Cooper, V. Meterko, Cognitive bias research in forensic science: A systematic review, *Forensic Science International* 297 (2019) 35–46.
5. I.E. Dror, G. Hampikian, Subjectivity and bias in forensic DNA mixture interpretation, *Science & Justice: Journal of the Forensic Science Society* 51(4) (2011) 204–208.

6. I.E. Dror, J. Kukucka, Linear Sequential Unmasking-Expanded (LSU-E): A general approach for improving decision making as well as minimizing noise and bias, *Forensic Science International: Synergy* 3 (2021) 100161.

7. C. Aitken, A. Nordgaard, The roles of participants' differing background information in the evaluation of evidence, *Journal of Forensic Science* 63(2) (2018) 648–649.

8. R.D. Stoel, C. Berger, W. Kerkhoff, E.J.A.T. Mattijssen, I.E. Dror, M. Hickman, K. Strom, Minimizing contextual bias in forensic casework. Forensic science and the administration of justice, *Critical Issues and Directions* 67 (2014) 67–86.

9. S.M. Willis, L. McKenna, S. McDermott, G. O'Donell, A. Barrett, B. Rasmusson, A. Nordgaard, C.E.H. Berger, M.J. Sjerps, J.-J. Lucena-Molina, G. Zadora, C. Aitken, T. Lovelock, L. Lunt, C. Champod, A. Biedermann, T.N. Hicks, F. Taroni, *ENFSI Guideline for Evaluative Reporting in Forensic Science*, European Network of Forensic Science Institutes. http://enfsi.eu/sites/default/files/documents/external _publications/m1_guideline.pdf, 2015.

10. Association of Forensic Science Providers, Standards for the formulation of evaluative forensic science expert opinion, *Science & Justice* 49(3) (2009) 161–164.

11. D. Taylor, B. Kokshoorn, T. Hicks, Structuring cases into propositions, assumptions, and undisputed case information, *Forensic Science International: Genetics* 44 (2020) 102199.

12. I.W. Evett, G. Jackson, J.A. Lambert, More on the hierarchy of propositions: Exploring the distinction between explanations and propositions, *Science & Justice* 40(1) (2000) 3–10.

13. T. Hicks, A. Biedermann, J.A.d. Koeijer, F. Taroni, C. Champod, I.W. Evett, The importance of distinguishing information from evidence/observations when formulating propositions, *Science & Justice* 55 (2015) 520–525.

14. FSR, *The Forensic Science Regulator guidance on DNA mixture interpretation FSR-G-222, The Forensic Science Regulator, 5 St Philip's Place, Colmore Row*, Birmingham, 2018, 1–63. https://www .gov.uk/government/publications/dna-mixture-interpretation-fsr-g-222 [last accessed 08 Janury 2023].

15. B. Kokshoorn, B.J. Blankers, J. de Zoete, C.E.H. Berger, Activity level DNA evidence evaluation: On propositions addressing the actor or the activity, *Forensic Science International* 278 (2017) 115–124.

16. D. Taylor, The evaluation of exclusionary DNA results: A discussion of issues in R v. Drummond, *Law, Probability and Risk* 15(1) (2016) 175–197.

5 Basic Mechanisms of Evaluation

Duncan Taylor

CONTENTS

DOI: 10.4324/9781003273189-5

5.1   INTRODUCTION

In this chapter a process for deriving an *LR* formula for an evaluation will be demonstrated. The *LR* derivation for the first three example cases, introduced in Chapter 2, will be stepped through. These same cases will then be used as examples to develop Bayesian networks (BNs) in Chapter 6, and it will be shown how BNs are simply a graphical way of representing a probabilistic formula (the formula derived here). During the evaluation of the observations in the first case, the steps of deriving a formula for the *LR* are explained. In the remaining two cases the same process is carried out but at an accelerated rate.

Even if the intention is to always carry out evaluations using a graphical tool, such as a BN, it is still important to have a conceptual understanding of the underlying construction of the probabilistic formulae. With this fundamental knowledge, it is easier to understand the manner in which the evaluation behaves and to avoid illogical construction. Some readers, who do aim to use BNs for their evaluation may find it useful to read this chapter simultaneously with Chapter 6. To do this it is suggested to the reader to go through the evaluation of the first case in Chapter 5 (where an *LR* formula is derived) and then that same case in Chapter 6 (where the BN is constructed), before coming back to Chapter 5 to move on to the next case.

5.2   DERIVING *LR* FORMULA BY HAND

In generality a number of steps are given below that can be used to derive an *LR* formula. These are likely to become more informative once they are applied in the running case examples. This is of course not the only way from which the *LR* formula can be derived as there are many different methods of structuring information in such a way that elements are not missed. The key is to have *some* kind of structure for thinking, whether that be graphical, tabulated or other.

5.2.1   STEP 1: PROPOSITION, BACKGROUND INFORMATION AND ASSUMPTIONS

Starting an evaluation (whether it is going to be done by deriving an *LR* formula explicitly or constructing a BN) the analyst must consider all the relevant information they have about the case. With this they develop their propositions, identify the key background information that is important to the case, and explicitly state any assumptions that are going to be made in the evaluation of the observations. Propositions, background information and assumptions for the case examples were generated in Chapter 4. If the reader has not yet read Chapter 4 then it is suggested that this is required before reading Chapter 5.

5.2.2   Step 2: Identify Pathways of Transfer

The analyst goes through a process of abductive reasoning, that is the analyst will consider the propositions in the case, and the framework of circumstances surrounding the alleged offence and develop a series of scenarios that could account for the observations. It may be that quite detailed information about the positions of prosecution and defence are given, and so the list will be constrained to consider possibilities that fall within the bounds of these positions. Some of these scenarios will likely be too similar to explanations and have a very low prior probability of having occurred and hence can be eliminated from the evaluation. As with many aspects of evaluation, there are no specific rules or thresholds for which one scenario need not be included. It is best to think about the issue in relative terms, i.e. if there are some scenarios being considered under a particular proposition that would seem to be quite reasonable (in statistical terms they would have a prior that was not close to 0) then there may be no need to include less reasonable (lower prior) options. However, those same less reasonable options that were excluded from the evaluation might be included as the most reasonable in other evaluations. An example which demonstrates this idea is the possibility of contamination occurring. In one scenario where the prosecution and defence both concede that the defendant touched an item (and the contention comes about the timing of the touch) then it might be considered that the prior probability of contamination is so low (compared to transfer probabilities) and it would not be meaningful to add to the evaluation. But in other situations where the defendant is claiming to have never had any direct contact with the item (and their claimed circumstances mean they are also disputing the possibility of any indirect transfer) then contamination may be the most reasonable explanation for the presence of the defendant's DNA (with the only other alternative being a chance DNA profile match with an alternate offender).

At the end of this abductive brainstorming session, the scientist will be left with a series of activities for each proposition that will be considered in the evaluation. Note that it is not always DNA profiling results that make up the observations as the observations can also be the presence of staining or the outcomes of body fluid testing.

5.2.3   Step 3: Identifying Temporal Order

Having the activities defined, it is helpful to draw the route of DNA transfer or to reason from the propositions to the observations. The pathways show all the factors that must be considered in getting to the observations. It is important to identify the order of events in time, as this will become essential when sequentially applying the relevant factors to the derivation. This step is not strictly necessary and is likely to be dropped as you become more familiar with derivations. It may seem very much like the beginnings of building a BN and in essence this is what it could be considered, although it is not taken as far as developing conditional probability tables and considering all possible outcomes. At this stage it may help to draw out the process like a branching tree of possibilities, and then eliminate the leaves in accordance with case observations.

5.2.4   Step 4: Start the Basic *LR* Formula

There are two ways in which this can be done, the first is to consider the probability of the observations given the propositions and then sequentially introduce terms (using the laws of probability) carrying out as much simplification as possible in each step. This leads to longer derivation, but one which is potentially more explicit in its assumptions. The second option is to use the pathways that were drawn in step 2 and assign probabilistic terms to each pathway (summing the result from each pathway within each proposition). This is often the preferred derivation approach as it keeps the formula smaller (with practice it can be carried out without graphically drawing out the pathways, but instead just making a mental tally of the pathways and their factors). However, it does come with the risk that certain conditional dependencies are not considered because the pathways are drawn linearly, and it is easy to overlook dependencies. Conditional dependencies are explored in more depth later in this chapter and Chapter 6.

5.2.5   STEP 5: COMPLETE THE *LR* DERIVATION

Once started as per step 4 the *LR* formula is expanded and simplified to its simplest form, within the assumptions being made. It is important to consider conditional dependencies between factors, which can become complex when there are multiple factors that are dependent on one element, or one factor that is dependent on multiple others. In the parlance of BNs these are referred to as convergent and divergent nodes and will be explained in Chapter 6.

5.2.6   STEP 6: ASSIGN PROBABILITIES

Once the *LR* formula is derived then the process shifts to assigning probabilities. Depending on the way that the data is considered, the probabilities may come from distributions that have modelled observational data or they may come directly from observation counts. The data will update prior beliefs in the probabilities (which can be either informed or uniform prior beliefs) in order to assign posterior probabilities in the evaluation. If properly done, the application of data to priors will prevent the occurrence of probability of 0 or 1 and hence avoid complete certainty in the evaluation model for one of the propositions. It is important to be transparent on the assignment of probabilities, with respect to any data used, the sources of that data, the way that it was used and any prior beliefs or subjective assignments made.

5.2.7   STEP 7: CALCULATE THE *LR*

Once the *LR* formula has been derived and the probabilities assigned for each factor within the formula then the process is completed by substituting the probabilities into their appropriate position in the *LR* and calculating the resulting value. In having derived an *LR* the advantage is that it is relatively easy to see how the *LR* would change if different probabilities were assigned. This process is more formally known as conducting a sensitivity analysis and is discussed later in Chapter 7.

The following sections progress to the case examples to demonstrate how to apply the steps described above to derive *LRs* for an evaluation.

5.3   CASE 1: THE ATTEMPTED KIDNAP

5.3.1   CASE 1: *LR* DERIVATION STEP 1

The *LR* derivation is started by setting the propositions, background and assumptions. This was talked about in Chapter 4 for all cases, but some additional discussion is provided here to demonstrate some points of theory. A series of propositions that ascend the hierarchy, going from sub-source to offence, can be seen in Table 5.1 (adapted from [1]). A contracted reference to the defendant and the complainant will be adopted as D and C, respectively (for Defendant and Complainant, although in this case the D could also stand for Drummond).

It may seem that when looking at the activity level propositions given in Table 5.1 they appear to be very close to describing the offence. Recall from Chapter 1 that any offence level proposition

TABLE 5.1
Propositions and Examples that Could Be Considered in Case 1

Hierarchy	Notation	Proposition
Offence	$^{Offense}H_p$	D attempted to kidnap C
	$^{Offense}H_d$	A male other than D attempted to kidnap C
Activity	$^{Activity}H_p$	C and D struggled which included D grabbing C's arm and C hitting D's chest
	$^{Activity}H_d$	A male other than D struggled with C which included him grabbing C's arm and C hitting him on the chest
Source	$^{Source}H_{p1}$	D has contributed skin cells to the top of C
	$^{Source}H_{d1}$	D has not contributed skin cells to the top of C
	$^{Source}H_{p2}$	C has contributed skin cells to the top of D
	$^{Source}H_{d2}$	C has not contributed skin cells to the top of D
Sub-source	$^{Sub\text{-}Source}H_{p1}$	D has contributed DNA to the sample from the top of C
	$^{Sub\text{-}Source}H_{d1}$	D has not contributed DNA to the sample from the top of C
	$^{Sub\text{-}Source}H_{p2}$	C has contributed DNA to the sample from the top of D
	$^{Sub\text{-}Source}H_{d2}$	C has not contributed DNA to the sample from the top of D

must have a legal component to it that speaks to the state of mind of the actors. The act of grabbing and hitting in a struggle scenario is not in itself an offence, and there may be numerous explanations by which these activities could occur in a legal context (such as play-fighting or an organised wrestling match). The fact that makes the hitting and grabbing an offence is that the complainant was not consenting to the activities and the intent behind the activities of the offender was presumably to kidnap (making the offence).

In Table 5.1 the sub-source propositions given are in a very general format. Typically, the propositions explicitly state the number of contributors and the contributors being considered. For example, the result from T's top may be evaluated with sub-source level propositions:

$^{Sub\text{-}Source}H_{p1}$: The sources of DNA are C, C's friend and D
$^{Sub\text{-}Source}H_{d1}$: The sources of DNA are C, C's friend and an unknown male

Also note that the source level propositions given in Table 5.1 are somewhat artificial as there is little need to specify the biological source of the DNA. In other cases the source is more important, particularly when the biological material is in contention (recall the proposition setting in Chapter 4 and the issues around source level propositions). Chapter 9 describes the manner in which biological sources can be addressed in the evaluation and the manner in which the results from different biological screen tests can be incorporated.

In the instance of this case the scientist excluded C from the DNA obtained from the sample of D's top and D from the DNA obtained from the sample of C's top. By making this interpretation the scientist has made the decision that the probability of obtaining the DNA profiling results if any of $^{Source}H_{p1}$, $^{Source}H_{p2}$, $^{Sub\text{-}Source}H_{p1}$ or $^{Sub\text{-}Source}H_{p2}$ is true, is zero. This would mean that any *LR* calculated from the pairs of sub-source propositions in Table 5.1 would also be zero regardless of the probability of the observations given the defence propositions. Note that as the sub-source propositions are exhaustive and mutually exclusive when the probability of the observations given one proposition in the pair is zero then the probability of the observations given the other proposition in the pair cannot also be zero. In Table 5.1 it would also have been possible to develop sub-sub-source level propositions where the references were considered as potential donors to components of the mixtures. Sub-sub-source level propositions are not provided for case 1 as the scientist manually excluded the individuals from the profiles and so in fact no sub-sub-source or sub-source propositions were explicitly defined. The sub-source level propositions are the only propositions that the scientist is able to comment on only from the given DNA profiling results. Recall there is no dispute over the cellular source of the DNA and in fact given the scientist excluded C from the top of D and D from the top of C, it is not clear how any evaluation of the observations given source level propositions could proceed. At the source level it may only be possible here to repeat the sub-source evaluation of an exclusion of the possibility of the prosecution proposition being true.

The court was clearly interested in the activity level propositions:

$^{Activity}H_p$: C and D struggled which included D grabbing C's arm and C hitting D's chest
$^{Activity}H_d$: A male other than D struggled with C which included him grabbing C's arm and C hitting him on the chest

There is sometimes a fine line between what is considered as background information that is relevant to the evaluation and what is considered an assumption. As explained earlier, background information is usually based on the framework of circumstances known to surround the case by testimony, admissions, witnesses or interviews, recorded footage, etc. Other factors for which there are no casework circumstances, but which are important to state explicitly as assumed for the evaluation are considered as assumptions. As with many aspects of evaluation, there is a slow transition from background information to assumptions and whereabouts along that line the particular factor sits may be based on personal preference as to whether it is listed as an assumption or information. In Chapter 4 the assumptions and background information being used in the evaluation were given. These are only the assumptions that are being made prior to the commencement of the evaluation. There are almost always further assumptions made during the derivation of the *LR* formula, which deal with simplification of the formula, or the assignment of probabilities, which relate to the applicability of the data to the case at hand. These will be listed during later steps of the derivation.

5.3.2   CASE 1: *LR* DERIVATION STEP 2

At this stage the various ways through which the activities could lead to the observations are brainstormed. In case 1 it may seem strange that scenarios being considered ultimately end up in no DNA being transferred; however, the process is no different to when DNA transfer has been observed. The main practical difference when considering an absence of DNA is that there is usually only a limited number of (or single) ways (compared to the presence of biological materials from persons-of-interest) in which the observation is explained. Starting first with the prosecution propositions that:

C and D struggled which included D grabbing C's arm and C hitting D's chest
And the first observation of:

- DNA from C, C's friend and an unknown female on C's top
  This could occur if:
  1. Background DNA was present on C's top. C transferred DNA to her own top from wearing it. C hugged her friend, and DNA from her friend was transferred to her top. During the attempted kidnap no DNA was transferred from D's hand to C's top when he grabbed her
  2. Background DNA was present on C's top. C transferred DNA to her own top from wearing. C hugged her friend, and DNA from her friend transferred to her top. During the attempted kidnap DNA was transferred from D's hand to C's top when he grabbed her arm but did not persist until the top being sampled at the forensic laboratory. Note that at this point we are not taking into account recovery at the laboratory (starting with the assumption that the correct area of the top was sampled)

And the second observation of:

- DNA from D and an unknown individual were found on D's top

  This could occur if:

3. Background DNA was present on D's top. D transferred DNA to his own top from wearing it. During the attempted kidnap no DNA was transferred from C's hand to D's top when she hit his collarbone area

4. Background DNA was present on D's top. D transferred DNA to his own top from wearing it. During the attempted kidnap DNA was transferred from C's hand to D's top when she hit his collarbone area but did not persist until (or was not recovered during) the top being sampled at the forensic laboratory

When it is specified that DNA material was not recovered, this means the DNA was not collected from the top with the tapelift or not visualised in subsequent analysis (e.g., a very minor component that is not observed in the electropherogram).

Now considering the defence proposition that:

A male other than D struggled with C which included him grabbing C's arm and C hitting him on the chest

And the first observation of:

- DNA from C, C's friend and an unknown female on C's top (given the assumption that the attacker was male, no DNA of an unknown offender has been recovered).

  This could occur if:

1. Background DNA was present on C's top. C transferred DNA to her own top from wearing it. C hugged her friend, and DNA from her friend transferred to her top. During the attempted kidnap no DNA was transferred from the alternate offender's (AO) hand to C's top when he grabbed her arm

2. Background DNA was present on C's top. C transferred DNA to her own top from wearing it. C hugged her friend, and DNA from her friend transferred to her top. During the attempted kidnap DNA was transferred from AO's hand to C's top when he grabbed her arm but did not persist until (or was not recovered during) the top being sampled at the forensic laboratory

There are numerous other possibilities that have not been considered because their prior probability is low compared to the scenarios being included in the evaluation. For example:

3. Background DNA was present on C's top. C transferred DNA to her own top from wearing it. C hugged her friend, and DNA from her friend transferred to her top. C's friend was her attacker and during the attempted kidnap DNA was transferred from C's friend's hand to C's top when he grabbed her arm. This possibility is ruled out due to the many issues it poses against the framework of circumstances known in the case, for example, C would be expected to recognise her friend, C's friend is not old enough to drive a car, etc.

4. Background DNA was present on C's top. C transferred DNA to her own top from wearing it. C hugged her friend, and DNA from her friend was not transferred to her top. During the attempted kidnap, DNA of AO was transferred from AO's hand to C's top when he grabbed her arm. The AO had the same DNA profile as C's friend. This option could be considered sensibly within the framework of circumstances of the case but is ruled out due to the very low probability of chance matching profiles between an AO and C's friend compared to non-transfer probability. If this chance matching possibility were included in the evaluation it would therefore have no noticeable effect on the *LR*

And the second observation of:

- DNA from D and an unknown individual was found on D's top
    This could occur if:
    5. Background DNA was present on D's top. D transferred DNA to his own top from wearing it

There is no mention of the attempted kidnapping under point 5 of the defence proposition because under this proposition D was not involved in any attempted kidnapping. Also, note that any events mentioned in the scenarios above which are the same for an item for both the prosecution and the defence propositions are not relevant to the alleged crime. This will be seen when the *LR* formula is being derived as these terms will cancel out and ultimately be absent in the final formula.

5.3.3 CASE 1: *LR* DERIVATION STEP 3

Having considered possible explanations for the observed DNA profile results given the proposition, step 3 maps the temporal order in which they have occurred within the framework of circumstances. Figure 5.1 shows this mapping for the points raised in step 2.

The boxes in Figure 5.1 show the factors that involve the potential for DNA transfer. Some of the boxes start with an action which then leads to an effect on the DNA transfer or persistence. The

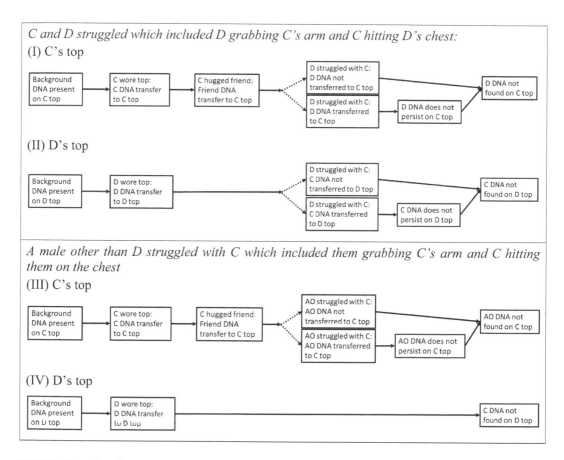

FIGURE 5.1   Timeline traces of DNA transfer in case 1. The top panel represents the timeline of C's and D's tops given the prosecution proposition and the bottom panel represents the timeline of the defence proposition.

dotted arrows in Figure 5.1 represent an uncertain outcome, i.e. that DNA may or may have not been transferred to C's top under either of the propositions and may or may not have been transferred to D's top under the prosecution proposition. If DNA has not been transferred, then this explains the absence of DNA found in laboratory observations. If DNA has been transferred then it must not have persisted (or been recovered; however, recovery is not explicitly modelled separately from persistence in this example).

5.3.4   CASE 1: *LR* DERIVATION STEP 4

There are two ways to proceed with the *LR* formula derivation. One is to start with the observations given the propositions and then sequentially add in factors and simplify when possible. The second method is to start with all factors present and then expand and simplify.

5.3.4.1   Starting with a Demonstration of the First Method

To calculate the *LR*, the probability of the observations, '*O*' given the activity level propositions and the background information (represented by '*I*') are given formulaically by:

$$LR = \frac{\Pr\left(O \mid {}^{Activity}H_p, I\right)}{\Pr\left(O \mid {}^{Activity}H_d, I\right)}$$

From this point onwards the background information term will be omitted, purely for visual clarity in the formula, but it must be remembered that it is always present in the conditioning information when carrying out an evaluation. The observations can now be broken apart into their separate components. A capital 'D' is used to represent DNA and subscripts represent the person of interest's DNA on an item with an arrow. As mentioned in Chapter 1, a line drawn over the top of a term represents its negation:

$\overline{D_{D \to C}}$ – The absence of D's DNA on C's top

$\overline{D_{C \to D}}$ – The absence of C's DNA on D's top

$\overline{D_{AO \to C}}$ – An absence of the alternate offender's reference (i.e. an unknown male) in the profile detected on C's top

$D_{C \to C}$ – The presence of C's DNA on C's top

$D_{F \to C}$ – The presence of C's friend's DNA on C's top

$D_{D \to D}$ – The presence of D's DNA on D's top

B_C – The presence of background DNA on C's top

B_D – The presence of background DNA on D's top

Before proceeding, note that many derivations start from the terminology of transfer and persistence and not from the DNA results themselves. Doing this is a short-cut, which can be used with experience. When starting out with *LR* derivation (or if carrying out a derivation for a complex scenario) it often helps to start at the actual observations (which is a lack of DNA rather than a lack of transfer).

Splitting the observations into their separate terms transforms the earlier *LR* equation to:

$$LR = \frac{\Pr\left(\overline{D_{D \to C}}, \overline{D_{C \to D}}, D_{C \to C}, D_{D \to D}, D_{F \to C}, B_C, B_D \mid {}^{Activity}H_p\right)}{\Pr\left(\overline{D_{D \to C}}, \overline{D_{C \to D}}, \overline{D_{AO \to C}}, D_{C \to C}, D_{D \to D}, D_{F \to C}, B_C, B_D \mid {}^{Activity}H_d\right)}$$

Using the third law of probability, the components of the *LR* can be broken into whatever parts are desired. The most convenient first demarcation is to split the factors that are not dependent on the propositions from all others. In this case these are the presence of the background DNA on the tops of C and D, the presence of C on her own top, the presence of C's friend on her top and the presence of D on his own top. All of these observations of DNA are from events that are not in dispute, hence they are the same in both propositions and therefore are independent of them.

$$LR = \frac{\Pr\left(\overline{D_{D \to C}}, \overline{D_{C \to D}} \mid D_{C \to C}, D_{D \to D}, D_{F \to C}, B_C, B_D, {}^{Activity}H_p\right) \Pr\left(D_{C \to C}, D_{D \to D}, D_{F \to C}, B_C, B_D \mid {}^{Activity}H_p\right)}{\Pr\left(\overline{D_{D \to C}}, \overline{D_{C \to D}}, \overline{D_{AO \to C}} \mid D_{C \to C}, D_{D \to D}, D_{F \to C}, B_C, B_D, {}^{Activity}H_d\right) \Pr\left(D_{C \to C}, D_{D \to D}, D_{F \to C}, B_C, B_D \mid {}^{Activity}H_d\right)}$$

In equation terms the concept of independence states is represented by

$$\Pr\left(D_{C \to C}, D_{D \to D}, D_{F \to C}, B_C, B_D \mid {}^{Activity}H_p\right) = \Pr\left(D_{C \to C}, D_{D \to D}, D_{F \to C}, B_C, B_D \mid {}^{Activity}H_d\right)$$

And therefore, these terms cancel in the *LR* equation. Also,

$$\Pr\left(\overline{D_{D \to C}}, \overline{D_{C \to D}} \mid D_{C \to C}, D_{D \to D}, D_{F \to C}, B_C, B_D, {}^{Activity}H_p\right) = \Pr\left(\overline{D_{D \to C}}, \overline{D_{C \to D}} \mid {}^{Activity}H_p\right)$$

$$\Pr\left(\overline{D_{D \to C}}, \overline{D_{C \to D}}, \overline{D_{AO \to C}} \mid D_{C \to C}, D_{D \to D}, D_{F \to C}, B_C, B_D, {}^{Activity}H_d\right) = \Pr\left(\overline{D_{D \to C}}, \overline{D_{C \to D}}, \overline{D_{AO \to C}} \mid {}^{Activity}H_d\right)$$

So that the *LR* quickly simplifies to

$$LR = \frac{\Pr\left(\overline{D_{D \to C}}, \overline{D_{C \to D}} \mid {}^{Activity}H_p\right)}{\Pr\left(\overline{D_{D \to C}}, \overline{D_{C \to D}}, \overline{D_{AO \to C}} \mid {}^{Activity}H_d\right)}$$

There have not yet been any events of transfer or persistence considered in the equation. This simplification of the *LR* from all the observations to just a few observations is the outcome of the effect mentioned earlier of including factors that are not relevant to the alleged crime.

Again, using the third law of probability to break apart the terms in the conditional probability statement into their individual components:

$$LR = \frac{\Pr\left(\overline{D_{D \to C}} \mid \overline{D_{C \to D}}, {}^{Activity}H_p\right) \Pr\left(\overline{D_{C \to D}} \mid {}^{Activity}H_p\right)}{\Pr\left(\overline{D_{D \to C}} \mid \overline{D_{C \to D}}, \overline{D_{AO \to T}}, {}^{Activity}H_d\right) \Pr\left(\overline{D_{C \to D}} \mid \overline{D_{AO \to C}}, {}^{Activity}H_d\right) \Pr\left(\overline{D_{AO \to C}} \mid {}^{Activity}H_d\right)}$$

In words, the numerator of the likelihood ratio is the probability of an absence of D's DNA on C's top if they struggled and given that there was an absence of C's DNA on D's top, multiplied by the probability of finding an absence of C's DNA on D's top (again if they struggled). The denominator is the same but considering the probability of the results in light of the defence proposition that C did not struggle with D, but AO did which also resulted in an absence of their DNA profile from C's top.

Further simplifying assumptions can still be made before any aspects of transfer and persistence are introduced into the derivation. That is the assumption that under the prosecution proposition the probability of a no correspondence with D's reference in the profile from C's top does not depend on whether there was no correspondence with C's reference in the profile from D's top. Note that there are potentially high-order dependencies that are not being taken into account when making this assumption. For example, if there is no correspondence with C's reference in the profile from D's top then this might suggest that the struggle was brief, which would then increase the chance of there being no correspondence with D's reference in the profile from C's top. Another example of a higher-order dependency would be the recovery efficiency of the forensic process. This potential dependency is ignored as (if it indeed exists) it is likely to be slight and doing so maximises the favour of the evidence for the defence (and hence could be described as a conservative but fair assumption). Under the defence propositions $\overline{D_{D\to C}}$, $\overline{D_{C\to D}}$ and $\overline{D_{AO\to C}}$ are independent because D and C have not been in contact with each other. Therefore, the *LR* is:

$$LR = \frac{\Pr\left(\overline{D_{D\to C}} \mid {}^{Activity}H_p\right)\Pr\left(\overline{D_{C\to D}} \mid {}^{Activity}H_p\right)}{\Pr\left(\overline{D_{D\to C}} \mid {}^{Activity}H_d\right)\Pr\left(\overline{D_{C\to D}} \mid {}^{Activity}H_d\right)\Pr\left(\overline{D_{AO\to C}} \mid {}^{Activity}H_d\right)}$$

The equation is now simplified to the point that factors of transfer and persistence are convenient to introduce.

Before continuing more terminology for transfer is introduced:

$T_{D\to C}$ – Transfer of D's DNA to C's top
$T_{C\to D}$ – Transfer of C's DNA to D's top
$T_{AO\to C}$ – Transfer of AO's DNA to C's top
$\overline{T_{D\to C}}$ – No transfer of D's DNA to C's top
$\overline{T_{C\to D}}$ – No transfer of C's DNA to D's top
$\overline{T_{AO\to C}}$ – No transfer of AO's DNA to C's top

and for persistence:

$P_{D\to C}$ – Persistence of D's DNA on C's top
$P_{C\to D}$ – Persistence of C's DNA on D's top
$P_{AO\to C}$ – Persistence of AO's DNA on C's top
$\overline{P_{D\to C}}$ – No persistence of D's DNA on C's top
$\overline{P_{C\to D}}$ – No persistence of C's DNA on D's top
$\overline{P_{AO\to C}}$ – No persistence of AO's DNA on C's top

---

BOX 5.1   THE ORDER OF INTRODUCING FACTORS INTO THE EVALUATION

As these terms are added, it tends to be done in the order in which they occur in the real world. Note must be made of what occurs downstream when carrying out simplifications. For example, take just the first term in the numerator of the *LR*, $\Pr\left(\overline{D_{D\to C}} \mid {}^{Activity}H_p\right)$. Consider what would occur if the factor of persistence was introduced first (i.e. before transfer):

$$\Pr\left(\overline{D_{D\to C}} \mid {}^{Activity}H_p\right) = \Pr\left(\overline{D_{D\to C}} \mid P_{D\to C}, {}^{Activity}H_p\right)\Pr\left(P_{D\to C} \mid {}^{Activity}H_p\right)$$

$$+ \Pr\left(\overline{D_{D\to C}} \mid \overline{P_{D\to C}}, {}^{Activity}H_p\right)\Pr\left(\overline{P_{D\to C}} \mid {}^{Activity}H_p\right)$$

It is now intuitively difficult to assign probabilities to terms such as $\Pr\left(\overline{D_{D\to C}} \mid P_{D\to C}, {}^{Activity}H_p\right)$, which in words is the probability of not observing any DNA of D on C's top if it persists. But it is not clear if this makes sense, i.e. it is not known if any DNA was on C's top to begin with, before considering whether it may have persisted. It should be noted here that given the same assumptions, the *LR* will ultimately be the same, regardless of the order of the introduction of terms. This is known to be the case from the third law of probability which shows:

$$\Pr\left(AB\right) = \Pr\left(A\mid B\right)\Pr\left(B\right) = \Pr\left(B\mid A\right)\Pr\left(A\right)$$

However, the derivation and the simplifications are easier in practice by introducing them in a natural order of occurrence, as the terms will possess natural intuitiveness.

The order in which terms are added should be the same as they are thought of occurring in real life (see Box 5.1). Also, the knowledge of downstream factors must be taken into account when carrying out the simplifications. Again, start with the term $\Pr\left(\overline{D_{D \to C}} \mid {}^{Activity}H_p\right)$, but this time a consideration of transfer is introduced which is given by

$$\Pr\left(\overline{D_{D \to C}} \mid {}^{Activity}Hp\right) = \Pr\left(\overline{D_{D \to C}} \mid T_{D \to C}, {}^{Activity}H_p\right)\Pr\left(T_{D \to C} \mid {}^{Activity}H_p\right)$$
$$+ \Pr\left(\overline{D_{D \to C}} \mid \overline{T_{D \to C}}, {}^{Activity}H_p\right)\Pr\left(\overline{T_{D \to C}} \mid {}^{Activity}H_p\right)$$

It would be easy to make the simplification that the probability of none of D's DNA on C's top if a transfer has not occurred is 1, i.e.:

$$\Pr\left(\overline{D_{D \to C}} \mid \overline{T_{D \to C}}, {}^{Activity}H_p\right) = 1$$

And the probability of none of D's DNA on C's top if a transfer has occurred is 0, i.e.:

$$\Pr\left(\overline{D_{D \to C}} \mid T_{D \to C}, {}^{Activity}H_p\right) = 0$$

which would lead the equation above to:

$$\Pr\left(\overline{D_{D \to C}} \mid {}^{Activity}H_p\right) = \Pr\left(\overline{T_{D \to C}} \mid {}^{Activity}H_p\right)$$

and there is no meaningful way in which persistence can be introduced, as there are no DNA observations left to be probabilistically described. The mistake here was not recalling that in the transfer route as there is still a need to consider that the DNA may or may not have persisted (as shown in the pathways in Figure 5.1), i.e. the simplification of $\Pr\left(\overline{D_{D \to C}} \mid T_{D \to C}, {}^{Activity}H_p\right) = 0$, was incorrect.

The introduction of transfer and persistence to the formula:

$$LR = \frac{\Pr\left(\overline{D_{D \to C}} \mid {}^{Activity}H_p\right)\Pr\left(\overline{D_{C \to D}} \mid {}^{Activity}H_p\right)}{\Pr\left(\overline{D_{D \to C}} \mid {}^{Activity}H_d\right)\Pr\left(\overline{D_{C \to D}} \mid {}^{Activity}H_d\right)\Pr\left(\overline{D_{AO \to C}} \mid {}^{Activity}H_d\right)}$$

will now be shown using a logical order of introduction and taking into account the full transfer pathway when making simplifications. Consider first the potential for transfer to have occurred. The probabilistic law of extending the conversation can be used to introduce this term into each of the elements of the *LR* shown above. It has already been shown that the introduction of transfer to the first term in the earlier example:

$$\Pr\left(\overline{D_{D \to C}} \mid {}^{Activity}H_p\right) = \Pr\left(\overline{D_{D \to C}} \mid T_{D \to C}, {}^{Activity}H_p\right)\Pr\left(T_{D \to C} \mid {}^{Activity}H_p\right)$$
$$+ \Pr\left(\overline{D_{D \to C}} \mid \overline{T_{D \to C}}, {}^{Activity}H_p\right)\Pr\left(\overline{T_{D \to C}} \mid {}^{Activity}H_p\right)$$

In Figure 5.1, within the considerations under ${}^{Activity}H_p$ for the top of C, there are no factors that sit between a non-transfer and the observations. Therefore, this term can be simplified (via making assumptions about it), without the risk of missing a term that had originally been intended to include. The assumption being made (which was one of the two made in the earlier example) is that the probability of none of D's DNA on C's top if a transfer has not occurred is 1, i.e.:

$$\Pr\left(\overline{D_{D \to C}} \mid \overline{T_{D \to C}}, {}^{Activity}H_p\right) = 1$$

so that result is the simplified term:

$$\Pr\left(\overline{D_{D\to C}} \mid {}^{Activity}H_p\right) = \Pr\left(\overline{D_{D\to C}} \mid T_{D\to C}, {}^{Activity}H_p\right)\Pr\left(T_{D\to C} \mid {}^{Activity}H_p\right) + \Pr\left(\overline{T_{D\to C}} \mid {}^{Activity}H_p\right)$$

Now the concept of persistence can be introduced. The persistence of DNA only makes sense to consider in observation terms. In the example we have, there is no point adding persistence to the transfer terms. While persistence could be introduced into the transfer terms, making the assumption that DNA transferred was independent of persistence, which means the terms would cancel back to their original form. For example, take the term $\Pr\left(\overline{T_{D\to C}} \mid {}^{Activity}H_p\right)$, if persistence were introduced:

$$\Pr\left(\overline{T_{D\to C}} \mid {}^{Activity}Hp\right) = \Pr\left(\overline{T_{D\to C}} \mid \overline{P_{D\to C}}, {}^{Activity}H_p\right)\Pr\left(\overline{P_{D\to C}} \mid {}^{Activity}H_p\right)$$

$$+ \Pr\left(\overline{T_{D\to C}} \mid P_{D\to C}, {}^{Activity}H_p\right)\Pr\left(P_{D\to C} \mid {}^{Activity}H_p\right)$$

And making the assumption of independence of transfer from persistence:

$$\Pr\left(\overline{T_{D\to C}} \mid {}^{Activity}H_p\right) = \Pr\left(\overline{T_{D\to C}} \mid {}^{Activity}H_p\right)\Pr\left(\overline{P_{D\to C}} \mid {}^{Activity}H_p\right)$$

$$+ \Pr\left(\overline{T_{D\to C}} \mid {}^{Activity}H_p\right)\Pr\left(P_{D\to C} \mid {}^{Activity}H_p\right)$$

$$\Pr\left(\overline{T_{D\to C}} \mid {}^{Activity}H_p\right) = \Pr\left(\overline{T_{D\to C}} \mid {}^{Activity}H_p\right)\left[\Pr\left(\overline{P_{D\to C}} \mid {}^{Activity}H_p\right) + \Pr\left(P_{D\to C} \mid {}^{Activity}H_p\right)\right]$$

$$\Pr\left(\overline{T_{D\to C}} \mid {}^{Activity}H_p\right) = \Pr\left(\overline{T_{D\to C}} \mid {}^{Activity}H_p\right)$$

With the last step coming from the fact that the terms of persistence or non-persistence are exhaustive and so their sum is 1. Hence, it can be seen that while persistence could be introduced to each term of the *LR*, it will only have an effect (i.e. will not cancel back to the original term) for probabilities of observations.

Going back to the term:

$$\Pr\left(\overline{D_{D\to C}} \mid {}^{Activity}H_p\right) = \Pr\left(\overline{D_{D\to C}} \mid T_{D\to C}, {}^{Activity}H_p\right)\Pr\left(T_{D\to C} \mid {}^{Activity}H_p\right) + \Pr\left(\overline{T_{D\to C}} \mid {}^{Activity}H_p\right)$$

And introducing persistence gives:

$$\Pr\left(\overline{D_{D\to C}} \mid {}^{Activity}H_p\right) = \Pr\left(\overline{D_{D\to C}} \mid T_{D\to C}, \overline{P_{D\to C}} {}^{Activity}H_p\right)\Pr\left(T_{D\to C} \mid {}^{Activity}H_p\right)\Pr\left(\overline{P_{D\to C}} \mid {}^{Activity}H_p\right) +$$

$$\Pr\left(\overline{D_{D\to C}} \mid T_{D\to C}, P_{D\to C} {}^{Activity}H_p\right)\Pr\left(T_{D\to C} \mid {}^{Activity}H_p\right)\Pr\left(P_{D\to C} \mid {}^{Activity}H_p\right) + \Pr\left(\overline{T_{D\to C}} \mid {}^{Activity}H_p\right)$$

Now all the intended factors have been introduced, simplifying assumptions can be made. The assumption is made that the probability of observing no DNA of D on C's top if DNA has transferred and persisted is 0, i.e.

$$\Pr\left(\overline{D_{D\to C}} \mid T_{D\to C}, P_{D\to C} {}^{Activity}H_p\right) = 0$$

And so:

$$\Pr\left(\overline{D_{D\to C}} \mid T_{D\to C}, P_{D\to C} {}^{Activity}H_p\right)\Pr\left(T_{D\to C} \mid {}^{Activity}H_p\right)\Pr\left(P_{D\to C} \mid {}^{Activity}H_p\right)=0$$

The assumption is also made that the probability of seeing no DNA from D on C's top if it has transferred, but not persisted is 1, i.e.:

$$\Pr\left(\overline{D_{D\to C}} \mid T_{D\to C}, \overline{P_{D\to C}} {}^{Activity}H_p\right)=1$$

Bringing the simplified term:

$$\Pr\left(\overline{D_{D\to C}} \mid {}^{Activity}H_p\right) = \Pr\left(T_{D\to C} \mid {}^{Activity}H_p\right)\Pr\left(\overline{P_{D\to C}} \mid {}^{Activity}H_p\right)+\Pr\left(\overline{T_{D\to C}} \mid {}^{Activity}H_p\right)$$

In words this can be described as the probability of not detecting the DNA of D on C's top is equal to the probability that DNA was transferred from D to C's top but did not persist plus the probability that no DNA was transferred from D to C's top. This represents a real-world intuitive sensibleness. The overall *LR* equation, incorporating this updated first term is:

$$LR = \frac{\left[\Pr\left(T_{D\to C} \mid {}^{Activity}H_p\right)\Pr\left(\overline{P_{D\to C}} \mid {}^{Activity}H_p\right)+\Pr\left(\overline{T_{D\to C}} \mid {}^{Activity}H_p\right)\right]\Pr\left(\overline{D_{C\to D}} \mid {}^{Activity}H_p\right)}{\Pr\left(\overline{D_{D\to C}} \mid {}^{Activity}H_d\right)\Pr\left(\overline{D_{C\to D}} \mid {}^{Activity}H_d\right)\Pr\left(\overline{D_{AO\to C}} \mid {}^{Activity}H_d\right)}$$

In this particular example all the terms in the *LR* are very similar in structure, i.e. they are all probabilities of not detecting some DNA, given a proposition. The two terms that are partly different from the rest are $\Pr\left(\overline{D_{D\to C}} \mid {}^{Activity}H_d\right)$ and $\Pr\left(\overline{D_{C\to D}} \mid {}^{Activity}H_d\right)$, as these terms have a proposition that specifies no contact between the people of focus in the observational terms. It is debatable that these terms could have been assigned probabilities of 1 directly, before even any consideration of transfer and persistence being introduced. Philosophically, it is not the probability that DNA was not observed that is one, but more so the probability for transfer of DNA to have occurred that is zero (and the complement, the probability of no transfer of DNA having occurred being 1). The terms have been left present in the equation and it will be seen that they are simplified away at the next point.

Introducing the factors of transfer and persistence into the other observational terms of the *LR*, in exactly the same way as was shown for the first term yields the *LR*:

$$LR = \frac{\begin{aligned}&\left[\Pr\left(T_{D\to C} \mid {}^{Activity}H_p\right)\Pr\left(\overline{P_{D\to C}} \mid {}^{Activity}H_p\right)+\Pr\left(\overline{T_{D\to C}} \mid {}^{Activity}H_p\right)\right]\times\\&\left[\Pr\left(T_{C\to D} \mid {}^{Activity}H_p\right)\Pr\left(\overline{P_{C\to D}} \mid {}^{Activity}H_p\right)+\Pr\left(\overline{T_{C\to D}} \mid {}^{Activity}H_p\right)\right]\end{aligned}}{\begin{aligned}&\left[\Pr\left(T_{D\to C} \mid {}^{Activity}H_d\right)\Pr\left(\overline{P_{D\to C}} \mid {}^{Activity}H_d\right)+\Pr\left(\overline{T_{D\to C}} \mid {}^{Activity}H_d\right)\right]\times\\&\left[\Pr\left(T_{C\to D} \mid {}^{Activity}H_d\right)\Pr\left(\overline{P_{C\to D}} \mid {}^{Activity}H_d\right)+\Pr\left(\overline{T_{C\to D}} \mid {}^{Activity}H_d\right)\right]\times\\&\left[\Pr\left(T_{AO\to C} \mid {}^{Activity}H_d\right)\Pr\left(\overline{P_{AO\to C}} \mid {}^{Activity}H_d\right)+\Pr\left(\overline{T_{AO\to C}} \mid {}^{Activity}H_d\right)\right]\end{aligned}}$$

At this point it is now clear that given the defence scenario that D and C have had no contact then the assumption can be made that there has been no transfer of DNA between them, i.e.:

$$\Pr\left(T_{D\to C} \mid {}^{Activity}H_d\right)=0$$

$$\Pr\left(\overline{T_{D\to C}} \mid {}^{Activity}H_d\right) = 1$$

$$\Pr\left(T_{C\to D} \mid {}^{Activity}H_d\right) = 0$$

$$\Pr\left(\overline{T_{C\to D}} \mid {}^{Activity}H_d\right) = 1$$

Giving:

$$LR = \frac{\left[\Pr\left(T_{D\to C} \mid {}^{Activity}H_p\right)\Pr\left(\overline{P_{D\to C}} \mid {}^{Activity}H_p\right) + \Pr\left(\overline{T_{D\to C}} \mid {}^{Activity}H_p\right)\right] \times \left[\Pr\left(T_{C\to D} \mid {}^{Activity}H_p\right)\Pr\left(\overline{P_{C\to D}} \mid {}^{Activity}H_p\right) + \Pr\left(\overline{T_{C\to D}} \mid {}^{Activity}H_p\right)\right]}{\Pr\left(T_{AO\to C} \mid {}^{Activity}H_d\right)\Pr\left(\overline{P_{AO\to C}} \mid {}^{Activity}H_d\right) + \Pr\left(\overline{T_{AO\to C}} \mid {}^{Activity}H_d\right)}$$

It is now time to turn to the meaning of the probabilistic terms themselves in order to take forward the *LR* formula simplification/derivation. Specifically, consideration is given to whether any terms within the equation, all currently distinguishable by unique subscripts, could be reasonably assumed to be equal. For example, in pre-derivation it was assumed that the complainant, defendant and alternate offender possessed roughly equal shedding ability (made through the assumption that none had a skin condition that would mean they donated unusually high levels of DNA, had similar shedding rates, personal hygiene, were not suffering from a cold, etc). Also, there was an assumption that they all had a similar DNA load available at the time of the alleged offence (made through the assumption that none had recently washed their hands). Given these points it can be assumed that the probability of DNA being transferred from D to C's top if D grabbed C would be the same as the probability of DNA being transferred from AO to C's top if AO grabbed C. This assumption is represented formulaically, by replacing the donor subscript of the transfer terms with a dot, i.e.:

$$\Pr\left(T_{D\to C} \mid {}^{Activity}H_p\right) = \Pr\left(T_{AO\to C} \mid {}^{Activity}H_d\right) = \Pr\left(T_{.\to C} \mid {}^{Activity}H_d\right)$$

From the laws of probability, and given the mutual exclusivity and exhaustive nature of transfer terms, it can also be stated:

$$\Pr\left(\overline{T_{D\to C}} \mid {}^{Activity}H_p\right) = \Pr\left(\overline{T_{AO\to C}} \mid {}^{Activity}H_d\right) = \Pr\left(\overline{T_{.\to C}} \mid {}^{Activity}H_d\right)$$

Note that once DNA has been transferred to C's top, the source of that DNA does not matter, its probability of persistence is assumed to be the same, so that:

$$\Pr\left(\overline{P_{D\to C}} \mid {}^{Activity}H_p\right) = \Pr\left(\overline{P_{AO\to C}} \mid {}^{Activity}H_d\right) = \Pr\left(\overline{P_{.\to C}} \mid {}^{Activity}H_d\right)$$

Giving the *LR*:

$$LR = \frac{\left[\Pr\left(T_{.\to C} \mid {}^{Activity}H_d\right)\Pr\left(\overline{P_{.\to C}} \mid {}^{Activity}H_d\right) + \Pr\left(\overline{T_{.\to C}} \mid {}^{Activity}H_d\right)\right] \times \left[\Pr\left(T_{C\to D} \mid {}^{Activity}H_p\right)\Pr\left(\overline{P_{C\to D}} \mid {}^{Activity}H_p\right) + \Pr\left(\overline{T_{C\to D}} \mid {}^{Activity}H_p\right)\right]}{\Pr\left(T_{.\to C} \mid {}^{Activity}H_d\right)\Pr\left(\overline{P_{.\to C}} \mid {}^{Activity}H_d\right) + \Pr\left(\overline{T_{.\to C}} \mid {}^{Activity}H_d\right)}$$

And having made this simplification, it can be seen that there is a common element of $\left[\Pr\left(T_{.\to C} \mid {}^{Activity}H_d\right)\Pr\left(\overline{P_{.\to C}} \mid {}^{Activity}H_d\right) + \Pr\left(\overline{T_{.\to C}} \mid {}^{Activity}H_d\right)\right]$ in the numerator and denominator (in fact this term is the entire denominator of the LR). Cancelling this common term yields an LR of:

$$LR = \Pr\left(T_{C\to D} \mid {}^{Activity}H_p\right)\Pr\left(\overline{P_{C\to D}} \mid {}^{Activity}H_p\right) + \Pr\left(\overline{T_{C\to D}} \mid {}^{Activity}H_p\right)$$

This cancellation is intuitive as it states that the results of the absence of relevant DNA on C's top is neutral with respect to the alleged offence. Either D's DNA did not transfer or persist under ${}^{Activity}H_p$ or AO's DNA did not transfer or persist under ${}^{Activity}H_d$, both with equal probability. The only item to which the LR is sensitive is D's top, specifically to probabilities of C's DNA being transferred and C's DNA persisting (or their complements).

Note that prior to the derivation of the LR a number of assumptions were made about the circumstances of the case so that the evaluation could proceed. It was stated at that time that the LR derivation itself would lead to a number of additional assumptions required for the simplification of the formula (and the general practicality of being able to carry out the evaluation). These assumptions, termed 'derivational assumptions', were given throughout the derivation but are also listed below in one block:

- The presence of the background DNA on the tops of C and D, the presence of C on her own top, the presence of C's friend on her top and the presence of D on his own top are the same under H_p and H_d
- Under the prosecution proposition, the probability of an absence of D's reference in the profile from C's top does not depend on whether there was an absence of C's reference in the profile from D's top
- Under the defence propositions $\overline{D_{D\to C}}$, $\overline{D_{C\to D}}$ and $\overline{D_{AO\to T}}$ are conditionally independent because D and C have not been in contact with each other
- The probability of none of D's DNA on C's top if transfer has not occurred is 1
- The probability of observing no DNA of D on C's top or no DNA of C on D's top or no DNA of AO on C's top, if DNA has transferred and persisted is 0 (assuming the probability of recovery is 1)
- The probability of seeing no DNA from D on C's top or no DNA of C on D's top or no DNA of AO on C's top, if it has transferred but not persisted is 1
- Given the defence scenario that D and C have had no contact there has been no direct or indirect transfer of DNA between them (no other interaction between them is known or assumed to have occurred)
- The probability of DNA being transferred from D to C's top if D grabbed C would be the same as the probability of DNA being transferred from AO to C's top if AO grabbed C
- Once DNA has been transferred to C's top, it does not matter what is the source of that DNA, its probability of persistence is assumed to be the same (note that this assumption may or may not be valid depending on the moment of transfer and activities of C between transfer event and seizure of the clothing)

Note that the way through which the derivation was carried out required each of these assumptions to be stated explicitly in order for simplifications of the LR formula to occur. The same LR derivation will now be carried out again but using a short-hand method that makes direct use of the pathways drawn out in Figure 5.1.

5.3.5   CASE 1: *LR* DERIVATION STEP 5: SHORT-HAND VERSION OF *LR* DERIVATION

There are two items relevant to the alleged crime, D's top and C's top. In Figure 5.1 a series of events were drawn out that lead to the observed DNA results. For the top of D these events were:

1. Background DNA, not relevant to the alleged crime, was present on D's top
2. D's DNA was transferred to his own top from wearing it (which is not relevant to the alleged crime activity)
3. C's DNA was not transferred to D's top during a struggle between D and C

These ultimately result in C's DNA not being observed on D's top. Any terms that appear in both the H_p and H_d pathways (here events 1 and 2, both not relevant to the proposed struggle) can be ignored, and probabilistic terms applied to all that remains, in this case event number 3 from above, which is assigned as $\Pr\left(\overline{T_{C \to D}} \mid {}^{Activity}H_p\right)$. Alternatively, for D's top the pathway is (this time not listing out points 1 and 2 from the previous):

4. C's DNA was transferred to D's top during a struggle between D and C
5. C's DNA did not persist on D's top from the time of the struggle to the time of sampling

For point 4 the term $\Pr\left(T_{C \to D} \mid {}^{Activity}H_p\right)$ is assigned, and point 5 the term $\Pr\left(\overline{P_{C \to D}} \mid {}^{Activity}H_p\right)$, meaning that the observations from D's top can be probabilistically represented under H_p events 3, 4 and 5 as:

$$\Pr\left(T_{C \to D} \mid {}^{Activity}H_p\right)\Pr\left(\overline{P_{C \to D}} \mid {}^{Activity}H_p\right) + \Pr\left(\overline{T_{C \to D}} \mid {}^{Activity}H_p\right)$$

Carrying out the same process for C's top (again not including the agreed actions in this case), requires the consideration that DNA from D either didn't transfer to C's top or did transfer but did not persist, yielding:

$$\Pr\left(T_{D \to C} \mid {}^{Activity}H_p\right)\Pr\left(\overline{P_{D \to C}} \mid {}^{Activity}H_p\right) + \Pr\left(\overline{T_{D \to C}} \mid {}^{Activity}H_p\right)$$

As the proposition specifies that both items are relevant then the results of C's top and D's top must be probabilistically described together. The 'and' translates to multiplication in mathematics so that the numerator of the *LR* is therefore:

$$\Pr\left(E \mid {}^{Activity}H_p\right) = \left[\Pr\left(T_{C \to D} \mid {}^{Activity}H_p\right)\Pr\left(\overline{P_{C \to D}} \mid {}^{Activity}H_p\right) + \Pr\left(\overline{T_{C \to D}} \mid {}^{Activity}H_p\right)\right] \times$$

$$\left[\Pr\left(T_{D \to C} \mid {}^{Activity}H_p\right)\Pr\left(\overline{P_{D \to C}} \mid {}^{Activity}H_p\right) + \Pr\left(\overline{T_{D \to C}} \mid {}^{Activity}H_p\right)\right]$$

Under the defence proposition the same considerations of transfer and persistence of DNA on C's top are required, but with the AO being the potential donor rather than D. This leads to:

$$\Pr\left(T_{AO \to C} \mid {}^{Activity}H_d\right)\Pr\left(\overline{P_{AO \to C}} \mid {}^{Activity}H_d\right) + \Pr\left(\overline{T_{AO \to C}} \mid {}^{Activity}H_d\right)$$

In the case of D's top, it can be seen in Figure 5.1 that under H_d there are no activities that are undisputed (i.e. the top is not relevant to the alleged crime) and so there are no probabilistic terms to apply. This leaves the denominator of the *LR* only addressing the observations on C's top as:

$$\Pr(E \mid {}^{Activity}H_d) = \Pr\left(T_{AO \to C} \mid {}^{Activity}H_d\right)\Pr\left(\overline{P_{AO \to C}} \mid {}^{Activity}H_d\right) + \Pr\left(\overline{T_{AO \to C}} \mid {}^{Activity}H_d\right)$$

And the *LR* is:

$$\left[\Pr\left(T_{D\to C} \mid {}^{Activity}H_p\right)\Pr\left(\overline{P_{D\to C}} \mid {}^{Activity}H_p\right)+\Pr\left(\overline{T_{D\to C}} \mid {}^{Activity}H_p\right)\right]\times$$

$$LR = \frac{\left[\Pr\left(T_{C\to D} \mid {}^{Activity}H_p\right)\Pr\left(\overline{P_{C\to D}} \mid {}^{Activity}H_p\right)+\Pr\left(\overline{T_{C\to D}} \mid {}^{Activity}H_p\right)\right]}{\Pr\left(T_{AO\to C} \mid {}^{Activity}H_d\right)\Pr\left(\overline{P_{AO\to C}} \mid {}^{Activity}H_d\right)+\Pr\left(\overline{T_{AO\to C}} \mid {}^{Activity}H_d\right)}$$

which is the same point as was reached in the previous method of derivation (albeit much more quickly) and could then be simplified (with the same assumptions) to:

$$LR = \Pr\left(T_{C\to D} \mid {}^{Activity}H_p\right)\Pr\left(\overline{P_{C\to D}} \mid {}^{Activity}H_p\right)+\Pr\left(\overline{T_{C\to D}} \mid {}^{Activity}H_p\right)$$

It is worth pointing out again here that the issue of recovery has not been considered in the evaluation. This is often the case, as the recovery component can be covered by the data used to assign probabilities (i.e. the results of the study being relied on will have an in-built recovery element).

5.3.5.1 Using the Pathways to Obtain the *LR* Formula

The pathways that were drawn in step 3 can be used to generate the same formula. Take the pathways drawn in Figure 5.1 and the following rules:

1) Start at the terms just prior to the observations (the right-hand side of the pathways) and work back to the left
2) For each box assign a probabilistic term
3) For each term in the series multiply the probabilistic terms together
4) For each set of terms that run in parallel add them together. This occurs at their junction in the pathway
5) For each proposition and each exhibit, you end up with a probabilistic equation
6) Multiply the probabilistic equations for each exhibit within a proposition
7) Divide the result of point 6 for H_p by the result of point for H_d

Diagrammatically this is shown for case 1 in Figure 5.2. Note the short-hand terms for the probabilistic elements have been used, i.e. $\Pr\left(T_{D\to C} \mid {}^{Activity}H_p\right) \equiv T_{D\to C}$

And the same process for the two exhibits under H_d using the pathways gives:

$$\left[B_C\times T_{C\to C}\times T_{F\to C}\left(\overline{T_{AO\to C}}+T_{AO\to C}\times\overline{P_{AO\to C}}\right)\right]\left[B_D\times T_{D\to D}\right]$$

so that the *LR* can be calculated by:

$$LR = \frac{\left[B_C\times T_{C\to C}\times T_{F\to C}\left(\overline{T_{D\to C}}+T_{D\to C}\times\overline{P_{D\to C}}\right)\right]\left[B_D\times T_{D\to D}\left(\overline{T_{C\to D}}+T_{C\to D}\times\overline{P_{C\to D}}\right)\right]}{\left[B_C\times T_{C\to C}\times T_{F\to C}\left(\overline{T_{AO\to C}}+T_{AO\to C}\times\overline{P_{AO\to C}}\right)\right]\left[B_D\times T_{D\to D}\right]}$$

And by cancelling common terms we get:

$$LR = \frac{\left(\overline{T_{D\to C}}+T_{D\to C}\times\overline{P_{D\to C}}\right)\left(\overline{T_{C\to D}}+T_{C\to D}\times\overline{P_{C\to D}}\right)}{\left(\overline{T_{AO\to C}}+T_{AO\to C}\times\overline{P_{AO\to C}}\right)}$$

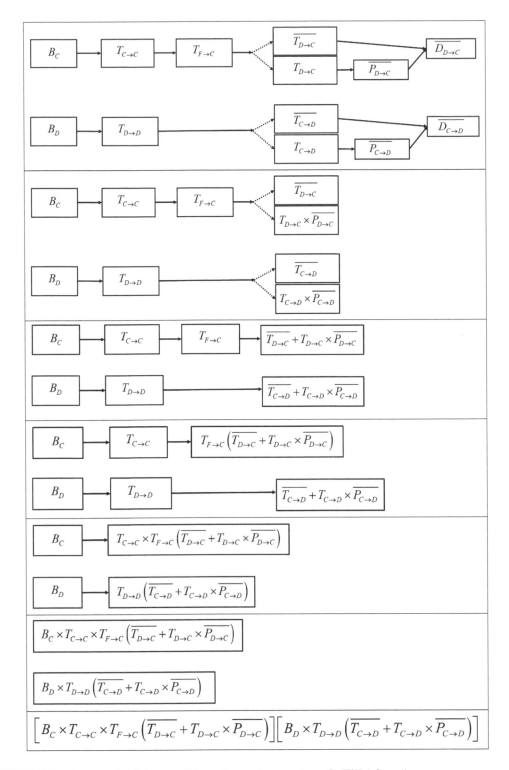

FIGURE 5.2   An example of the use of the pathways to generate an Pr(E|Hp) formula.

With the same assumptions as previously described regarding the probability of transfer and persistence between AO and D, simplification of the *LR* equation gives:

$$LR = \overline{T_{C \to D}} + T_{C \to D} \times \overline{P_{C \to D}}$$

which again is the same formula obtained as in two other times. Note that in the example just provided all the terms were included, even those that were present in both the H_p and H_d pathways (for example the transfer of DNA from F to C). Previously it was stated that these terms can be ignored, and by including them in the example it can be seen why this is the case. These terms (being present in both H_p and H_d pathways) will simply cancel each other out in the *LR* formula. Whichever method was used to obtain the *LR* formulae, it leads to step 6, assigning the probabilities.

5.3.6   CASE 1: *LR* DERIVATION STEP 6: ASSIGNING PROBABILITIES

Having derived the *LR* formula, there will be one or more probabilistic terms that require assignment in order to carry out the evaluation. The manner in which the observations are discretised (broken up into discrete brackets) for the evaluation, or indeed if the data is being treated as continuous, will dictate the way in which raw data from studies is used. It is also likely to affect the way in which the derivation is carried out in the first place. If bespoke in-house experiments are being performed, then it may dictate aspects of how the experiments are carried out. If data is being used from literature, then it may dictate which studies are applicable (i.e. can provide the information needed). In this way the type of data obtained can be thought of as adhering to the Bauhaus tenet of 'form follows function', i.e. the form of the data being used will be determined by its function in the evaluation. The spectrum of data discretisation is shown in Figure 5.3, with the far left being the least discretised (i.e. continuous, which is not discretised) and the right side being the most discretised (presence or absence).

5.3.6.1   Incorporating Prior Belief with Data to Assign Probabilities

A discussion on various factors that need to be considered when assigning probabilities is given in Section 5.6. From earlier the *LR* formula derived was:

$$LR = \Pr\left(T_{C \to D} \mid {}^{Activity}H_p\right)\Pr\left(\overline{P_{C \to D}} \mid {}^{Activity}H_p\right) + \Pr\left(\overline{T_{C \to D}} \mid {}^{Activity}H_p\right)$$

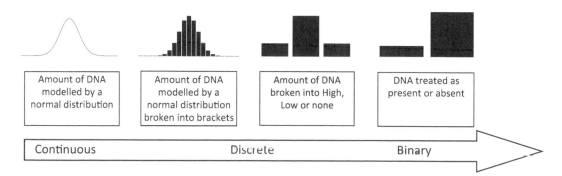

FIGURE 5.3   Ways in which DNA profile data can be discretised, going from left being the least discretised (continuous) and the right being the most discretised.

Meaning that three probabilities must be assigned in order to carry out the evaluation:

- $\Pr\left(T_{C \to D} \mid {}^{Activity}H_p\right)$ – the probability that DNA was transferred from C's hand to D's top when she hit him
- $\Pr\left(\overline{P_{C \to D}} \mid {}^{Activity}H_p\right)$ – the probability that C's DNA would not persist on D's top for 2 days (the time between the alleged offence and when the top from D was tested by the forensic laboratory
- $\Pr\left(\overline{T_{C \to D}} \mid {}^{Activity}H_p\right)$ – the probability that DNA was not transferred from C's hand to D's top when she hit him

As the probability of transfer not occurring and the probability of transfer occurring are mutually exclusive and exhaustive, it follows that $\Pr\left(\overline{T_{C \to D}} \mid {}^{Activity}Hp\right) = 1 - \Pr\left(T_{C \to D} \mid {}^{Activity}Hp\right)$ and so there is no need to assign three different probabilities, but instead just two; one for the probability of transfer and one for the probability of DNA not persisting.

First consider the assignment of a probability of DNA transfer from a hand to a top from hitting. There is little data available that directly addresses this issue. Some options that are available at the time of writing this book are Fonneløp et al. [2] or Breathnatch [3], but for this particular example Daly et al. [4] will be used. In Daly et al. cloth samples were held by volunteers for 60 s before sampling using tapelifts (the same sampling technique as used in the samples taken for case 1). Table 1 of the Daly study shows that 53 out of 100 samples produced less than 0.01 ng/L of DNA and for the example this value will be used as an approximate cut-off for obtaining DNA profile information for Profiler Plus™ (used to type the traces case 1).

Another study is that of Bowyer [5], who in their PhD thesis asked volunteers to punch and slap an acetate sheet. They found that one-hour post handwashing punching an acetate sheet yielded detectable DNA 9 out of 15 swabs of the sheet (Table 4-6 of their thesis).

Both studies diverge in aspects from the case circumstances. The Daly study replicates the materials (i.e. hand contacting cloth) but not the activity (i.e. holding in the Daly study as opposed to hitting in the case). The Bowyer study replicates the activity (i.e. hitting) but not the materials (i.e. hitting an acetate sheet rather than cloth). Luckily in this instance both studies show a broadly similar result, i.e.:

- In Daly et al., DNA transferred $47/100 = 0.47$ of the time
- In Bowyer, DNA transferred $9/15 = 0.6$ of the time

This gives some confidence in an assignment of a probability around the 0.5–0.6 range and a value could be assigned that aligned with either study or an average from the studies or a weighted average (weighted by the number of samples in the studies). If the results from the two studies had been quite different from each other then further investigation would be required. For example, the study by Goray et al. [6] looked at the amount of DNA transfer under controlled conditions and investigated differences in material type (plastic or cotton) and contact type (passive, pressure or friction). Note that the Goray study is not being used directly to assign a probability but is being used in the general body of knowledge in the assignment (specifically the choice of which study to use).

For the probability $\Pr\left(T_{C \to D} \mid {}^{Activity}H_p\right)$ a probability is assigned using the data from the Daly study, as the analytical techniques most closely align with that of the case. The data in Daly et al. revealed that 47 out of 100 experiments showed a DNA transfer (defined for this evaluation as more than 0.01 ng of DNA being detected). One option would be to use these values to calculate a proportion 47/100 = 0.47 and use this directly as the probability assignment. There are instances where doing this is reasonable, but issues arise in two situations:

1) When the sample size of the study is small, then it is more subject to sample variation and so unrealistically high or low probabilities may be assigned, due to stochastic effects of sampling variation
2) If the experiment shows the same result for all samples, i.e. if in the Daly study all 100 experiments showed that DNA had been transferred, then an issue would arise when assigning a probability of non-transfer. Using the unadjusted sample numbers this probability would be assigned a value of 0, which can cause issues in many evaluations (i.e. those that would require a transfer event under one of the propositions) whereby one of the propositions becomes certain to have occurred

The extreme manifestation is an instance where data is available from a single experiment that has been carried out to determine whether DNA transfer has occurred or not. With a single observation the probabilities, if calculated using the unadjusted count, can only be assigned 0 and 1. But this is clearly too strong an assignment to make based on a single observation. However, there is one data-point, and whilst not much, it does provide some information, which should be utilised to update prior knowledge. In order to overcome these problems, the data learned from the literature must be considered within a Bayesian framework. Specifically, prior belief in the experimental outcome should be updated with the observations from the study in order to assign a posterior probability to the evaluation.

There are various ways in which prior probabilities can be assigned, but the important point is that they are values assigned before the knowledge of experimental data is obtained. They can be informed (i.e. there is some prior belief spread unequally between possible outcomes) or uninformative (i.e. all outcomes are considered a priori equally probable). Imagine a situation where an experiment is conducted that seeks to determine the probability of DNA being transferred by contact. There are two possible outcomes: DNA is transferred or DNA is not transferred, and initially there is no information about which is most probable. One option in this instance is to represent that lack of knowledge with an uninformative prior, i.e. initially assign an equal prior probability to either occurrence. If there were many possible categorical outcomes being considered (e.g. if the experiment sought to capture DNA amount data in categories of 'none', 'low', 'medium' and 'high' amounts) then the idea of a non-informative prior can be extended to consider that, without any information, any one of these outcomes is considered equally probable.

The mechanical manner in which experimental counts can be updated is to consider the prior as providing one count to each category and then add the experimentally observed counts on to these priors.

Consider the extreme example posed earlier where a single experiment was carried out to determine whether DNA transfer would occur. Say that in this experiment a DNA transfer was observed, and the desire is to use this data in the evaluation. Assigning from raw counts would see the probabilities for being assigned as 1 for transfer and 0 for no transfer. However, if uniform prior counts are applied then the posterior counts and probabilities are calculated as shown in Table 5.2. For the 'DNA transfer' option a prior count of 1 is added to the experimental observation of 1 to obtain a posterior count of 1 + 1 = 2. For the 'No DNA transfer' scenario a posterior count of 1 is obtained by the calculation 1 + 0 = 1. The posterior count can then be used to obtain the posterior probability; i.e. for DNA transfer the posterior count is 2, the total posterior count (the sum of all the posterior counts) is 3 and therefore the posterior probability is 2/3 ~ 0.66.

Note now in this example the posterior probabilities have been shifted significantly from their position when based on counts alone. The probability of DNA transfer has shifted from 1 to 0.66 and the probability of no transfer has shifted from 0 to 0.33. These posterior

TABLE 5.2

Example of Calculating Posterior Probabilities from a Single Observational Count

Experimental Outcome Category	Prior Count	Experimental Observation Count	Posterior Count	Posterior Probability
DNA transfer	1	1	2	2/3 ~ 0.66
No DNA transfer	1	0	1	1/3 ~ 0.34
TOTAL	2	1	3	1

probabilities still provide some support for a transfer over no transfer (as would be expected from the one transfer observed experimentally) but the level of support is relatively weak (as expected from the fact that the experimental sample size is so small). This is the expected and desired behaviour of prior probabilities; that is, they have a large effect when the sample size is small and a small effect when the sample size is large. Intuitively this can be thought of as the fact that a very small experiment doesn't provide much evidence to shift the analyst from their prior beliefs, whereas a large experiment provides ample evidence that is more likely to be convincing.

Returning to case 1 and the task of using the data from Daly *et al.* to assign a value for $\Pr\left(T_{C \to D} \mid {}^{Activity}H_p\right)$. Consider that data comes from an experiment that had two possible outcomes, transfer occurred (defined as an observation of DNA greater than 0.01 ng/sL) or transfer did not occur (defined as an observation of DNA less than 0.01 ng/sL). The observation of transfer and no transfer is added to the priors in order to obtain a posterior count. For transfer the posterior count is $1 + 47 = 48$ and for no transfer $1 + 53 = 54$. The posterior probabilities are then calculated form these adjusted counts:

$$\Pr\left(T_{C \to D} \mid {}^{Activity}H_p\right) = 48 / (48 + 54) \approx 0.47$$

$$\Pr\left(\overline{T_{C \to D}} \mid {}^{Activity}H_p\right) = 1 - \Pr\left(T_{C \to D} \mid {}^{Activity}H_p\right) = 1 - 0.47 = 0.53$$

Table 5.3 shows the construction of the posterior counts and then probabilities.

Note that in the example of the data from Daly *et al.* [4] there is virtually no change in the posterior probability to the probability calculated from the unadjusted counts. In fact, the reporting of results to two significant figures hides the change completely. The actual changes are from probabilities of 0.47 and 0.53 using the raw counts to 0.4706 and 0.5294 if displayed out to four significant figures. A description of the mathematics behind the described addition of a prior count is given in Box 5.2.

TABLE 5.3

Example of Calculating Posterior Probabilities from Observational Counts

Experimental Outcome Category	Prior Count	Experimental Observation Count	Posterior Count	Posterior Probability
DNA transfer	1	47	48	48/102 ~ 0.47
No DNA transfer	1	53	54	54/102 ~ 0.53
TOTAL	2	100	102	1

BOX 5.2   MODELLING COUNTS WITH A DIRICHLET DISTRIBUTION

In the context of the event being modelled, the categories are considered to be mutually exclusive and exhaustive. This means that the summed probabilities of all categories must be one. To describe multiple probabilities in statistical terms a distribution having these properties must be chosen, i.e. one in which multiple categorical outcomes have probabilities that sum to one. The technical method that has been described is to model the data with a Dirichlet distribution. A Dirichlet distribution provides a set of probabilities, $\alpha_1,...,\alpha_n$, that sum to one, given a set of positive real counts, $x_1,...,x_n$, for the N possible categories. The general form of a Dirichlet distribution is:

$$f\left(\alpha_1,...,\alpha_N\right) = \frac{\Gamma\left(\sum_{i=1}^{N}\alpha_i\right)}{\prod_{i=1}^{N}\Gamma\left(\alpha_i\right)} \prod_{i=1}^{N} x_i^{\alpha_i-1}$$

when $\alpha = \alpha_1,...,\alpha_n$ the above distribution is written as $D\left(\alpha\right)$.

When an uninformative prior distribution with **K** elements is utilised, the prior $(1,...,1)$ count or $D(1)$ distribution is updated with experimental counts to produce a posterior $\left(x_1+1, x_2+1,...,x_N+1\right)$ count or $D\left(\mathbf{a}+\mathbf{1}\right)$ distribution. The mean of this posterior Dirichlet distribution for any particular category, 'i', can be calculated by:

$$mean\left(\alpha_1\right) = \frac{\alpha_i}{\sum_{i=1}^{N}\alpha_i} = \frac{\left(x_i+1\right)}{\left(x_1+1\right)+...+\left(x_N+1\right)} = \frac{x_i+1}{T+N}$$

where T is the total count of all experimental observations and N is the number of categories into which the observations are broken. In the case of the data from Daly et al., in order to calculate the posterior probability of DNA transfer occurring, take, $x_i = 47$, $N = 2$ (i.e. transfer or no transfer) and $T = 100$, so that:

$$\Pr\left(T_{C\to D} \mid {}^{Activity}Hp\right) = \frac{x_i+1}{T+N} = \frac{47+1}{100+2} = \frac{48}{102} \approx 0.47$$

which is as previously calculated.

5.3.6.2   The Difference between Proportions and Rates

Now turn to the second probability that requires assignment, which is the probability that C's DNA would not persist on the top of D over 2 days. As with DNA transfer there are numerous factors likely to affect persistence, such as:

- The surface type of the object the DNA is present on
- The length of time between deposition and sampling
- The conditions the item is exposed to during the time in question

One example of a DNA persistence study for contact DNA is the work by Raymond et al. [7]. In this study biological material was deposited on items and these items were either kept outdoors or

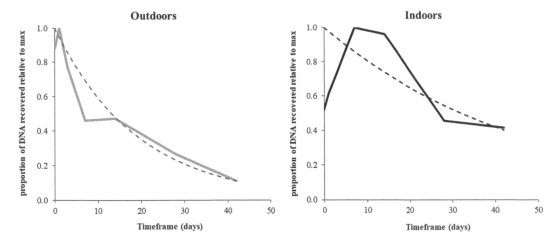

FIGURE 5.4 Graphs showing the data from Raymond *et al.* grouped into two categories, outdoors (grey, left) and indoors (black, right) with the trends modelled with an exponential curve (dashed lines).

indoors over the course of approximately 40 days. The amount of DNA recovered was examined at daily intervals. While there is a high degree of variability in the amount of DNA that was detected, an examination of the observations in their study shows that very low (or even no detectable) levels of degradation occur within the first few days (the timeframe that is relevant in case 1). In fact, during the first few days the amount of DNA detected was larger than the initially deposited amount. This of course is not possible and must be due to sampling or recovery effects within the experiment. Figure 5.4 shows the data from the Raymond *et al.* study are summarised, graphed and with a downwards (exponential) curve fitted to the two datasets. It is a modified version of Figure 3 from Taylor *et al.* [8].

The exponential curves shown in Figure 5.4 are examples of fitting a model to the data, which becomes important, particularly when dealing with DNA amounts that are transferred or persist. For example, it is known that DNA cannot increase on an item, and so the raw results of the experimental data could not sensibly be used directly (i.e. otherwise the model would show an increase in DNA on items held indoors for the first 8 days!). Modelling also allows data to be extrapolated out to areas which the study did not cover or interpolated to positions within the range of values covered by the study, but not specifically trialled. As such, modelling becomes extremely useful for applying data to case examples when some differences must be taken into account between the circumstances of the case and the study. There are a number of considerations that must be taken into account when modelling data, and a more detailed discussion is left in Chapter 8.

One further important point is dealing in detail with the manner in which data is presented (and how it can then be used in evaluations). The data that informed the modelling in Figure 5.4 was based on DNA amounts. By looking at the exponential distribution that has been fit to the indoors scenario (and assuming that the indoors scenario is the most closely aligned with the case circumstances in case 1) then it can be seen that after 2 days the proportion of DNA recovered is approximately 0.95. This means that the DNA amount will have decreased by 0.05, which does not mean that the probability that DNA and a detection threshold will persist on an item is 0.95 (i.e. it cannot be assumed that $\Pr\left(\overline{P_{C\rightarrow D}} \mid {}^{Activity}Hp\right) = 0.05$). The difference here is between an amount and a rate and the two cannot be interchanged. The rate depends on additional information, namely the amount of starting, which must be taken into account before the model can be used to inform the rate model. For example, imagine that the timeframe of interest in case 1 had been 35 days so that it is expected that the amount of DNA could have dropped by approximately half. If it was found that contacts from hitting tended to transfer a lot of DNA, say for example 10 ng, then after 35 days it would be expected to see DNA on 100% of tops, but all around 5 ng, i.e. the rate of persistence would be 100% but the amount of DNA persisting is 50%. If, on the other hand, the amount of DNA transferred from hitting was only just above the detection threshold then it would be expected for a persistence rate of close to 0% and, although it wouldn't be detected, it would still be expected that the amount of DNA transfer would be around 50%. So, it can be seen that depending on the starting amount of DNA the rate of persistence, which is the value required for the evaluation, could range from 0 to 1.

This demonstrates the reason that studies on amounts of transfer or persistence cannot be directly applied to rates of transfer or persistence. If there was a need (or desire) to use the results of the Raymond *et al.* [7] study, it would need to be paired with the Daly *et al.* [4] study by considering the number of experiments in the Daly study that would have fallen below the threshold of 0.01 ng/uL had they reduced by 5%. This could then be used as a probability of persistence in the form required for evaluation. As it stands, given the fact that it appears from the persistence study that the probability of persistence is high, the assumption can be made that $\Pr\left(\overline{P_{C \to D}} \mid {}^{Activity}Hp\right) \approx 0$ so that the *LR* formula simplifies to:

$$LR = \Pr\left(\overline{T_{C \to D}} \mid {}^{Activity}Hp\right)$$

with a value for the only probability assignment required already obtained. Note that while this is a further simplification of the *LR* formula (and hence an assumption), it is technically a data-driven probability assignment, based on literature about DNA persistence.

The reader may be disappointed that even after so many pages of derivation and explanation that the *LR* has simplified to a single probabilistic term. However, carrying out the evaluation in this detail has covered many important topics. It also shows that even the seemingly simplest of evaluations has an enormous amount of background planning, thought and assumption that goes into the evaluation. Given the level of thought that has gone into this evaluation, the analyst would be well versed to answer questions about the effect of various components of the evaluation or to justify the assumptions made. This leads to the last step of the *LR* derivation, which is to calculate the *LR* using the derivation and the probability assignments.

5.3.7   CASE 1: *LR* DERIVATION STEP 7: CALCULATING THE *LR*

The *LR* calculation case 1 is relatively straight-forward in this step, having a simple formula and a single probability. From the previous section the *LR* formula was:

$$LR = \Pr\left(\overline{T_{C \to D}} \mid {}^{Activity}Hp\right)$$

And data from the Daly *et al.* [4] study was used, with an uninformative prior, to calculate:

$$\Pr\left(\overline{T_{C \to D}} \mid {}^{Activity}Hp\right) = 0.53$$

Putting these two components together the result of the evaluation is:

$$LR = 0.53$$

As this is less than one, the observations support the defence proposition over the prosecution proposition. In these instances, it is usual to invert the *LR* and report the result in support of the defence proposition (which in this case is 1/0.53 ~ 2). This could then be reported as:

There is an absence of D's DNA obtained from the tapelift of the top of C. There is an absence of C's DNA obtained from the tapelift of the top of D. In the evaluation of the evidence I have considered two possible propositions for these results;

- The prosecution proposition I have considered is that C and D struggled which included D grabbing C's arm and C hitting D's chest
- The defence proposition I have considered is that C had no prior contact, direct or indirect, with D. C struggled with an unknown male alternate offender (AO), who grabbed her arm and whom she hit

When I consider the probability that DNA would be transferred in such an encounter, the DNA observations are in the order of 2 times more probable to have been obtained if the defence proposition had occurred rather than the prosecution proposition. This provides slight (or weak or limited) support to the proposition that C had no prior contact, direct or indirect, with D and that C struggled with an unknown male, compared to the proposition that C and D struggled.

Much time has been spent going through the steps in deriving an *LR* for case 1. It is a good case to learn from given the relative simplicity of the evaluation itself (in terms of the formula, rather than the level of thought that must go into it). The proceeding sections detail the *LR* derivation for two further cases (cases 2 and 3). These will be worked through in much-abbreviated form, only spending time to explain in detail the new aspects that arise during their evaluation.

5.4 CASE 2: THE THREE BURGLARS

This section deals with the steps needed to derive an *LR* formula for case 2, the case of the three burglars.

5.4.1 CASE 2: STEP 1: PROPOSITION, BACKGROUND INFORMATION AND ASSUMPTIONS

Recall from Chapter 4 that the scenario described provided the following:
Background information:

- D lent his gloves to people who commit burglaries

Propositions:

- H_1: D held the knife and grabbed the hair of the victim
- H_2: AO held the knife and grabbed the hair of the victim, after wearing the gloves of D

Assumptions:

- The knife submitted for examination was used by the perpetrator
- The perpetrator did not wear gloves while handling the knife or while interacting with the victim
- Under H_2 the AO took off the suspect's gloves immediately prior to handling the knife and grabbing the hair of the victim
- Samples were taken, secured, packaged and transported according to standard operating procedures
- None of C, D, H or AO have conditions that would make them unusually high DNA shedder (i.e. they are representative of the typical spread of shedding ability within the population)
- The sample from the knife contains DNA of C, H and D, no DNA of other individuals
- The sample from the hair contains DNA of C and D, no DNA of other individuals

5.4.2 CASE 2: STEP 2: IDENTIFY PATHWAYS OF TRANSFER

Under the prosecution proposition of:
D held the knife and grabbed the hair of the victim
The observation of:
C, H and D's DNA on the knife handle could occur by:

- C and/or H using the knife in everyday activity and transferring DNA, no background DNA being present on the knife handle, D transferring DNA during the offence, and C, H and D DNA persisting throughout the incident and subsequent time period until sampling

The observation of:
   C and D's DNA on the hair could occur by:

- C's DNA being present on her own hair, no background DNA being present on the hair, D transferring DNA during the offence, and C and D DNA persisting during the time period between incident and sampling

Under the defence proposition of:
AO held the knife and grabbed the hair of the victim, after wearing gloves of D
The observation of:
C, H and D's DNA on the knife handle could occur by:

- C and H using the knife in everyday activity and transferring DNA (and that DNA persisting through the crime), no background DNA being present on the knife handle, D's DNA being present on his own gloves, D's DNA transferring to AO's hands from AO wearing D's gloves, AO not transferring any of his own DNA to the knife handle while handling it during the offence, AO transferring D's DNA from his hands to the knife handle during the offence, and C and D DNA persisting during the time period between incident and sampling

The observation of:
   C and D's DNA on the hair could occur by:
- C's DNA being present on her own hair (and that DNA persisting through the crime), no background being DNA present on the hair, D's DNA being present on his own gloves, D's DNA transferring to AO's hands from AO wearing D's gloves, AO not transferring any of his own DNA to the hair while handling it during the offence, AO transferring D's DNA from his hands to the hair during the offence, and C and D's DNA persisting during the time period between incident and sampling

5.4.3 CASE 2: STEP 3: IDENTIFYING TEMPORAL ORDER

The timeline of events shown in a graphical format is seen in Figure 5.5. Note that in this example we do not include a consideration of persistence. This is done to keep the example as simple as possible, and could be included (and would be particularly important if some time had elapsed between the offence and sampling of the exhibits).

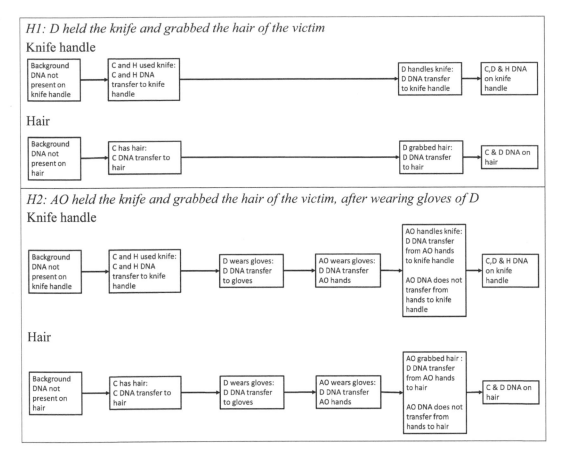

FIGURE 5.5 Timeline traces of DNA transfer in case 2. The top panel represents the timeline of the items given the prosecution proposition and the bottom panel represents the timeline of the items given the defence proposition.

5.4.4 CASE 2: STEP 4: START THE BASIC *LR* FORMULA

The derivation of the *LR* formula is started by assigning terms to each of the events in the timeline shown in Figure 5.5. Two rules are applied when working directly from the timeline:

a) If activities and transfer events are listed 'on an item in both propositions', then it will cancel out in the *LR* and so can be omitted from the derivation (for example the transfer of DNA of C and H to the knife handle from regular use)

b) If an activity or transfer is listed 'within a proposition on multiple items', then it is only assigned a probability once (for example the wearing of D's gloves by the AO and the associated DNA transfer of D's DNA to AO's hands)

Using these guidelines the components of the *LR* formula can be derived. A short-hand notation is used in this example (compared to case 1), which is in a form commonly seen in literature, whereby the probability of transfer is represented by the letter '*t*'.

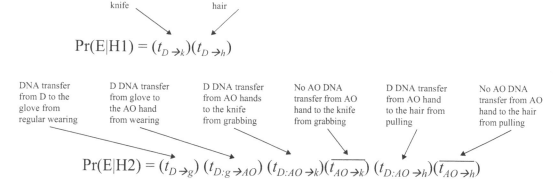

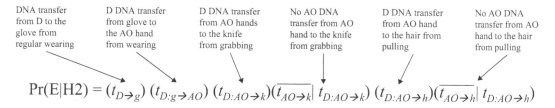

Note that by constructing the components of the *LR* formula in this manner there have been many assumptions made about the independence of transfer events that are not as explicitly obvious as when the full derivation (as in case 1) is carried out. Perhaps in this example, the most impactful is the assumption of independence under H_2 between the transfer of D's DNA from AO's hands to the knife or hair and the transfer of AO's DNA from AO's hands to the knife or hair. The likely dependence between these transfer events can be seen in Figure 5.5 by the fact that they are transfers arising from a single event, and hence both are drawn in the same activity box. The dependence could be included in the derivation, by modification of the above formula for $\Pr(E|H_2)$ to:

5.4.5   Case 2: Step 5: Complete the *LR* Derivation

In this case, as is often the case for relatively simple derivations, the *LR* formula is completed by taking the ratio of the two *LR* components given in step 4.

$$LR = \frac{\left(t_{D \to k}\right)\left(t_{D \to h}\right)}{\left(t_{D \to g}\right)\left(t_{D:g \to AO}\right)\left(t_{D:AO \to k}\right)\left(\overline{t_{AO \to k}} \mid t_{D:AO \to k}\right)\left(t_{D:AO \to h}\right)\left(\overline{t_{AO \to h}} \mid t_{D:AO \to h}\right)}$$

Further simplification can occur only by making assumptions about the transfer probabilities. For this example, it is assumed that:

- the probability of DNA being transferred from someone's hand to a knife is the same regardless of whether it is their own DNA or another person's DNA being transferred. This has the effect of assuming $t_{D:AOk} = t_{Dk}$
- the probability of DNA being transferred from someone's hand to hair is the same regardless of whether it is their own DNA or another person's DNA being transferred. This has the effect of assuming $t_{D:AOh} = t_{Dh}$

This simplifies the *LR* to:

$$LR = \frac{1}{\left(t_{D \to g}\right)\left(t_{D:g \to AO}\right)\left(\overline{t_{AO \to k}} \mid t_{D:AO \to k}\right)\left(\overline{t_{AO \to h}} \mid t_{D:AO \to h}\right)}$$

5.4.6   CASE 2: STEP 6: ASSIGN PROBABILITIES

After *LR* derivation it becomes apparent that there is a need to assign probabilities for:

- t_{Dg}: the probability of transferring DNA to gloves from regular wearing
- $t_{D:gAO}$: the probability of transfer of DNA from gloves to hands from once-off wearing
- $\overline{t_{AO \to k}} \mid t_{D:AO \to k}$: the probability of not transferring self DNA from hands to a knife from a once-off hold when non-self DNA on the hands has been transferred from that same hold
- $\overline{t_{AO \to h}} \mid t_{D:AO \to h}$: the probability of not transferring self DNA from hands to hair from a once-off hold when non-self DNA on the hands has been transferred from that same hold

There are numerous sources of data that could be used to inform the probability assignments in this case. It is also likely that as this book is being read, further research that was not available at the time of writing might have been conducted and published. Due to this fact one example is provided of assigning probability from data, for t_{Dg}.

Also, while it has not been explicitly stated, as it was not yet required, the decision is now made to consider just the presence or absence of DNA (rather than DNA amounts broken up into finer-scale brackets). Consider the work of van den Berge *et al.* [9] who examined the presence of DNA on various public and private objects. Examination of winter gloves found that of 20 samples taken, all showed the presence of DNA of the owner. Using the general formula for calculating a mean posterior probability using a Dirichlet (**1**) prior:

$$t_{Dg} = (20+1)/(20+2) \sim 0.95$$

Without giving literature justification for the remaining probability assignments, they are set as:

- $t_{D:gAO}$: 0.75
- $\overline{t_{AO \to k}} \mid t_{D:AO \to k}$: 0.1
- $\overline{t_{AO \to h}} \mid t_{D:AO \to h}$: 0.1

5.4.7   CASE 2: STEP 7: CALCULATE THE *LR*

The *LR* value calculated using the probability assignments from step 6 and the *LR* formula from step 5 yields:

$$LR = \frac{1}{(0.95)(0.75)(0.1)(0.1)} \approx 140$$

In other words, the probability of obtaining the DNA results in this case is 140 times higher if D held the knife and grabbed the hair of the victim, rather than if AO held the knife and grabbed the hair of the victim, after wearing gloves of D.

5.4.8 CASE 2: SUPPLEMENTAL

Before moving to the final case, an evaluation of the two evidence items individually is shown. Consider first just the knife handle and the *LR* formula that would be derived from the terms that are only related to this item:

$$LR = \frac{1}{\left(t_{D \to g}\right)\left(t_{D:g \to AO}\right)\left(\overline{t_{AO \to k}} \mid t_{D:AO \to k}\right)}$$

Substituting the assigned probability values from step 6 yields:

$$LR = \frac{1}{\left(0.95\right)\left(0.75\right)\left(0.1\right)} \approx 14$$

Similarly, considering only the results from the hair sample:

$$LR = \frac{1}{\left(t_{D \to g}\right)\left(t_{D:g \to AO}\right)\left(\overline{t_{AO \to h}} \mid t_{D:AO \to h}\right)} = \frac{1}{\left(0.95\right)\left(0.75\right)\left(0.1\right)} \approx 14$$

Note that if the *LR*s were multiplied for the two evidence items when they were considered individually, the result is $14 \times 14 = 196$, i.e. higher than 140, the *LR* resulting from the evaluation of both items jointly.

This demonstrates that when multiple items are to be evaluated they cannot always be considered independent (and this assumption must be justified) and the total *LR* will not necessarily be a multiplication of the individual *LR*s for each item. In this case the effect comes from the fact that in both individual evaluations the transfer of DNA from D to D's gloves and then to AO's hands must be taken into account. But by multiplying both *LR* together it is being formulaically stated that this chain of transfers has occurred twice, when in reality (and as correctly specified in the joint evaluation) it only occurs once. Recall from Chapter 4 that when dependencies exist between the different activities that it is often better to consider them together as a 'package deal'.

There are other dependencies in the scenario that are not taken into account within the evaluation, doing so would take the complexity of the formula deviation beyond practical levels. For example, there may exist some dependency in the order in which the events took place, i.e. if the knife handle was touched first and some DNA transferred, then less DNA would be available to transfer from the hands of the offender to the C's hair. Additionally, there could be back transfer (i.e. DNA transferring from the knife handle to the hand of the offender) which would then be available to transfer to the second touched object. Not including these other factors is equivalent to making assumptions that they do not occur. In practice, these types of assumptions are commonly made, partly due to the complexity they bring to the evaluation when they are not made, partly due to the availability of data that could inform such transfer probabilities and partly due to the belief that they would not make a significant difference to the overall evaluation.

5.5 CASE 3: THE FAMILY ASSAULT

The next example introduces the use of DNA amounts (in the form of high/low/none) and incorporates the results of body fluid testing.

Recall that the circumstances of this case are that C has stayed for a week at her older brother's (D) house and accuses her brother of biting her on the vagina, over her underwear. On the underwear is faecal staining, and an Rapid Stain Identification (RSID) test for saliva gives a positive reaction. A tapelift of the underwear produced DNA from C, and then Y-chromosome Short Tandem repeat (Y-STR) testing produces a profile matching the brother.

5.5.1   Case 3: Step 1: Proposition, Background Information and Assumptions

Recall from Chapter 4:
Background information

- C and D have been staying together in D's house for a week
- C normally lives with her biological father (F)
- C and D are full siblings
- C's underwear has faecal stains

Propositions

- H_1: D bit C on the vagina over the underwear
- H_2: C was not bitten

Assumptions

- No mutations have occurred between D and D's father that would affect the Y-STR profile
- C's underwear has not been washed since the alleged offence (given the timeframe and activities stated by C and the witness, this is highly probable)
- C's underwear was not bought less than a week prior to the alleged offence (and so has been worn/washed at her normal residence)

5.5.2   Case 3: Step 2: Identify Pathways of Transfer

Under the prosecution proposition of:
D bit C on the vagina over the underwear
The observation of:
C's DNA on C's underwear could be explained by:

- C wearing her own underwear (note that this result will not continue to be considered in the evaluation as it does not affect the evaluation since the activity is accepted as occurring under both H_1 and H_2)

Y-STR profile matching the reference of D on the outer crotch of C's underwear could be explained by:

- D's DNA transferring to C's underwear due to cohabitation
- D's father's (F) DNA transferring to C's underwear due to cohabitation
- D transferring DNA when he bit the underwear

RSID positive reaction to the outer crotch of C's underwear
Could be explained by:

- background levels of saliva present on C's underwear
- a positive reaction of RSID test to faeces on C's underwear
- cross-reactivity with other body fluids or non-human sources of amylase (however it is assumed this probability to be much lower than the probability of a positive reaction to faeces and background levels of saliva on underwear and thus this is omitted)
- D transferring saliva when he bit the underwear

Under the defence proposition of:
  C was not bitten
The observation of:
  Y-STR profile matching the reference of D on the outer crotch of C's underwear
  could be explained by:

- D's DNA transferring to C's underwear due to cohabitation
- D's father's DNA transferring to C's underwear due to cohabitation

RSID positive reaction to the outer crotch of C's underwear
  could be explained by:

- background levels of saliva present on C's underwear
- a positive reaction of RSID test to faeces on C's underwear

5.5.3   CASE 3: STEP 3: IDENTIFYING TEMPORAL ORDER

In this case there are multiple events that can lead to a DNA result. In fact any one or a combination of these events can lead to a result. For example, under H_1 a positive reaction of the crotch of C's underwear to the RSID saliva test could be caused by the presence of background saliva on C's underwear or from a positive reaction to faeces or because D's saliva was transferred during biting. Anyone one of these events (or any combination of them) will lead to a positive RSID test (according to the assumptions and the model). Unlike for separate items, it is not always possible to separate test results from the same item, depending on the assumptions in the model. For example, under H_1 the pathways for the Y-STR result and the RSID results could be considered in separate pathways as shown in Figure 5.6. Remember also that when the probabilities of the downstream event are exhaustive (i.e. the RSID test either did or did not give a positive reaction to saliva from biting, once it has already given a positive reaction to faeces) then summing across pathways will result in only the first event that leads to a transfer having an effect on the final *LR* formula. In this case, only those events leading up to a transfer need to be shown in the pathway.

The issue with the pathways shown in Figure 5.6, whereby the RSID and Y-STR tests have been separated, is that it may be desired to make the assumption that biting will result in a transfer of DNA and saliva. If the *LR* is formulated in separate parts using the pathways in Figure 5.6 there is no intuitive manner to do so. Instead for the single sample with dual results, the final result must be considered as the combination of results (i.e. matching Y-STR profile and RSID saliva positive) and then draw pathways that lead to both outcomes. The timeline of events shown in a graphical format is seen in Figure 5.7.

The graphical display of the pathways of transfer shown in Figure 5.7 is fine if the presence or absence of DNA is considered. In this example though, DNA amount in brackets of 'high/low/none' classification is required. In order to do this, using the graphical method of *LR* formula construction, means that these amounts must be shown in the pathways. Consider that in case 3, low amounts of D's DNA were detected on C's underwear. For example, a simple transfer of DNA from F to C's underwear from cohabitation cannot by itself account for the result. The transfer must be of low amounts of DNA and cannot be paired with a transfer of high levels of DNA from D to C's underwear from cohabitation (because, in a pure accumulation model, that would result in high levels of DNA being observed).

Including DNA amounts represents an increase in the complexity of the pathways figure, and the resulting pathway is shown in Figure 5.8, but note that at some point of complexity, the graphical pathway shortcut method becomes too cumbersome, and the full *LR* derivation (as shown for the case 1) will become preferable. The pathways shown in Figure 5.8 might already suggest this point has been reached. In addition, a number of decisions will need to be made as to how to deal with combinations of DNA transfer; for example, does the accumulation of two low levels of DNA result in a total amount of DNA that is still low, or is it now high? In the model it will be considered that two low amounts have a 50% chance of resulting in a low total and a 50% chance in a high level.

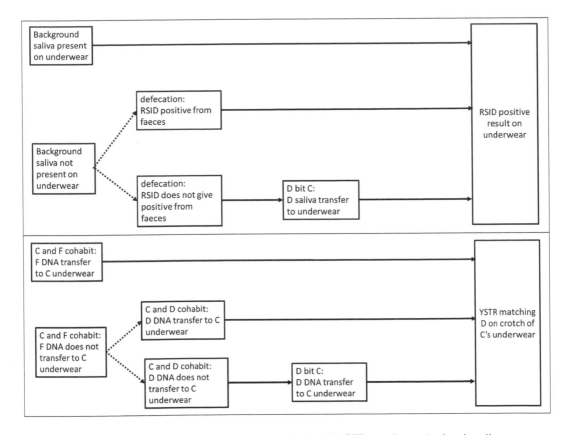

FIGURE 5.6 Pathway diagram for RSID saliva test (top) and Y-STR test (bottom), showing divergent possibilities through dashed arrows.

This additional probability is specified in Figure 5.8 by an asterisk against the DNA transfers that are affected by the modelled accumulation. Note that in the probability assignment, whenever an asterisk is present in Figure 5.8 there will be an associated factor of ½.

5.5.4 CASE 3: STEP 4: START THE BASIC *LR* FORMULA

5.5.4.1 Case 3: Using Presence/Absence of DNA

Again, making sure that if activities and transfer events are listed on an item in both propositions, then it can be omitted from the derivation and the following probabilities can be derived:

$$\Pr(E \mid H1) = t_{F \to C}^{cohabit} \left[b_s + \overline{b_s} \left(R^{deficate} + \overline{R^{defecate}} t_{D \to C}^{biting} \right) \right] +$$

$$\overline{t_{F \to C}^{cohabit}} \left\{ \begin{array}{l} t_{D \to C}^{cohabit} \left[b_s + \overline{b_s} \left(R^{deficate} + \overline{R^{defecate}} t_{D \to C}^{biting} \right) \right] + \\ \overline{t_{D \to C}^{cohabit}} \left[b_s \times t_{D \to C}^{biting} + \overline{b_s} \left(R^{defecate} \times t_{D \to C}^{biting} + \overline{R^{defecate}} \times t_{D \to C}^{biting} \right) \right] \end{array} \right\}$$

Note that practice Question 7 assigns the task of using the pathways shown in Figure 5.8 to produce the equation given. This may be a good time to attempt that question if a deeper understanding of the derivation is desired. The final term can be simplified to give:

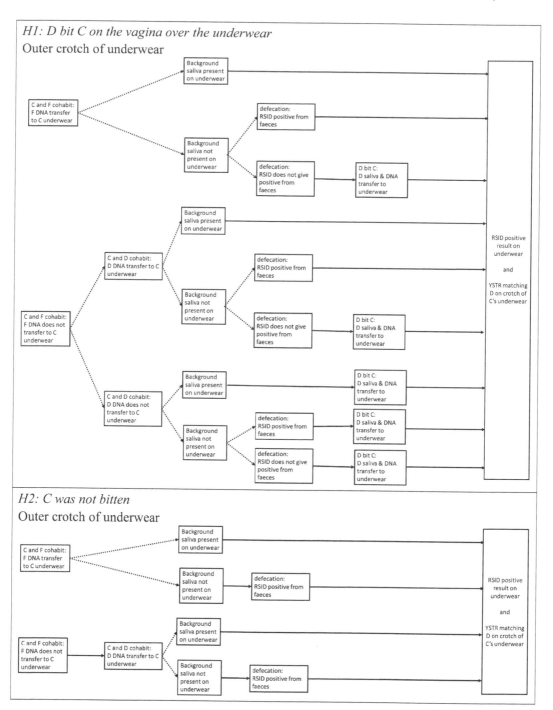

FIGURE 5.7   Timeline traces of DNA transfer in case 3. The top panel represents the timeline of the items given the prosecution proposition and the bottom panel represents the timeline of the items given the defence proposition.

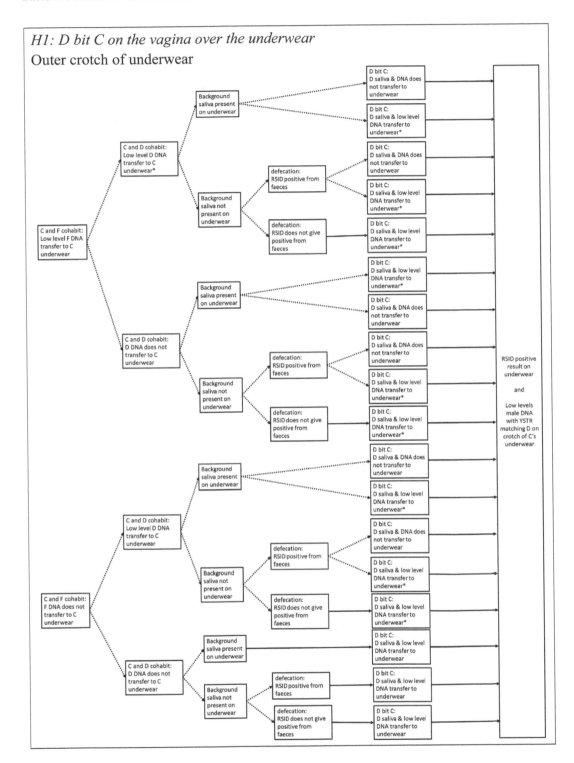

FIGURE 5.8 Timeline traces of DNA transfer in case 3 considering DNA amounts. The top panel represents the timeline of the items given the prosecution proposition and the bottom panel represents the timeline of the items given the defence proposition.

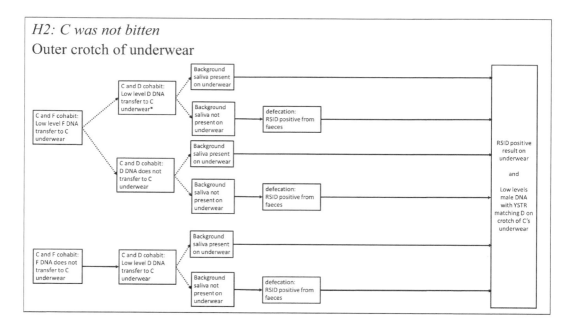

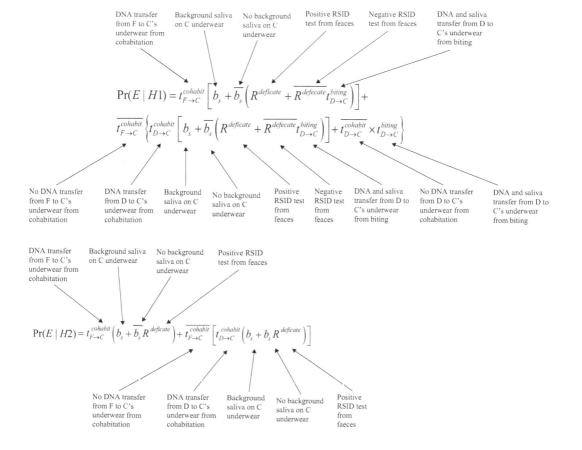

Note that in the pathways shown in Figures 5.7 and 5.8 there are boxes that describe the deposition of both saliva and DNA, for example, 'D bit C: D saliva & DNA does not transfer to underwear'. As in the example for the previous case, these two factors (the deposition of saliva and DNA) are in the same box as they are dependent on each other, i.e. if there was a high level of saliva deposition then there this is expected to be associated with a high level of DNA deposition also. If the dependency is written out in full for the above example it would take the form:

$$t_{D\to C}^{biting} = \left(t_{saliva:D\to C}^{biting}\right)\left(t_{DNA:D\to C}^{biting} \mid t_{saliva:D\to C}^{biting}\right)$$

The simplifying assumption is made that if saliva is transferred then so too will DNA, i.e. $\left(t_{DNA:D\to C}^{biting} \mid t_{saliva:D\to C}^{biting}\right)=1$, and if saliva is not transferred then DNA will not be transferred, i.e. $\left(\overline{t_{DNA:D\to C}^{biting}} \mid \overline{t_{saliva:D\to C}^{biting}}\right)=1$. A mild abuse of terminology can be carried out here by the use of the ambiguous term $t_{none:D\to C}^{biting}$ to denote no transfer of saliva and DNA.

5.5.4.2 Case 3: Using High/Low/None Categories of DNA

The pathways are quite complex in Figure 5.8, and this translates to quite a complex formula. Some visual simplicity can be achieved by blocking out repeated parts of the equation. For example, consider just the component:

$$t_{none:D\to C}^{biting} + 0.5\left(t_{low:D\to C}^{biting}\right)$$

Note that as explained in Section 5.5.4.1 the generic term is used such as $t_{low:D\to C}^{biting}$ signifies a transfer of low levels of saliva and DNA (and so assume the levels of DNA and saliva to be completely dependent, therefore the dependency is not written out explicitly). In words the above equation is the probability of no DNA transferring from D to C's underwear from biting plus the probability of low levels of DNA transferring from D to C's underwear from biting. Recall the associated modification of 0.5, as explained earlier to account for the probability of an accumulation of two low amounts of DNA leading to a low amount of total DNA. Previously, a line over the top of a transfer term was used to specify a non-transfer. Now, as there are more than two categories (none, just being one of them) an additional component is introduced in the subscript of the term.

Define this equation fragment as t^{bite}. In a similar manner, consider the formula fragment (which includes the last abbreviated definition):

$$R^{defecate} \times t^{bite} + \overline{R^{defecate}} \times 0.5\left(t_{low:D\to C}^{biting}\right)$$

And define this as $t^{bite/defecate}$. Now define:

$$t^{bite/defecate/background} = b_s \times t^{bite} + \overline{b_s} \times t^{bite/defecate}$$

Then, using these simplified terms the probabilities sought are:

$$\Pr(E\mid H1) = t_{low:F\to C}^{cohabit}\left[0.5\times t_{low:D\to C}^{cohabit}\times t^{bite/defecate/background} + t_{none:D\to C}^{cohabit}\times t^{bite/defecate/background}\right]+$$

$$t_{none:F\to C}^{cohabit}\begin{Bmatrix} t_{low:D\to C}^{cohabit}\times t^{bite/defecate/background} + \\ t_{none:D\to C}^{cohabit}\times\left[b_s\times t_{low:D\to C}^{biting} + \overline{b_s}\left(R^{defecate}\times t_{low:D\to C}^{biting} + \overline{R^{defecate}}\times t_{low:D\to C}^{biting}\right)\right]\end{Bmatrix}$$

Note that the final terms of the *LR* were not able to use the defined terms above as they did not incorporate the 0.5 element within them. Some simplification is therefore possible:

$$\Pr(E\mid H1) = t_{low:F\to C}^{cohabit}\left[0.5\times t_{low:D\to C}^{cohabit}\times t^{bite/defecate/background} + t_{none:D\to C}^{cohabit}\times t^{bite/defecate/background}\right] +$$

$$t_{none:F\to C}^{cohabit}\left(t_{low:D\to C}^{cohabit}\times t^{bite/defecate/background} + t_{none:D\to C}^{cohabit}\times t_{low:D\to C}^{biting}\right)$$

and:

$$\Pr(E\mid H2) = t_{low:F\to C}^{cohabit}\left[0.5\times t_{low:D\to C}^{cohabit}\times t^{defecate/background} + t_{none:D\to C}^{cohabit}\times t^{defecate/background}\right] +$$

$$t_{none:F\to C}^{cohabit}\times t_{low:D\to C}^{cohabit}\times t^{defecate/background}$$

where the term was defined:

$$t^{defecate/background} = b_s + \overline{b_s}\times R^{defecate}$$

As an example, note what occurs if all terms are expanded within $\Pr(E\mid H_2)$

$$\Pr(E\mid H2) = t_{low:F\to C}^{cohabit}\times 0.5\times t_{low:D\to C}^{cohabit}\times t^{defecate/background} +$$

$$t_{low:F\to C}^{cohabit}\times t_{none:D\to C}^{cohabit}\times t^{defecate/background} +$$

$$t_{none:F\to C}^{cohabit}\times t_{low:D\to C}^{cohabit}\times t^{defecate/background}$$

$$= t_{low:F\to C}^{cohabit}\times 0.5\times t_{low:D\to C}^{cohabit}\times\left(b_s + \overline{b_s}\times R^{defecate}\right) +$$

$$t_{low:F\to C}^{cohabit}\times t_{none:D\to C}^{cohabit}\times\left(b_s + \overline{b_s}\times R^{defecate}\right) +$$

$$t_{none:F\to C}^{cohabit}\times t_{low:D\to C}^{cohabit}\times\left(b_s + \overline{b_s}\times R^{defecate}\right)$$

$$= t_{low:F\to C}^{cohabit}\times 0.5\times t_{low:D\to C}^{cohabit}\times b_s +$$

$$t_{low:F\to C}^{cohabit}\times 0.5\times t_{low:D\to C}^{cohabit}\times\overline{b_s}\times R^{defecate} +$$

$$t_{low:F\to C}^{cohabit}\times t_{none:D\to C}^{cohabit}\times b_s +$$

$$t_{low:F\to C}^{cohabit}\times t_{none:D\to C}^{cohabit}\times\overline{b_s}\times R^{defecate} +$$

$$t_{none:F\to C}^{cohabit}\times t_{low:D\to C}^{cohabit}\times b_s +$$

$$t_{none:F\to C}^{cohabit}\times t_{low:D\to C}^{cohabit}\times\overline{b_s}\times R^{defecate}$$

The law of total probability can be used to expand out the b_s terms with respect to defecation:

$$b_s = b_s(R^{defecate} + \overline{R^{defecate}}) = b_s R^{defecate} + b_s\overline{R^{defecate}}$$

And the very lengthy formula is obtained:

$$= t^{cohabit}_{low:F \to C} \times 0.5 \times t^{cohabit}_{low:D \to C} \times b_s \times R^{defecate} +$$

$$t^{cohabit}_{low:F \to C} \times 0.5 \times t^{cohabit}_{low:D \to C} \times b_s \times \overline{R^{deeicate}} +$$

$$t^{cohabit}_{low:F \to C} \times 0.5 \times t^{cohabit}_{low:D \to C} \times \overline{b_s} \times R^{defecate} +$$

$$t^{cohabit}_{low:F \to C} \times t^{cohabit}_{none:D \to C} \times b_s \times R^{defecate} +$$

$$t^{cohabit}_{low:F \to C} \times t^{cohabit}_{none:D \to C} \times b_s \times \overline{R^{defecate}} +$$

$$t^{cohabit}_{low:F \to C} \times t^{cohabit}_{none:D \to C} \times \overline{b_s} \times R^{defecate} +$$

$$t^{cohabit}_{none:F \to C} \times t^{cohabit}_{low:D \to C} \times b_s \times R^{defecate} +$$

$$t^{cohabit}_{none:F \to C} \times t^{cohabit}_{low:D \to C} \times b_s \times \overline{R^{defecate}} +$$

$$t^{cohabit}_{none:F \to C} \times t^{cohabit}_{low:D \to C} \times \overline{b_s} \times R^{defecate}$$

The result is a sum where each summed element is a different combination of events that could explain the observations and each term within a sum is a probability associated with an event. This brings about an alternate way of laying out formula components that do not use the graphical pathways, and that is to list out and tabulate all pathways. Taking $Pr(E|H_2)$, which has just been expanded, the data from which it could have arisen is shown in Table 5.4.

TABLE 5.4
Tabulate Form of *LR* Construction for $Pr(E|H_2)$

Events That Account for the Low Levels of Make DNA		Events that Account for the Positive RSID Test		
DNA Transfer F→C from Cohabitation	DNA Transfer D→C from Cohabitation	Background Saliva Present on C Underwear	Faecal Staining	Product
$t^{cohabit}_{low:F \to C}$	$0.5 \times t^{cohabit}_{low:D \to C}$	b_s	$R^{defecate}$	$t^{cohabit}_{low:F \to C} \times 0.5 \times t^{cohabit}_{low:D \to C} \times b_s \times R^{defecate}$
$t^{cohabit}_{low:F \to C}$	$0.5 \times t^{cohabit}_{low:D \to C}$	b_s	$\overline{R^{defecate}}$	$t^{cohabit}_{low:F \to C} \times 0.5 \times t^{cohabit}_{low:D \to C} \times b_s \times \overline{R^{defecate}}$
$t^{cohabit}_{low:F \to C}$	$0.5 \times t^{cohabit}_{low:D \to C}$	$\overline{b_s}$	$R^{deficate}$	$t^{cohabit}_{low:F \to C} \times 0.5 \times t^{cohabit}_{low:D \to C} \times \overline{b_s} \times R^{defecate}$
$t^{cohabit}_{low:F \to C}$	$t^{cohabit}_{none:D \to C}$	b_s	$R^{defecate}$	$t^{cohabit}_{low:F \to C} \times t^{cohabit}_{none:D \to C} \times b_s \times R^{defecate}$
$t^{cohabit}_{low:F \to C}$	$t^{cohabit}_{none:D \to C}$	b_s	$\overline{R^{defecate}}$	$t^{cohabit}_{low:F \to C} \times t^{cohabit}_{none:D \to C} \times b_s \times \overline{R^{defecate}}$
$t^{cohabit}_{low:F \to C}$	$t^{cohabit}_{none:D \to C}$	$\overline{b_s}$	$R^{defecate}$	$t^{cohabit}_{low:F \to C} \times t^{cohabit}_{none:D \to C} \times \overline{b_s} \times R^{defecate}$
$t^{cohabit}_{none:F \to C}$	$t^{cohabit}_{low:D \to C}$	b_s	$R^{defecate}$	$t^{cohabit}_{none:F \to C} \times t^{cohabit}_{low:D \to C} \times b_s \times R^{defecate}$
$t^{cohabit}_{none:F \to C}$	$t^{cohabit}_{low:D \to C}$	b_s	$\overline{R^{defecate}}$	$t^{cohabit}_{none:F \to C} \times t^{cohabit}_{low:D \to C} \times b_s \times \overline{R^{defecate}}$
$t^{cohabit}_{none:F \to C}$	$t^{cohabit}_{low:D \to C}$	$\overline{b_s}$	$R^{defecate}$	$t^{cohabit}_{none:F \to C} \times t^{cohabit}_{low:D \to C} \times \overline{b_s} \times R^{deficate}$

The sum of the product column in Table 5.4 provides the formula for $\Pr(E|H_2)$, with some simplification by contraction of like terms, that was previously given.

The advantage of the table is that the routes can be more intuitively listed, but the disadvantage is that it starts the *LR* derivation as a sum of basic elements that then must be simplified to get the final *LR* formula, whereas the graphical form tends to produce *LR* formulae that are already in simplified form (as has been seen).

5.5.5 CASE 3: STEP 5: COMPLETE THE *LR* DERIVATION

5.5.5.1 Presence/Absence of DNA

The *LR* is calculated by division of the two individual probabilities in step 4.

$$
LR = \frac{
\begin{array}{l}
t_{F\to C}^{cohabit}\left[b_s + \overline{b_s}\left(R^{defecate} + \overline{R^{defecate}}\,t_{D\to C}^{biting}\right)\right] + \\[4pt]
\overline{t_{F\to C}^{cohabit}}\left\{t_{D\to C}^{cohabit}\left[b_s + \overline{b_s}\left(R^{defecate} + \overline{R^{defecate}}\,t_{D\to C}^{biting}\right)\right] + \overline{t_{D\to C}^{cohabit}}\times t_{D\to C}^{biting}\right\}
\end{array}
}{
t_{F\to C}^{cohabit}\left(b_s + \overline{b_s}R^{defecate}\right) + \overline{t_{F\to C}^{cohabit}}\left[t_{D\to C}^{cohabit}\left(b_s + \overline{b_s}R^{defecate}\right)\right]
}
$$

Because the system by which the *LR* formula was derived was already quite efficient, there is little additional simplification that can be done, and the process can move straight to step 6.

5.5.5.2 High/Low/None Categories of DNA

Again, the ratio of probabilities is given from the previous step and there is little further simplification to be done.

$$
LR = \frac{
\left\{
\begin{array}{l}
t_{low:F\to C}^{cohabit}\left[0.5\times t_{low:D\to C}^{cohabit}\times t^{bite/defecate/background} + t_{none:D\to C}^{cohabit}\times t^{bite/defecate/background}\right] + \\[4pt]
t_{none:F\to C}^{cohabit}\left(t_{low:D\to C}^{cohabit}\times t^{bite/defecate/background} + t_{none:D\to C}^{cohabit}\times t_{low:D\to C}^{biting}\right)
\end{array}
\right\}
}{
\left\{
\begin{array}{l}
t_{low:F\to C}^{cohabit}\left[0.5\times t_{low:D\to C}^{cohabit}\times t^{defecate/background} + t_{none:D\to C}^{cohabit}\times t^{defecate/background}\right] + \\[4pt]
t_{none:F\to C}^{cohabit}\times t_{low:D\to C}^{cohabit}\times t^{defecate/background}
\end{array}
\right\}
}
$$

5.5.6 CASE 3: STEP 6: ASSIGN PROBABILITIES

5.5.6.1 Presence/Absence of DNA

Again, not much time will be spent on identifying specific sources of data for probability assignment as the field moves so quickly and new data becomes available each day. For examples of data being used for assigning probabilities in this case the reader is directed to the supplemental material of Taylor *et al.* [10]. This exemplar case will show an example of using DNA amounts in the form of a categorical high/low/none. High is defined as greater than 2 ng of DNA (in total), low as greater than zero (or technically greater than the analytical detection capability of the laboratory) but less than 2 ng and none as 0 ng. In the case at hand the amount of male DNA detected on the underwear was low (approximately 0.5 ng). Therefore, in the following definitions DNA amounts are used in the descriptions of the probability meanings. For saliva a presence/absence discretisation scheme is used.

The following probabilities require values to be assigned:

- $t_{F\to C}^{cohabit}$: The probability of transferring DNA from F to C's underwear from the cohabitation of F and C. This probability is assigned a value of 0.52. Note that the assumption is made that $t_{F\to C}^{cohabit} = t_{D\to C}^{cohabit} = 1 - \overline{t_{F\to C}^{cohabit}} = 1 - \overline{t_{D\to C}^{cohabit}}$

- b_s: The probability of background saliva being present on the underwear of C. This probability is assigned a value of 0.07. Note the assumption is made that $b_s = 1 - \overline{b_s}$
- $R^{defecate}$: The probability that faeces will yield a positive RSID saliva reaction. This probability is assigned a value of 0.67. Note that the assumption is made that $R^{defecate} = 1 - \overline{R^{defecate}}$
- $t_{D \to C}^{biting}$: The probability of transferring DNA from D to C from biting. This probability is assigned a value of 0.92

5.5.6.2 High/Low/None Categories of DNA

The probabilities for the RSID saliva positive from faeces and the presence of background saliva do not change from the presence/absence evaluation and so those probabilities are not specified here. New or different probabilities are:

- $t_{low:F \to C}^{cohabit}$: The probability of transferring low amounts of DNA from F to C's underwear from cohabitation of F and C. This probability is assigned a value of 0.43. Note that the assumption is made that $t_{low:F \to C}^{cohabit} = t_{low:D \to C}^{cohabit}$
- $t_{none:F \to C}^{cohabit}$: The probability of transferring no DNA from F to C's underwear from cohabitation of F and C. This probability is assigned a value of 0.48. Note that the assumption is made so that $t_{none:F \to C}^{cohabit} = t_{none:D \to C}^{cohabit}$
- $t_{low:D \to C}^{biting}$: The probability of transferring low amounts of DNA from D to C's underwear from D biting C. This probability is assigned a value of 0.08
- $t_{none:D \to C}^{biting}$: The probability of transferring no DNA from D to C's underwear from D biting C. This probability is assigned a value of 0.08

5.5.7 CASE 3: STEP 7: CALCULATE THE LR

5.5.7.1 Presence/Absence of DNA

The LR is now calculated from the formula as given in step 5 and substituting the probabilities as given in step 6.

$$LR = \frac{t_{F \to C}^{cohabit}\left[b_s + \overline{b_s}\left(R^{defecate} + \overline{R^{defecate}}t_{D \to C}^{biting}\right)\right] + }{t_{F \to C}^{cohabit}\left(b_s + \overline{b_s}R^{defecate}\right) + \overline{t_{F \to C}^{cohabit}}\left[t_{D \to C}^{cohabit}\left(b_s + \overline{b_s}R^{defecate}\right)\right]}$$

$$\overline{t_{F \to C}^{cohabit}}\left\{t_{D \to C}^{cohabit}\left[b_s + \overline{b_s}\left(R^{defecate} + \overline{R^{defecate}}t_{D \to C}^{biting}\right)\right] + \overline{t_{D \to C}^{cohabit}} \times t_{D \to C}^{biting}\right\}$$

$$LR = \frac{0.52\left[0.07 + 0.93\left(0.67 + 0.33 \times 0.92\right)\right] + 0.48\left\{0.52\left[0.07 + 0.93\left(0.67 + 0.33 \times 0.92\right)\right] + \left(0.48 \times 0.92\right)\right\}}{0.52\left(0.07 + 0.93 \times 0.67\right) + 0.48\left[0.52\left(0.07 + 0.93 \times 0.67\right)\right]}$$

$$LR \approx 1.8$$

5.5.7.2 High/Low/None Categories of DNA

Starting with the defined terms:

$$t^{bite} = t_{none:D \to C}^{biting} + 0.5\left(t_{low:D \to C}^{biting}\right) = 0.08 + 0.5 \times 0.08 = 0.12$$

$$t^{bite/defecate} = R^{defecate} \times t^{bite} + \overline{R^{defecate}} \times 0.5\left(t^{biting}_{low:D \to C}\right) = 0.67 \times 0.12 + 0.33 \times 0.5 \times 0.08 = 0.0936$$

$$t^{bite/defecate/background} = b_s \times t^{bite} + \overline{b_s} \times t^{bite/defecate} = 0.07 \times 0.12 + 0.93 \times 0.0936 \approx 0.0954$$

$$t^{defecate/background} = b_s + \overline{b_s} \times R^{defecate} = 0.07 + 0.93 \times 0.67 = 0.6931$$

And now the *LR*:

$$LR = \frac{\left\{ \begin{array}{l} t^{cohabit}_{low:F \to C}\left[0.5 \times t^{cohabit}_{low:D \to C} \times t^{bite/defecate/background} + t^{cohabit}_{none:D \to C} \times t^{bite/defecate/background}\right] + \\ t^{cohabit}_{none:F \to C}\left(t^{cohabit}_{low:D \to C} \times t^{bite/defecate/background} + t^{cohabit}_{none:D \to C} \times t^{biting}_{low:D \to C}\right) \end{array} \right\}}{\left\{ \begin{array}{l} t^{cohabit}_{low:F \to C}\left[0.5 \times t^{cohabit}_{low:D \to C} \times t^{defecate/background} + t^{cohabit}_{none:D \to C} \times t^{defecate/background}\right] + \\ t^{cohabit}_{none:F \to C} \times t^{cohabit}_{low:D \to C} \times t^{defecate/background} \end{array} \right\}}$$

$$LR = \frac{0.43\left[0.5 \times 0.43 \times 0.0954 + 0.48 \times 0.0954\right] + 0.48\left(0.43 \times 0.0954 + 0.48 \times 0.08\right)}{0.43\left[0.5 \times 0.43 \times 0.6931 + 0.48 \times 0.6931\right] + 0.48 \times 0.43 \times 0.6931}$$

$$LR \approx 0.19$$

5.5.8 Case 3: Supplemental

There are several points of interest that this final case derivation demonstrates. Firstly, note how the transition from considering DNA as being present/absent to considering DNA amounts of high/low/none changes the direction of support from supporting H_1 (with an *LR* of approximately 2 in support of H_1) to supporting H_2 (with an *LR* of approximately 5 in support of H_2). This shift comes from the difference between:

- Observing male DNA and expecting biting to transfer DNA with a higher probability than cohabitation, and
- Observing low levels of male DNA and expecting biting to transfer high levels of DNA, but importantly, expecting biting to transfer low amounts of DNA with a lower probability than cohabitation.

This is a good example of the increased discrimination power of using DNA amount, even if the transition is only from two categories to three.

Also, note that in H_1 the transfers of DNA and saliva are considered through all routes (and not simply the route that corresponds to the disputed alleged activity). The importance here is that the transfers from the undisputed activities can accumulate with the disputed activity, and this has an important effect on the evaluation. As explained with a toy example by Taylor *et al.* in [11]:

Imagine that Mr A has been invited over to Ms B's home to perform an amateur dental check-up, which Mr A does without wearing any gloves. This is consensual and accepted by both prosecution and defence as having occurred legally and consensually. After the exam it is alleged that Mr A picked up Ms B's wallet and took out some money (without permission) and left the house. Swabs are taken of Mr A's hands two hours after the incident (before he washed his hands) to help with the issue of whether or not he has picked up Ms B's wallet. The DNA profile obtained from the swabs of Mr A's hand can be considered as a mixture of two persons corresponding to both DNA profiles of A and B. For our example we need not elaborate further, but one could imagine that the mixture proportions (or DNA amounts) would also likely be informative. We simply consider that the DNA profiling of Mr. A's hand swabs reveal the presence of Ms B's DNA. The propositions are then:

H_p: Mr A took money from Ms B's wallet
H_d: Mr A didn't touch Ms B's wallet
I: The accused performed a dental check-up on Ms. B and did not wash his hands afterwards; swabs taken two hours after the alleged facts.

There are two ways that DNA from Mr B could have come to be in Mr A's hands. Either directly transferred from the dental exam or indirectly transferred from Mr A to Mr A's wallet and then to Mr B's hands. The important point demonstrated with this example is that it would almost certainly be expected to see DNA from Mr B on Mr A's hands, not due to a secondary transfer from the wallet (which is associated with the alleged crime and has only occurred under H_p), but from the prior dental exam, which has occurred under both propositions.

While this example shows the importance of considering both mechanisms for DNA transfer under H_p, there are many examples where the importance is not so obvious, and there are numerous examples of when it has not been taken into account. For example, imagine a situation where the defendant and complainant were at a social event together and then later an alleged non-consensual digital penetration of the complainant's vagina by the defendant's fingers occurred. In this situation it is less obvious than the dentistry example, there are two mechanisms from DNA transfer from the complainant to the defendant that can be considered under Hp; from the social activity and from the alleged rape.

5.6   FACTORS TO CONSIDER WHEN ASSIGNING PROBABILITIES

5.6.1   ASSIGNING PROBABILITIES FROM CONTINUOUS DNA AMOUNT TO BINARY DNA PRESENCE

Any step away from fully continuous results in a discretised breakup of DNA, and all options shown in Figure 5.3 (apart from the continuous option on the left) can be thought of as just variants of where, and how many breaks there are. For example, even the presence or absence is simply a choice to discretise the data with a single breakpoint at the laboratory threshold of DNA detection. In this way the presence/absence category could just as validly be labelled as ≥ 0.01 ng and <0.01 ng (as an example where the limit of detection in the laboratory was 0.01 ng). For derivations it is likely the case that only the probability associated with the transfer/persistence/etc bracket that relates to the case observations will be required.

5.6.2   OBSERVATIONS OTHER THAN DNA AMOUNT

Observations besides DNA also need to be discretised. For example, consider the results of a presumptive test for a particular body fluid, which is indicated by a colour change within a certain time. The observations of this test could be considered in a continuous manner by modelling the time-to-positive distribution for the test in the presence of the body fluid or the time-to-positive could be broken into categories (i.e. quick positive, slow positive, negative) or it could be binary (positive or negative).

When dealing with DNA data, another consideration when setting up the evaluation is how the observations will be interpreted. So far, the observations described have always been amounts of DNA detected, but this does not need to be the case. Observations could be properties of the DNA profile or the DNA profile combined with an interpretation scheme, for example, the observations could be discretised into categories:

- Single source matching defendant
- Mixture with defendant as major

- Mixture with defendant as minor
- Mixture without defendant
- Uninterpretable mixture
- Single source not defendant
- No profile

There are some advantages and some disadvantages to categorising the observations in this manner. The advantage is that the observations themselves can be more stable than DNA amounts, in situations where DNA amount is highly variable and observation numbers are limited. For example, many experiments dealing with the amount of DNA transferred during a contact scenario lead to highly variable results, which depend on a number of factors (shedder status of the person, nature of the surface, duration and vigour of the contact, etc.). Even when taking into account these factors, a study looking at the amount of DNA transferred will require many data points before the amount can be sufficiently modelled. Then the solution employed is to fall back to modelling the data on a presence or absence basis (through the use of a logistic regression, which is written about in Chapter 7). While this has less power to discriminate contact types than DNA amounts it may be able to make use of more variable DNA amount data by simplifying the response variable. The same thinking applies when categorising the DNA profile results into interpretational categories, e.g. while the amount of DNA transferred may be highly variable, there may be some trends of consistency in the number of times the person contacting is the major donor. Again, this loses discrimination power compared to DNA amounts, but less so than simplifying the outcome to a binary presence/absence. The disadvantage to breaking up the observations into DNA profile interpretation categories (rather than in some other manner) is that it makes the data specific to the performance of the laboratory and its interpretation scheme. For example, another laboratory wanting to use the results of a study in their evaluation, could only do so (or at least would best align) if their laboratory had the same sampling procedure, utilised the same DNA extraction process, used the same DNA profiling kit and carried out interpretation of profiles in the same way.

The further back in the DNA profiling process the results can be considered, the more generally applicable they will be. Consider that instead of using DNA profile interpretation categories quantified DNA amounts were used. Then a laboratory would only need to have the same DNA sampling and DNA extraction methodology in order to be able to use the results. If knowledge of the extraction efficiency of the laboratory was known, then the quantified DNA amounts could be extrapolated back to the amount of DNA on the sampling device. Now a second laboratory could use those results only requiring an alignment of sampling methodology. One final step could be to also model the sampling efficiency of a particular device and then apply that on top of the extraction efficiency model in order to extrapolate right back to the amount of DNA on the item. With these results any laboratory would then be able to make use of the observations of the study regardless of the methodology they used (and as they would just need to apply their own sampling and extraction efficiencies to the study findings in order to use them).

There is also a limitation to breaking up the results in the DNA profile interpretation categories in that it is only practical to use these results once in an evaluation, when considering the final results. For example, imagine the scenario being modelled was one where the defendant was said to have shaken the hand of the alternate offender, and it was the alternate offender who then touched an object that had been sampled. If the interest lies in the potential for DNA of the defendant being on the object, then there are two transfer events here to consider, DNA transfer from handshaking, and then DNA transfer from touching the object. If experimental data was available that investigated DNA transfer during a handshake, and the results were broken up by profile interpretation categories, then it is not clear how these results would be used as conditional categories to be considered in probability assignments for outcomes of the second transfer. In fact, the only way to break up the data according to profile interpretation categories in this instance would be to find a study (which becomes quite unlikely as the specificity of the scenario increases) or carry out a study of the DNA profiles resulting from the full transfer route (i.e. handshake and then touching an object).

Alternatively, if DNA amounts were modelled then probabilities could be assigned to different DNA amounts of the defendant being transferred to the hands of the alternate offender. Then these DNA amounts become inputs to the second transfer in the overall route, i.e. each transfer in a string of transfer events can be informed by its own experiments and pieced together as required (as long as appropriate consideration of dependencies is taken into account during the evaluation). Note that differences in methodology between the study and the case at hand will still need to be accounted for. The advantages and disadvantages of the different options are discussed in later chapters.

5.6.3 How to Obtain Data to Inform Probabilities

Once the manner in which the treatment of the factors in the evaluation has been decided then the task is to assign probabilities to these factors. There are a number of ways in which probabilities can be assigned, ranging from pure intuition (i.e. a probability based purely on experience and an individual knowledge base without having any experimental data) to purely data-driven (i.e. based completely on experimental data that aligns with the case circumstances). Given that it is never the case that all relevant details of case circumstances are known, assumptions will always need to be made. Hence the assignment of the probability still holds a measure of subjectivity (it always will). Even if the assumption is made that the data perfectly align with the case, this is still an assumption with underlying uncertainty. Therefore, the purely data-driven is a theoretical extreme, rather than an obtainable position. Often the information used to assign the probability will fall somewhere between these extremes when some data is available, but an experience-based numerical adjustment must be made for the fact that it does not exactly align with some aspect of the evaluation. Recall the well-known statement of de Finetti [12]:

Probability does not exist

Meaning that probabilities are states of mind rather than states of nature [13]. There exists a spectrum of data sources (or decisions about data) that can be used to assign probabilities (or decline to do so). From most to least robust these are:

1. Bespoke experiments
2. Data from literature
3. Considering a range of reasonable values
4. Assign a value based on the expert's experience or knowledge
5. Do not carry out an evaluation

Bespoke experiments are the gold standard in producing data that most closely aligns with the case circumstances. This is because the experiments can be set up to mimic the case circumstances as closely as possible within the bounds of ethical science and practicality. They will also have the advantage that they are produced under the same laboratory processes and conditions as the case samples were analysed and so are directly applicable without further adjustments for any laboratory-based differences. When using bespoke experiments then any option of data collection (i.e. DNA amounts, presence or absence, DNA profile interpretation categories) is available to use by the laboratory.

Bespoke experiments are often not possible, due to the time and resources required to carry them out. This level of resourcing is not common in forensic laboratories, and particularly not if they were required on a regular basis in a laboratory that regularly evaluates observations given activity level propositions. Also, just because bespoke experiments are able to be carried out it does not mean that data will be available that exactly mimics the framework of circumstances surrounding the case. There are limitations to practicality:

- Impracticalities of reproduction – There are numerous examples of how the circumstances of a case would make it impractical to exactly replicate in practice. For example, one such situation might be in a cold case where a sample has been tested after 20 years and has degraded. The evaluation may require a probability assigned to the level of degradation that is expected over this length of time, but of course it is not practical to set up a series of 20-year-long experiments in order to inform the evaluation for that case
- Ethics – As some aspects of evaluations involve illegal activities, it is not ethical to reproduce them exactly in bespoke experiments. For example, the amount of DNA transferred to a screwdriver handle if it was used to stab a person or the amount of DNA shed from a badly burnt individual could not be ethically reproduced
- Unknown aspects of case specifics – Studies into DNA transfer from a contact have shown that there are many factors that play a part in whether (and how much) DNA is transferred. These can include the shedder status of the person, nature of the surface, duration and vigour of the contact, etc. Some of these aspects are not known, or not known with precision; for example, the level of pressure applied to an object that was handled would not be known, even to a person who admits to touching an object
- Unknown aspects of framework – This is similar to the previous point except rather than relating to a specific aspect of an event it relates to unknown aspects of the activities. For example, if a probability must be assigned to DNA transferring between two people who attended a party, there may be little to no information regarding the specifics of how (or if) they directly interacted or the level of shared items that could account for indirect transfers

With regard to the last two points, the lack of knowledge on the specifics surrounding an activity or a lack of specifics about the activities performed within the framework of circumstances does not mean that probabilities cannot be assigned. In these instances, the lack of knowledge must be considered within the evaluation and data chosen/generated that mirrors the uncertain factors. Alternatively, the choice could be made for the model to explicitly specify possible activities and data found for each activity, along with a prior probability that reflects the uncertainty of those activities occurring in the case. To explain further with an example, it may be the case that a probability must be assigned to whether DNA will be transferred from a person grabbing a knife handle and using it to stab a person. It is known that the length of time an object is held will have an effect on the probability that DNA will be transferred. One option would be to set up a series of experiments that ask people to pick up and use a knife to stab an object (roughly holding the properties of the victim in the offence) without giving very specific instructions on how long the knife should be handled. In doing this, variability in the results is naturally obtained which encompasses the uncertainty surrounding the distribution of times that people hold a knife to stab someone. In mathematical terms, this distribution of DNA transfer values can be considered to have integrated out the effect of holding time, and by making the assumption that the offender in the case is not abnormal in comparison to the general population with respect to how long they hold knives to stab people, the data will appropriately suit the case situation.

The second option would be to carry out a set of experiments that asked people to hold the knife for specific times. Having done this for a range of different hold times a model could be built that assigns a probability of DNA transfer occurring given different hold times. A separate experiment can then be carried out asking people to stab an object without any direction on the length of time to hold the knife. The distribution of lengths of time that people hold knives whilst stabbing can then be modelled and incorporated into the evaluation along with the model of DNA transfer given a hold time. If the hold time in a case is not known, then all possible hold times in the distribution are considered according to their probability of occurrence. In this example it can be seen more directly how the evaluation is integrating out the effect of hold time. The second method is a more complex manner to carry out experimentation and evaluation. However, it has the advantage that if the knife hold time somehow becomes available (say for example mobile phone footage of the assault is uncovered) then the model already has the data that can make use of this new information. In the first model, there is no way to incorporate this new information. The more complex model is also more likely to be usable in other cases.

While the final point (lack of knowledge about the activities themselves) may be more difficult to account for, it is dealt with in the same way as just described for a lack of specifics about an event. Take the example of DNA transfer during a party. One option is that experimentation can be carried out on people who are not given specific instructions on how to interact at a party (and specific events will occur with what is assumed to be their general prior probability at any party). The other option is that people can be given specific instructions to carry out certain events, DNA transfer is measured during these events, and then this is paired with the probabilities of those events occurring.

If bespoke experimentation is not available, then the next best source of informative data is from literature. This may be published scientific papers but could also include theses or in-house validation reports. Unlike bespoke experimentation, when data is sought from the literature it often has some divergence from the specific aspects of the case (see Box 5.1). This may be in the manner of transfer, the objects studied, the timeframes of the activity and sampling, the samples taken, the laboratory process or the manner of interpretation of the observations. When a divergence occurs this needs to be acknowledged and the ramifications of the difference are accounted either statistically or subjectively, or the potential evaluative impact should be explored (by way of the sensitivity of the LR to this value). For example, consider the scenario that the difference between the study and the circumstances of the case is that the laboratory process in the study uses a different profiling kit than the laboratory wishing to use the data. This may be able to be taken into account statistically, by looking back at the DNA amount results rather than the profiling results (or it may be possible to extrapolate back to the DNA amounts from the DNA profiling results if the right information is available). While the results of the DNA profiling system of the study are used to determine such elements as the mixture proportions, and when an individual has contributed DNA, this aspect of the data interpretation does not need to align with a laboratory wishing to use the study data as the data itself is not a description of the performance of the profiling kit. It is similar to the fact that it is somewhat irrelevant what system is used to quantify the amount of DNA. As long as that system is accurate in its determination then DNA amounts from the study can be used, without the need for quantification systems between the study and the laboratory to be aligned.

If the DNA amounts are not available in the literature and the information required to extrapolate profiling kit results back to DNA amounts is not present, then a subjective adjustment might be possible. For example, imagine a situation where a study claims that a certain proportion of the time a person's DNA profile is observed after contact, and they have used a modern profiling kit. You wish to use the observations of this study, but the case you wish to apply it to uses an older generation profiling kit that has a lower sensitivity than the modern kit used in the study. You may wish to take the proportions of DNA profile observation from the study and reduce that proportion by some amount that you believe would approximate the difference in DNA profiling kit sensitivities. A follow-up can then be done that tests how sensitive the LR is to the data being used. This is explained in detail in Chapter 8 but mentioned here as it is a mechanism by which divergence between data and case circumstances can be investigated. In brief the sensitivity analysis will trial a range of values for the probability being investigated and will calculate the LR at each point. Then the amount by which the LR changes when the probability (and hence underlying data leading to that probability assignment) changes is investigated. If the LR changes dramatically with only small changes in the probability, and the data being used to inform the probability assignment is divergent from the case circumstances in a number of ways then it may be decided that the evaluation is lacking robustness and more investigation is required into finding/producing data to inform that probability assignment.

BOX 5.1: HOW CLOSE MUST EXPERIMENTAL DATA BE TO THE
CASE CIRCUMSTANCES?

There is no specific cut-off or delineation at which a study becomes too different to the case circumstances in order to use. It is often the case that some probabilities will simply have no direct data that can inform the probability assignment. In these cases, the best the analyst can do is finding one or more pieces of information from less directly applicable studies that will assist in the assignment. For example, the analyst might be required to assign a probability to DNA being transferred by:

- Holding a bicycle handlebar, and must use a study that looks at DNA transfer to a knife handle when it is being held
- Slapping someone in the face, and must use a study that looks at DNA transfer during handshaking
- Having fingers inside someone's mouth, and must use a study that looks at the amount of DNA transferred to fingers during digital penetration of the vagina, and another study that looks at the amount of DNA from spitting onto someone's face

With a lack of any other information, all of the above studies can provide some guidance in assigning a probability. All would need to have the divergences from the case circumstances and the study circumstances clearly explained in the evaluation. As exemplified in the final scenario, there may be more than one study available to guide the probability assignment in different ways. This leads to the third step in the spectrum of probability assignments, which is to consider a range of sensible values, and test the sensitivity of the *LR* to the extremes in that range (and potentially at various points along it). Note that the purist view of probability assignment is that a probability reflects one's state of mind about their belief in an event occurring, and to test multiple probabilities (as in the extremes of a range of probability) in effect is to admit to having multiples states of mind (which is not logical). The theory behind this point is valid, but perhaps a pragmatic middle ground can be adopted, in which it is recommended to carry out an exploration of the outcome of the evaluation for a range of values assigned to a probability, which have been conservatively set by one or more semi-relevant data sources.

The above discussion brings about thoughts on a situation that occurs during evaluations when a probability assignment can be informed from multiple sources. Is it better to use a smaller-sized study that is well aligned with the circumstances of the case or a larger study, but one that is less well aligned? As before there are no rules-based answers that can be provided to this point. The answer to this question will depend on the circumstances of the case, the specifics of the studies themselves, how close or divergent the studies' circumstances are to the case, whether divergences can be accounted for in a justifiable and sensible manner, and many other nuanced factors.

Moving through the spectrum of probability assignment methods from purely data-driven to pure intuition, the fourth point is reached which is to assign based on experience or knowledge alone. These types of probabilities (i.e. not based on any identifiable literature source) are often referred to as 'soft' assignments. An example of such assignments can be seen in [14] which makes use of expert elicitation. In this study the probability values were assigned by taking an average of a set of values given by multiple experts. Such a regime (expert elicitation) utilises the concept of the 'wisdom of the crowd', whereby collective answers will tend to be more consistently accurate than that of a single expert (trial by jury itself makes use of this concept). If a purely experienced-based assignment must be made, then a collective assignment from multiple experts may be preferable than the evaluator alone assigning a value based only on their experience.

In reality assigning a probability from experience is simply an extreme form of any of the previous forms of data collection. If a probability is being assigned from experience, then there must have been some experience in similar or related matters on which to draw (otherwise the assignment is doing little more than choosing a random value). While a specific source of data may not be able to be identified in such an assignment, it will be the conglomeration of readings on a range of peripherally related topics, and observations from casework, that will be in the subconscious of the analyst's mind when providing a value. It is preferable that if the background knowledge being relied on is previous casework then it is supported by structured analysis of similar case files, which can be justified by an argument, and be disclosed for review, (as required, for example, by the ENFSI guideline [15]).

Whether the probability assignment is done by an experience-based method or from literature, the use of previous casework is a common source of data. To use such a source is acceptable, as long as the limitations are recognised (i.e. the assumptions that are made by using casework, for instance about the ground truth of the data, need to be explicitly stated). One type of study that has classically used casework is time since intercourse investigations. In these studies, sexual assault casework observations of sperm counts on vaginal swabs are compared against victim statements detailing the time between the offence and the vaginal swabs being taken. The use of results from criminal cases is based on the victim's account of events and requires the assumption that they are true and accurate. This will not necessarily always be the case; putting aside simply untruthful testimony, the account of victims of sexual assault could be distorted by the trauma they have endured. Also, during rape, the offender will often not ejaculate [16]. This will not always be known to the victim and will affect the seminal components observed afterwards (some aspects can be partially mitigated through data screening, for example only choosing cases where the victim is confident that the offender ejaculated). When using data from casework sources, it needs to be stated that it is assumed the information provided by the victims is accurate with respect to time and events. However, if controlled experiments are used then there are limitations to the cases that the results can be applied. For instance, those where it is assumed intercourse and ejaculation have taken place. Even then, it has been noted that consensual intercourse is usually followed by a period of inactivity (sleep), whereas after rape the victims are usually more active and agitated [17], which is known to affect the persistence of seminal components in the vagina [18]. Using controlled data may better be considered as time since ejaculation studies rather than time since intercourse. As can be seen, there are always considerations that must be taken into account when applying data to a case, regardless of the source.

The final, and least preferred, option when assigning probabilities is to decline to carry out an evaluation, citing a lack of data leading to the inability to provide a robust opinion. It may surprise the reader that this is the least preferred option. This view comes from the culture of forensic DNA evaluations, where it is considered conservative to decline to evaluate a DNA profile if there exists a level of uncertainty around the evaluation (for example, not being able to determine the number of contributors with confidence is often a cause of decline to evaluate). In the sub-source proposition framework, this is probably a conservative behaviour as declining to provide an evaluation simply does not provide the court with any information. This is not so for evaluations where the propositions are at the activity level, particularly when the sub-source level evaluations (i.e. on the individual DNA profiles) have already been given. The risk of not providing an evaluation of the observations given activity level propositions is that the *LR* given for the evaluations of DNA profiles (given sub-source level propositions) will be incorrectly translated in the minds of the court or jury to issues that apply to higher levels in the hierarchy of issues. This was called the 'association fallacy' by Gill in his book 'misleading DNA evidence' [19] and then later specified to 'source level fallacy' and activity level fallacy' by Meakin *et al.* [20]. Even if this fallacy does not occur, then a jury or judge who has heard an evaluation considering sub-source level propositions, will then need to decide how the defendant has interacted with that item (e.g. directly contacting perhaps in disputed ways or indirectly contacting through one or more intermediaries) and then whether the defendant is guilty or innocent. The jury or judge does not have the luxury of deciding that this line of decision-making is too complicated to perform (i.e. they must ultimately decide on a verdict). A forensic scientist is better equipped with the knowledge of transfer, persistence, prevalence and recovery of DNA from an object than a member of the court or jury, and the responsibility rightly sits then with the scientist to provide advice on these matters. If the scientist declines to carry out an evaluation due to the complexity of the factors involved, then this shifts the responsibility to the factfinder to perform a mental evaluation, unaided by the knowledge that scientific investigative study and logical evaluation can provide. It has to be asked that if a scientist feels this is too complex then how one could expect that a jury perform the task. This is the responsibility of the scientist in court and the basis for the reason that declining to carry out an evaluation is the least preferred option (particularly when the sub-source level evaluation results have been reported). Even when the knowledge about the case is incomplete (which is always the case) and the data to inform the probability assignment diverges from case circumstances (which it always does) the scientist can still provide some opinion to the court. This opinion will be based on the best knowledge that forensic science has to offer, and the court will be better equipped to understand the significance of the evidence than they would have been without an opinion being offered. If the scientist declines to provide an evaluation of the observations given activity level propositions (and does not provide any guidance) and the sub-source observations alone have a strong potential to misleading, then it may be preferable to decline to provide an evaluation of observations given any level of proposition, in order not to unintentionally mislead the court.

5.7   FINAL WORDS

The final point to make is that (as seen in this chapter) the complexity in deriving the *LR* formula when considering DNA amounts approaches a level where even using a guidelines and a step-wise system, approached a level of impracticality. The rise in complexity that occurs when multiple items and multiple tests are performed is rapid, and the derivation quickly becomes unwieldy and prone to derivational errors. This situation then calls for the next chapter of this book, where graphical tools can be used to represent aspects of the *LR* formula without the need for derivation. This tool is the Bayesian network.

5.8   PRACTICE QUESTIONS

Q1) You have the following data on DNA transfers from your laboratory. It catalogues the number of observations of each type of result for three contact types:

	Type of contact		
Transfer category	Passive	Pressure	Friction
High amount of DNA (> 1 ng)	2	9	13
Low levels of DNA (>0–1 ng)	7	15	0
No DNA (0 ng)	11	3	0

You are carrying out an evaluation where you believe the most appropriate category to consider (i.e. that aligns with the circumstances of your case) is 'friction'. You are going to evaluate the data in a binary fashion, considering DNA to be present or absent. Using the data in the table above, what posterior probabilities are you going to assign for the probability of DNA transfer and no DNA transfer.

Q2) In Section 5.3.2 a consideration was outlined that was being omitted from the evaluation of case 1 due to the negligible effect it would have. The consideration was:

Background DNA was present on C's top. C transferred DNA to her own top from wearing. C hugged her friend, and DNA from her friend was not transferred to her top. During the attempted kidnap DNA was transferred from AO's hand to C's top when he grabbed her arm. The AO had the same DNA profile as C's friend.

Draw out the new transfer pathways that would need to be considered to include this possibility.

Q3) Following on from Question 2, carry out an *LR* derivation and show the effect is negligible.

Q4) In case 1 it was considered that if the defendant was not the offender, then an alternate offender must be the attacker. Consider now that the defence changed their proposition to challenge the activity itself, i.e. that no attack occurred. Draw out the transfer pathways that would now apply under H_d.

Q5) Following on from Question 4, derive an *LR* formula considering the new defence proposition. What effect does it have on the evaluation?

Q6) In the derivation of the *LR* for the case 1 an assumption was made that the shedding ability of D, C and AO are all approximately the same. Imagine now that information was given that the defendant has a skin condition which means he is much more prone to shedding his DNA than other people. This means the assumption can no longer be made that the defendant and an AO have a similar shedding ability. How does this affect the derivation? What is the new *LR* formula?

Q7) In Section 5.3.5.1 it was shown how the pathways (produced in step 3 of a formulaic derivation) could be used directly to derive the *LR* in a short-hand manner for case 1. Carry out the same exercise using the pathways in Figures 5.5 and 5.8 for cases 2 and 3 to show how the *LR* formula is obtained for both cases.

Q8) In Table 5.4 an example was given whereby instead of deriving an *LR* formula using pathways (or by sequential consideration of factors) a table was used that laid out all the combinations of events (and their respective probabilities) for case 3 under H_2 that could lead to the observations. Carry out this same task for H_1 and derive the *LR* formula for the case in this tabular manner.

Q9) At the end of the chapter in Section 5.4.8, an example was given to demonstrate how it was important to consider all pathways of DNA transfer that could occur under both propositions. Derive a formula for the *LR* that would relate to this evaluation. You have seen several different ways of deriving an *LR* formula and worked through examples of each in the questions so far. Use whatever method you find most intuitive for this question.

Q10) Following on from Question 9, now derive an *LR* formula that only takes into consideration the DNA transfer from the money to the hands of the defendant under H_p (i.e. under H_p the evaluation will not also consider the earlier dentistry work that occurred and the potential for DNA transfer arising from that activity). Using this derivation, and comparing it to the derivation from Question 9, show how omitting the earlier activity under H_p has led to an illogical result.

5.9 REFERENCES

1. D. Taylor, The evaluation of exclusionary DNA results: A discussion of issues in R v. Drummond, *Law, Probability and Risk* 15(1) (2016) 175–197.
2. A.E. Fonneløp, M. Ramse, T. Egeland, P. Gill, The implications of shedder status and background DNA on direct and secondary transfer in an attack scenario, *Forensic Science International: Genetics* 29 (2017) 48–60.
3. M. Breathnach, L. Williams, L. McKenna, E. Moore, Probability of detection of DNA deposited by habitual wearer and/or the second individual who touched the garment, *Forensic Science International: Genetics* 20 (2016) 53–60.
4. D.J. Daly, C. Murphy, S.D. McDermott, The transfer of touch DNA from hands to glass, fabric and wood, *Forensic Science International: Genetics* 6 (2013) 41–46.
5. V.L. Bowyer, *Investigation into DNA Transfer during Physical Child Abuse*, University of Leicester, 2009.
6. M. Goray, R.J. Mitchell, R.A.H. van Oorschot, Investigation of secondary transfer of skin cells under controlled conditions, *Legal Medicine* 12 (2010) 117–120.
7. J.J. Raymond, R.A.H. van Oorschot, P.R. Gunn, S.J. Walsh, C. Roux, Trace evidence characteristics of DNA: A preliminary investigation of the persistence of DNA at crime scenes, *Forensic Science International: Genetics* 4 (2009) 26–33.
8. D. Taylor, A. Biedermann, L. Samie, K.-M. Pun, T. Hicks, C. Champod, Helping to distinguish primary from secondary transfer events for trace DNA, *Forensic Science International: Genetics* 28 (2017) 155–177.
9. M. van der Berge, G. Ozcanhan, S. Zijlstra, A. Lindenbergh, T. Sijen, Prevalence of human cell material: DNA and RNA profiling of public and private objects after activity scenarios, *Forensic Science International: Genetics* 21 (2016) 81–89.
10. D. Taylor, A. Biedermann, T. Hicks, C. Champod, A template for constructing Bayesian networks in forensic biology cases when considering activity level propositions, *Forensic Science International: Genetics* 33 (2018) 136–146.
11. D. Taylor, B. Kokshoorn, T. Hicks, Structuring cases into propositions, assumptions, and undisputed case information, *Forensic Science International: Genetics* 44 (2020) 102199.
12. B. deFinetti, *Theory of Probability*. Vol 1, John Wiley and Sons, New York, 1974.
13. F. Taroni, P. Garbolino, A. Biedermann, C. Aitken, S. Bozza, Reconciliation of subjective probabilities and frequencies in forensic science, *Law, Probability and Risk* 17(3) (2018) 243–262.
14. T.R.D. Wolff, A.J. Kal, C.E.H. Berger, B. Kokshoorn, A probabilistic approach to body fluid typing interpretation: An exploratory study on forensic saliva testing, *Law, Probability and Risk* 14(4) (2015) 323–339.
15. S.M. Willis, L. McKenna, S. McDermott, G. O'Donell, A. Barrett, B. Rasmusson, A. Nordgaard, C.E.H. Berger, M.J. Sjerps, J.-J. Lucena-Molina, G. Zadora, C. Aitken, T. Lovelock, L. Lunt, C. Champod, A. Biedermann, T.N. Hicks, F. Taroni, *ENFSI Guideline for Evaluative Reporting in Forensic Science*, European Network of Forensic Science Institutes. http://enfsi.eu/sites/default/files/documents/external_publications/m1_guideline.pdf, 2015.
16. A.N. Groth, A.W. Burgess, Sexual dysfunction during rape, *The New England Journal of Medicine* 297(14) (1977) 764–766.
17. J.E. Allard, The collection of data from findings in cases of sexual assault and the significance of spermatozoa on vaginal, anal and oral swabs, *Science & Justice* 37 (1997) 99–108.
18. A. Davies, E. Wilson, The persistence of seminal constituents in the human vagina, *Forensic Science* 3 (1974) 45–55.
19. P. Gill, *Misleading DNA Evidence: Reasons for Miscarriages of Justice*, Elsevier, 2014.
20. G.E. Meakin, B. Kokshoorn, R. van Oorschot, B. Szkuta, Evaluating forensic DNA evidence: Connecting the dots, *Forensic Science* 3 (4) (2021) e1404.

6 Fundamentals of Bayesian Networks

Duncan Taylor

CONTENTS

DOI: 10.4324/9781003273189-6

	6.8.4	Step 4	230
	6.8.5	Step 5	230
	6.8.6	Step 6	231
	6.8.7	Step 7	232
	6.8.8	Evaluation	232
6.9	Practice Questions		233
6.10	References		235

6.1 INTRODUCTION

In the previous chapter three examples were worked through with the goal of deriving an *LR* formula considering activity-level propositions. It was seen how even for relatively simple scenarios the evaluation could quickly become complex when issues of dependence, multiple items or multiple tests on an item were involved. Even for the case 3 example where a single item was tested, the inclusion of DNA amount in the result category required the consideration of many factors. A potential solution to having to manually derive complex *LR* formulas is the use of a tool known as a Bayesian network (also sometimes called a belief network, or a directed acyclic graph). Bayesian networks are a graphical tool that represents probabilistic relationships and dependencies by the use of symbols and arrows. In this way a Bayesian network is similar to the pathways of transfer that were used in the *LR* derivations of the previous chapter. Unlike those pathways, BNs have tables of probability that underpin each factor, thereby allowing for a more formal and structured form. Due to their ability to represent general probabilistic inferences, BNs have wide-ranging uses, and there have been numerous applications of BN within science ranging from quality control monitoring [1], preparation for legal challenges [2] and forensic science in complex pedigree evaluation [3–5], to DNA profile mixture evaluation [6], helping to address activity-level propositions [7] (see e.g. [8] for a review) and for crime-level investigation [9, 10].

In this chapter the fundamental basics of BNs will be shown, how they can be formally structured for evaluations that consider activity-level propositions and the ways they can be extended to accommodate changing information. BNs are simply a graphical reflection of the *LR*s that can be derived by hand, and this will be demonstrated by developing BNs in the same three cases for which *LR*s were derived in the previous chapter. Given BNs are a graphical representation, full advantage is taken of the visual aspect of BN construction so that they become more than simply a means to assign an *LR*, but also through the architecture of their layout alone the lines of thought become apparent.

6.2 WHAT ARE BAYESIAN NETWORKS?

The term 'Bayesian Network' was first coined by Judea Pearl in 1985 [11] to represent the idea that in the use of these graphs for probabilistic inference there are aspects of subjectivity in assignments and that the process for updating information makes use of Bayesian conditioning. A Bayesian network has two main components, a directed acyclic graph (DAG) and a set of conditional probability tables. A DAG is a graph (with foundations in the mathematics of graph theory) where the vertices represent some object. In Bayesian network terms these vertices are called 'nodes' and represent random variables for which there exists some uncertainty. Nodes are connected to one another by the fact that effects on the random variable represented by one node can influence our belief in the properties of a random variable represented in another node. In formal terms this is a causal probabilistic dependence and is graphically represented in a DAG by an arrow between two nodes, the direction of which indicates the direction of causation (called an 'edge' or an 'arc'). Figure 6.1 shows an example of a DAG that has two nodes, A and B, where B is dependent on A as represented by an arc. A biological example might be that the alleles possessed by the parent (A) of a child (B) will influence our belief in the alleles that the child will possess. In fact, this is the terminology that

FIGURE 6.1 Two-node DAG showing the dependence of node B on node A.

is commonly used with BNs, i.e. a parent node has an arc leaving it and pointing to its child node (or multiple children nodes). If a node in a BN has no parents, then it is considered to be a 'founder' node.

A fundamental property of a DAG is that it contains no loops (hence acyclic), i.e. if starting at node X, then there is no pathway of arcs that can be followed which will return to node X. Figure 6.2 shows some examples of DAGs and a non-DAG.

The overall placement of nodes and arcs within a BN is referred to as the BN architecture. There are different types of connections that can be constructed within the architecture of a BN, and at the simplest level, these are serial connections, convergent connections and divergent connections as shown in Figure 6.3.

A serial connection represents a situation where (as shown in Figure 6.3 left) node C is dependent on node B and node B is dependent on node A. A convergent connection (such as that shown in Figure 6.3 mid) is one where there are multiple parents of a single child node. The diagram in Figure 6.3 mid shows two parents, but there can be any number. A divergent connection (such as that shown in Figure 6.3 right) is one where a single parent has multiple child nodes (again any number). As will be seen during the construction of BNs for casework, each of these types of connection has its own place representing different real-world events.

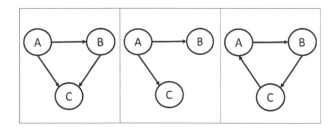

FIGURE 6.2 Left and mid panels show three-node DAGs, and the right panel shows a graph that is cyclic and hence not a DAG.

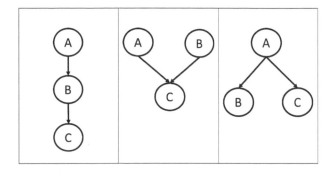

FIGURE 6.3 The left panel shows a serial connection, the mid panel shows a convergent connection and the right panel shows a divergent connection.

Note that there is no connection in Figure 6.3 left from node A directly to node C. This represents a situation, where if information about the state of B is known, then node A will not provide any information that can further inform us about node C. For example, imagine that A, B and C represent the genotypes of a grandparent, parent and child within a family, respectively. If knowledge of the grandparent's genotype is available, then this can be used to make predictions (with associated probabilities) on the genotypes of the parent and the child. If knowledge of the genotype of the parent is also available, then this gives more direct information about the possible genotypes of the child, and in this case the genotype of the grandparent can be removed without any loss of information about the genotype of the child. In formal terms the nodes A and C are 'd-separated' by node B. This concept becomes very important in BN evaluation as it tells us that the probability of the BN (and the states of any nodes for which observations are available) can be calculated by the probability of each node and that the probability of each node depends only on the possible states of its parent(s). This property is returned to later.

Continuing with the real-world analogy of a pedigree containing a grandparent, parent and child then if the genotype of the child is known, it is also possible to draw inferences about what the genotypes of the parent might be, and what the genotype of the grandparent might be. So too this is true of information propagation throughout nodes of a BN, i.e. information about any of the nodes can affect others, even in the opposite direction that the arrows point. This is due to the laws of probability and Bayesian inference, and such inference will be demonstrated with an example in the coming section.

6.3   HOW DO THEY WORK?

6.3.1   A Two-Node Paternity Example

As mentioned earlier, a BN, as well as being a DAG, also has conditional probability tables for each node. If each node represents a factor, say the genotype of a parent, then consider that there are multiple states which that factor can take. For example, the node 'genotype of a parent' might contain states that represent combinations of all alleles observed in a population. Each of these states will have an associated prior probability (that is the probability expected for the state to take before any specific information is given about the node or related nodes). In the case of genotypes, the probability might be the probability expected based on allele frequencies known to exist in a population. Imagine a population with alleles X, Y and Z with frequencies 0.5, 0.49 and 0.01, respectively. A person in this population (i.e. a parent) could possess genotypes, G_P, which make up any combination of alleles X, Y and Z with the following prior probabilities:

Genotype $[X,X]$: $\Pr(G_P = [X,X]) = p_X^2 = (0.5)^2 = 0.25$

Genotype $[X,Y]$: $\Pr(G_P = [X,Y]) = 2p_X p_Y = 2(0.5)(0.49) = 0.49$

Genotype $[X,Z]$: $\Pr(G_P = [X,Z]) = 2p_X p_Z = 2(0.5)(0.01) = 0.01$

Genotype $[Y,Y]$: $\Pr(G_P = [Y,Y]) = p_Y^2 = (0.49)^2 = 0.2401$

Genotype $[Y,Z]$: $\Pr(G_P = [Y,Z]) = 2p_Y p_Z = 2(0.49)(0.01) = 0.0098$

Genotype $[Z,Z]$: $\Pr(G_P = [Z,Z]) = p_Z^2 = (0.01)^2 = 0.0001$

Note that logically the sum of the prior probabilities of all states of the node must sum to 1 (as they do in this instance). Consider now the very simple, one-node BN as shown in Figure 6.4, which has the values shown in Table 6.1 for the different states this node can take.

The desire is to model the way that the genotype of the parent can be used to update belief in the genotype of a child, G_C (or indeed how knowledge of a child's genotype can update belief in the parent's genotype). A node is added to the BN shown in Figure 6.4 to create a two-node BN seen in Figure 6.5. There is an arc drawn from the 'Genotype of parent' node to the 'Genotype of

FIGURE 6.4   Single-node BN representing the genotype of a parent.

TABLE 6.1

Probabilities for Genotype States of 'Genotype of Parent' Node

Genotype of parent		
	XX	0.25
	XY	0.49
	XZ	0.01
	YY	0.2401
	YZ	0.0098
	ZZ	0.0001

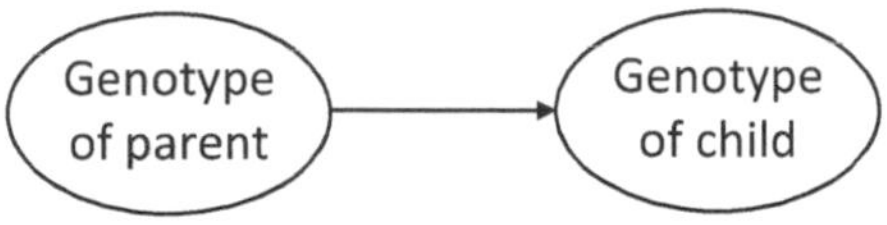

Genotype of parent ────▶ Genotype of child

FIGURE 6.5   Two-node BN representing the genotype of a parent and their child.

child' node which represents a causal probabilistic dependence, i.e. there is a natural order which indicates that the genotype of the parent will dictate what the child's genotype will be; however, the genotype of the child cannot dictate what the parent's genotype will be as that is already set. Note an important distinction here between the dependency of one event on another in only one direction; however, our knowledge can be updated in either direction.

Logically, the child can also possess any of the genotypes that the parent can possess, but unlike the 'genotype of parent' node the 'genotype of child' node has a dependency, which must be taken into account when assigning prior probabilities. For example, if the parent has genotype $[X,X]$, then the child will also certainly possess an X allele (discounting the possibility of mutation for the time being) and so could be $[X,X]$, $[X,Y]$ or $[X,Z]$, but nothing else. The child will possess these genotypes with probabilities 0.5, 0.49 and 0.01, respectively. These probabilities are based on the probability that an X allele has been passed on to the child from the parent (i.e. a probability of 1), multiplied by the probability of the other allele that the child possesses coming from someone at random in the population. Probabilistically this is written as:

$$\Pr(G_C = [X,X] \mid G_P = [X,X]) = 1 \times 0.5 = 0.5$$

$$\Pr(G_C = [X,Y] \mid G_P = [X,X]) = 1 \times 0.49 = 0.49$$

$$\Pr(G_C = [X,Z] \mid G_P = [X,X]) = 1 \times 0.01 = 0.01$$

$$\Pr(G_C = [Y,Y] \cup G_C = [Y,Z] \cup G_C = [Z,Z] \mid G_P = [X,X]) = 0$$

Again, logically given a state for the parent genotype, the probabilities for all possible genotypes that a child can take must equal 1 (but this probability will not be spread across genotypes in the

TABLE 6.2

Conditional Probability Table Showing Probabilities for Genotype of Child Given the States of the Genotype of the Parent

Genotype of parent		XX	XY	XZ	YY	YZ	ZZ
Genotype of child	XX	0.5	0.25	0.25	0	0	0
	XY	0.49	0.495	0.245	0.5	0.25	0
	XZ	0.01	0.005	0.255	0	0.25	0.5
	YY	0	0.245	0	0.49	0.245	0
	YZ	0	0.005	0.245	0.01	0.25	0.49
	ZZ	0	0	0.005	0	0.005	0.01

same way as if it were not conditioned on any information). If the same logic is followed through for all possibilities, then the conditional probabilities are obtained for the 'genotype of child' node as shown in Table 6.2. Table 6.2 can be read as the probability that a child will possess a specific genotype (listed in the vertical genotype column), given the genotype of the parent (listed in the horizontal genotype row) is equal to the value in the table that corresponds to the row and column of interest. For example, the probability of a child possessing the genotype [X,Y] given the parent is [Y,Z] is 0.25 (that comes from a probability of 0.5 for inheriting the [Y] and 0.5 for a random person in the population providing the [X]), or:

$$\Pr(G_C = [X,Y] \mid G_P = [Y,Z]) = 0.25$$

Note how in Table 6.2 the sum of the conditional probabilities of the genotype of the child node for each state of the genotype of the parent node is 1. If the values are entered for the prior probabilities of a parent as shown in Table 6.1 and for the child as shown in Table 6.2, the prior probabilities for either individual possessing any genotype are shown in Figure 6.6.

Note that the BN shown in Figure 6.6 has been produced with the software HUGIN. There are a number of software programs that can be used to construct BNs and some of these are discussed later in Section 6.4.1. In HUGIN BNs are represented in the way that has been described, with ovals representing nodes and dependencies (or arcs) displayed as arrows between nodes. When a BN is being 'run' in HUGIN, the probabilities for the states at any node can be displayed in a 'monitor window'. In Figure 6.6 these are the small windows (or boxes) below the nodes. The values shown by the software HUGIN are percentages (hence will be the probability value multiplied by 100). As well as for the numerical value for the probability there is also a coloured bar which provides a graphical representation. If the bars are green, then this is a state in a node that has not been informed by an observation. If an observation at a particular node is made (or assumed to have been made), then the bar will turn red, and the observed state will have 100% probability. Providing a node in a BN with information about an observation is referred to as 'instantiating' the BN. If the

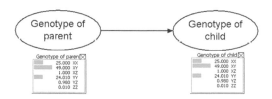

FIGURE 6.6   Two-node network showing prior probabilities for states of each node.

reader is unfamiliar with BNs, then it is suggested that they replicate the examples shown in this chapter as they are being described.

In Figure 6.6 it can be seen that the prior probability of the parent possessing a genotype is simply a reflection of the expected frequency of that genotype in the population. However, it can also be seen that this is the case for the genotype of the child. As the genotype of the child is dependent on the genotype of the parent, each genotype of the child must consider all possible genotypes the parent can possess and the prior that the parent will possess it. This is an example of the law of total probability being applied. In explanation, consider the genotype $[X,X]$ for the child. The probability be sought is:

$$\Pr(G_C = [X,X])$$

But in order to evaluate this term all the possible genotypes of the parent must be considered:

$$\Pr(G_C = [X,X]) =$$
$$\Pr(G_C = [X,X] \mid G_P = [X,X]) \Pr(G_P = [X,X]) +$$
$$\Pr(G_C = [X,X] \mid G_P = [X,Y]) \Pr(G_P = [X,Y]) +$$
$$\Pr(G_C = [X,X] \mid G_P = [X,Z]) \Pr(G_P = [X,Z]) +$$
$$\Pr(G_C = [X,X] \mid G_P = [Y,Y]) \Pr(G_P = [Y,Y]) +$$
$$\Pr(G_C = [X,X] \mid G_P = [Y,Z]) \Pr(G_P = [Y,Z]) +$$
$$\Pr(G_C = [X,X] \mid G_P = [Z,Z]) \Pr(G_P = [Z,Z])$$

which, when enumerated (using values from Table 6.2) is:

$$\Pr(G_C = [X,X]) =$$
$$0.5 \times 0.25 +$$
$$0.25 \times 0.49 +$$
$$0.25 \times 0.01 +$$
$$0 \times 0.2401 +$$
$$0 \times 0.0098 +$$
$$0 \times 0.0001$$
$$= 0.25$$

In compact notation the equation for the child possessing genotype j, $G_{C,j}$ can be written as:

$$\Pr(G_{C,j}) = \sum_i \Pr(G_{C,j} \mid G_{P,i}) \Pr(G_{P,i})$$

where $G_{P,i}$ is genotype i in the parent. The same process can be carried out for all other genotypes of the child to reveal the probabilities displayed in the 'Genotype of child' table in Figure 6.6. Intuitively this makes sense as the system has not been provided with any information and so anyone in the BN will have a prior probability of possessing genotypes in the frequency that they are expected in the population. As a further demonstration of this point, it could be imagined that the parent also has a parent, whose genotype is unknown. Modelling this additional person in the BN should therefore not affect the prior probabilities for the genotype of the parent from what they are shown in Table 6.1.

The desire is now to supply the BN with some information. This is where the real usefulness of BN is realised, as information can be supplied to any node by 'telling' it what is known (or what has been observed). This process of specifying the state of a node that represents an observation

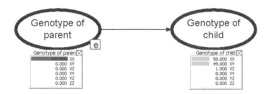

FIGURE 6.7 Two-node network showing posterior probabilities for states of each node when the genotype of the parent node has been instantiated.

is called instantiation, and when instantiation occurs, the information is passed to other nodes (or propagated) in accordance with the laws of probability and Bayesian updating. The mechanical way that this is achieved is through propagation algorithms (the reader is directed to [12] for more information on these algorithms). Information is either propagated through the network in the direction of arcs, or back propagated in order for all nodes to be updated.

In Figure 6.7 the results of instantiating the genotype of parent node to value [X,X] are shown. In other words the BN is being told that the parent genotype is known to be [X,X] and the interest lies in the probabilities of different genotypes being observed in a child, given this information. As the BN started with prior probabilities for the genotypes the child could possess and instantiation is updating those probabilities with information, the probabilities shown in Figure 6.7 are now considered posterior probabilities.

Instantiating the genotype of the parent to [X,X] is effectively working in the first column of Table 6.2 that corresponds to that genotype. In Figure 6.7 the genotypes that the child can possess have posterior probabilities as they are shown in that column of Table 6.2. Intuitively it is expected to obtain the same posterior probabilities for the genotype of the parent if the genotype [X,X] were instantiated in the child, as was obtained in the child when genotype [X,X] was instantiated in the parent. In other words, the same updating of genotype probabilities is expected regardless of which individual in the BN is instantiated. From a biological point of view, the BN is modelling the relationship between a child and parent, and this can equivalently be considered to be that a parent passes one allele to their child, or that a child inherits one allele from their parent. This effect is shown in Figure 6.8.

In order to show how the back-propagation has occurred in Figure 6.8 for the parental genotype to be updated, recall the formula for Bayesian inference:

$$Pr(A \mid B) = \frac{Pr(B \mid A)Pr(A)}{Pr(B)}$$

Let A be that the parent possesses genotype [X,X] and B be that the child possesses genotype [X,X]. The Bayesian inference formula above can then be written as:

$$Pr(G_P = [X,X] \mid G_C = [X,X]) = \frac{Pr(G_C = [X,X] \mid G_P = [X,X])Pr(G_P = [X,X])}{Pr(G_C = [X,X])}$$

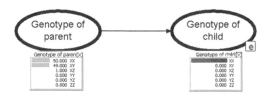

FIGURE 6.8 Two-node network showing posterior probabilities for states of each node when the genotype of the child node has been instantiated.

The probability of the genotype of the child being [X,X] can be determined using the law of total probability so that:

$$\Pr(G_P = [X,X] \mid G_C = [X,X]) = \frac{\Pr(G_C = [X,X] \mid G_P = [X,X])\Pr(G_P = [X,X])}{\sum_i \Pr(G_C = [X,X] \mid G_{P,i})\Pr(G_{P,i})}$$

From Table 6.2, the value for $\Pr(G_C = [X,X] \mid G_P = [X,X])$ is 0.5 and it is known from Table 6.1 that the value of $\Pr(G_P = [X,X])$ is 0.25. The value for the sum of conditional probabilities for the genotype of the child, given all possible genotypes for the parent, was already enumerated during the calculation of the posterior probabilities of the genotypes of the child when the parent was instantiated, and its value was found to be 0.25. This gives the posterior probability of the parent possessing genotype [X,X] as:

$$\Pr(G_P = [X,X] \mid G_C = [X,X]) = \frac{0.5 \times 0.25}{0.25} = 0.5$$

as seen in Figure 6.8. So too could the posterior probabilities of the other genotypes for the parent be worked through and it would be found that they possessed the values shown in Figure 6.8.

6.3.2 A CONVERGENT THREE-NODE BN EXAMPLE

Rather than continuing with the parent/child genotypes (which could be modelled with a convergent three-node network with two parents and one child), a convergent BN will be demonstrated for a classic logic problem known as the Monty Hall problem. This example extends the Bayesian network to a three-node example as well as demonstrating the required calculations for a convergent dependency structure. It also moves away from pure biological issues to show how BNs can be used for standard logic problems with everyday issues.

Imagine you are on a game show and given the choice of three doors. You will win whatever is behind the door you end up choosing. Behind one door is a car and behind the other doors are goats and, for the purpose of this example, it is assumed you wish to win the car. You pick a door, say number 1, and the host (Monty Hall), who knows what's behind the doors, opens another door, say number 2, which has a goat. He then asks you, 'Do you want to pick door number 3 or stay with door number 1?'. The problem is to determine whether:

1. You are more likely to win the car if you switch doors,
2. You are more likely to win the car if you stick with your original door or
3. It makes no difference to your chance of winning the car whether you switch or stay.

Most people when posed with this problem will provide answer number 3, that it makes no difference to your chance of winning the car whether you stick with your original choice of doors or whether you switch doors. This belief comes from the thinking that at that point where you must choose to switch or stay there are two doors left, one with a car and the other with a goat, and you currently have one of these chosen. This appears then to be a simple 50:50 chance of whether you already have the correct door, and so there is no benefit (but equally no loss) from switching. However, this thinking does not take into account one important piece of information: The door that Monty has revealed. Using the information that Monty has revealed, a door that held a goat, allows us to update our knowledge of where the car might be. In fact, it turns out to be advantageous to always switch doors when given the opportunity and it doubles your chance of winning the car.

This problem was famously known as being printed in *Parade* magazine in a column called 'Ask Marilyn' (authored by Marilyn vos Savant, who at one point held the world record for highest IQ, quoted as 228 in the Guinness Book of World Records in the 1980s). Many readers of the column refused to believe that switching doors was beneficial and around 10,000 readers (including 1000 with PhDs) wrote to the magazine to dispute it, claiming vos Savant was wrong.

In order to intuitively see the benefit of switching, imagine that instead of 3 doors there were 100 doors, behind which were 99 goats and 1 car. You randomly choose a door and have a probability of 0.01 of having chosen the one with the car. The car is behind one of the other 99 doors with a probability of 0.99. Monty now reveals what's behind 98 of the doors, all being goats, and leaves a single unopened door (in addition to the unopened one you originally chose), asking if you would like to switch. In simplistic terms the probability that you chose the right door to begin with (by chance) is still 0.01, but the probability of the car being behind another door of 0.99 has now all been concentrated down to a single door. It is clear to see here that your probability of winning the car is dramatically improved (by 99 times) if you switch.

This is a good example of a seemingly simplistic logic problem that is in reality quite difficult to intuit. It is exactly the sort of situation where working through the problem probabilistically becomes important to understand the effects of information on the significance of results. Let's think of this problem in terms of a BN. There are three variables that must be considered:

- The door holding the car
- The door that the contestant chooses
- The door that Monty reveals

and therefore, there will be three nodes to the BN. Start with the BN seen in Figure 6.9:

Thinking about the dependencies between these factors, the following can be considered:

- Is the door I choose dependant on the door which the car sits behind? – No, it cannot be as I am not aware of the door that the car sits behind.
- Is the door I choose dependant on the door that Monty chooses? – No, it cannot be as I make my choice before Monty.
- Is the door that the car sits behind dependent on my choice? – No, it cannot be as the car is placed behind a door before I even make my choice.
- Is the door that the car sits behind dependent on Monty's choice? – No, it cannot be as the car is placed behind a door before Monty makes a choice.
- Is the door that Monty chooses dependent on my choice? – Yes, my choice will affect Monty's as Monty cannot reveal a door I have already chosen.
- Is the door that Monty chooses dependent on the position of the car? – Yes, as Monty will not reveal a door that the car sits behind.

and so, there are two dependencies in the BN, one each from the door that holds the prize and the door I choose to the child node representing the door Monty reveals. This represents a simple convergent BN. While this is a non-forensic example of a BN construction, the same process of identifying factors that are important in the evaluation (and so will become nodes in a BN) and identifying dependencies in the dataset (which will become arcs in the BN) is carried out.

FIGURE 6.9   The nodes in the BN that will be used to evaluate the Monty Hall problem.

Consider now the prior probabilities that will populate the conditional probability tables sitting behind the nodes. In this case all three nodes will have the same possible states:

- Door 1
- Door 2
- Door 3

The two nodes, 'Door I choose' and 'Door that holds prize', are founder nodes (i.e. they have no parents) and so probabilities can be assigned to the three door states without having to consider any other factors. For these nodes, it is most appropriate to assign equal prior probabilities of 1/3, i.e. prior to the game being played (when no information is available other than the rules and structure of the game) there is an equal probability that the car will be sitting behind any of the doors, and there is an equal probability that any of the doors will be chosen by the contestant. The probability tables are given in Table 6.3.

The node 'Door Monty reveals' depends on both nodes 'Door I choose' and 'Door that holds prize'. There are two general possibilities that need to be modelled:

1) If I choose a door initially that holds a goat, then Monty will certainly have to reveal the one remaining door that holds the other goat. In this case the choice for Monty is fixed.
2) If I happen to initially choose the door that holds the car, then Monty will have two doors to choose from to reveal, both with goats, and he will choose between them with equal probability.

Apply this thinking to produce a conditional probability table for node 'Door Monty reveals' as shown in Table 6.4. Seen in Table 6.4 for the first time the effect of having a convergent relationship in a BN, namely that each possible combination of states for each of the parent nodes is considered. The order that the parent nodes are listed in Table 6.4 does not matter (i.e. in Table 6.4 the node 'Door I choose' could be above 'Door that holds the prize'). All that will change from a different

---

TABLE 6.3

Conditional Probability Tables for Nodes 'Door I Choose' (Left) and 'Door That Holds Prize' (Right) in the Monty Hall Game

Door I choose	1	1/3	Door that holds the prize	1	1/3
	2	1/3		2	1/3
	3	1/3		3	1/3

TABLE 6.4

Conditional Probability Tables for Nodes 'Door Monty Reveals' in the Monty Hall Game

Door that holds the prize		1			2			3		
Door I choose		1	2	3	1	2	3	1	2	3
Door Monty Reveals	1	0	0	0	0	0.5	1	0	1	0.5
	2	0.5	0	1	0	0	0	1	0	0.5
	3	0.5	1	0	1	0.5	0	0	0	0

ordering of parent nodes is the order of the entries in the cells so that the probabilistic outcome remains the same for the same combination of parental states.

Reading down the columns of Table 6.4 identifies the probability of the states shown on the left-hand side. For example, in the right-hand (last) column of Table 6.4 the door that holds the prize is door 3, and the door I chose was also 3. Given this information, according to the rules of the game Monty cannot also choose door 3 and so the row relating to door 3 for the door that Monty reveals (i.e. the last row of Table 6.4) must be 0. This can be seen in the bottom right value of Table 6.4. This leaves Monty with a choice of revealing the goat behind either door 1 or door 2, which he does with equal probability. This can be seen in the remaining two conditional probability values in the right-hand column of Table 6.4.

Taking into account the nodes shown in Figure 6.9, the dependencies between the nodes as discussed above, and incorporating the conditional probabilities as shown in Tables 6.3 and 6.4 lead to the BN shown in Figure 6.10. Figure 6.10 shows the prior probabilities for the states of each node before any information (i.e. any door choices by either the contestant or Monty) is obtained. As expected by simple intuition (but also possible to calculate using the probability values in Tables 6.3 and 6.4), the initial state is that there are equal probabilities for any of the door states in any of the nodes.

Note that our knowledge represents that of an outside observer and not any actor within the evaluation, i.e. we do not have the knowledge of the person who has placed the items behind the doors, nor do we know what Monty knows, nor do we know which door the contestant is going to choose before they make the choice. The BN construction however allows any combination of possible outcomes, so that it is possible to provide the information to the BN as it is obtained and see what probabilistic inferences are possible if that, or different, information had been obtained. This concept is important when moving to constructing BNs for casework examples and is one of the fundamental differences in the way the evaluation proceeds for a case compared to the derivation of the *LR* in the previous chapter. When an *LR* is derived, it is done with the knowledge of the results and this knowledge becomes a fundamental part of the evaluation (i.e. all pathways are considered that could lead to that result). When constructing a BN, the outcomes are usually an exhaustive set of all possible observations. In fact, the BN can be constructed when only the case circumstances, and not the forensic observations, are known (this is how the practice of case assessment and interpretation can be carried out as explained in Chapter 1). Of course, all possible outcomes could also be considered with *LR* derivations; however, it is likely to come at the cost of multiple derivations (one for each set of outcomes).

Having constructed the BN to represent the circumstances of the Monty Hall example, it is now possible to provide the BN with some information, specifically the choice of a door by the

FIGURE 6.10   Initial construction of a BN that models the Monty Hall problem, showing the conditional probability tables, before any information is known.

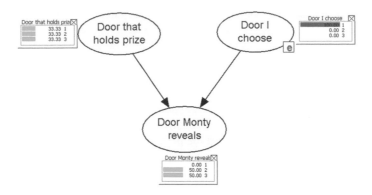

FIGURE 6.11 The Monty Hall BN, showing the conditional probability tables, when the contestant choice is known and instantiated.

contestant. Figure 6.11 shows the updated probabilities for each node within the BN, having propagated the information throughout the network. In Figure 6.11 the example is that the contestant has chosen door number 1 and this is identified by the probability bar for the 'Door I choose' node being instantiated (red), and all the probability being placed on the door 1 state. In other words, node 1 has been instantiated to the door 1 state.

Note that in Figure 6.11 it can be seen how the choice of the door by the contestant has changed the probabilities of Monty's choice. Because in the rules of this example, Monty cannot choose the same door as the contestant, and this is ensured by the probabilistic setup in Table 6.4 Monty is now restricted to choosing one of door 2 or 3. Again, because our knowledge represents that of an outside observer and not Monty, it is not known which door the car is behind, and the choice of the door by the contestant does not provide any information about that fact. In the BN it can be seen how the probabilities of the states in the 'Door that holds the prize' node are unchanged from their initial state. To use earlier terminology the 'Door that holds the prize' node is d-separated from the 'Door I choose' node by the 'Door Monty reveals' node. Given the knowledge of contestant door choice, it would not be expected for the probabilities to have changed in the 'Door that holds the prize' node from their prior state.

Finally, information is provided to the BN about the door that Monty reveals, by instantiating one of the two possible options that do not have 0 probability in Figure 6.11. The resulting propagation of this information through the BN to ultimately update the probabilities in the states of the 'Door that holds the prize' node is shown in Figure 6.12. Here 'Door 2' has been instantiated for Monty's

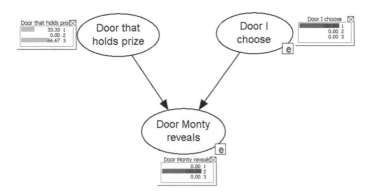

FIGURE 6.12 The Monty Hall BN, showing the conditional probability tables, when the contestant choice and Monty's choice are known and instantiated.

choice but could have also been 'Door 3' and the resulting intuition regarding the advantage of door switching by the contestant would still be seen. The reader is invited to construct the BN for the Monty Hall problem and prove this to themselves. This may also be a good time to work through the first three practice questions so that some familiarity with conditional probabilities and BN construction is gained before continuing.

Figure 6.12 demonstrates the solution to the Monty Hall problem, specifically that the door the contestant originally chose has half the probability of revealing a car than the other unrevealed door. Start first by considering the probability that the contestant (C) has chosen door 1 ($D1$), Monty (M) has chosen door 2 ($D2$) and the prize (P) is behind door 1 ($D1$). Using the law of Bayesian inference, the probability that door 1 holds the car is sought, given Monty has revealed door 2:

$$\Pr(P = D1 \mid M = D2, C = D1) = \frac{\Pr(M = D2 \mid P = D1, C = D1) \times \Pr(P = D1 \mid C = D1)}{\Pr(M = D2 \mid C = D1)}$$

Note that in the formula above the contestant's choice is a conditional piece of information throughout. The marginalisation constant (the denominator on the right-hand side) in the formula above is expanded using the law of total probability to consider the probability of Monty choosing door 2, given that the prize could be behind any door.

$$\Pr(P = D1 \mid M = D2, C = D1) = \frac{\Pr(M = D2 \mid P = D1, C = D1) \times \Pr(P = D1 \mid C = D1)}{\sum_{i=1}^{3} \Pr(M = D2 \mid C = D1, P = Di) \Pr(P = Di)}$$

which expands to:

$$\Pr(P = D1 \mid M = D2, C = D1) = \frac{\Pr(M = D2 \mid P = D1, C = D1) \times \Pr(P = D1 \mid C = D1)}{\begin{pmatrix} \Pr(M = D2 \mid C = D1, P = D1)\Pr(P = D1) + \\ \Pr(M = D2 \mid C = D1, P = D2)\Pr(P = D2) + \\ \Pr(M = D2 \mid C = D1, P = D3)\Pr(P = D3) \end{pmatrix}}$$

The values from the equation above can be substituted with their known quantities (i.e. all the terms on the right-hand side of the equality sign), giving:

$$\Pr(P = D1 \mid M = D2, C = D1) = \frac{\frac{1}{2} \times \frac{1}{3}}{\left(\frac{1}{2} \times \frac{1}{3} + 0 \times \frac{1}{3} + 1 \times \frac{1}{3}\right)} = \frac{1}{3}$$

which is the value shown in the 'door 1' state of the 'Door that holds the prize' node in Figure 6.12. Now the same process can be carried out considering the probability that the prize is behind door 3. Formulaically this is expressed as:

$$\Pr(P = D3 \mid M = D2, C = D1) = \frac{\Pr(M = D2 \mid P = D3, C = D1) \times \Pr(P = D3 \mid C = D1)}{\sum_{i=1}^{3} \Pr(M = D2 \mid C = D1, P = Di) \Pr(P = Di)}$$

And enumerating in the same way as before yields:

$$\Pr(P = D3 \mid M = D2, C = D1) = \frac{1 \times \frac{1}{3}}{\left(\frac{1}{2} \times \frac{1}{3} + 0 \times \frac{1}{3} + 1 \times \frac{1}{3}\right)} = \frac{2}{3}$$

which is the value shown in 'door 3' state of the 'Door that holds the prize' node in Figure 6.12. In other words, there is twice the probability that the car is behind the door not initially chosen by the contestant, so they should switch (given the assumption that they do wish to win the car).

6.3.3 A DIVERGENT BN EXAMPLE

The final example of a simple BN to explore is that of a divergent BN. The example returns to a forensic example and again will use a biological example, but note that this scenario could apply to any evidence type that has matching traces. Imagine that a crime has been committed and at the scene are two blood spots, which are believed to have been left during the commission of the crime. The two blood spots are sampled, and DNA profiles obtained have single source DNA profiles that match each other (i.e. possess the same alleles at each locus). A suspect is arrested, and a reference sample is generated that matches the profile seen in the two blood spot samples taken from the scene. The frequency of the profile from the reference of the suspect (and in both blood spot samples) is 1 in 1 million in the relevant population (i.e. that of potential alternate offenders). The suspect denies ever being at the scene and says that the blood must be from someone else.

The BN construction is started considering three nodes. The first node is a source-level propositional node with the states *Hp* and *Hd*, where:

Hp – relates to the prosecution proposition that the suspect is the source of the blood
Hd – relates to the defence proposition that the suspect is not the source of the blood

The second node represents the first scene's blood spot, yielding a DNA profile that matches the suspect. The third node represents the second scene's blood spot, yielding a DNA profile that matches the suspect.

In the BN construction for this scenario, the probabilities of either of the scene's blood having a profile that matches the suspect will depend on whether the suspect has donated the blood spots. These are the only dependencies in the BN and so there are only two arcs starting with the propositional node and entering the two matching nodes. The prior probabilities for the two states of the propositional node, *Hp* and *Hd*, are each set to 0.5. This does not imply that these are the prior probabilities that should be used by the recipient of this evaluation. The equal assignment of prior probabilities represents an assignment of convenience, i.e. it allows for the *LR* to be equal to the posterior odds:

$$\frac{\Pr(Hp \mid E)}{\Pr(Hd \mid E)} = \frac{\Pr(E \mid Hp)}{\Pr(E \mid Hd)} \text{ when } \Pr(Hp) = \Pr(Hd)$$

This fact becomes useful when dealing with casework evaluations, as will be demonstrated later on in the chapter.

The probabilities of the matching nodes are based on the frequency of the profile in the population and are shown in Table 6.5.

TABLE 6.5

Conditional Probability Table for the Scene Sample Having a Profile Matching the Reference of the Suspect

Source of blood spots		*Hp*	*Hd*
Scene sample profile matches suspect	Match	1	0.000001
	No match	0	0.999999

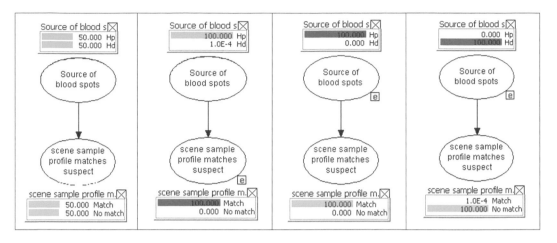

FIGURE 6.13 BN construction for a scene sample with DNA profile matching a suspect (left) and instantiations of either the results node (middle left) or the proposition node (middle right and right).

Starting by considering just one of the scene stains first, the BN in Figure 6.13 can be constructed. The far-left panel of Figure 6.13 shows the BN with probabilities entered as described and no information instantiated. In the middle-left panel of Figure 6.13 is the BN with the results node 'Scene sample profile matches suspect' instantiated to show that the scene profile matches the reference of the suspect. When propagated through the BN, the result is that the probabilities for *Hp* and *Hd* states in the propositional node 'Source of blood spots' are approximately 1 and 0.000001, respectively. Two aspects to note here are that the software displays percentages and so the values are 100 times the probability, and the number of significant figures displayed is restricted. In the case of Figure 6.13 the values are shown to three significant figures and so the exact *Hp* probability of 99.9999 has been rounded to 100.00. The ratio of these two values is 1 million.

The values shown in the BN are posterior probabilities, given whatever information has been instantiated in the BN. This means that technically the values seen in the propositional node of Figure 6.13 middle left are the posterior probabilities of the propositions, given the evidence Pr(*Hp*|*E*) and Pr(*Hd*|*E*). Recalling the third principle of evidence evaluation, that scientists can only comment on the evidence and not the propositions, it seems that the principle has been violated by the way that the BN is being used. Indeed, technically it has. To obtain the values for Pr(*E*|*Hp*) and Pr(*E*|*Hd*), the instantiations could occur for each of the *Hp* and *Hd* states in the propositional node (as shown in Figure 6.13 middle right and right, respectively) and the values in the results node taken, which corresponds to the results in the case, i.e. 1 from Figure 6.13 middle right and 0.000001 from Figure 6.13 right. Again, the ratio of these two values is 1 million.

However, because equal prior probabilities have been assigned to the two states of the propositional node the equality $\frac{\Pr(Hp\,|\,E)}{\Pr(Hd\,|\,E)} = \frac{\Pr(E\,|\,Hp)}{\Pr(E\,|\,Hd)}$ holds true. This means it is acceptable to instantiate just the results node, and the ratio of the probabilities for the two states in the propositional node will be the same as the ratio of the probabilities of the state in the results node that corresponds to the evidence, given sequential instantiations of the *Hp* and *Hd* states of the propositional node. The instantiation of the results nodes (such as seen in Figure 6.13 middle left) is a more intuitive way of entering results into a BN, more easily allows for exploration of the effects of different outcomes on the *LR* and is typically the method used in casework evaluations. It will be the way that results of BN instantiations are shown throughout this book. Note that an adjustment will need to be made if unequal priors have been used in the propositional node, or if dealing with propositional nodes that have more than two states.

For completeness, the final way in which the *LR* could be calculated is to consider the probability of the entire BN when the results node is instantiated, and the *Hp* and *Hd* states of the propositional node are instantiated in turn. Remember that the probability of the BN is the probability of the children given their direct parents. In this case, if both the match state of the results node and the *Hp* state of the propositional node were instantiated, then the probability would be $\Pr(Match \mid Hp) \times \Pr(Hp) = 1 \times 0.5 = 0.5$. If the *Hd* state of the propositional node was instantiated, then the probability would be $\Pr(Match \mid Hd) \times \Pr(Hd) = 0.000001 \times 0.5 = 0.0000005$. The ratio of these two probabilities is then 1 million. This method is not normally conducted due to the additional work required to calculate the probabilities but can be useful to identify a situation where the probability of the evidence given either proposition is very low, and may be suggestive of a non-exhaustive set of propositions.

The BN for the example described is shown in Figure 6.14 in its instantiated form (i.e. both nodes for the two evidence profiles instantiated to match).

In Figure 6.14 the propositional node shows probabilities that lead to $LR = 1 \times 10^{12}$ (obtained by the division of the probability of the '*Hp*' state by the probability of the '*Hd*' state in the 'Source of blood stain' node), despite the frequency of the suspect profile only 1 in 1 million. The reason for the large *LR* is that the architecture of the BN shown in Figure 6.14 implies that if the suspect is not the source of the two scene samples, then they must have independently (i.e. in two different people) matched the suspect's reference. In equations:

$$\Pr(Hp \mid Match1, Match2) = \frac{\Pr(Match1, Match2 \mid Hp) \times \Pr(Hp)}{\Pr(Match1, Match2)}$$

And as the matches are independent (according to the BN structure in Figure 6.14) given the propositional node:

$$\Pr(Hp \mid Match1, Match2) = \frac{\Pr(Match1 \mid Hp) \times \Pr(Match2 \mid Hp) \times \Pr(Hp)}{\Pr(Match1, Match2)}$$

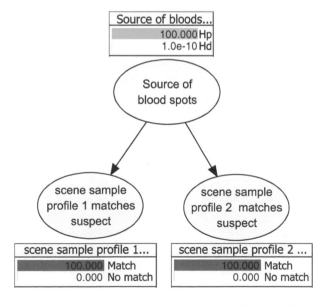

FIGURE 6.14   BN construction for two scene samples with DNA profiles matching a suspect.

Similarly:

$$\Pr(Hd \mid Match1, Match2) = \frac{\Pr(Match1 \mid Hd) \times \Pr(Match2 \mid Hd) \times \Pr(Hd)}{\Pr(Match1, Match2)}$$

So that the *LR* shown in the BN can be calculated by the ratio of these two terms:

$$LR = \frac{\Pr(Hp \mid Match1, Match2)}{\Pr(Hd \mid Match1, Match2)} = \frac{\left[\dfrac{\Pr(Match1 \mid Hp) \times \Pr(Match2 \mid Hp) \times \Pr(Hp)}{\Pr(Match1, Match2)} \right]}{\left[\dfrac{\Pr(Match1 \mid Hd) \times \Pr(Match2 \mid Hd) \times \Pr(Hd)}{\Pr(Match1, Match2)} \right]}$$

$$LR = \frac{\Pr(Match1 \mid Hp) \times \Pr(Match2 \mid Hp)}{\Pr(Match1 \mid Hd) \times \Pr(Match2 \mid Hd)} \times \frac{\Pr(Hp)}{\Pr(Hd)}$$

Noting the fact that in the BN equal prior probabilities were set for the two propositions:

$$LR = \frac{1 \times 1}{0.000001 \times 0.000001} = 1 \times 10^{12}$$

This is unlikely to be the evaluation that is intended. It is more likely that the defence proposition is that a single alternate offender exists who has left both stains. In order to consider a single offender, the BN must be built with an additional dependency. This can either be done by drawing an arc from the first matching node to the second (with probabilities that reflect that if the first node matches then the second one will also match with certainty) or by adding a node that considers the probability that an alternate offender will have a matching profile to the suspect and making the matching nodes both children of this new founder node. Such an architecture is shown in Figure 6.15, with Table 6.6 showing the new conditional probabilities. The 'suspect profile matching AO profile' node is a founder and has two states, 'matching' and 'not matching' with prior probabilities of 0.000001 and 0.999999, respectively (recall AO is shorthand for an alternate offender).

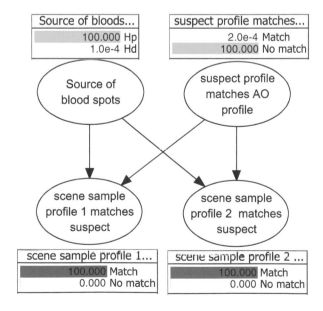

FIGURE 6.15   BN construction for two scene samples with DNA profiles matching a suspect and only a single alternate offender.

TABLE 6.6

Conditional Probability Table for Matching Nodes in the BN Shown in Figure 6.15

	Source of blood spots		Hp		Hd	
Suspect profile matching AO profile		Matching	Not matching	Matching	Not matching
Scene sample profile matches suspect	Match	1	1	1	0
	No match	0	0	0	1

Note that in Table 6.6 there are no probabilities that need to be assigned from data, they are now all 0 or 1 depending on the circumstances of the case and assumptions being made. The final point to make is that while in the example provided, it is relatively clear that the defence proposition would sensibly be that a single alternate offender has left both blood spots, there are situations where these types of decisions are not straightforward. In these instances, it is possible to include the uncertainty in this aspect of the evaluation within the BN. One final BN construction is shown for the evaluation of the two-stain scenario in Figure 6.16, with the left panel showing the results of instantiating the two matching nodes but not specifying whether the same AO has contributed both stains. The middle panel specifies that the same AO has contributed both stains. The right panel specifies that two different AOs have contributed the stains. Whether or not there are one or two AOs is only important under *Hd*. Note that additional consideration is required in the evaluation, i.e. prior probabilities for whether the same or two different AOs are the sources of the stains must be assigned. These probabilities are often (as with this example scenario) the province of the judiciary and not the scientist. The number of offenders has therefore been classically considered as an offence-level consideration [13]. Care must be taken when including different uncertainties in the evaluation, particularly whether their priors should be set by the analyst or the judiciary. If they make up part of a reported evaluation, it should be clear how these values have been set, as well as the effect different values will have on the evaluation. An alternative to including the prior probabilities in the evaluation is to instantiate both states of the new node and determine the *LR* given both possibilities. This can be seen in the middle and right panels of Figure 6.16, which show the same *LR*s as Figures 6.14 and 6.15. The same *LR*s are obtained in Figure 6.16 (middle) as in Figure 6.15, and Figure 6.16 (right) as in Figure 6.14 (right). This is due to the fact that these pairs of BN are making the same assumptions, just explicitly in Figure 6.16 through instantiation, rather than implicitly in Figures 6.14 and 6.15 through the BN architecture.

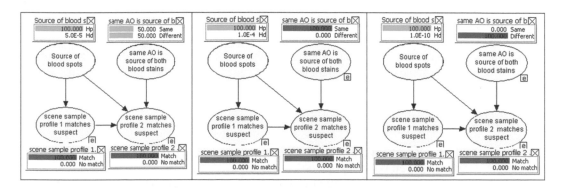

FIGURE 6.16   BN construction for the two-stain scenario showing the results of instantiating the two matching nodes but not specifying whether the same AO has contributed both stains (left), specifying the same AO has contributed both stains (middle) and specifying the two different AOs have contributed the stains (right).

TABLE 6.7

Conditional Probability Table for Matching Nodes in the BN Shown in Figure 6.16

Source of blood spots		Hp				Hd			
Same AO source of both blood stains		Same		Different		Same		Different	
Suspect profile matching AO profile		match	no match	match	no match	match	no match	match	no match
Scene sample profile matches suspect	Match	1.0	1.0	1.0	1.0	1.0	0.0	0.000001	0.000001
	No match	0.0	0.0	0.0	0.0	0.0	1.0	0.999999	0.999999

The conditional probability table for the second matching node of the BN seen in Figure 6.16 is given in Table 6.7, and it can be seen from this that the profile frequency information has been shifted back into the matching nodes.

6.4 HOW TO CONSTRUCT A BN FOR A CASE?

Section 6.2 showed the basic construction of BNs with a serial, convergent and divergent BN example, and their associated *LR* derivations. The aim of this exercise was to show that BNs are simply graphical representations of manipulations of the laws of probability. There are many different ways that BNs can be constructed, and if they represent the same set of assumptions and use the same data, then they are carrying out the same evaluation and will therefore produce the same *LR*. The implication of this is that as long as the data and assumptions are agreed upon, then there are many different ways of constructing a BN, and none are wrong. BN architecture is somewhat of a personal preference, and it can be designed with different goals in mind. A common goal in computing, and when dealing with very large BNs, is to make the architecture as efficient as possible, and not to include redundant nodes or arcs. Such issues of efficiency are not a concern in typical forensic evaluations as the complexity of the BNs does not approach the level where computational costs become a limiting factor. This section will demonstrate one possible method for BN construction and apply it to the three cases for which *LR*s were manually derived in Chapter 5. The method described only provides a general framework, and there are still many architectural decisions that are made during construction. However, it provides a consistent approach so that anyone looking at a BN constructed in this way will be able to see the lines of thought and reasoning that have gone into the evaluation (even without looking at the conditional probability tables). Full advantage is taken of the graphical nature of BNs, and in some cases redundant nodes are added because their presence better graphically represents assumptions in the evaluation.

6.4.1 BN CONSTRUCTION SOFTWARE

There are a range of software programs available, which range from interface-driven BN construction to command-line-driven BN construction.

These include freeware such as GeNIe (https://www.bayesfusion.com/) (which is free for academia but must be paid for in other private or public applications), R libraries, gRain [14] and BNlearn [15] and commercial software such as HUGIN (www.hugin.com) or AgenaRisk (http://www.agenarisk.com). Many of these software options allow the user to store BN graphs in a generic

(.net) format. This allows compatibility of use between the different software options. These are just three of many possible options for constructing BNs chosen as from the author's experience, they are most commonly used in forensic sciences (for further information on a range of possible options, see [16]).

Throughout this book, the software HUGIN is predominantly used, although this should not be taken as an endorsement for that product over any other. Different software packages should be explored, and the right package matched the needs of the individual or organisation when carrying out evaluations.

6.4.1.1 GeNIe

GeNIe from Bayes Fusion LLC (https://www.bayesfusion.com/genie/) is a graphic interface-driven software for construction and interrogation. GeNIe is compatible with Excel and has many graphical features for displaying the results of analyses or aspects of the data. GeNIe uses a suite of graphical tools called SMILE so that other software can be developed around a GeNIe 'engine'. The developers of GeNIe promote the fact that they spend large amounts of development effort on the intuitive useability of the software. GeNIe allows manual or formulaic entry of probability values. GeNIe is free for academic uses but requires payment for public or private use.

6.4.1.2 BNlearn/gRain

BNlearn (www.bnlearn.com) is a powerful R package that can be used to learn BN structures from raw data (although this is not a feature that will be utilised in BN construction for activity-level evaluations in this book). It can also be used to create BN structures and then accessed from other R packages such as gRain used to propagate information through the network. BNlearn is predominantly command driven and can be used for setting up bespoke analyses to test the functioning of the BN architecture and its sensitivity to data (see [17] for an example). There are graphical front-end additions that can use the BNlearn engine such as BayesianNetwork (https://cran.r-project.org/web/packages/BayesianNetwork/vignettes/BayesianNetwork.html).

While powerful and free, BNlearn may not be as accessible to those who are not familiar with R or programming in general. For example, the R code to recreate the BN and calculate the instantiations shown in Figure 6.15 is:

```
#load libraries
library(bnlearn)
library(gRain)
#builds up the BN
dag <- empty.graph(nodes = c("source_of_blood_spots", "suspect_
profile_matches_AO_profile", "SS1_matches_suspect",
"SS2_matches_suspect"))
#draw arcs
dag <- set.arc(dag, from = "source_of_blood_spots", to =
"SS1_matches_suspect")
dag <- set.arc(dag, from = "source_of_blood_spots", to =
"SS2_matches_suspect")
dag <- set.arc(dag, from = "suspect_profile_matches_AO_profile", to
= "SS1_matches_suspect")
dag <- set.arc(dag, from = "suspect_profile_matches_AO_profile", to
= "SS2_matches_suspect")
#sets levels in nodes
source_of_blood_spots.lv <- c("Hp", "Hd")
suspect_profile_matches_AO_profile.lv <- c("same", "different")
SS1_matches_suspect.lv <- c("match", "no_match")
SS2_matches_suspect.lv <- c("match", "no_match")
#adds probabilities to nodes
```

```
source_of_blood_spots.prob <- array(c(0.5, 0.5), dim = c(2,1),
dimnames = list(source_of_blood_spots = source_of_blood_spots.lv))
suspect_profile_matches_AO_profile.prob <- array(c(0.000001,
0.999999), dim = c(2,1), dimnames = list(suspect_profile_matches_AO_
profile = suspect_profile_matches_AO_profile.lv))
SS1_matches_suspect.prob <- array(c(1, 0, 1, 0, 1, 0, 0, 1), dim =
c(2, 2, 2), dimnames = list(SS1_matches_suspect = SS1_matches_
suspect.lv, source_of_blood_spots = source_of_blood_spots.lv,
suspect_profile_matches_AO_profile = suspect_profile_matches_AO
_profile.lv))
SS2_matches_suspect.prob <- array(c(1, 0, 1, 0, 1, 0, 0, 1), dim =
c(2, 2, 2), dimnames = list(SS2_matches_suspect = SS2_matches_
suspect.lv, source_of_blood_spots = source_of_blood_spots.lv,
suspect_profile_matches_AO_profile = suspect_profile_matches_AO
_profile.lv))
#set up conditional probability tables
cpt <- list(source_of_blood_spots = source_of_blood_spots.prob,
suspect_profile_matches_AO_profile = suspect_profile_matches_AO
_profile.prob, SS1_matches_suspect = SS1_matches_suspect.prob,
SS2_matches_suspect = SS2_matches_suspect.prob)
#create the BN
bn <- custom.fit(dag, cpt)
junction <- compile(as.grain(bn))
#instantiates nodes
instdag <- setEvidence(junction, nodes = c("SS1_matches_suspect",
"SS2_matches_suspect"), states = c("match", "match"),
propagate=TRUE)
#gets the query of interest
HpHd_result <- querygrain(instdag, nodes = "source_of_blood_spots")
#calculate LR
LR <- HpHd_result$source_of_blood_spots[["Hp"]]/HpHd_result$source
_of_blood_spots[["Hd"]]
#display the LR
LR
```

which gives the value for *LR* as 1 million.

6.4.1.3 HUGIN

HUGIN (www.hugin.com) is a graphical interface-driven BN construction tool. Figures that display BNs in this chapter are screen captures from the HUGIN software. Conditional probability tables are set through tables displayed on the screen, and the interface can be switched between Edit mode (for constructing the BN) and Run mode (for instantiating information and viewing posterior probabilities). HUGIN also includes a basic tool for parameter sensitivity analysis.

HUGIN comes in different versions, which includes a 'lite' (free) version, and an app for tablet or smartphone. HUGIN also has APIs that allow other software to be built that are based on the HUGIN 'engine'. HUGIN allows manual or formulaic entry of probability values.

6.4.1.4 AgenaRisk

AgenaRisk (http://www.agenarisk.com) is another graphical user interface-driven Bayesian network construction software that also boasts features of spreadsheet software. AgenaRisk has the ability to blend discrete and continuous models, with a number of inbuilt continuous distributions available and has a number of advanced features dealing with risk analysis. AgenaRisk has a trial version that can be downloaded and used for a limited time, and then requires payment for continued use.

6.4.2   STEPS TO CONSTRUCTING A BN FOR AN EVALUATION, GIVEN ACTIVITY-LEVEL PROPOSITIONS

The steps that will be worked through were first published by Taylor et al. [18] and the general template has since been used for BN construction in other published examples of case evaluations in biology [19–21], and fingermarks [22]. Some additional advice on BN node naming and state ordering will be provided that, while not required for the BN to function, make the construction and review more standard, and hence streamlined. In Chapter 11 validation requirements for BNs will be discussed.

As with the process of *LR* derivation, in order to start a BN construction, the specifics of the case, the factors that are being taken into consideration, the assumptions that will be made, the availability of data to inform probability assignment and the intended manner of breaking up the possible values of states into discrete groups (discretisation scheme) will all need to have been thought through first. This should not be a surprise, as the construction of a BN and the derivation of an *LR* are fundamentally the same task being carried out in different forms. With the mentioned information in mind, the steps for BN construction are:

Step 1: Define the main proposition node
Step 2: Define activity node(s)
Step 3: Group similar observations
Step 4: Define observations node(s)
Step 5: Define transfer and persistence node(s)
Step 6: Define root nodes(s)
Step 7: Checking for absolute support within the BN

Each of these steps will be explained in detail during the first evaluation case example. In Figure 6.17 a general BN structure is shown that considers a single result in relation to two competing activities. The BN shown can be constructed using the steps above and is in the general form that it will be used in the following example. Note that in the BN in Figure 6.17 (and all examples used in the book) the nodes are coloured in the following way:

- Black nodes – refer to the propositional node,
- Blue nodes – refer to the activity nodes,
- Yellow nodes – refer to the transfer, persistence and accumulation nodes (an accumulation node fuses two or more variables into a single node. It is the opposite of a technique called parent divorcing [23], where the complexity of a node is broken down by reformulating it into one or several layers of intermediate, more basic nodes),
- Red nodes – refer to the observation nodes and
- Grey nodes – refer to the root nodes.

This colouring has no other purpose than providing quick comprehension of the BN at a glance.

As shown in Figure 6.17, the convention used is that all arcs point downwards, so that the proposition node is at the top of the BN and the results nodes are at the bottom of the BN. While the node names in Figure 6.17 are very general (as they do not relate to any specific case), when specific case information is available one form of naming convention that can be used (and which are used throughout this book) are:

- Proposition nodes are called '*Hp/Hd*'.
- Activity nodes are named: '[PERSON] [ACTIVITY] [PERSON] [LOCATION] at [TIMEFRAME]' or '[PERSON] and [PERSON] [ACTIVITY]'. For example 'D kissed C

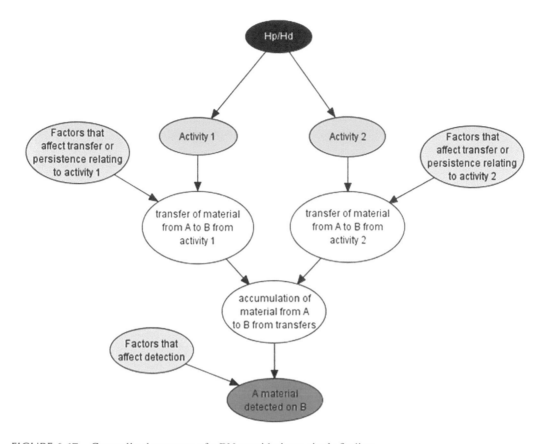

FIGURE 6.17   Generalised structure of a BN considering a single finding.

in the bedroom after the party' or 'D and C socially interacted'. Note that the timeframe
aspect can be dropped from the name if it is not being considered within the evaluation.
- For transfer mechanisms, the nodes are named: '[PERSON] [MATERIAL] transfer from
  [LOCATION] to [LOCATION] from [ACTIVITY]'. For example 'D DNA transfer from
  D hands to C underwear from digital penetration'.
- For the persistence of DNA, the nodes are named: '[PERSON] [MATERIAL] persisted
  on [LOCATION] through/for [EVENT]'. For example 'D DNA persisted on shirt through
  washing'.
- Accumulation nodes that combine transfer events from different activities are named:
  '[PERSON] [MATERIAL] on [PERSON] [LOCATION]'. For example 'D DNA on C
  underwear'.
- Background nodes are named: 'Background [MATERIAL] on [LOCATION]'. For exam-
  ple 'Background DNA on underwear'.
- Result nodes describe the observation, DNA profile or presumptive test that is being
  used in the evaluation and are named: 'Results of [TEST] on [ITEM]', or '[PERSON]
  [MATERIAL] on [ITEM]'. For example 'results of Hemastix test on knife blade' or 'D
  DNA on C's shirt'.

Again, the node naming above is only a suggestion, and there are cases where the naming convention must be deviated from in order to make clear what the node represents. Some of the conventions
above will seem overly descriptive, but the benefit is seen when complex chains of transfer events
are being described. The naming avoids having multiple nodes all named ambiguously with a title

such as 'DNA transfer'. In addition to the naming convention, all node names in the final BN are numbered. This allows for easy reference to any node by its name or number during the BN description. The numbers start at the top (with the propositional node always being node 1) and move from top to bottom, row by row from left to right.

In any BN construction the goal is to have a single probability being considered in each node and to avoid having multiple transfer, persistence, prevalence or recovery considerations either all being combined in one node or having values combined outside the BN with a final value being assigned to states in a node within the BN. Again, this makes it simple to see where probabilities are being used, to reassign a probability value for a particular event, or to test the effect of one probability on the overall evaluation. It should be noted that it is not always easy to separate the factors of transfer, persistence and recovery. For example, in experiments carried out to inform a transfer probability, the results of the transfer experiments have to be sampled, DNA extracted and (at the least) quantified. Therefore, the results from the study are actually showing data that relates to transfer and recovery. The only way to obtain data on the transfer alone is to have separately modelled recovery and then apply that model to the experimental transfer data to extrapolate back to the effects of transfer alone. The method for such extrapolations is beyond the information given in this book, but an example can be seen in [24], where results from recovery (and in fact two aspects of recovery; the sampling efficiency and the extraction efficiency) are applied to observations of DNA on hands in order to extrapolate back to purely the amount of DNA that was on hands.

Often the effects of recovery from experimental data obtained from literature are assumed to be similar to the effects of recovery in the laboratory applying them and so recovery is not separately taken into account in the evaluation.

The final point before beginning the case examples is that nodes will often be populated with probability assignments of either 0 or 1. Results nodes in particular will often have one state that relates to each of the possible combinations of incoming arcs. The convention used in this book is that the states in the nodes will be ordered in such a way that the 1's will start from the top left and work their way down to the bottom right of the conditional probability table. Once again, this is simply a measure to make the BN construction as standard as possible, which assists greatly in construction and review. It is much quicker to troubleshoot a BN that appears to be functioning counter-intuitively when as many elements as possible are standardised.

6.5   CASE 1: THE ATTEMPTED KIDNAP

The first of the cases for which the *LR* derivation was demonstrated in the case of *R v Drummond* [25]. The case details are not reiterated here, and it is suggested that if the reader has chosen to skip Chapter 2, they may wish to go back and read the case details for case 1 before continuing with this chapter.

6.5.1   CASE 1 – STEP 1: DEFINE PROPOSITION NODE

Determine the competing propositions that reflect what each party is putting forward and all associated activities with each. Remember that when one party is suggesting that an individual has not been involved in an activity, this may mean that they are stating the activity did not occur, or it may be that they are stating the activity occurred, but with someone other than the individual. These two options will result in different BN constructions (recall the discussions on this in Chapter 4).

In case 1 the following activity-level propositions were considered:

$_{Activity}Hp$: C and D struggled which included D grabbing C's arm and C hitting D's chest
$_{Activity}Hd$: A male other than D struggled with C which included them grabbing C's arm and C hitting them on the chest

FIGURE 6.18   BN for step 1 of case 1.

TABLE 6.8

Conditional Probability Table for Propositional Node in the BN Shown in Figure 6.18

Hp/Hd		
	Hp	0.5
	Hd	0.5

In each step the resulting BN is given, on its journey to completion. Figure 6.18 shows the case 1 BN construction after step 1, but as step 1 relates mainly to defining the proposition the BN is a single propositional node. The prior probabilities for the two states of this node are 0.5, which are shown in Table 6.8.

6.5.2   STEP 2: DEFINE ACTIVITY NODE(S)

Draw one node for each activity (it is important that each of these nodes represents a real activity and not an explanation of the phenomenon, such as saying 'secondary transfer'). Make the propositional node the parent for all the activity nodes. This structure is recommended, even for activities that are not in question (they are true under either proposition), i.e. the activities that are important to the evaluation of the observations, but are common in the description of events by both prosecution and defence (e.g. the victim and suspect had dinner together, prior to an alleged assault). This creates a more populated BN, but the advantage of the added complexity is that all activities that may impact the forensic results, disputed or not, are clearly identifiable in the BN structure. This is an example of the way that full advantage can be taken of the graphical nature of BNs. The ideal is for a full understanding of the evaluation to be gained by anyone looking at the BN from its architecture alone, without having to delve into the probability assignments or node explanations.

As with many evaluations, it is a personal preference when some items are included in the BN structure, particularly when they do not influence the final *LR*. For example, in case 1 there was DNA on C's top that was identified as coming from C's friend (F), whom she hugged shortly before the alleged attack. This interaction between C and F is accepted in both propositions and will not play a part in the evaluation. It was seen in the *LR* derivation that any terms relating to F's DNA were cancelled out in the *LR*. Therefore, a decision could be made in the BN construction to consider only the results of C's, D's or unknown DNA on the tops of C and D. Conversely, if the DNA results from the top of C were specified fully, including the presence of DNA from F, then an activity could be included in the BN that specified C hugging F. In step 2 of the BN construction for case 1 (shown in Figure 6.19) the hug between C and F has been included, and the associated potential for DNA transfer to occur in the evaluation, but note the *LR* is insensitive to the presence of DNA from F.

As a demonstration of a decision not to include an activity in the BN construction, the presence of D on their own top or C on their own top is not considered. They could be considered by including activity nodes such as 'D wore their own top' and 'C wore their own top', with probabilities assigned for transferring DNA to tops from wearing, but again the *LR* is insensitive to these factors and in

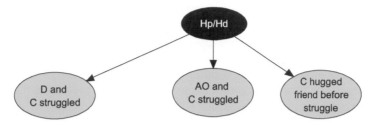

FIGURE 6.19   BN for step 2 of case 1.

TABLE 6.9

Conditional Probability Table for Activity Nodes in the BN Shown in Figure 6.19

Hp/Hd		*Hp*	*Hd*	*Hp/Hd*		*Hp*	*Hd*	*Hp/Hd*		*Hp*	*Hd*
D and C	Yes	1	0	AO and C	Yes	0	1	C hugged	Yes	1	1
struggled	No	0	1	struggled	No	1	0	friend before	No	0	0
								struggle			

this case, and they are not included in this example. The consequence of this decision is that in the results nodes there will be no accounting of C on their own top or D on their own top, and the nodes will represent DNA results other than the top owners.

The conditional probability tables for the activity nodes are given in Table 6.9. In Table 6.9 it is clear that the 'C hugged friend before struggle' node is not disputed as the probability assignments are the same, given a *Hp* or *Hd* state of the propositional, parent node. For the other two activity nodes, the probability assignments are different, given the different states of the propositional node.

6.5.3   STEP 3: GROUP SIMILAR RESULTS

To simplify the task, group the results of similar samples e.g. palm swab/finger swab/fingernail swab or tapelift of inner front, crotch and front waistband of underpants. The term 'similar' is vague. In this context, it means samples that are taken from closely related areas of the same item, where it might be expected that due to the propensity for DNA to transfer, a presence of DNA in one area and an absence in an adjacent are likely to have little evaluative difference to simply considering the presence of DNA in the general area. The propensity of DNA to transfer and the sensitivity of DNA analysis mean that the collected trace material can effectively be considered a single item. Considering multiple items separately leads to a complex set of dependencies, for which their full consideration will not bring more insight into the evaluation of the observations in this case. At this point it is also important to realise the impact of the forensic process on the integrity of traces. What is the impact of handling, packaging and transport on the re-distribution of traces on an item? How does the examination strategy impact on the definition of trace 'areas' on an item? If the impact is unknown or high, this may be another argument to combine results from multiple traces into a single 'meta-trace' to consider in the evaluation.

It will not always be appropriate to consider multiple items together. For example, if one of the propositions considers the presence of DNA to be from a single event, while the other considers it from multiple events, then the results cannot be combined. For example, in case 1 there are not any samples that could be considered for combination. However, if there had been multiple tapelifts taken from the top of C, then the choice may have been made to consider these as a single sample that demonstrates the presence or absence of D's DNA on C's top (that all came from the activity of

C hitting D's top). But, if multiple instances of D's DNA had been found on C's top, then the defence position may be that each occurrence was due to independent contamination events of the extraction tubes during DNA profiling, in which case this constituted multiple events and hence the samples would not be able to be combined for evaluation.

These two options could be considered extremes of a general evaluation setup where the items are considered dependent to some degree. This would require an evaluation that accounts for an effect that is referred to as cross-transfer, for which a simple BN setup and probability derivation is shown in [26] and a more complex example for a DNA example shown in [21]. Consider the example just given, where multiple areas of C's top were sampled and all returned D's DNA. At one extreme, these are considered to all be completely independent contamination events. Alternatively, it could be considered that perhaps one contamination event occurred which deposited D's DNA onto C's top, and then from site-to-site intra-exhibit contact occurring within the exhibit packaging (as explored in [27]). Therefore, there is a probability that the contaminating DNA was transferred to other areas of C's top, or they may be independent contamination events. Such a treatment would require modelling of cross-transfer. And at the other extreme one contamination event of C's top with D's DNA could be considered and that site-to-site cross-transfer is then guaranteed to occur, in which case the point has been reached where the dependence between samples is completely gone and they can be grouped as a 'similar result' as per the practice of this step.

For the running example, however, the BN for step 3 does not change from the BN in step 2 seen in Figure 6.19.

6.5.4   STEP 4: DEFINE OBSERVATIONS NODE(S)

Add 'observations' nodes (at this stage unlinked to the activity or proposition node) below the activity nodes. There will be one observations node for each group of observations relevant to the propositions. It is advisable not to ignore or leave out results from the BN as they can have an impact on the relative support given to each proposition. One should for example not leave out 'negative' results from items or traces that did not yield DNA of the defendant, as such results (much like the presence of DNA) may support one proposition over the other. This is true even for screening tests for body fluid that found no indication of that fluid and so no samples were taken for DNA analysis. The finer the resolution between node states (for example DNA amounts could be expressed in 1 ng brackets, or at a finer resolution in 100 pg brackets) that can be used in an observations node, the better the BN will be able to use the information in the observations. The offset for this is that sometimes the availability of data will be such that few (or often only a binary presence/absence) delineations are possible.

Observations nodes can be displayed in different ways. The results of DNA profiling a sample in a case may be a single node, or it may be broken up as the presence or absence (or amount) of DNA from individual contributors to a single sample, i.e. for the sample taken from C's top, the results could be represented in a single node 'result of DNA profiling from C's top' or it could be considered in several results nodes 'D DNA on C's top' and 'U DNA on C's top' and 'F DNA on C's top'. It is personal preference as to which of these methods is used, as the evaluation results will be the same either way. It may also be that the type of data available will lend itself to a particular architecture.

A balance exists between being practical and limiting the number of results included in a BN and ensuring that the evaluation is not constructed in a way that is observations led. A good rule of thumb is to include a node for each type of result that is known to have been tested for but not limit the states of those nodes to the results in the case. For example, in case 1 it is known that DNA profiling results are available for samples taken from C's top and D's top and so there would be results nodes included for these two aspects. Nodes would not be added for presumptive tests for blood or saliva or fingernail swabs from C because none of these samples were taken or tests done. Even

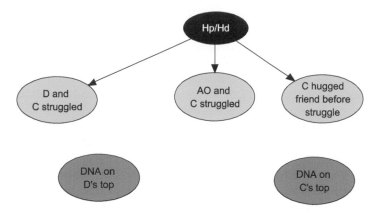

FIGURE 6.20 BN for step 4 of case 1 (note the large white gap between activity and results nodes appears odd but will be filled in during the coming steps).

if built into the BN architecture, there would be nothing to instantiate in these nodes and so they would not add to the evaluation. For the DNA results nodes that are included (DNA results from D's top and C's top), the states included should reflect the breadth of results that might be obtained in the case. Having said this, choices can be made regarding results that are likely irrelevant to the crime. For example, in case 1, whether or not to include the presence of C on her own top, D on his own top or F on either top will not add to the evaluation. Note that technically there is a potential pathway under *Hp* for F's DNA to be transferred to C's top, then C's hand and finally on to D's top; however, that complexity of transfer pathway is not modelled in this example, which technically makes the assumption that the presence/absence of C's DNA to D's top will be vastly more informative towards the evaluation. This is discussed in more detail in Section 6.6.

In case 1, the results of DNA profiling are modelled as a single node for each sample, as shown in Figure 6.20, that is, one node for the sample taken from the top of C and one node for the sample taken from the top of D. The states for these samples will need to encompass all combinations of DNA being considered as potentially transferring to these items. Conditional probability tables are provided later on when the BN is more progressed and connected.

6.5.5 Step 5: Define Transfer and Persistence Node(s)

Add transfer and persistence nodes that describe the mechanisms by which the activities would lead to the observations. The following should be considered:

- There may be multiple activities that all contribute to a single result.
- Some pathways will require multiple steps (nodes).
- The order of activities, and therefore transfers, may be important and could affect the way in which transfers are mapped.
- There may be nodes which are purely present as 'accumulation nodes' that combine the results of multiple transfers to the same object. These may not be strictly necessary (in that the BN could be constructed without them) but can help with comprehensibility.

In case 1, there is a need to consider how the activities of:

- D struggling with C, or
- AO struggling with C, and
- C hugging F

could lead to DNA transfer and ultimately the DNA results from the sample of C's top. In each case the BN structure is relatively simple, and a single node holding transfer probabilities can be added, i.e. 'D DNA transfer from D to C's top from grabbing', 'AO DNA transfer from AO to C's top from grabbing' and 'F DNA transfer from F to C's top hugging'. For D's top consideration is given to the probability of DNA transfer from C as a result of the struggle (Figure 6.21). There is no consideration of AO for D's top because if D is not the offender, then D and the AO have had no contact and so DNA transfer is not expected (or in fact assumed not to occur).

In each case the conditional probability table will take the same value form of a transfer occurring (or not occurring), given the states of the parent activity occurring or not. Table 6.10 shows the conditional probability table for these initial transfer nodes. The same study is used for assigning transfer probabilities for the hitting or grabbing of shirts as was used in the *LR* derivation chapter for case 1, i.e. the study by Daly et al. [28] where cloth samples were held by volunteers for 60 s before sampling using tapelifts. If any of these activities did not take place, then it is certain that a DNA transfer could not have occurred and so the probabilities of DNA transfer given an activity state of no are 0 and 1 for yes and no respectively. For the hugging activity, the same DNA transfer probability again is used as for the other activities. Within the evaluation, the probability assigned for DNA transfer from F to C from hugging does not matter, i.e. the *LR* is completely insensitive to the value

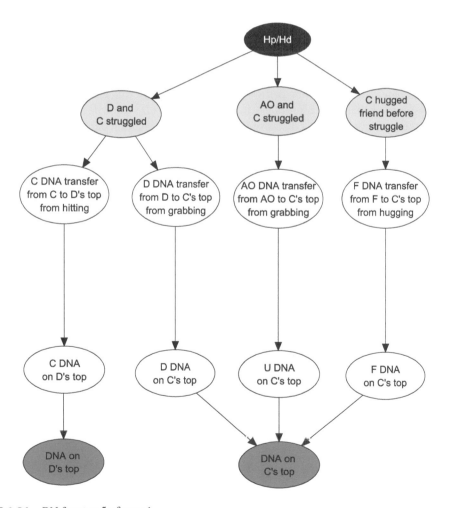

FIGURE 6.21   BN for step 5 of case 1.

TABLE 6.10

Conditional Probability Table for Transfer Nodes in the BN Shown in Figure 6.21

D and C struggled		Yes	No	D and C struggled		Yes	No
C DNA transfer from C to	Yes	0.47	0	D DNA transfer from D to C's	Yes	0.47	0
D's top from hitting	No	0.53	1	top from grabbing	No	0.53	1
AO and C struggled		**Yes**	**No**	**C hugged friend before struggle**		**Yes**	**No**
AO DNA transfer from AO	Yes	0.47	0	F DNA transfer from F to C's	Yes	0.47	0
to C's top from grabbing	No	0.53	1	top from hugging	No	0.53	1

and any value assigned will give the same *LR*. This was seen in the *LR* derivation with the terms relating to DNA transfer from F to C cancelling between the numerator and denominator of the *LR*.

The second row of yellow nodes seen in Figure 6.21 relates to the presence or absence of individual components of DNA on the two samples taken. If the construction of the BN was desired where each component of the DNA profiles was separate, then these nodes would be considered the results nodes (and hence be red) and would possess no children nodes. In the provided example of building a BN for case 1, they are intermediaries between the transfer nodes and the result nodes. The second line of yellow nodes could be skipped altogether, and the transfers could lead directly into the results, but in the next step root nodes are added, and having these additional transfer nodes as separate entities is more convenient.

6.5.6   STEP 6: DEFINE ROOT NODES(S)

Add in 'root' nodes. These are nodes that do not refer to any activities but have a relevant parental relationship with either the transfer steps or the observations nodes. Examples of this type of node are:

- Background levels of saliva on underpants,
- Background levels of DNA on hands or
- Contamination of the exhibit.

As with many other steps in the BN construction, the number of root nodes added depends on the desired richness of the BN. Many root nodes can be added but will not have any noticeable effect on the evaluation as they do not represent relevant pieces of information or present probabilities that are tiny compared to other nodes. For example, in an evaluation where a suspect's DNA may have come to be on an item due to their proximity with it, a root node could be added for an alternate offender having a matching DNA profile to D. However, this would be a highly improbable explanation for observing a profile matching D on the item compared to a transfer-by-proximity explanation. Hence it is usually assumed, when evaluating observations given activity-level propositions, that the source of the DNA is not in question.

In case 1, nodes are added for background DNA being present on the tops of D and C, but note that the results of the case were such that there was no relevant background DNA seen in the DNA profiles. Once again, these nodes are added to make the BN more generalised and to allow an exploration of what different results would have supported in the evaluation framework.

Root nodes are also added for the contamination of D's top with C's DNA in the forensic laboratory and for the contamination of C's top with D's DNA in the forensic laboratory. As the samples were processed separately, these contaminations represent two independent events and hence are represented by two separate nodes. For the states of 'yes' and 'no' in the contamination nodes, probabilities of 0.001 and 0.999 are assigned, respectively. The values used by an analyst for contamination nodes are specific to the laboratory in which the results were generated, and will be calculated from values such as those compiled by Kloosterman et al. [29]. Because C's DNA was not detected on D's top and D's DNA was not detected on C's top, the assumption is that contamination has not occurred (although it could be done, it is usual to consider contamination of a sample with unknown DNA in an evaluation). Despite this contamination, nodes are added, for reasons that will be explained in the final BN construction step. After step 6, the BN obtained is that shown in Figure 6.22 (which now also has nodes numbered).

The states of 'yes' and 'no' for background DNA being present on the tops of D and C are assigned values of 0.8 and 0.2, respectively. These values are based on a study by Szkuta et al. [30] where individuals were asked to wear shirts for one day either on a workday or on a non-workday. Out of 448 samples taken, only 86 were single profiles from the wearer and 10 produced no DNA profile. The remainder of the samples (78%) showed signs of DNA other than the wearer of the clothing.

Table 6.11 shows the conditional probability tables for the results nodes (17 and 18). Note that the order of categories in these nodes is such that the probability assignments of 1 work from top left to bottom right.

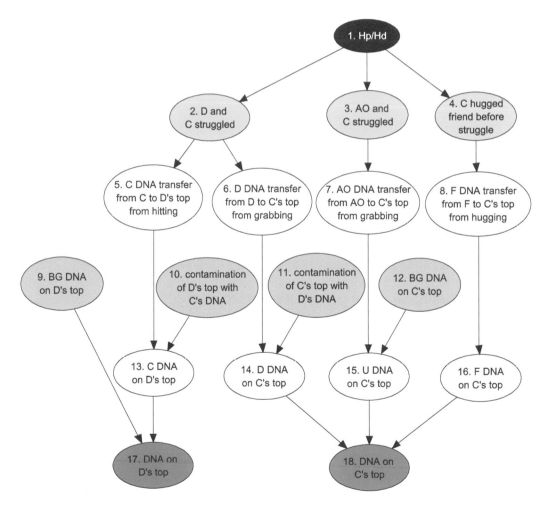

FIGURE 6.22   BN for step 6 of case 1.

TABLE 6.11

Conditional Probability Table for Results Nodes (17 Top and 18 Below) in the BN Shown in Figure 6.22

9. BG DNA on D's top		Yes		Yes	No		
13. C DNA on D's top		Yes	No	Yes	No		
17. DNA on	C + U	1	0	0	0		
D's top	U	0	1	0	0		
	C	0	0	1	0		
	None	0	0	0	1		

16. F DNA on C's top		Yes				No			
15. U DNA on C's top		Yes		No		Yes		No	
14. D DNA on C's top		Yes	No	Yes	No	Yes	No	Yes	No
18. DNA on	D + F + U	1	0	0	0	0	0	0	0
C's top	F + U	0	1	0	0	0	0	0	0
	D + F	0	0	1	0	0	0	0	0
	F	0	0	0	1	0	0	0	0
	D + U	0	0	0	0	1	0	0	0
	U	0	0	0	0	0	1	0	0
	D	0	0	0	0	0	0	1	0
	None	0	0	0	0	0	0	0	1

6.5.7 STEP 7: CHECKING FOR ABSOLUTE SUPPORT WITHIN THE BN

The final step relates partially to the architecture of the BN but also to the probabilities entered into conditional probability tables that underlie each node. Try to avoid specifying BNs where the instantiation of any single finding will lead to all the posterior probability being placed on a single proposition, because this would amount to a categorical conclusion. This can be avoided in two ways.

Firstly, in the probabilities associated with transfer and persistence used in the conditional probability tables that underlie each node, don't use values of 0 or 1. For nodes that rely on counts of experimental observations, a Dirichlet (1, …, 1) prior is applied to calculate the posterior probability of each count for use in the BN, as explained in Chapter 5. Note that when considering non-TPPR nodes, such as accumulation nodes, activity nodes or results nodes values of 0 or 1 are fine to assign.

Secondly, all results should be possible to be observed under either proposition (put technically, the posterior probability of any result given either proposition should never be 0). This will typically mean either that the competing activities will lead to two separate pathways for the finding to occur or that a 'root node' will present some alternate account of the observations. In some cases, this may be in the form of a coincidental DNA profile match or a contamination event. This is the reason that the contamination nodes were included in the BN seen in Figure 6.22. As previously stated, the evaluation of the results in some cases (such as case 1) does not require these nodes in order to evaluate the observations in the case, given the competing activity-level propositions. It is a general matter of making the evaluation more flexible to other situations. In some cases, for instance, if the BN is being constructed as part of a case assessment and interpretation (see Chapter 1), then the results will not be known yet. Then it will be important to ensure that complete support is not provided to either proposition, regardless of what results are obtained.

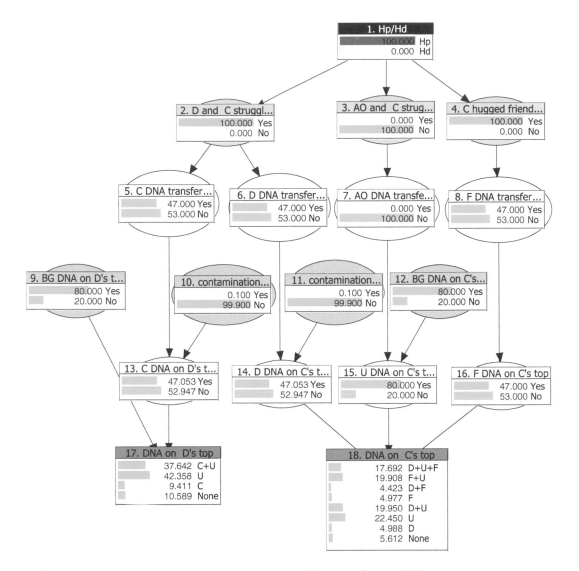

FIGURE 6.23 Instantiation of the *Hp* state in the proposition node for case 1 BN.

One way to test whether the BN has been constructed to satisfy this point is to first instantiate one state of the proposition node and check that none of the observations nodes have all their probability on a single state (the red nodes should not have a state equal to 100% or 0%). Then instantiate the other state in the propositional node and carry out the same check. A check of the BN as described above is shown in Figure 6.23 for the instantiation of the *Hp* state in the propositional node and Figure 6.24 for the instantiation of the *Hd* state in the propositional node.

Note that none of the states in results nodes 17 or 18 of either Figures 6.23 or 6.24 have a posterior probability of 0 or 1, and so the BN will not give absolute support for either proposition, regardless of the results instantiated. In the BN in Figure 6.22, the prosecution pathway that leads to C's DNA being on D's top is through nodes 2 → 5 → 13 → 17 and the defence pathway is 10 → 13 → 17. For D's DNA on C's top, the prosecution pathway is 2 → 6 → 14 → 18 and the defence is 11 → 14 → 18.

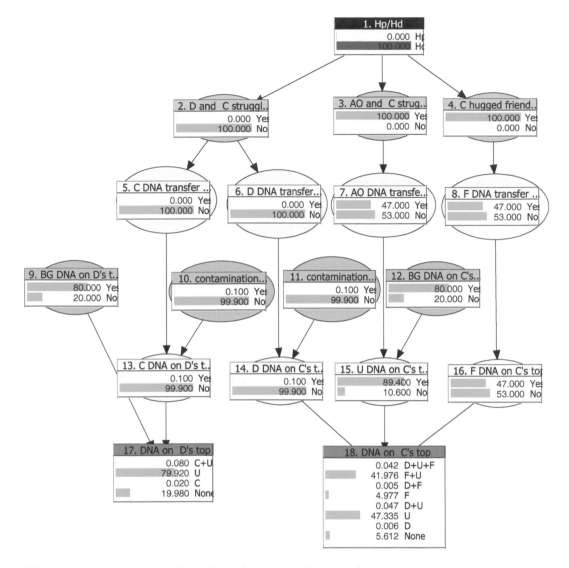

FIGURE 6.24 Instantiation of the *Hd* state in the proposition node for case 1 BN.

6.6 FINAL WORDS ON BN CONSTRUCTION

While the steps provided here provide a pathway to follow when constructing a BN for an evaluation that considered activity-level propositions, in reality the process can be quite iterative. Rarely is the first BN that is constructed for a case going to be the same as the final product used for court. Not specified in the steps of construction given above are when different aspects of the BN construction, other than the general architecture, are carried out:

- When to assign states to nodes
- When to order states
- When to assign probabilities

Some of these tasks will be able to be carried out as BN construction is occurring, while others will require the BN to have been completed before they can be attempted. It may be that the availability of data from literature will dictate the discretisation of states, and hence the BN architecture. In this instance the probability assignment may come before the BN construction (or may alter the BN construction if an initial structure has already been attempted).

There is also usually a period of 'sanity checking' the performance of the BN. In this period, it is good practice to predict whether the *LR* should move in favour of supporting *Hp* or *Hd*, given different combinations of results instantiation. If your predictions are correct, then this gives confidence in the structure of the BN and values for probability assignment. If the evaluation does not behave in an intuitive manner (and this is likely to occur in the early stages of BN construction), then the reason must be tracked down. It could be:

- An error with a probability assignment (i.e. a typo in the tables),
- An assumption that is being made implicitly by the architecture of the BN that was not intended, or realised,
- A product of combining results from different studies, i.e. when very similar transfer events are informed by two different sources of literature, and these sources present markedly different conclusions from their study, then the use in two separate nodes can cause evaluative issues. In these instances, it may make for a more logical BN behaviour to use one study in the BN for both nodes or combine the two studies in some way and use them in the two nodes. The scenario of informing two similar nodes with different studies can occur when pooling fragments of BNs used in different cases that have been evaluated over a long period of time (so that availability of literature has changed between them).

6.6.1 The Choice of BN Complexity

If the evaluation of forensic observations is being carried out after the observations have already been obtained (i.e. not as part of a case assessment and interpretation exercise), then results are known prior to the commencement of BN construction. In such a situation it is possible to make the BN construction simpler, with the disadvantage that it will only be suitable for the results obtained in this specific case, and will be limited in that no exploration can be carried out. A BN construction that suits only the case and specific set of results known to have been obtained is similar to an *LR* derivation in that if new results are obtained (i.e. through additional amplification of the DNA sample), then a new BN is required. It was mentioned in Chapter 5 that BNs are more expansive than *LR* derivations. *LR* derivation is for the observations obtained, while BNs are set up to handle any results outcome, e.g. the effect of observing D's DNA on C's shirt on the evaluation could be explored. Figure 6.25 shows the evaluation of the results obtained in case 1.

Note the instantiations of the results nodes in Figure 6.25 and recall the results in case 1:

- On the sample from the top of the complainant: A mixed DNA profile was obtained, explainable by a contribution of DNA from three people; the complainant, the friend of the complainant, and a low-level female, who is assumed to be unrelated to the alleged offence (as the attacker in the scenario was definitely male).
- On the sample from the top of D: A mixed DNA profile was obtained, explainable by a contribution of DNA from two people; D and an unidentified source, assumed to be unrelated to the alleged offence.

Recall the decision was made during BN construction not to include the presence of DNA on tops from their owners. This means that while, for example, the top of D yielded a DNA profile from D and an unknown, the instantiation will only be for the presence of DNA for an unknown (as seen in node 17 of Figure 6.25). The presence of D is not important to the evaluation and so the name of

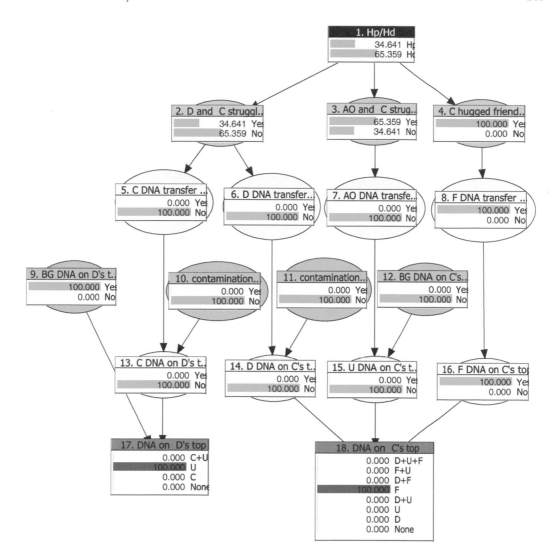

FIGURE 6.25 Instantiation of the BN shown in Figure 6.22 with the results obtained in case 1.

node 17 could be more aptly 'non-D DNA present on D's top'. In a similar vein, the presence of C on her own top is not important to the evaluation and node 18 could be named 'non-C DNA present on C's top'. The other aspect of the results of the sample from C's top is that there was the presence of DNA from an unknown female, who is assumed to be irrelevant to the offence. The unknown DNA modelled in the BN for C's top relates to relevant unknown DNA (or in this case, unknown male DNA). Therefore, the instantiation of 'F' in node 18 of Figure 6.25 occurs as there was no relevant unknown DNA and C's DNA is not being evaluated. There are several alternative approaches to this, which people may prefer:

1) Explicitly model male and female background DNA separately, and then expand the states of node 18 to include combinations to include male or female unknown DNA.
2) Relabel the background node to 'male BG DNA on top' and the states in node 18 to talk about male (or relevant) unknown DNA.
3) Structure the BN so that the results nodes are the individual components of the DNA profile and again talk about male unknown DNA (or relevant unknown DNA).

Compared to the *LR* derivation for case 1, given in Chapter 5, the only additional assumption that has been made in the BN construction is that contamination can occur (by not including contamination in the *LR* derivation, it was implicitly assumed that it cannot occur). However, the probability of contamination is small, and because DNA from D was not found on C's top and DNA from C was not found on D's top, then the posterior probability of contamination having occurred will be 0. The *LR* obtained from the instantiations shown in Figure 6.25 can be calculated by the ratio of the posterior probabilities of the two states in the propositional node, i.e. 34.641/65.359 = 0.53, i.e. the probability of non-transfer from the Daly et al. [28] study, and the same as the *LR* obtained in Chapter 5.

This should provide some comfort that the *LR* derivation and the BN construction are reflections of the same evaluation (as they have made the same assumptions and used the same underlying data to assign probabilities).

6.6.2 VARIANTS ON THE BN FOR CASE 1

The BN shown in Figure 6.22 is just one possible construction that would lead to exactly the same *LR*. The BN can be made simpler to handle only the results obtained (similar to the way that the *LR* is only derived for the results obtained), but this limits their use for other cases/flexibility to answer questions. Figure 6.26 shows an example of simplified BN for case 1, along with the instantiation, leading to the same *LR*.

Again, the instantiations shown in Figure 6.26 yield the same *LR* (and again this is because the same assumptions and data are being used).

In the *LR* derivation for case 1, given in Chapter 5, the effects of persistence on the evaluation were also considered. So too persistence can be included in the BN construction. The manner in which persistence is considered will dictate the architecture of the BN. For example, one way that persistence could be added to the BN shown in Figure 6.22 is shown in Figure 6.27. There are several important features of the way persistence is incorporated into the BN in Figure 6.27 of which to take note:

- Nodes 14–18 have been added in Figure 6.27, which represent whether DNA has persisted from the time between the offence and sampling of the items.
- Temporal ordering is important. The persistence considered is the time between the offence and the sampling of items for forensic testing. The implication of this is that the persistence of the potential contaminations is not considered (or more correctly, they are assumed to persist if present). This is built into the BN architecture by the fact that the persistence node acts on nodes in the chain between activities and results after the transfers from struggles and hugging, but before the contaminations are considered.
- A single persistence node (node 12 in Figure 6.27) has been added and made a parent of all the individual DNA persistence nodes (14–18). The implication for setting up the BN in this manner is that persistence is considered as having occurred to all or none of the DNA, i.e. if DNA has not persisted on C's top, then it will not have persisted on D's top. Alternatively, if any states of nodes 23 or 24 are instantiated to something other than 'none' (i.e. indicating that DNA is present), then persistence will obtain a posterior probability of 1 and non-persistence 0. This will then carry over to the other results node. Put simply, DNA cannot persist on one item and not persist on the other.

This last point is somewhat unrealistic in the context of the case. The two items in the case were seized and sampled at different times, they are made from different materials, and they may have been held in different conditions. In addition, the DNA transferred from the hug between F and C occurred at a different time to the offence and persistence may be different for this component.

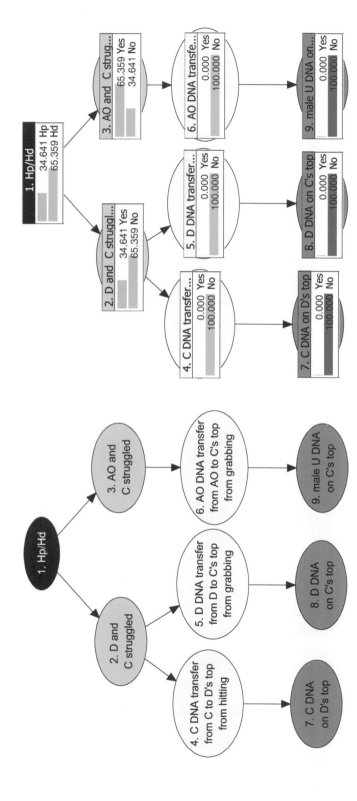

FIGURE 6.26   A simplified version of the BN, left, (compared to the worked example) case 1 and the instantiated form, right, showing posterior probabilities.

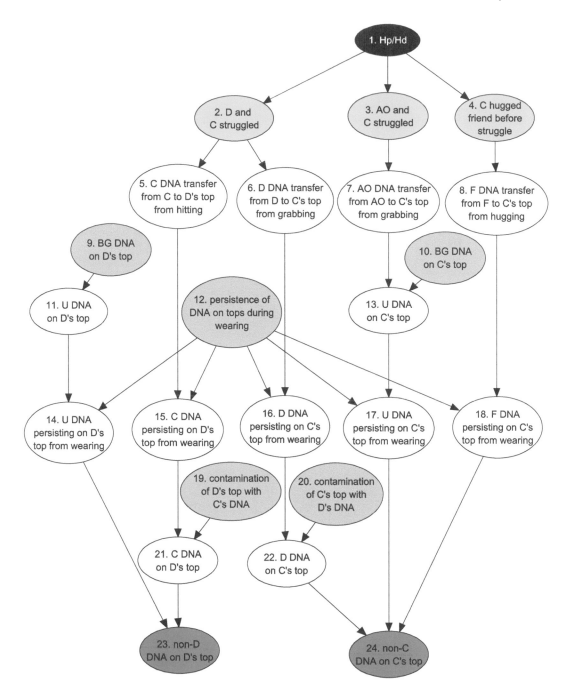

FIGURE 6.27   BN from Figure 6.22 updated to include persistence.

In order for the independent persistence events to be considered as such, multiple persistence nodes need to be made and applied to their respective DNA components. Such a BN construction is shown in Figure 6.28, which is likely to better reflect persistence in reality.

The complexity of the BN used to evaluate observations in case 1 between the 9-node BN seen in Figure 6.26 and the 26-node BN seen in Figure 6.28 is a matter of personal preference. Although it has not been shown, the *LR* for the BN seen in Figure 6.28 is still 0.53, so it can be seen that the

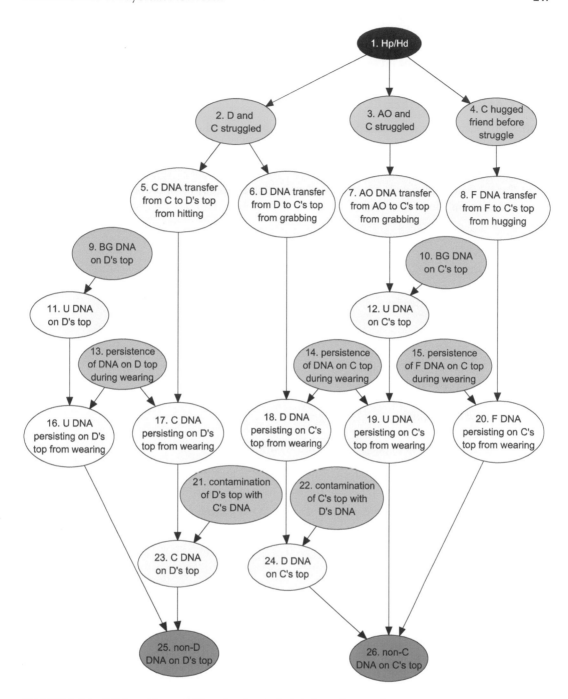

FIGURE 6.28 BN from Figure 6.27 updated to include independent persistence events.

exact architecture does not matter as long as the assumptions are properly expressed within the BN architecture. The construction of the BN for case 1 has explained a number of key concepts, and the remaining two cases will be presented at a much quicker pace.

6.7   CASE 2: THE THREE BURGLARS

Case 2 introduces some new concepts that were not encountered in case 1:

- Case 2 deals with the presence of DNA from the defendant (unlike case 1 where an absence of DNA was being dealt with).
- Case 2 deals with new types of dependencies that are required in order for the evaluation to behave in a logical manner.

The following abbreviations for individuals are used in the case:

D – the defendant
C – the complainant
AO – the alternate offender
H – the husband of C

6.7.1   STEP 1

Recall from Chapter 4 the propositional setup of the case is:

Background information:

- D lent his gloves to people that commit burglaries

Propositions:

- H1: D held the knife and grabbed the hair of the victim
- H2: AO held the knife and grabbed the hair of the victim, after wearing the gloves of D

Assumptions:

- The knife submitted for examination was used by the perpetrator.
- The perpetrator did not wear gloves while handling the knife or while interacting with the victim.
- Under H2, the AO took off the suspect's gloves immediately prior to handling the knife and grabbing the hair of the victim.
- Samples were taken, secured, packaged and transported according to standard operating procedures.
- None of C, D, H or AO have conditions that would make them unusually high DNA shedders, i.e. they are representative of the typical spread of shedding ability within the population.
- The sample from the knife contains DNA of C, H and D, but no DNA of other individuals.
- The sample from the hair contains DNA of C and D, but no DNA of other individuals.

This provides all the information required for the setup of the evaluation, and Figure 6.29 gives the standard beginnings of the BN setup that is the same for all BN construction.

FIGURE 6.29   BN for step 1 of case 2.

6.7.2   STEP 2

Case 2 is another example of a case where the actor is in dispute rather than the activity. This means that it must be considered that if D did not commit the activities specified in *Hp*, then an AO did. Therefore, there are four activity nodes to cover these activities:

- D handled knife (which has occurred under *Hp* and not under *Hd*)
- AO handled knife (which has occurred under *Hd* and not under *Hp*)
- D grabbed C's hair (which has occurred under *Hp* and not under *Hd*)
- AO grabbed C's hair (which has occurred under *Hd* and not under *Hp*)

A node is also added for the activity of the AO wearing D's gloves, which has occurred under *Hd* and not under *Hp* (although it would not matter if this also occurred under *Hp* as under *Hp* the AO is not involved in the offence). Adding these activity nodes and making them all children of the propositional node yield the partial BN seen in Figure 6.30. The decision is made not to model the presence of DNA for C or H on the knife or C's hair as they do not impact the evaluation and so there are no associated activity nodes.

6.7.3   STEP 3

There are two results obtained that are included in the evaluation, the DNA profiling of the knife sample and the DNA profiling of the hair sample. Nodes are added for these two results (but not connected to any other nodes at this stage) giving the BN seen in Figure 6.31.

The states for both of the results nodes will be:

- D + U
- D
- U
- None

6.7.4   STEP 4

In this case there are no results for which it is appropriate to combine.

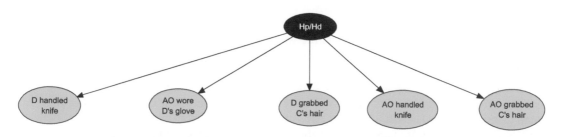

FIGURE 6.30   BN for step 2 of case 2.

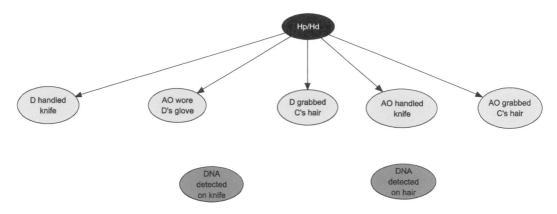

FIGURE 6.31   BN for step 3 of case 2.

6.7.5 STEP 5

This is the most involved step of any BN construction. For each of the activities, there is an initial transfer node:

- D DNA transfer to knife from handling, for the activity of D handling the knife.
- AO DNA transfer to knife from handling, for the activity of AO handling the knife.
- D DNA transfer to C's hair from grabbing, for the activity of D grabbing C's hair.
- AO DNA transfer to C's hair from grabbing, for the activity of AO grabbing C's hair.
- D DNA transfer from D's glove to AO's hands from wearing, for the activity of AO wearing D's gloves.

From this point, there are two further considerations:

- Additional transfer considerations, specifically that if D's DNA has transferred to AO's hands from AO wearing D's gloves, then the AO could further transfer D's DNA to C's hair if the AO grabbed it and to the knife if AO handled it. Therefore, two additional transfer nodes are required as children of the node considering D DNA transfer from D's glove to AO's hands from wearing.
- It is expected that the AO has their own DNA on their hands. If the AO also has D's DNA on their hands, then if the AO has touched an object the transfer of D's DNA and AO's DNA are not independent. In other words, if there are multiple sources of DNA on someone's hands, it is expected that they are transferred together or not transferred, and it is much less likely that only one of the sources of DNA would be transferred and not the others. Recall in Chapter 5 a probability of 0.9 was assigned to transfer together and 0.1 to a transfer of only one source of DNA.

Finally, nodes are added that accumulate the various sources of DNA on the knife or hair, which can then be passed down to the results nodes. Adding all these nodes into the BN results in the architecture seen in Figure 6.32.

In Figure 6.32 it can be seen that the dependency between multiple sources of DNA on AO's hands being transferred together is accounted for by the arc between 'D DNA transfer from AO hand to knife from AO handling' and 'AO DNA transfer to knife from handling' and an arc between 'D DNA transfer from AO hands to C hair from AO grabbing' and 'AO DNA transfer to hair from grabbing'.

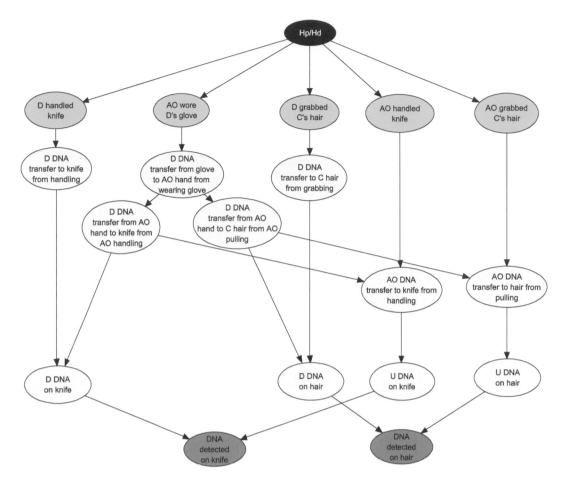

FIGURE 6.32   BN for step 5 of case 2.

6.7.6   STEP 6

Because the defence proposition invokes the existence of an alternate offender for which there is no reference, there must be an account of finding unknown DNA on the items. A common source of unknown DNA is background DNA, and so two root nodes are added that take into account the possibility of background DNA on the two sampled items in the case.

Also added is a root node for the presence of D's DNA on their own gloves. There are several ways this situation could be dealt with:

- Add in a root node as described to capture the probability that someone's DNA will be on their own regularly used gloves, as just described.
- Assume that D's DNA will be on his own gloves and not add a node (then the presence of his DNA on his own gloves becomes an assumption that is implicit in the evaluation from the structure of the BN in the same way that has been done for the presence of AO's DNA on their own hands).
- Add another activity node 'D regularly wore gloves' rather than a root node and create a transfer node that feeds into the 'D DNA transfer from glove to AO hand from wearing' node.

The final BN structure for case 2 is given in Figure 6.33.

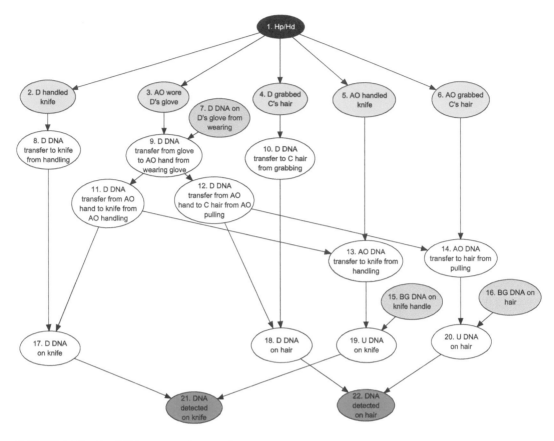

FIGURE 6.33 final BN structure for case 2.

6.7.7 STEP 7

The check for absolute support in the BN comes partly from the BN architecture and partly from the probability assignments. At this stage, then it is required to provide the conditional probability tables for the 22 nodes in the BN in Figure 6.33. These are given in Tables 6.12–6.34, and the results

TABLE 6.12
Conditional Probability Table for Node 1 in Figure 6.22

Hp/Hd		
	Hp	0.5
	Hd	0.5

TABLE 6.13
Conditional Probability Table for Node 2 in Figure 6.22

1. Hp/Hd		Hp	Hd
2. D handled knife	Yes	1	0
	No	0	1

TABLE 6.14

Conditional Probability Table for Node 3 in Figure 6.22

1. *Hp/Hd*		*Hp*	*Hd*
3. AO wore D's glove	Yes	0	1
	No	1	0

TABLE 6.15

Conditional Probability Table for Node 4 in Figure 6.22

1. *Hp/Hd*		*Hp*	*Hd*
4. D grabbed C's hair	Yes	1	0
	No	0	1

TABLE 6.16

Conditional Probability Table for Node 5 in Figure 6.22

1. *Hp/Hd*		*Hp*	*Hd*
5. AO handled knife	Yes	0	1
	No	1	0

TABLE 6.17

Conditional Probability Table for Node 6 in Figure 6.22

1. *Hp/Hd*		*Hp*	*Hd*
6. AO grabbed C's hair	Yes	0	1
	No	1	0

TABLE 6.18

Conditional Probability Table for Node 7 in Figure 6.22

7. D DNA on D's glove from wearing	Yes	0.95
	No	0.05

TABLE 6.19

Conditional Probability Table for Node 8 in Figure 6.22

2. D handled knife		Yes	No
8. D DNA transfer to knife from handling	Yes	0.8	0
	No	0.2	1

TABLE 6.20
Conditional Probability Table for Node 9 in Figure 6.22

3. AO wore D's glove		Yes		No	
7. D DNA on D's glove from wearing		Yes	No	Yes	No
9. D DNA transfer from glove to AO hand from wearing glove	Yes	0.75	0	0	0
	No	0.25	1	1	1

TABLE 6.21
Conditional Probability Table for Node 10 in Figure 6.22

4. D grabbed C's hair		Yes	No
10. D DNA transfer to C hair from pulling	Yes	0.47	0
	No	0.53	1

TABLE 6.22
Conditional Probability Table for Node 11 in Figure 6.22

9. D DNA transfer from glove to AO hand from wearing glove		Yes	No
11. D DNA transfer from AO hand to knife from AO handling	Yes	0.8	0
	No	0.2	1

TABLE 6.23
Conditional Probability Table for Node 12 in Figure 6.22

9. D DNA transfer from glove to AO hand from wearing glove		Yes	No
12. D DNA transfer from AO hand to C hair from AO pulling	Yes	0.47	0
	No	0.53	1

TABLE 6.24
Conditional Probability Table for Node 13 in Figure 6.22

5. AO handled knife		Yes		No	
11. D DNA transfer from AO hand to knife from AO handling		Yes	No	Yes	No
13. AO DNA transfer to knife from handling	Yes	0.9	0.25	0	0
	No	0.1	0.75	1	1

TABLE 6.25
Conditional Probability Table for Node 14 in Figure 6.22

6. AO grabbed C's hair		Yes		No	
12. D DNA transfer from AO hand to C hair from AO pulling		Yes	No	Yes	No
14. AO DNA transfer to hair from pulling	Yes	0.9	0.34	0	0
	No	0.1	0.66	1	1

TABLE 6.26
Conditional Probability Table for Node 15 in Figure 6.22

15. BG DNA on knife handle	Yes	0.5
	No	0.5

TABLE 6.27
Conditional Probability Table for Node 16 in Figure 6.22

16. BG DNA on hair	Yes	0.5
	No	0.5

TABLE 6.28
Conditional Probability Table for Node 17 in Figure 6.22

8. D DNA transfer to knife from handling		Yes		No	
11. D DNA transfer from AO hand to knife from AO handling		Yes	No	Yes	No
17. D DNA on knife	Yes	1	1	1	0
	No	0	0	0	1

TABLE 6.29
Conditional Probability Table for Node 18 in Figure 6.22

10. D DNA transfer to C hair from pulling		Yes		No	
12. D DNA transfer from AO hand to C hair from AO pulling		Yes	No	Yes	No
18. D DNA on hair	Yes	1	1	1	0
	No	0	0	0	1

TABLE 6.30
Conditional Probability Table for Node 19 in Figure 6.22

13. AO DNA transfer to knife from handling		Yes		No	
15. BG DNA on knife handle		Yes	No	Yes	No
19. U DNA on knife	Yes	1	1	1	0
	No	0	0	0	1

TABLE 6.31
Conditional Probability Table for Node 20 in Figure 6.22

14. AO DNA transfer to hair from pulling		Yes		No	
16. BG DNA on hair		Yes	No	Yes	No
20. U DNA on hair	Yes	1	1	1	0
	No	0	0	0	1

TABLE 6.32

Conditional Probability Table for Node 21 in Figure 6.22

17. D DNA on knife		Yes		No	
19. U DNA on knife		Yes	No	Yes	No
21. DNA detected on knife	D + U	1	0	0	0
	D	0	1	0	0
	U	0	0	1	0
	None	0	0	0	1

TABLE 6.33

Conditional Probability Table for Node 22 in Figure 6.22

18. D DNA on hair		Yes		No	
20. U DNA on hair		Yes	No	Yes	No
22. DNA detected on hair	D + U	1	0	0	0
	D	0	1	0	0
	U	0	0	1	0
	None	0	0	0	1

TABLE 6.34

Conditional Probability Table for Node 11 in Figure 6.41

Faeces on underwear		Yes	No
Faecal stains observed on C's underwear	Yes	1	0
	No	0	1

of instantiating the *Hp* and *Hd* states of the propositional node are shown in Figures 6.34 and 6.35. None of the states of the results nodes (21 or 22) are given probabilities of 0 or 1 and so there is no combination of results that can lead to absolute support for a proposition within the BN.

6.7.8   EVALUATION CASE 2

The state 'D' can be instantiated in node 21 (relating to the knife) and state 22 (relating to the hair) in order to carry out the evaluation for the observed observations. Doing so results in the posterior probabilities in the *Hp* and *Hd* state of the propositional node of 0.99293 and 0.00707, respectively, as seen in Figure 6.36. The *LR* is therefore approximately 140, just as was obtained from the manual *LR* derivation for this case in Chapter 5.

6.8   CASE 3: THE FAMILY ASSAULT

Case 3 has three different features from cases 1 and 2:

- It is a scenario where the activity itself is being disputed (rather than disputing who carried out the activity),
- It introduces the use of a multi-level discretisation of the DNA results (high/low/none) rather than modelling the presence or absence of DNA, and
- It introduces results other than DNA profiles, specifically visual stain identification and presumptive body fluid tests.

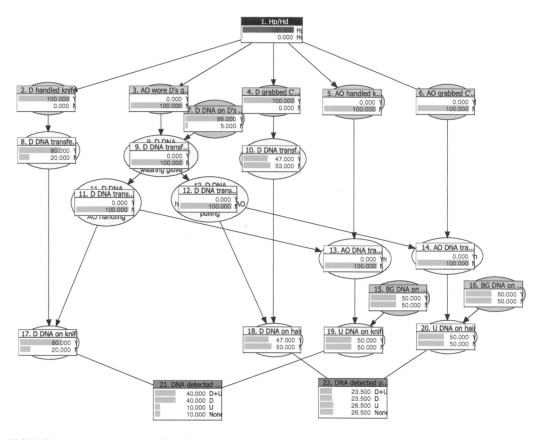

FIGURE 6.34 Instantiation of the *Hp* state in the propositional node for case 2 BN.

The BN construction for this case was demonstrated by Taylor et al. [18] and the BN is supplied as supplementary material to that publication.

6.8.1 Step 1

Recall from Chapter 5 the following propositional information:

> *Background information:*
> - C and D have been staying together in D's house for a week
> - C normally lives with her biological father (F)
> - C and D are full siblings
> - C's underwear has faecal stains
>
> *Propositions:*
> - H1: D bit C on the vagina over the underwear
> - H2: C was not bitten
>
> *Assumptions:*
> - No mutations have occurred between D and D's father that would affect the Y-STR profile
> - C's underwear has not been washed since the alleged offence
> - C's underwear was not bought less than a week prior to the alleged offence (and so has been worn/washed at her normal residence)

The starting of the BN is shown in Figure 6.37.

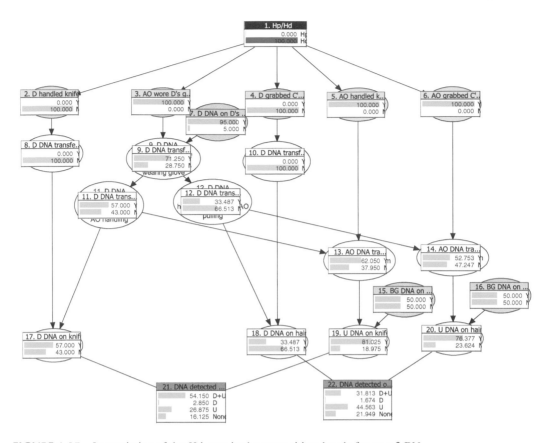

FIGURE 6.35   Instantiation of the *Hd* state in the propositional node for case 2 BN.

6.8.2   Step 2

For this case scenario, the activity itself is in dispute, and so a single activity node is required to capture this activity. Any other activity nodes are added to include non-disputed activities that affect the evaluation. For case 3, two additional activity nodes are added, one that considers that C and D were cohabiting at the time of the alleged assault, and one that specifies that C and F cohabited shortly before the alleged assault. The fact that C and F cohabited shortly before the assault is relevant as the DNA profiling results were Y-STR profiles and D and F were patrilineally related (so that they are expected to share the same Y-STR profile). The arrows between the proposition node and non-disputed activities are not required. Their presence is a matter of choice and removing them should not affect the evaluation. Adding these three activity nodes gives rise to the BN seen in Figure 6.38.

6.8.3   Step 3

Now results nodes to cover the observations from the case are added to the BN. For case 3, one of the available results are the Y-STR profile obtained from C's underwear, which matches the reference Y-STR profiles of both D and F. As the Y-STR profile will be present in all male members of the family who have an unbroken patrilineal line to D and F (unless mutation has occurred), the profile will be referred to as the 'family Y-STR profile'.

Another result comes from testing the underwear for the presence of saliva, with the RSID saliva test, and a node is added for the results of this test. Note that the DNA results are not associated

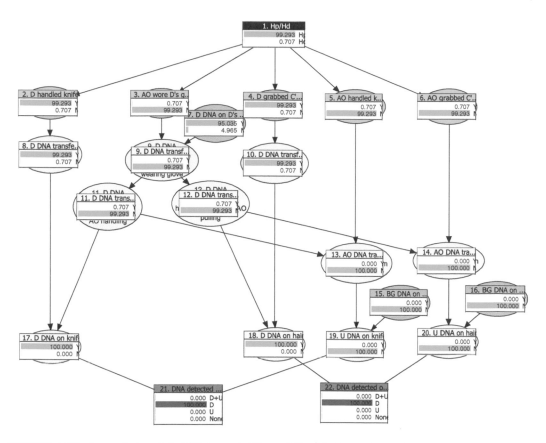

FIGURE 6.36   instantiation of the BN shown in Figure 6.33 with the results obtained in case 2.

FIGURE 6.37   BN for step 1 of case 3.

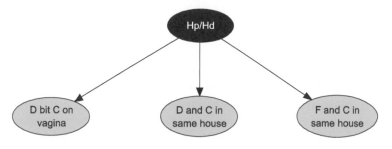

FIGURE 6.38   BN for step 2 of case 3.

with the saliva results, i.e. the distinction of whose saliva is not made. The results node is simply the positive or negative reaction of the RSID test.

Finally, the result of the visual examination of the underwear is available, which has revealed the presence of faecal staining. These three results nodes are added to obtain the BN shown in Figure 6.39.

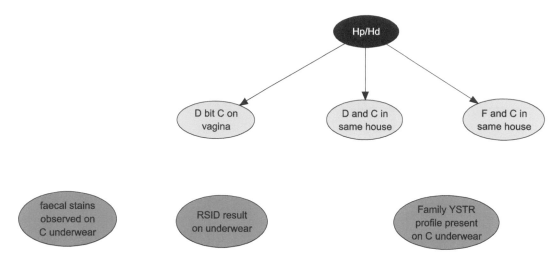

FIGURE 6.39   BN for step 3 of case 3.

Note the slight deviation in BN construction from the original publication from which it is derived [18], in which the visual result was not explicitly given its own node. This decision is discussed later.

6.8.4   Step 4

In case 3, the observations of the two tapelifts of the outer surface of the underwear are combined. In both cases the results were that high levels of C's DNA were present (this is expected as they are her underwear and would be assumed during DNA profile interpretation). On one of the tapelifts, low levels of male DNA were detected, and the subsequent Y-chromosome profiling generated a Y-STR profile. The two tapelifts are considered as one large tapelift of the entire outer front and crotch of the underwear and that low levels of male DNA were present.

This decision is made with the assumption that the presence of DNA on any one part of the outer crotch of the underwear is evaluatively no different to the presence of DNA on all parts of the outer crotch of the underwear. This is because the two locations are close together and, based on the case information, it is uncertain whether the biting occurred at one of the two locations or both. Additionally, cross-transfer of material from these locations on the underwear due to packaging and transport may have occurred given the way the item was packaged. Hence the observations from the two samples can be considered a single result.

6.8.5   Step 5

Both cohabitation activities have a chance of transferring DNA to the outer crotch of the underwear and so will have an associated transfer node that sits between the activity and the results nodes. The biting activity node will have a child transfer node that is specific to the probability of saliva being transferred. It is assumed that if saliva is transferred, then so too will DNA be transferred and so the saliva transfer nodes will diverge into two chains, one that leads to the RSID saliva test results node and the other that ultimately leads to the presence of the family Y-STR node. The result can be seen in Figure 6.40.

In a similar vein to previous BN constructions, there are different ways that the presence of faeces could be included in the BN. One method was given in [18], where the presence of faeces was not explicitly given its own node, and rather it was a root node that was later instantiated. Another method could have been to include another activity node that specified the defaecation occurring, which could then have led straight into the RSID results node.

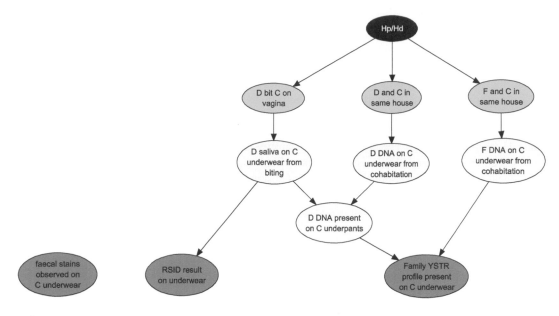

FIGURE 6.40   BN for step 5 of case 3.

A node for the visual observation of faeces is included, but it is not connected to the remainder of the BN at this stage.

6.8.6   STEP 6

Two root nodes are added to the BN, one that considers the probability of background levels of saliva being present on C's underwear and the other considers the presence of faecal stains on C's underwear. The latter of these nodes connects the two results nodes of observing faecal staining and the RSID saliva test result. The resulting, final BN architecture is seen in Figure 6.41.

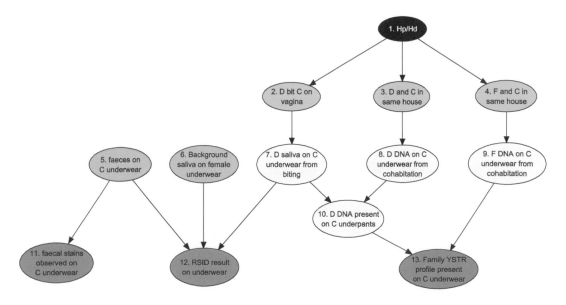

FIGURE 6.41   BN for step 6 of case 3.

Other possible root nodes that could have been considered is one for the presence of chance matching background male DNA on the underwear of C; however, this was not considered as the presence of the DNA from either D or F was accepted. Also, the probability of matching Y-STR profiles is much less than the probability of transfer of DNA to underwear from cohabitation and so including the chance matching root nodes would not have an observable effect on the evaluation.

To see the conditional probability tables (and probability assignments) for the BN in Figure 6.41, the reader is directed to the supplementary material of [18], which is a mock report for this case. The only difference between the BN in the supplementary material and the BN in Figure 6.41 is node 11 from Figure 6.41. The conditional probability table for this node is provided in Table 6.34.

As seen from the probability assignments in Table 6.34, there is no account for the possibility that faecal stains are present but not observed, or absent but thought to be present. In effect this node does not provide any additional functionality to the BN from that seen in [18], and its presence is simply so that visually only the results nodes are being instantiated.

6.8.7 Step 7

The check for absolute support within the BN is shown in Figures 6.42 and 6.43.

As can be seen, none of the results have posterior probabilities of 0 or 1 when either *Hp* or *Hd* are instantiated and so absolute support for one proposition will not occur regardless of results instantiation.

6.8.8 Evaluation

Instantiating the results in case 3:

- Faecal staining observed in the underwear of C
- A positive RSID saliva test result
- The presence of low-level male DNA with the family Y-STR profile

gives posterior probabilities as seen in Figure 6.44.

The *LR* can be obtained by the ratio of the posterior probabilities of *Hp* and *Hd* in the propositional node, and yield *LR* = 0.19, just as derived in Chapter 5 for this case.

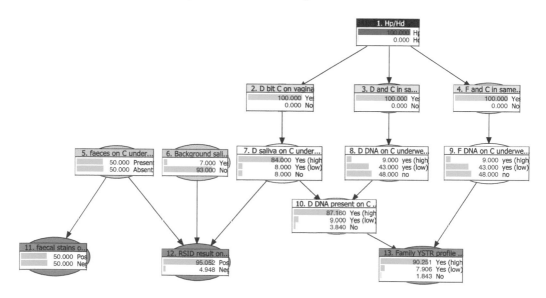

FIGURE 6.42 Instantiation of the *Hp* state in the propositional node for case 3 BN.

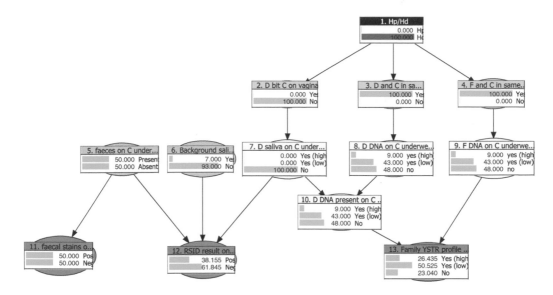

FIGURE 6.43   Instantiation of the *Hd* state in the propositional node for case 3 BN.

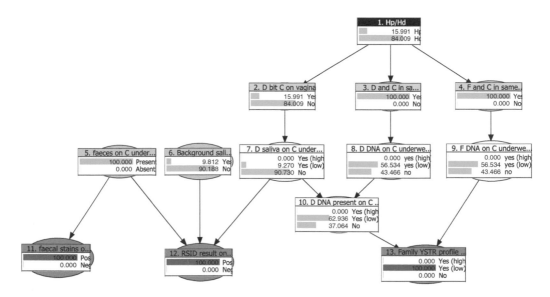

FIGURE 6.44   Instantiation of the BN shown in Figure 6.41 with the results obtained in case 3.

6.9   PRACTICE QUESTIONS

Q1)   The values of the first column of Table 6.2 are derived from the text preceding the table. Show that the values in the rest of the table are correct by deriving their value yourself. Recall that the population has alleles *X*, *Y* and *Z* with frequencies 0.5, 0.49 and 0.01, respectively.

Q2)   In Figure 6.6, the value for the probability of the child having genotype [*X*,*X*] is derived and shown to be 0.25. Show how the remaining probabilities for the genotype of the child in Figure 6.6 can be determined by derivation of their value. Recall that the population has alleles *X*, *Y* and *Z* with frequencies 0.5, 0.49 and 0.01, respectively.

Q3) In Chapter 1 an example was given where a patient gives a positive result to a test for a disease that has a prevalence in the population of 1 in 1 million people. The test is known to have a 99.9% true positive rate and a 0.001% false positive rate. It was shown in that chapter that the positive test result means that the patient has an approximately 9.1% chance of being truly positive. Draw out the structure of the BN and provide the setup of the conditional probability tables. Now build a BN that demonstrates this example. If you don't have software that can build BNs, then now is the perfect time to start! Look at the BN software described in Section 6.4.1 and download a free/trial/academic version of one of the programs described. For any of these software programs, there are manuals, tutorials and online help available that will guide you through the mechanics of using the programs.

Q4) Construct a BN for case 1 considering the defence proposition that there was no attacker at all, i.e. the complainant made up the event.

Q5) How would you extend the BN for case 1 (the attempted kidnapping case) if C and D had both been present at a party together the night before the alleged assault?

Q6) In the evaluation framework for case 2, it was considered that if the AO had DNA from D on his hands, then there was a dependency between the transfer of AO DNA and the transfer of D DNA from AO's hands. What part of the architecture of the BN seen in Figure 6.33 accounts for this dependency? What do you think would be the result of the evaluation without these considerations of independence? If you have BN software, construct the BN without these dependency assumptions and see if your predictions were correct.

Q7) Nodes 15 and 16 from the BN constructed for case 2 (the three burglars) required probability for background DNA being present on items. What sort of factors might be important in assigning this probability? What sources of data could be used? Is there one or more literature sources that you can identify that would be applicable here? What are the limitations of using those sources?

Q8) Consider the BN in case 3 (Figure 6.41), the family assault. How would that BN structure change if the alternate proposition had been that someone other than D bit C?

Q9) Again considering the BN in case 3 (Figure 6.41), add to the BN the possibility of laboratory contamination of the underwear of C with the DNA of D. Also add in the possibility that the Y-STR profile inherited by D from F had a 1 in 1000 chance of a mutation occurring, in which case their Y-STR profiles would not match. How do these two factors affect the evaluation? Should they have been included in the BN construction? Why/why not?

Q10) In Chapter 5, in the supplemental information for case 3 (Section 5.4.8) the following scenario was given:

Mr A has been invited over to Ms B's home to perform an amateur dental check-up, which Mr A does without wearing any gloves. This is accepted by both prosecution and defence as having occurred legally and consensually. After the exam, it is alleged that Mr A picked up Ms B's wallet and took out some money (without permission) and left the house. Swabs are taken from Mr A's hands two hours after the incident (before he washed his hands) to help with the issue of whether or not he has picked up Ms B's wallet. The DNA profile obtained from the swabs of Mr A's hand can be considered as a mixture of two persons corresponding to both DNA profiles of A and B.

Draw out the general structure of this BN using the steps described in this chapter. What probabilities will you need to assign to carry out the evaluation? If you have BN software, construct a BN to evaluate these observations. What is the *LR* according to your evaluation?

6.10 REFERENCES

1. Q.A. Le, G. Strylewicz, J.N. Doctor, Detecting blood laboratory errors using a Bayesian network: An evaluation on liver enzyme tests, *Medical Decision Making* 31 (2011) 325–337.

2. W. Edwards, Influence diagrams, Bayesian imperialism, and the Collins case: An appeal to reason, *Cardozo Law Review* 13 (1991) 1025–1074.

3. A.P. Dawid, et al., Probabilistic expert systems for forensic inference from genetic markers, *Scandinavian Journal of Statistics* 29 (2002) 577–595.

4. C. van-Dongen, et al., Bonaparte: Application of new software for missing persons program, *Forensic Science International: Genetics Supplement Series* 3(1) (2011) e119–e120.

5. P.J. Green, J. Mortera, Sensitivity of inferences in forensic genetics to assumptions about founding genes, *The Annals of Applied Statistics* 3(2) (2009) 731–763.

6. J. Mortera, A.P. Dawid, S.L. Lauritzen, Probabilistic expert system for DNA mixture profiling, *Theoretical Population Biology* 63 (2003) 191–205.

7. I.W. Evett, et al., Interpreting small quantities of DNA: The hierarchy of propositions and the use of Bayesian networks, *Journal of Forensic Sciences* 47(3) (2002) 520–530.

8. A. Biedermann, F. Taroni, Bayesian networks for evaluating forensic DNA profiling evidence: A review and guide to literature, *Forensic Science International: Genetics* 6 (2012) 147–157.

9. J. de Zoete, M. Sjerps, R. Meester, Evaluating evidence in linked crimes with multiple offenders, *Forensic Science International: Genetics* 57(3)(2017) 228–238.

10. C. Vlek, et al., Building Bayesian networks for legal evidence with narratives: A case study evaluation, *Artificial Intelligence and Law* 22(4) (2014) 375–421.

11. J. Pearl, Bayesian networks: A model of self-activated memory for evidential reasoning, in: Proceedings of the 7th Conference of the Cognitive Science Society, University of California (1985) 15–17.

12. F. Taroni, et al., *Bayesian Networks and Probabilistic Inference in Forensic Science* (2nd ed.), John Wiley & Sons, Ltd, Chichester, 2014.

13. I.W. Evett, Establishing the evidential value of a small quantity of material found at a crime scene, *Journal of the Forensic Science Society* 33(2) (1993) 83–86.

14. S. Højsgaard, Graphical independence networks with the gRain package for R, *Journal of Statistical Software* 46(10) (2012) 1–26.

15. M. Scutari, Learning Bayesian networks with the bnlearn R package, *Journal of Statistical Software* 35(3) (2010) 1–22.

16. M.A. Mahjoub, K. Kalti, Software comparison dealing with Bayesian networks, in: Liu D., Zhang H., Polycarpou M., Alippi C., He H. (Eds.) *Advances in Neural Networks. ISNN 2011. Lecture Notes in Computer Science*, Springer, Berlin, Heidelberg, 2011.

17. D. Taylor, T. Hicks, C. Champod, Using sensitivity analyses in Bayesian networks to highlight the impact of data paucity and direct future analyses: A contribution to the debate on measuring and reporting the precision of likelihood ratios, *Science & Justice* 56(5) (2016) 402–410.

18. D. Taylor, et al., A template for constructing Bayesian networks in forensic biology cases when considering activity level propositions, *Forensic Science International: Genetics* 33 (2018) 136–146.

19. L. Mayuoni-Kirshenbaum, et al., How did the DNA of a suspect get to the crime scene? A practical study in DNA transfer during lock-picking, *Australian Journal of Forensic Sciences* 54(1) (2020) 15–25.

20. P. Gill, et al., DNA commission of the international society for forensic genetics: Assessing the value of forensic biological evidence: Guidelines highlighting the importance of propositions. Part I: Evaluation of DNA profiling comparisons given (sub)source propositions, *Forensic Science International: Genetics* 36 (2018) 189–202.

21. D. Taylor, L. Samie, C. Champod, Using Bayesian networks to track DNA movement through complex transfer scenarios, *Forensic Science International: Genetics* 42 (2019) 69–80.

22. A. de Ronde, et al., The evaluation of fingermarks given activity level propositions, *Forensic Science International* 302 (2019) 109904.

23. U. Kjaerulff, A. Madsen, *Bayesian Networks and Influence Diagrams, A Guide to Construction and Analysis*, Springer, New York, 2008.

24. D. Taylor, et al., Helping to distinguish primary from secondary transfer events for trace DNA, *Forensic Science International: Genetics* 28 (2017) 155–177.

25. D. Taylor, The evaluation of exclusionary DNA results: A discussion of issues in R v. Drummond, *Law, Probability and Risk* 15(1) (2016) 175–197.

26. F. Taroni, P. Juchli, C. Aitken, A probabilistic account of the concept of cross-transfer and inferential interactions for trace materials, *Law, Probability and Risk* 19(3-4) (2020) 221–233.

27. M. Goray, R. van Oorschot, J. Mitchell, DNA transfer within forensic exhibit packaging: Potential for DNA loss and relocation, *Forensic Science International: Genetics* 6 (2012) 158–166.

28. D.J. Daly, C. Murphy, S.D. McDermott, The transfer of touch DNA from hands to glass, fabric and wood, *Forensic Science International: Genetics* 6 (2013) 41–46.

29. A. Kloosterman, M. Sjerps, A. Quak, Error rates in forensic DNA analysis: Definition, numbers, impact and communication, *Forensic Science International: Genetics* 12 (2014) 77–85.

30. B. Szkuta, et al., Assessment of the transfer, persistence, prevalence and recovery of DNA traces from clothing: An inter-laboratory study on worn upper garments, *Forensic Science International: Genetics* 42 (2019) 56–68.

7 Advanced Mechanisms of Evaluation

Duncan Taylor

CONTENTS

7.1 INTRODUCTION

In the previous chapter, evaluations were carried out on three example cases using Bayesian networks as a graphical tool to construct the likelihood ratio. In Chapter 6 it was seen that the LRs obtained using the BNs were the same as those that had been previously derived by hand in Chapter 5. This was because the assumptions being made and the data used in the evaluation were the same, and the use of a BN was simply a tool to ease the technical component of the evaluation.

There are numerous ways in which the use of BNs can be extended from the relatively simple examples given in Chapter 6, to encompass more complex scenarios, use more efficient architecture or consider the observations in more nuanced ways. This chapter will go through the following extensions to the standard BNs from Chapter 6.

Section 7.2 demonstrates the use of numbered or interval node states in a BN. These are when the values of states of the BN are used in the calculation of values in child nodes, rather than the node states being considered simply as a label. If using software that allows the formulaic construction

DOI: 10.4324/9781003273189-7

of conditional probabilities, then there are streamlining advantages of using numbered or interval nodes when considering observations as continuous variables.

Section 7.3 reacquaints the reader with the fourth example case. This fourth case will make use of data to model distributions for various factors being considered in the evaluation. As such the method of assigning conditional probabilities will be distinctly different from the way that they have been dealt with up to this point.

In Section 7.4 the fourth case is used as a vehicle to demonstrate the modelling of variables in an evaluation. The idea of using formulaic means of populating conditional probabilities is used (in part because with a relatively fine-scaled discretisation of states for various parental factors, the accumulated numbers of conditional probabilities can quickly rise to thousands, or tens of thousands).

Section 7.5 provides some general thoughts on modelling data, and how this relates to their use within a BN, but more broadly within an evaluation. Examples are provided of different types of modelling and explain the uses of each.

Section 7.6 discusses the use of Object-Oriented Bayesian Networks (OOBN). These types of BN construction make use of efficient architecture, by identifying repeated components within a large BN and making it into an object. There are numerous advantages of using OOBNs when dealing with complex BNs. Apart from creating a visually clearer picture, they also have the advantage of only needing to create and populate a single set of reused BN nodes' conditional probabilities.

7.2   USING INTERVALS IN BNS

This next section will provide the building blocks that will lead to modelling continuously distributed data within a BN. Up until this point, the states assigned to any event have been given designations such as 'yes', 'no', 'high', 'low', 'none', 'presence', '$C + U$', etc. Each of these designations have meaning to us, as they are ascribed to particular events or states of events, but they have no inherent meaning within the BN itself. They are acting purely as labels for convenience, i.e. all state designations of 'yes' could be changed to something meaningless such as 'purple' and all state designations of 'no' to 'yellow', and as long as the meaning of each colour was remembered during probability assignment, the resulting evaluation would not be affected (except perhaps in the sense of comprehension).

While this section describes some mathematics that is more complex than other sections in this book, it is not necessary to understand the details here in order to use numbered or interval nodes. Often software will allow numbered or interval states to be used, with any probability assignments calculated formulaically from the parent nodes occurring automatically. All that is required is a conceptual understanding of the use and benefits of these node types. A final note before beginning is that the mathematical methods described in this chapter are not necessarily those that are applied in BN software tools. Software must often balance accuracy with speed and practicality when it comes to calculations, which means that they may give slightly different values (due to using simplifying assumptions or approximations) to those calculated using the methods described here.

7.2.1   Numbered Nodes

The next topic is introduced using a simplistic, non-forensic, scenario. The way that the concepts relate to forensic BN construction is explained later after the concepts are introduced. Consider that instead of assigning the designations of the states with labels, they were instead designated with a numerical value that had meaning with respect to the evaluation, specifically each state label represented the value the state was taking. For example, consider the number of fish that will be caught by a man on a fishing trip. The range may be from none to four, and therefore the states of the BN node are designated values 0, 1, 2, 3 and 4. These designations (as well as simply distinguishing the states from one another, which could be achieved with any labels) represent the number of fish. For example these states are assigned probabilities 0.1, 0.3, 0.3, 0.2, 0.1 (note that the states are considered to be exhaustive; maybe an assumption can be made that the time spent fishing could never result in catching more than four fish). The node that has just been set up is considered a numbered node (as opposed to a labelled node) and it is now possible to deal with probability assignments formulaically. For example, consider that the man took his son on the trip and that the son could also catch fish. As the son is quite young and inexperienced, they could only catch 0 or 1 fish (hence states 0 and 1 will be created), each with probability of 0.5. How many fish are likely to be caught in total by father and son? The BN shown in Figure 7.1 shows the relatively simple BN setup.

The total fish caught must have values that exactly match (and exhaustively cover) all possible combinations of the two parental nodes; in this case, the minimum is 0 and the maximum is 5. The conditional probabilities assigned to the 'Total fish caught' node did not need to be entered manually. They can be calculated formulaically with the equation (or expression):

$$\text{Total fish caught} = \text{Fish caught by father} + \text{Fish caught by son}$$

where the values for the father and son are the numerical values for the states. For example, the probability that a fishing trip will result in three fish being caught comes from the probabilities of:

The father catching 2 fish and the son catching 1 fish (0.3×0.5), or

The father catching 3 fish and the son catching 0 fish (0.2×0.5)

$$= 0.3 \times 0.5 + 0.2 \times 0.5$$

$$= 0.25$$

as seen in the total fish caught node for state 3. Numbered nodes can be very useful in situations where the states of a node take a set of finite and defined values. Usually, numbered nodes are used in root nodes, and less frequently as child nodes of other numbered nodes (as it is in Figure 7.1). The reason for this is that when the calculations become more complex, the need to capture all possible values in a numbered node can become difficult, and it is less intuitive to determine what the values in that node mean in a real-world context. For example, in the fishing example imagine that the father has two children and the option of taking neither, one or both of them on the fishing trip. As his children are quite noisy, he knows that taking them along means that they will scare the fish and result in him catching less. In fact, the father knows that each child he brings along will halve

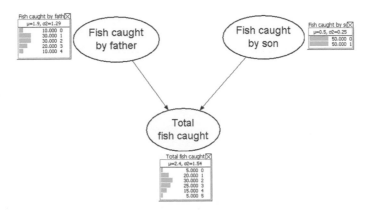

FIGURE 7.1   BN showing the number of fish caught by father and son, using numbered nodes. Both parent nodes and the child node are 'numbered nodes'.

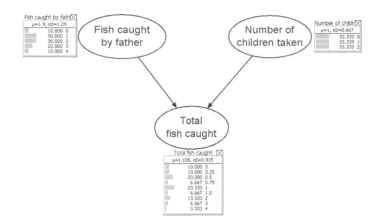

FIGURE 7.2   BN showing the number of fish caught by father by bringing some number of children, using numbered nodes. Both parent nodes and the child node are 'numbered nodes'.

the number of fish he will catch. Consider the BN in Figure 7.2, where the total fish caught node is now calculated by the equation:

$$\text{Total fish caught} = \frac{\text{Fish caught by father}}{2^{\text{(number of children taken)}}}$$

The BN in Figure 7.2 shows the architecture of the BN and the states of each node, along with their probabilities before any information is instantiated. As with the BN in Figure 7.1, all nodes in the BN in Figure 7.2 are numbered nodes. Note now how the values in the 'total fish caught' node are more complex than before. This is because the equation to calculate the total does not add to whole numbers. In fact, the values in the 'total fish caught' node are difficult to interpret as they don't have a real-world meaning, e.g. what does it mean that the father will catch 0.25 of a fish with probability 0.1?

7.2.2   INTERVAL NODES

To interpret the number of fish expected to be caught by the father from the total fish caught node, some post-processing of the data is required. Perhaps this would involve rounding the values to their nearest whole number. For example, 0.25 of a fish may round down to 0 fish so that there is a total probability of 0.2 that the father will come home with 0 fish. This same type of grouping of common values can be achieved in a BN by considering the states in a third form: As a collection of intervals. In the case of the fishing example, the BN from Figure 7.2 could be used, with intervals constructed around whole numbers that covered the entire range of possible values:

0–0.5
0.5–1.5
1.5–2.5
2.5–3.5
3.5–4.5

Any sums of values from the parent nodes falling into the first bracket are summed and displayed as the probability for that bracket, which can be interpreted as 0 fish. Carrying this process on for all combinations of parent values produces the BN shown in Figure 7.3.

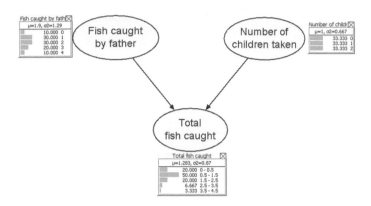

FIGURE 7.3 BN showing the number of fish caught by father by bringing some number of children, using numbered nodes and an interval node for the total fish caught. Both parent nodes are 'numbered nodes' and the child node is an 'interval node'.

The total fish caught can now be interpreted as there being a 0.2 probability that 0 fish will be caught, a 0.5 probability that 1 fish will be caught and so on. This is under the assumption that there is an equal probability that the father will take none, one or both of his children.

It is also possible to consider the parent nodes as interval nodes if the property being measured does not possess integer values. A common example could be the amount of DNA on an object. An experiment is carried out that investigates the amount of DNA obtained by swabbing. The amount of DNA on the object is modelled in discrete brackets of:

- 0 ng to 1 ng with the probability of 0.2
- 1 ng to 2 ng with the probability of 0.5
- 2 ng to 3 ng with the probability of 0.3

Although any break-up of the data could be used. The item can be swabbed once or twice. The amount of DNA collected by swabbing is determined by the equation:

$$\text{Sampled DNA} = \text{starting DNA} \times (1 - 0.5^{\text{number of swabs}})$$

which could, for instance, be based on a validation study that showed that each swab used recovers half of all DNA present or remaining on the surface. The BN shown in Figure 7.4 shows this scenario.

The probability table for the 'sampled DNA' node in Figure 7.4 is shown in Table 7.1.

The values in the cells of Table 7.1 are calculated by adjusting the range of the interval used for the starting DNA by the number of swabs. For example, if the range of starting DNA was 1–2 ng and two swabs were used, then a uniform distribution of expected DNA amount is obtained as shown in Figure 7.5.

The lower bound of the uniform distribution is adjusted from its original value of 1 to 0.75 when considering two swabs by:

$$\text{Adjusted lower-bound interval} = 1 \times (1 - 0.5^2)$$

$$= 1 \times (1 - 0.25)$$

$$= 0.75$$

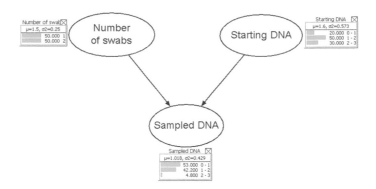

FIGURE 7.4 BN showing the amount of DNA obtained from one or two swabs of an item. Both the nodes specifying DNA amounts are interval nodes, and the swab node is a numbered node. One parent node is a 'numbered node' and the other parent node, and child node are 'interval nodes'.

TABLE 7.1

Probability Table for the 'Sampled DNA' Node in Figure 7.7

Starting DNA (ng)		0–1		1–2		2–3	
Number of swabs		1	2	1	2	1	2
Sampled DNA (ng)	0–1	1	1	1	0.33	0	0
	1–2	0	0	0	0.67	1	0.67
	2–3	0	0	0	0	0	0.33

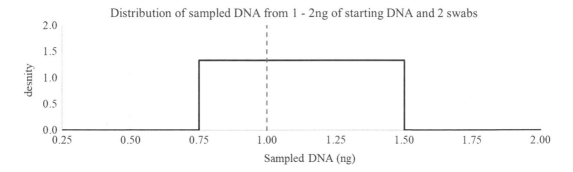

FIGURE 7.5 Distribution of DNA expected from 1 to 2 ng of starting DNA and using two swabs.

and similarly, the upper bound is adjusted to 1.5. The height of the distribution (its density) is determined by the fact that the area under the curve must equal 1. Therefore, the height is one divided by the width of the interval:

$$\text{density} = \frac{1}{(1.5 - 0.75)} = \frac{1}{0.75} = 1.33$$

The red line in Figure 7.5 shows the boundary of two states of the sampled DNA factor (i.e. the boundary between 0–1 ng and 1–2 ng). When the probability is assigned to the interval states of the sampled DNA factor, it is done so taking into account the area of the adjusted uniform distribution seen in Figure 7.5 that falls into the upper and lower bounds of the interval. For example, for the sampled DNA interval of 0–1 ng, the area of the adjusted uniform distribution within 0 and 1 is 0.33, which can be obtained by:

$$\text{Pr(sampled DNA} = 0 - 1 \mid \text{starting DNA} = 1 - 2, \text{swabs} = 2) = (1 - 0.75) \times 1.33 = 0.33$$

Similarly:

$$\text{Pr(sampled DNA} = 1 - 2 \mid \text{starting DNA} = 1 - 2, \text{swabs} = 2) = (1.5 - 1) \times 1.33 = 0.67$$

Which are the values shown in Table 7.1.

7.2.2.1 Summing Interval Nodes

The final BN construction shows the situation where both parents and the child are interval nodes. The first situation described below is when the probabilities assigned to the states of the child nodes result from a sum of the probability states of the parent nodes (such as when considering an accumulation of DNA from two objects). The second situation described is when the probabilities assigned to the states of the child node result from a product of the probability states of the parent nodes (such as when considering a transfer of DNA from one object to another).

First consider a scenario where there are two objects which both transfer an amount of DNA to a third object. All three nodes are represented by a series of three intervals set up with the prior probabilities as seen in Figure 7.6. The child node is the total DNA amount node and is calculated by:

$$\text{Total DNA amount} = \text{DNA transferred from object 1} + \text{DNA transferred from object 2}$$

Note that the intervals in the total DNA node in Figure 7.6 must span from the sum of the lower bound of the two smallest brackets possessed by its parents ($0 + 0 = 0$) to the sum of the upper bound of the two largest brackets possessed by its parents ($6 + 3 = 9$). The conditional probability table for the total DNA node is shown in Table 7.2.

In order to understand the distribution of probability to the states in the 'total DNA' node from any combination of states in the two parent nodes, the probability density function for the sum of two uniform random variables must be considered.

If X is a uniform variable between values of 'a' and 'b', i.e.

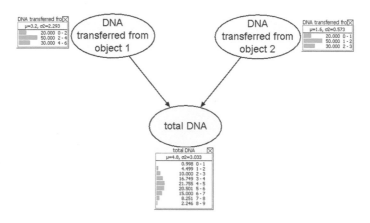

FIGURE 7.6 BN showing the amount of total DNA obtained from the accumulation of DNA transferred from two objects. All the nodes are interval nodes. Both parent nodes and the child node are 'interval nodes'.

TABLE 7.2

Probability Table for the 'Total DNA' Node in Figure 7.8

DNA transferred from object 1		0–2			2–4			4–6		
DNA transferred from object 2		0–1	1–2	2–3	0–1	1–2	2–3	0–1	1–2	2–3
Total DNA	0–1	0.25	0	0	0	0	0	0	0	0
	1–2	0.50	0	0	0.25	0	0	0	0	0
	2–3	0.25	0.25	0	0.50	0	0	0.25	0	0
	3–4	0	0.50	0	0.25	0.25	0	0.50	0	0
	4–5	0	0.25	0.25	0	0.50	0	0.25	0.25	0
	5–6	0	0	0.50	0	0.25	0.25	0	0.50	0
	6–7	0	0	0.25	0	0	0.50	0	0.25	0.25
	7–8	0	0	0	0	0	0.25	0	0	0.50
	8–9	0	0	0	0	0	0	0	0	0.25

$$X \sim U[a, b]$$

And similarly:

$$Y \sim U[c, d]$$

The sum of these variables is Z:

$$Z = X + Y$$

When X and Y are ordered so that $b + c \leq d + a$, the probability density distribution of Z, $f(z)$, can be given by:

$$f(z) = \begin{cases} 0 & z \leq a+c \\ (z-a-c)/\left[(b-a)(d-c)\right] & a+c < z < b+c \\ 1/(d-c) & b+c \leq z < d+a \\ (b+d-z)/\left[(b-a)(d-c)\right] & d+a \leq z < b+d \\ 0 & z \geq b+d \end{cases} \qquad (7.1)$$

Figure 7.7 shows this distribution graphically.

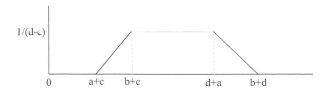

FIGURE 7.7 The distribution of Z, $f(z)$, the sum of uniform random variables $X \sim U[a, b]$ and $Y \sim U[c, d]$.

When using distributions in order to assign probability, it is the area under the distribution that is of importance, and for this the cumulative density function, $cdf(z)$, can be used, as given by the following equation for a sum of intervals:

$$cdf(z) = \begin{cases} 0 & z \le a+c \\ (z-a-c)^2 / \left[2(b-a)(d-c)\right] & a+c < z < b+c \\ (2z-a-b-2c)/\left[2(d-c)\right] & b+c \le z < d+a \\ 1-(b+d-z)^2 / \left[2(b-a)(d-c)\right] & d+a \le z < b+d \\ 1 & z \ge b+d \end{cases} \quad (7.2)$$

Consider the combination of states for the DNA transfer nodes where the DNA transferred from object 1 is 2–4 ng and the DNA transferred from object 2 is 1–2 ng. The distribution of DNA expected from the sum of these two states is the sum of:

$$X \sim U[1, 2]$$

$$Y \sim U[2, 4]$$

Using Equation 7.2, the probability density function, $f(z)$, for the sum $Z = X + Y$ is

$$f(z) = \begin{cases} 0 & z \le 3 \\ (z-3)/2 & 3 < z < 4 \\ 1/2 & 4 \le z < 5 \\ (6-z)/2 & 5 \le z < 6 \\ 0 & z \ge 6 \end{cases}$$

Graphically, this is shown in Figure 7.8.

As before, the area under the curve seen in Figure 7.8 between state boundaries can be used to determine the probability assignments for the total DNA states. The area taken up by the distribution between 3 and 4 ng of DNA is 0.25 (half the length times the height). This can be determined using the cumulative density function as:

Pr(3 ng < total DNA < 4 ng | transfer from object 1 = 2–4 ng, transfer from object 2 = 1–2 ng)

$$= cdf(z = 4) - cdf(z = 3)$$

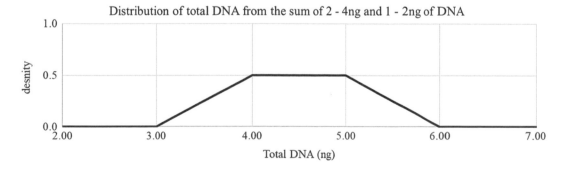

FIGURE 7.8 Distribution of total DNA expected from the sum of 2–4 ng and 1–2 ng of DNA.

$$\left(2\{z=4\}-a-b-2c\right)/[2(d-c)]-\left(\{z=3\}-a-c\right)^2/\left[2(b-a)(d-c)\right]\qquad =$$

When $a = 1$, $b = 2$, $c = 2$, $d = 4$,

$$\left(2\times4-1-2-2\times2\right)/[2(4-2)]-\left(3-1-2\right)^2/\left[2(2-1)(4-2)\right]\qquad =$$

$$= 1/4$$

Similarly, the area under the curve between 4 and 5 ng is 0.5 and the area under the curve between 5 and 6 ng is again 0.25. All other states have an area (and hence probability) of 0. This is seen in the probability assignments in Table 7.2.

7.2.2.2 Multiplying Interval Nodes

The final commonly used instance of intervals is when the child node is a product of the parent interval nodes. This may occur when considering scenarios of DNA transfer. For example, consider a situation where an object has some starting DNA, and the amount transferred to the second object is described by an interval node with states 0–33%, 33–66% and 66–100%. Such a BN is shown in Figure 7.9. The DNA-transferred node has states that have probabilities assigned using the equation:

DNA transferred = DNA on object 1 × proportion of DNA transferred

The conditional probability table for the DNA-transferred node is shown in Table 7.3.

In order to understand the distribution of probability to the states in the 'DNA transferred' node from any combination of states in the two parent nodes, the probability density function for the product of two uniform random variables must be considered.

If X is a uniform variable between values of 'a' and 'b', i.e.:

$$X \sim U[a, b]$$

And similarly:

$$Y \sim U[c, d]$$

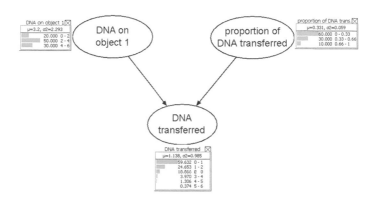

FIGURE 7.9   BN showing the DNA transferred from the starting amount of DNA on object 1 and the proportion of DNA transferred. All the nodes are interval nodes. Both parent nodes and the child node are 'interval nodes'.

TABLE 7.3

Probability Table for the 'DNA Transferred' Node in Figure 7.9

DNA on object 1		0–2			2–4			4–6		
Proportion of DNA transferred		0–0.33	0.33–0.66	0.66–1	0–0.33	0.33–0.66	0.66–1	0–0.33	0.33–0.66	0.66–1
DNA transferred	0–1	1	094	0.61	0.94	0.11	0	0.61	0	0
	1–2	0	0.06	0.39	0.06	0.76	0.23	0.39	0.22	0
	2–3	0	0	0	0	0.13	0.58	0	0.58	0.03
	3–4	0	0	0	0	0	0.19	0	0.20	0.41
	4–5	0	0	0	0	0	0	0	0	0.44
	5–6	0	0	0	0	0	0	0	0	0.12

The product of these variables is Z:

$$Z = X \times Y$$

When X and Y are ordered so that $ad \leq bc$, the probability density distribution of Z, $f(z)$, can be given by:

$$f(z) = \begin{cases} 0 & z < ac \\ \left[\ln(z) - \ln(ac)\right]/\left[(b-a)(d-c)\right] & ac \leq z < bc \\ \ln(b/a)/\left[(b-a)(d-c)\right] & bc \leq z < ad \\ \ln(bd/z)/\left[(b-a)(d-c)\right] & ad \leq z < bd \\ 0 & z \geq bd \end{cases} \qquad (7.3)$$

where 'ln' is the natural (base e) logarithm of the bracketed number. Figure 7.10 shows this distribution graphically.

Again, to assign probabilities the cumulative density function is utilised. The manner in which the cumulative density function (cdf) was created previously was not specified. The fact that the probability density function shown in Figure 7.8 was made from straight lines meant that the area could also be calculated with standard geometry. The cdf formula for the product scenario is not so straightforward. The area under the curve is obtained by integration. Specifically, for a value of z that is between ac and bc the cumulative area is obtained by:

$$\Pr(Z \leq z) = \int_{ac}^{z} f(z)dz = \int_{ac}^{z} \frac{\ln(z) - \ln(ac)}{(b-a)(d-c)} dz$$

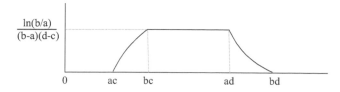

FIGURE 7.10   The distribution of Z, $f(z)$, the product of uniform random variables $X \sim U[a, b]$ and $Y \sim U[c, d]$.

which, without showing all integration steps, is:

$$\Pr(Z \leq z) = \frac{z\big[\ln(z) - \ln(ac) - 1\big] + ac}{(b-a)(d-c)}$$

The area under the curve from point ac to bc can be calculated by substituting in bc for z so that:

$$\Pr(Z \leq bc) = \frac{bc\big[\ln(bc) - \ln(ac) - 1\big] + ac}{(b-a)(d-c)}$$

This is useful for simplification as the area under the curve for some point of z between bc and ad can be calculated by:

$$\Pr(Z \leq z) = \int_{ac}^{z} f(z)dz = \int_{ac}^{bc} f(z)dz + \int_{bc}^{z} f(z)dz = \Pr(Z \leq bc) + \int_{bc}^{z} f(z)dz$$

$$= \Pr(Z \leq bc) + \int_{bc}^{z} \frac{\ln\left(\dfrac{b}{a}\right)}{(b-a)(d-c)} dz$$

$$= \Pr(Z \leq bc) + \frac{(z - bc)\ln\left(\dfrac{b}{a}\right)}{(b-a)(d-c)}$$

And in a similar manner if z is between ad and bd:

$$\Pr(Z \leq z) = \Pr(Z \leq ad) + \int_{ad}^{z} f(z)dz$$

$$\Pr(Z \leq z) = \Pr(Z \leq ad) + \frac{z\left[1 - \ln\left(\dfrac{z}{bd}\right)\right] - ad\left[1 - \ln\left(\dfrac{ad}{bd}\right)\right]}{(b-a)(d-c)}$$

So that the *cdf* for all values of z is given by:

$$cdf(z) = \begin{cases} 0 & z < ac \\[2mm] & ac \leq z < bc \\[2mm] \dfrac{z\big[\ln(z) - \ln(ac) - 1\big] + ac}{(b-a)(d-c)} & \\[4mm] \Pr(Z \leq bc) + \dfrac{(z - bc)\ln\left(\dfrac{b}{a}\right)}{(b-a)(d-c)} & bc \leq z < ad \\[6mm] \Pr(Z \leq ad) + \dfrac{z\left[1 - \ln\left(\dfrac{z}{bd}\right)\right] - ad\left[1 - \ln\left(\dfrac{ad}{bd}\right)\right]}{(b-a)(d-c)} & ad \leq z < bd \\[6mm] 1 & z \geq bd \end{cases} \qquad (7.4)$$

Consider the combination of states for the DNA on object 1 as 4–6 ng and the transfer proportion as 0.66–1. The distribution of DNA expected from the product of these two states is the product of:

$$X \sim U[4, 6]$$

$$Y \sim U[0.66, 1]$$

Graphically, this is shown in Figure 7.11.

Again, the cumulative distribution function can be used for the product of two uniform distributions to assign probabilities. For the example being considered, the cumulative distribution is shown in Figure 7.12. In Figure 7.12, the x-axis shows the DNA amount transferred and the y-axis shows a cumulative probability for that amount of DNA transferred (rather than a density as shown in previous graphs). For example, there is approximately a probability of:

- 0 that less than 2 ng will be transferred,
- 0.03 that less than 3 ng being transferred,
- 0.9 that less than 5 ng will be transferred and
- 1 that less than 6 ng will be transferred.

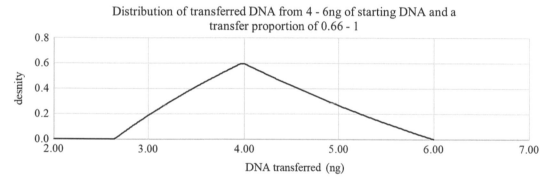

FIGURE 7.11   Distribution of transferred DNA expected when object 1 has 4–6 ng and the transfer proportion was 0.66–1.

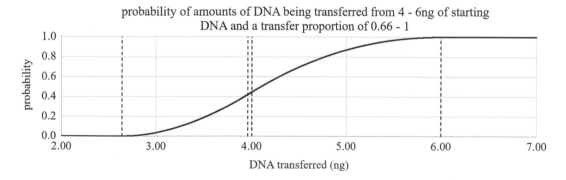

FIGURE 7.12   Cumulative probability of transferred DNA expected when object 1 has 4–6 ng and the transfer proportion was 0.66–1.

Because the y-axis in Figure 7.12 is a probability, it only takes values from 0 to 1. The grey vertical lines show the boundaries for the different components of the distribution i.e. in general terms the boundaries of ac, bc, ad and bd.

Using the cumulative distribution graph from Figure 7.12 (or more accurately the cdf equation), the probability of transferred DNA being between 2 and 3 ng is:

Pr(2 ng < transferred DNA < 3 ng | DNA on object 1 = 4–6 ng, transfer proportion = 0.66–1)

$$= cdf(z = 3) - cdf(z = 2)$$

In this case $a = 4$, $b = 6$, $c = 0.66$, $d = 1$. Only the areas starting after $ac = 2.64$ need be considered (as the cdf is zero up until this point). In fact, as the calculations below will show, the lower-bound calculation need not be enumerated at all as it will be zero:

$$= cdf(z = 3) - cdf(z = 2.64)$$

$$= \frac{3\left[\ln(3) - \ln(4 \times 0.66) - 1\right] + 4 \times 0.66}{(6 - 4)(1 - 0.66)} - \frac{2.64\left[\ln(2.64) - \ln(4 \times 0.66) - 1\right] + 4 \times 0.66}{(6 - 4)(1 - 0.66)}$$

$$= 0.034 - 0$$

$$= 0.034$$

which is the value seen in the corresponding cell of Table 7.3 (far-right column, shown up to two significant figures as 0.03). The other values within Table 7.3 can be assigned in a similar fashion.

An issue occurs when the interval bounds for the states are negative. From Equations 7.3 and 7.4, it can be seen that a problem arises with negative values as there is no defined value for the log of a negative number. Note that in these instances the issue can be solved by considering the bounds as its positive symmetrical counterpart and then translating the distribution of probability symmetrically around 0 to the negative bounds. For example, the product of an $X \sim U[-2, -1]$ and $Y \sim U[0.33, 0.66]$ provides a probability of 0.125 to the interval $[-2, -1]$ and a probability of 0.875 to the interval $[-1, 0]$. These values can be obtained by using Equation 7.4 and considering the product of $|X| \sim U[1, 2]$ and $Y \sim U[0.33, 0.66]$ which provides a probability of 0.125 to the interval $[1, 2]$ and a probability of 0.875 to the interval $[0, 1]$. Note that the probability assigned to the brackets in the positive domain is symmetrical around 0 to the probabilities assigned when X is in the symmetrical, negative domain. From this explanation, it becomes apparent that in order to calculate probabilities for products of interval nodes there can be no node that has an interval crossing 0 (if these exist, they will need to be split around zero).

Requiring this type of calculation is unusual in the forensic domain as parameters being manipulated usually cannot take negative values. Even when modelling factors such as DNA amount on a log scale (where intervals with negative bounds will be present), we do not need this type of consideration as the application of factors such as transfer proportion must be applied on a natural scale, i.e. the log DNA amount will need to be converted to a DNA amount (on a natural scale) before the transfer proportion applies, which means that it will no longer take interval values in the negative domain. Even the addition of interval nodes in the negative domain is rare (again as adding DNA amounts is not carried out in a log scale, and the value must be converted to their, always positive, natural values). One instance where the addition of intervals that have negative bounds may occur is when multiplying values on a natural scale, but in a log scale so that the action becomes an addition. For example, if DNA amount was modelled on a log scale and transfer proportion was also modelled on a log scale, then the multiplication on a natural scale is equivalent to the sum of these terms on a log scale (and would lead to a child node that was also on a log scale).

7.3 CASE 4: THE GUN IN THE LAUNDRY

This case was not examined in earlier chapters due to the fact that the evaluation is too complex to reasonably be carried out using formulaic *LR* derivation. It also uses distributions to model DNA amounts and interval nodes within the BN and so goes beyond the BN setup described in Chapter 6.

Recall that the circumstances surrounding this case are that a defendant (D) is accused of robbing a service station armed with a handgun. The firearm is found in a laundry basket covered with D's unwashed laundry. D claims that he has never seen the firearm and that it must have been put in his laundry basket by someone else (an alternate offender, AO). D states that he last did his laundry yesterday, so whoever put the firearm in his laundry must have done so within the last 24 hours. The defendant states that he has not been home much in the last 24 hours and couldn't tell if anyone had been in the house.

There was 9.078 ng of total DNA detected on the swab from the firearm and D aligned with a contributor that accounts for 93.84% of the mixed DNA profile, which means that 8.52 ng has originated from D and 0.56 ng has originated from unknown sources. This is accepted by both parties.

Table 7.4 shows the results of the samples taken from the laundry. In all cases D was assumed to be a contributor of DNA to the sample.

Recall from Chapter 4 the following proposition setup:

Background information:

- The firearm was found in a basket of D's unwashed laundry.

Propositions :

- H1: D handled the firearm and placed it in his laundry basket
- H2: An AO handled the firearm and placed it in D's laundry basket

Assumptions:

- Whoever placed the firearm in the basket (i.e. either the defendant, D, or an alternate offender, AO) was the regular user of the firearm.
- Whoever (regularly) used the firearm did not wear gloves when doing so.
- The firearm had not been recently cleaned.
- Neither the defendant nor alternate offender had any skin or other conditions that would mean they shed DNA unusually more or less than an average person.

In the next section the BN is developed for the evaluation of these case observations.

TABLE 7.4

DNA Results from Laundry Items in Case 4

Sample	total DNA (ng)	Number of contributors	D mixture proportions	D DNA (ng)	Other DNA (ng)
1	1.5456	4	64.45%	0.996	0.5496
2	4.0368	4	92.43%	3.731	0.3058
3	0.6846	5	53.1%	0.364	0.3206
4	0.1955	2	79%	0.154	0.0415
5	15.396	5	87.3%	13.44	1.956
6	32.352	3	87.53%	28.32	4.032

7.4   MODELLING EFFECTS AND AMOUNTS IN BNS

7.4.1   BN CONSTRUCTION FOR CASE 4

The construction of the BN is deliberately kept simple for this case, as the main point being demonstrated is the use of nodes that represent distributions of DNA amounts, interval nodes and formulaic probability assignments for conditional probability tables (rather than exploring a rich BN
architecture).

The BN construction starts, as always, with the proposition node which is the parent of activity
nodes. In this case the activities that need to be considered are:

1. D handled the firearm,
2. An AO handled the firearm, and
3. The firearm was placed in the laundry bag with D's clothing.

The first of these activities has occurred under Hp, but not Hd, the second has occurred under Hd
but not Hp and the third has occurred under both propositions. The results nodes will be the amount
of D's DNA found on the firearm (including none) and the amount of unknown DNA found on the
firearm (including none). The DNA transfers that lead down from the activities to the results nodes
will consider the main routes:

- D's DNA transferring to the firearm from direct handling by D.
- D's DNA transferring to the firearm from contact with clothing in the laundry bag.
- Unknown DNA coming from the AO's DNA transferring to the firearm from direct han-
  dling by AO.
- Unknown DNA coming from background DNA on the firearm.

Given this, the structure of the BN is shown in Figure 7.13.

The following are not considered in the BN architecture (although some of these will be explored
in later chapters):

- Unknown DNA transferring from the items of clothing in the laundry bag to the firearm.
- Whether or not the unknown DNA on the items of laundry is the same as the unknown
  DNA on the firearm.
- Which items of laundry the firearm may have come into contact with.
- The fact that there are multiple sources of unknown DNA on the firearm (and only one of
  these could be an AO).
- The action of removing any DNA from an AO during contact with the items of laundry.

Note that in the BN construction that leads to the architecture shown in Figure 7.13, the process is
no different to constructing a BN for cases 1–3 in Chapter 6. In fact, at this stage of BN construction the choice could be made to model the DNA amounts in any manner from a binary presence/
absence to DNA amounts (although the node names shown in Figure 7.13 are indicative that DNA
amounts will be used).

Nodes 5–14 in the BN in Figure 7.13 are interval nodes. The extremes of the states in the nodes,
and the level of discretisation depend on the situation and the preference of the analyst. Discretisation
is the manner in which a continuous distribution can be broken into a finite number of discrete segments. There is little information to guide here other than there must be a balance between the
complexity of the BN (the consideration of which will generally drive the discretisation to courser
steps) and the resolution obtained in the results (the consideration of which will generally drive the
discretisation to finer steps). The extremes of the bounds tend to be set by the range of reasonable
values expected to have been accounted for. DNA amounts in nodes 5, 7, 8 and 10 are modelled on a

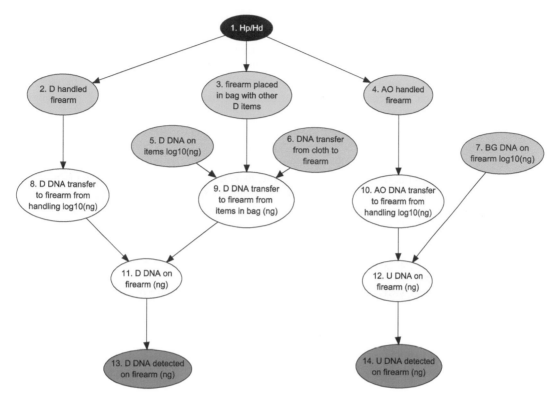

FIGURE 7.13BN for evaluations of observations in case 4.

$\log_{10}$ scale with normal distribution. Node 8 modelled the transfer proportions with a Beta distribution. Nodes 9, 11, 12, 13 and 14 have probabilities assigned formulaically. Specifically:

$$\text{Node } 9 = 10^{\text{node } 5} \times \text{node } 6$$

$$\text{Node } 11 = 10^{\text{node } 8} + \text{node } 9$$

$$\text{Node } 12 = 10^{\text{node } 7} + 10^{\text{node } 10}$$

$$\text{Node } 13 = \text{Node } 11$$

$$\text{Node } 14 = \text{Node } 12$$

Note that nodes 13 and 14 are not required in this BN as they just replicate nodes 11 and 12. Nodes 11 and 12 could have been made the results nodes. The reason for the separate nodes, 13 and 14, in this instance is simply to separate the formulaic assignments from the results nodes for visual clarity of the example. It also makes the BN more easily expandable if recovery considerations were later added.

There are several ways in which the DNA results from the samples taken from the laundry items could have been used in the evaluation.

- *Modelling a distribution of available DNA from the results* – This method, which is the method used for the example evaluation, uses the DNA results for the items in the laundry basket in order to build up a distribution of DNA that was available to transfer to the firearm. While this means that the results are quite generally used, it probably reflects our state of knowledge about the framework of circumstances that surround the firearm. It is not known which items of laundry the firearm came into contact with, or with what vigour or length of time, or how much area of the gun or clothing came into contact with each other, or whether subsequent transfers removed any DNA from earlier transfers. Therefore, it is only possible to model a general distribution for the amount of DNA that may have been available to transfer to the firearm.

The process itself involves choosing a distribution that reasonably reflects the spread of DNA amounts observed and then using that distribution to assign a probability to the intervals within node 6 in the BN. DNA amounts are most often distributed on a log scale with a normal distribution (this tends to describe the distribution of DNA amounts seen in many applications). Figure 7.14 shows the normal distribution fitted to the six DNA amounts of clothing in the laundry basket. The points along the x-axis are the DNA amounts (on a $\log_{10}$ scale $-0.002, 0.572, -0.439, -0.812, 1.128$ and 1.452). The black line shows the smoothed density of the points, and the red line shows the fitted normal distribution. In this case the distribution has a mean of 0.317 and a variance of 0.812 when fitted using maximum likelihood estimation.

In this example, several samples were taken from the item, and this led to several measurements of DNA amount from which a distribution could be modelled. It may not always be the case that there are multiple values from which to model a distribution in this manner. In such a situation there are two main options. The first is to use the mean of the single value obtained in the case and set a variance based on the general level of variance seen in DNA amounts in literature. For example, if a DNA amount of 1 ng was obtained, then the mean $\log_{10}(ng)$ amount could be set at 0, and the variance could be set from the case 4 example results, i.e. 0.812. Doing this makes the assumption that there is an approximately equal variance in DNA distributions between different items of clothing. Alternatively, a distribution for DNA on clothing could be obtained from the literature (or from previous casework) and used in the BN itself, which is discussed in the next point.

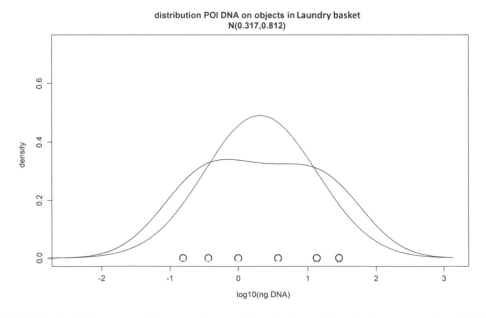

FIGURE 7.14　D DNA amounts of samples from clothing (black circles), with observed density (black line) and fitted normal distribution (red line).

- *Modelling the distribution parameters for available DNA within the BN* – The idea of this method is very similar to the previous one, except that the application occurs in the BN rather than outside the BN. This method is more akin to a process known as hierarchical Bayesian modelling (HBM), whereby the individual sample results are considered as observations from a population with a distribution. The hyper-prior distributions for the parameters in the distribution are then used to model the population distribution for DNA amount. Figure 7.15 shows node 5 from the BN seen in Figure 7.13 (with the rest of the BN removed) and the architecture that could be used to determine the distribution of D's DNA available on the items of laundry using the BN itself. In the BN shown in Figure 7.15 the variables of the normal distribution being used to describe the amount of D's DNA on the laundry are nodes labelled '5. mean' and '5. var'. As well as informing the distribution of DNA for node 5, these parameter nodes also inform the DNA amounts for six nodes, one for each of the samples taken from the laundry. All six laundry sample nodes and node 5 in Figure 7.15 have probability assigned using the formula:

  NORMAL(mean, var)

  When setting up an evaluation as part of an HBM, it is also required to set prior distribution for the parameter values as well. For example, the mean and variance nodes in Figure 7.15 are interval nodes that have upper and lower bounds, all of which must have probabilities assigned. The mean node is arguably easier to assign as it can be informed by knowledge from the literature on the amount of DNA expected on worn clothing. This distribution can be quite broad, and in fact this is desirable as it allows the observation

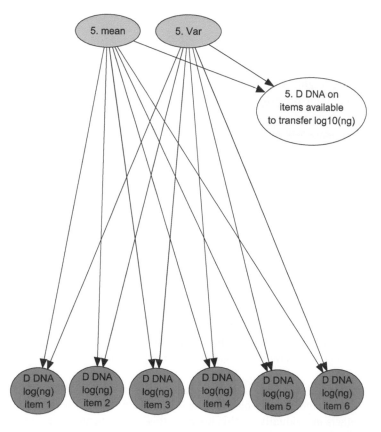

FIGURE 7.15   A hierarchical Bayesian modelling setup to inform the amount of D DNA available for transfer to the firearm.

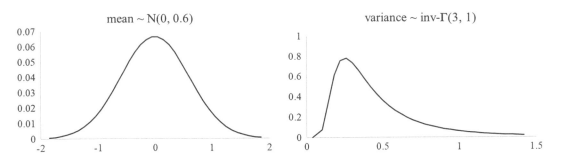

FIGURE 7.16   Prior distributions for the mean and variance.

to drive the posterior distribution for the mean. As an example, a normal distribution is used with hyper-prior mean of 0 and a variance of 0.6 (seen in Figure 7.16 left). A prior distribution for the variance is less intuitive. A variance must be greater than 0 and so the prior distribution must be bound at the lower end by zero. There is much literature about the choice of appropriate hyper-prior values for a variance distribution, but a commonly used choice of distribution is an inverse gamma, written as inv- (see the work of Gelman [1, 2] for in-depth discussions on the topic of variance parameter hyper-prior choices). The specifics of hyper-prior choices or their distributions are not explored here. Note that in some software there is a limitation to the type of distributions that are available formulaically, and probability assignment for more unusual distributions will have to be developed external to the software.

The extreme values of bounds used in interval nodes depend somewhat on the range of values that are reasonably expected to be encountered. For example, there is little value in including bounds up to 6 for log DNA as it is highly improbable that amounts up to 1 million ng of DNA will be relevant. Having said that, due to the bounds of distributions (for example the fact that normal distributions exist from −infinity to infinity) some software require the full possible ranges to be filled when probabilities are assigned formulaically. Figure 7.17 shows the BN from Figure 7.15 with the posterior probabilities after the values for the laundry samples have been instantiated (albeit the architecture of the BN is completely covered by the tables of probabilities). All DNA amount nodes (the six laundry item amount nodes and node 5 from the original BN) must have intervals that range from −infinity to infinity. This can be seen in the extreme value of the first and last node state ranges.

In each of the uninstantiated interval nodes, a mean and variance are provided. In Figure 7.17 the mean and variance of the D DNA amount available for transfer can be seen to be 0.236 and 0.639, respectively. Note that these values are close to the mean and variance seen when the DNA amounts from the laundry samples were modelled directly as shown in Figure 7.14. The difference in this case is due to the choice of priors in the HBM method, as well as the fact that the continuous normal and inverse-gamma distributions are modelled by a number of discrete chunks in the BN.

It was mentioned in the previous point that this method could be used if only a single observation from the case was available. If this is the case, then the architecture in Figure 7.15 could be used, but with a single item observation node rather than the 6 shown. The consequence of this would be to less strongly inform the general mean and variance hyper-parameters.

- *Modelling individual contacts* – Another manner in which the DNA transfer resulting from contact(s) between the laundry and the firearm could be treated within the BN is to consider the contacts individually. In this instance this would mean that a transfer probability would be assigned to each laundry item, with the same starting prior probabilities. There would then be a series of accumulations leading to a total amount of DNA transferred.

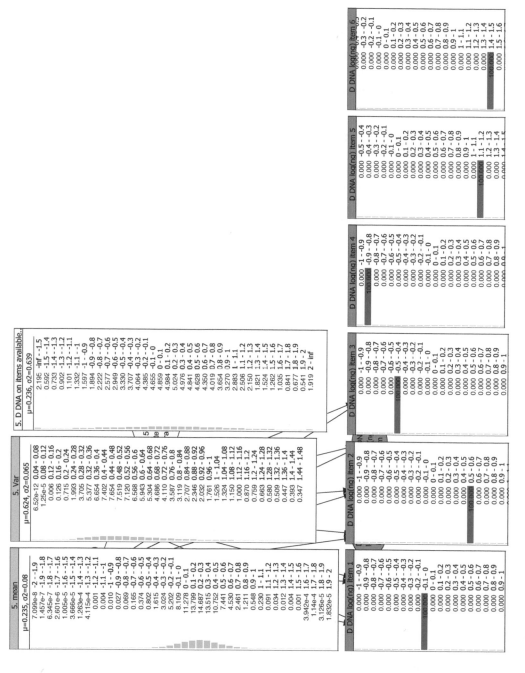

FIGURE 7.17 Posterior distribution for mean, variance and D DNA amount available for transfer after the sample results from the laundry items have been instantiated.

If the model was also taking into account the removal of DNA once transferred, from subsequent contacts then this style of BN architecture could handle such considerations. However, to do so it would be required to make some determination of the contact order. Clearly such a modelling setup is not applicable to case 4, but there may be other instances when it is warranted.

7.4.2   How to Model DNA Amounts as Distribution

When modelling a series of observations whose measurements are from a continuous range within a BN, there are a number of decisions that need to be made. The first is the decision as to what distribution to use to model the observations. This will largely be driven by the properties of the observations themselves, i.e. what are they observations of, and what are the properties of that factor. For example, if talking about DNA amounts, then the values must be positive (i.e. a negative DNA amount does not make sense) and in theory have no upper bound. There are a number of distributions that have these same properties, such as a gamma or log-normal distribution. It is common, however, to transform data in such a way that it makes the modelling easier, or more standard. Ideally transformations are made so that it adheres to a normal distribution. In the case of DNA amounts a $\log_{10}$ transformation is typically used and this allows modelling as a normal distribution.

Once the distribution is chosen, its parameters must be set so that the distribution is a reasonable reflection of the observations. There are a few ways in which this is typically done:

Least squares – this is most commonly used in many forensic science applications. If the distribution of data is normal, then the least squares method (i.e. obtaining the mean as the value that minimises the squared distance of the datapoints to that value) finds the mean of N observations x_1, ..., x_N by:

$$\mu = \frac{1}{N} \sum_{i=1}^{N} x_i$$

Using the mean, the variance can be calculated by:

$$\sigma^2 = \frac{\sum_{i=1}^{N} (x_i - \mu)^2}{N-1}$$

These properties can be calculated by many software applications such as excel:

```
=average(cells)
=var(cells)
```

Or R:

```
mean (values)
var(values)
```

Maximum Likelihood Estimation (MLE) – Maximum likelihood calculates the density of the proposed distribution at each of the points where the observations lie. The total likelihood of the distribution is the product of the densities at each of the observation values. The MLE method, as the name suggests, finds the parameter values that maximise the likelihood of the distribution (or in other words the probability density of the observations is maximised). Under certain circumstances (namely that the random elements in the data are normally distributed), then the MLE will provide the same value for the mean parameter as the least squares method. The variance will be a factor of

$N - 1/N$ compared to the variance previously given (that minimises the mean square error) and is therefore MLE is a bias estimator.

Whenever the random elements are not normally distributed, then the MLE can have more power to identify the parameters of the distribution than the least squares.

Hierarchical Bayesian Modelling – An example was shown of this method earlier using a BN (Figure 7.15). This method is similar to MLE, but the parameters in the distribution themselves have prior distribution with their own hyper-priors. This method is the most flexible, but also the most complex and requires explicit prior distributions and hyper-prior values to be chosen.

Whatever manner the modelling takes place, there is likely to be a requirement to break the continuous distribution into discrete chunks for the BN software. The choice of how many brackets and the span of their values is somewhat up to personal preference, with the guiding notes that there is a balance between many brackets (which add complexity and computation requirements to the BN evaluation) and few brackets (which risks losing too much fine-scale resolution in the evaluation). The extreme ranges of the brackets are likely to be guided by the extreme of the range of values expected to be encountered in real life (as there is no point going beyond these bounds).

When assigning the probabilities to brackets, using a distribution that has been modelled externally to the BN (such as the DNA amounts in nodes 8 and 10 in the BN shown in Figure 7.13) cumulative density functions will be required. This is carried out in the same manner as was earlier shown when explaining the functioning of interval nodes. For example, node 5 in the BN from Figure 7.13 has bracket widths of 0.1 ranging from −1.5 to 2. The lowest bracket extends down to −infinity. The modelled DNA amount of D's DNA on the items in the bag available for transfer was modelled by $N(0.317, 0.812)$ as shown in Figure 7.14. In order to assign the probability in node 5 for the bracket (as an example) 1–1.1, then the cumulative density at 1 is subtracted from the cumulative density at 1.1. In excel this is achieved by the formula:

```
=NORM.DIST(1.1,0.317,SQRT(0.812),TRUE)- NORM.DIST(1,0.317,SQRT(0.812),
TRUE)
```

which gives a value of 0.00333, or in R by the command:

```
pnorm(1.1,0.317,sqrt(0.812))-pnorm(1,0.317,sqrt(0.812))
```

which again yields 0.00333. In both software programs, spreadsheets or codes can be set up to calculate the probabilities for all the brackets in an automated way. Doing so for node 5 leads to the values shown in Figure 7.18.

The *x*-axis in Figure 7.18 shows the starting bracket value. Therefore, if the column labelled with value '1' is chosen, the probability being represented by the height of the bar (0.00333) represents the probability of possessing between 1 and 1.1 $\log_{10}$(ng) of DNA.

7.4.3   How to Model Transfer Proportion

Modelling of transfer proportions is similar in nature to modelling DNA amounts with respect to the different choices of modelling regime, discretisation of probability and application to a BN node. The choice of distribution is likely to be different, as a transfer proportion can only exist between 0 and 1. For this reason, it may be sensible to choose a distribution that is also bound by 0 and 1, such as a Beta distribution. This is the manner in which node 6 from the BN in Figure 7.13 was modelled.

The data used to inform transfer proportions is easiest to model when the data has been paired, i.e. the same sample is measured before and after the transfer experiment. This allows the data to be modelled directly, by considering the reduction in the data pairing. Without paired values, modelling transfer proportion can be more difficult. One option is to model the data before and after the transfer separately even if this comes from different sources. For example, the amount of DNA on

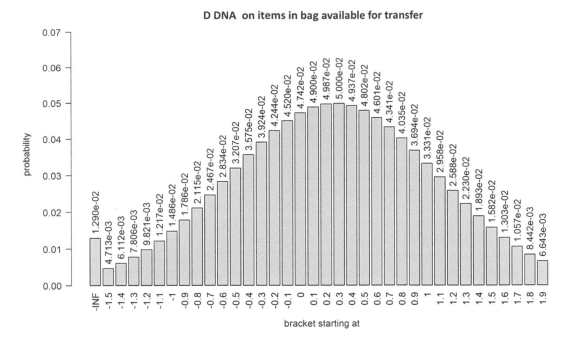

FIGURE 7.18   Node 5 probability assignments.

people's hands may come from one study and the amount on knife handles that have been grabbed may come from another study. Each of these DNA amounts could be modelled separately and then a simulation carried out that draws values from the two modelled distributions to mimic paired data. Often in this case some correction is needed to account for impossible values (such as transfer proportions above 1). This is the manner that the probabilities for node 6 can be obtained. Figure 7.19 shows the two separately modelled distributions for the amount of DNA present on cloth and the amount of DNA on a hard surface that had been in contact with the cloth, taken from the work of

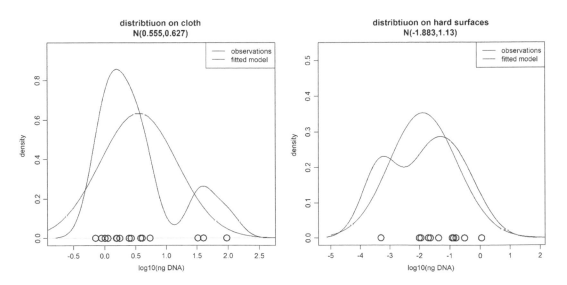

FIGURE 7.19   DNA distributions of DNA on cloth (left) and transferred to hard objects (right).

Carson et al. [3]. The distributions have been fitted in the same way as for producing Figure 7.14, with the observed values (black circles in Figure 7.19) representing individual DNA amounts from the work of Carson et al. Together these two distributions can be used to assign probabilities to the proportion of DNA transferred from cloth to a hard surfaced items when they come in contact.

Imagine a value is drawn from the normal distribution used to model DNA amount on cloth (Figure 7.19 left) and another value is independently drawn from the distribution of DNA on hard surfaces (Figure 7.19 right). By dividing the simulated DNA amount on a hard surface by the simulated DNA amount on the cloth a simulated transfer proportion is obtained. Carrying out this task numerous times simulates many transfer proportions, which can then be modelled. A sensible distribution to use for modelling DNA transfer proportion is a Beta distribution as it is bound by 0 and 1 (the same as a transfer proportion is bound) and the distribution can be skewed (again as may be expected that a transfer proportion distribution to be). Figure 7.20 shows the results of simulating paired data (by drawing from these separate distributions) and the Beta distribution fit by MLE.

Finally Figure 7.21 shows the break-up of data for use in node 6 of the BN. For this example, the discretisation was chosen to be carried out in ten steps, 0–0.1, 0.1–0.2, ..., 0.9–1. Again, it was produced by using the cumulative probability distribution of the Beta(0.242, 3.948) as shown in Figure 7.20.

As the modelled transfer proportion is continuous (i.e. the Beta distribution shown by the red line in Figure 7.20) the discretised distribution (such as that shown in Figure 7.21) can have any level of resolution, i.e. the choice could have been made to break the distribution into two brackets, 0–0.5 and 0.5–1, or it could have been broken into 1 million brackets. There is no rule for setting the size of brackets. A general guide is to use the most course break-up that does not significantly affect the evaluation compared to if the continuous distribution had been used. Discretisation is discussed further in Chapter 8.

The evaluation of the observations in case 4 with respect to the activity-level propositions is given in Chapter 8.

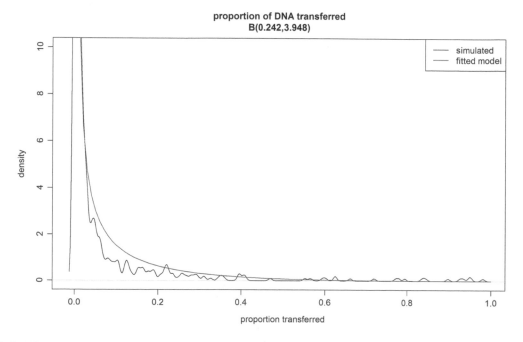

FIGURE 7.20   DNA transfer proportion showing simulated values (black) and modelled distribution (red).

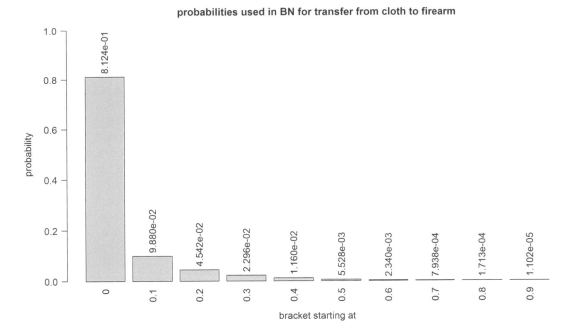

FIGURE 7.21   DNA transfer proportion brackets (states) used in node 6 of the BN along with the probability assigned to each.

7.5   GENERAL THOUGHTS ON MODELLING

7.5.1   How to Model

7.5.1.1   Linear Regression

There are generally two different ways that people think about modelling data. One of these is in the manner discussed in this chapter up to this point, taking a series of observations that are assumed to be from a single distribution of values and fitting the appropriate distribution to the observations. The other type of modelling is often associated with the descriptive or predictive modelling carried out by regressions. Such regression analyses would look at an independent variable given one or more dependent variables and assess the strength of the association between them. An example of this in the context of forensic biology could be looking at the amount of DNA deposited from the hand of an individual, given dependent variables like surface area, surface type, amount of friction etc.

In fact, these two seemingly different types of modelling are carrying out the same function, the first one is just working under a simplified model and with some assumptions. Some basic concepts of modelling are provided below, but not delved into in any great depth.

When modelling measurements with a distribution, then the model can be thought of as a regression, where the response variable, y, is modelled by:

$$y = \beta_0$$

where β_0 is a single parameter that does not depend on any other features of the data or experiment. While this provides a point value for the amount of DNA expected, it will also almost certainly be accompanied by residuals, i.e. the distance between each of the N observations, $x_1,...,x_N$, and the value of β_0. Assumptions can then be made about the distribution of the residuals, for example, that they follow a normal distribution, which then allows a complete model to be specified and used in evaluations. Given the simplicity of the model, a formal regression is usually forgone, and with the assumption of normality of residuals, the value of β_0 is calculated directly as the mean of $x_1,...,x_N$, and the standard deviation of the residuals is again calculated directly as the standard deviation of $x_1,...,x_N$.

If the data collected possesses some kind of stratification (with the groups having different distributions of the measured variable), then there are two ways in which data analysis can proceed. As an example, consider modelling the amount of DNA on an individual's hands. Data is collected by randomly choosing N people from the population, swabbing their hands and quantifying the amount of DNA. When plotted, the distribution of DNA amount appears to have two peaks (or modes) as shown in Figure 7.22 (for $N = 200$). This is referred to as a bi-modal distribution (or multi-modal to specify any number of modes more than one). In Figure 7.22 the DNA amount is displayed on a $\log_{10}$ scale, because (as previously discussed) this transforms the observations in such a way that they have a normal distribution.

The assumption of normality of this data would not be reasonable. There are tests that can be carried out to test for the normality of data (or residuals), such as the:

- Visual diagnostics, such as producing a box plot, or Quantile–Quantile (QQ) plots. These are intuitive as they are based on visual performance but will not give a formal statistic, i.e. they do not test a hypothesis.
- Kolmogorov–Smirnov test, which can be used for tests on any distribution type but is not very powerful.
- Shapiro–Wilk test, which is specific to a normal distribution, but more powerful than the Kolmogorov–Smirnov test.
- Various other tests, which are less commonly used such as the Anderson–Darling test, Cramer–von Mises test, D'Agostino skewness test, Anscombe–Glynn kurtosis test, D'Agostino–Pearson omnibus test and the Jarque–Bera test, which have various strengths and limitations.

Say that one of these tests for normality has been performed and suggests that a normal distribution is not a good description of the observations. One possibility is that the data represents two groups of individuals, each with their own normal distribution of the data. For example, gender could be tested as a contributing factor to the modelling of DNA amount, as studies have found that males tend to possess more DNA on their hands than females [4–6]. One possibility to deal with this data stratification would be to break the total dataset into two groups, one containing only males and the other containing only females and then model the distributions individually on the two subsets of data. This has

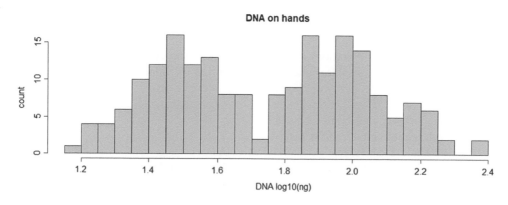

FIGURE 7.22 Example of a bi-modal distribution of DNA amounts on hands.

the advantage that the data analysis is simple, and there are no assumptions that the groups have equal variances. The disadvantages are that the data is divided and so the amount of information available to inform each distribution is less. In this given example, this is not such an issue, however in many studies, where there are multiple variables, the result of dividing into the smallest groups of specific variable values can mean each dataset is too small to provide meaningful information for modelling.

A second option, which retains the power of using all the datapoints is to carry out a single regression but include gender as a variable. As the grouping of males and females has no numerical meaning (and there is no natural order in which to place them), gender is considered a categorical variable and the model becomes:

$$y = \beta_0 + \beta_1 X$$

where $X = 1$ if the person is male and 0 if they are female. When carrying out the regression, the mechanics of setting up regression with a categorical variable that has multiple states is through the use of 'dummy variables'. These are variables created for each state of the variable that contains a 1 in the corresponding state and a 0 in all others. A regression such as this makes the assumption that the variances are the same between genders. If this is not the case, i.e. there are dependencies of the distribution of residuals with respect to one or more parameters (an effect called heteroscedasticity), then adjustments can be made (such as weighting the variance terms, or carrying out a generalised least squares regression). There are many good online texts and resources that can be accessed to assist in these areas. If the assumption of equal variance is able to be made, then the single regression treatment has more power in modelling DNA amounts than carrying out one regression on male-only data and separately one for female-only data.

It may be that one of the variables has a numerical meaning, for example consider that in addition to gender, age also plays a role in how much DNA is present on an individual's hands [7]. Age can be measured on a continuous spectrum, and as long as the (possibly unreasonable) assumption is made that there is a linear trend of DNA amount with respect to age (i.e. that DNA starts at its highest amounts being transferred in lower ages and steadily decreases as age increased) then DNA amount can be modelled by:

$$y = \beta_0 + \beta_1 X + \beta_2 A$$

where A is now the age of the person (using any unit is fine, as long as that unit is consistently used throughout the evaluation). Finally, consider that the decrease in DNA amount with age may occur at different rates in the different genders. In order to account for this, 'interaction terms' must be considered, which (in regression terms) is modelled by considering the multiplication of the variables being considered as potentially interacting:

$$y = \beta_0 + \beta_1 X + \beta_2 A + \beta_3 X \times A$$

When faced with many combinations of variables, there is somewhat of an artform to trialling different combinations of variables and their interactions in order to settle on a final model. Decisions on:

- How many terms to include in the model,
- Whether they need to be transformed,
- Whether they are considered categorical or continuous (e.g. primary/secondary/tertiary transfers could be modelled as categorical variables, or could be considered as 'number of transfers' in which case the model would be carried out using a continuous variable with a numerical value of 1, 2 or 3) or
- Which interactions should be taken into account

will all depend on the dataset, what factors present as being significantly important to the model, what is known about the variable being measured from other studies and some personal intuition. If too few variables are modelled, then they may appear to be insignificant to the modelling of the response variable simply due to the fact that they are very much confounded by other stratifications within the dataset that have not been accounted for. Too many variables and everything will appear insignificant as there is too much ability for the model to twist multiple parameters to explain the observations. John Von Neumann [8] famously once said, 'With four parameters I can fit an elephant, and with five I can make him wiggle his trunk', which was later explored in a paper by Mayer et al. [9].

7.5.1.2 Logistic Modelling

Up until this point, regression modelling has been described where the response variable was a numerical value with physical real-world property, for example an amount of DNA. This may not always be the case and the response variable may be a binary indicator, such as 'DNA detected' vs. 'DNA not-detected'. When carrying out a regression with a binary response variable, the typical course of action is to carry out a logistic regression. In a logistic regression the same consideration of dependant variables is considered as described above; however rather than these variables being combined to describe an amount of DNA, they would be combined to describe a probability that DNA will be detected. Taking the last regression equation from above, then the logistic regression is represented as:

$$\ln\left(\frac{p}{1-p}\right) = \beta_0 + \beta_1 X + \beta_2 A + \beta_3 X \times A$$

where p is the probability that DNA will be detected, and ln in the natural logarithm (i.e. the logarithm with base 'e').

There are different reasons why a logistic regression is carried out. One may be that this is the only reasonable way to measure the outcome. For example, both Zuidberg and Ramos [10, 11] carried out experiments to see what areas of items were contacted during certain activities. The items were divided into multiple areas, and a note was made whenever an activity led to contact with any of the areas (either through fluorescent dye, or chalk marking). Each area was measured as positive or negative and so a logistic regression was required to analyse the pattern of contacts. In other situations, the choice may be made to use logistic regression due to sample size and variability. For example, if the experimental sample size was not adequate to model DNA amounts in a situation where transferred DNA amounts are highly variable (such as a contact DNA activity), then there may not be enough power in the dataset to properly model the significance of variables. The response variable in the model can be simplified to a presence/absence (around a particular threshold value) which then may allow regression to proceed and have enough power to identify important variables to the probability of obtaining a result.

7.5.1.3 Modelling with Bayesian Statistics

The modelling that has been discussed so far is described in a classic frequentist method, with the regression providing the point values for parameters that best explain the data (by minimising the least square distance) and providing a p-value to test for significance. The p-value for a parameter represents the comparison of the model to the null hypothesis (typically that the data is not stratified in the way being suggested by the model using that parameter). The p-value is the probability that a dataset in which the stratification proposed by the parameter in question is not present, and would show the observations by chance. Typically, when the chance falls below 5% ($p < 0.05$), then the observed data is considered to be sufficiently improbable to have been obtained by chance (if the parameters were indeed not playing any role in describing the data) and so the null hypothesis is rejected.

There are Bayesian equivalents for carrying out regressions that perform the same function of assessing what factors are important to describe the observations. In a Bayesian linear regression model the response variable is drawn from a distribution, with parameters that dictate the mean and variance, i.e. using the last model described above:

$$y \sim N\left(\mu, \sigma^2\right),$$

where $\mu = \beta_0 + \beta_1 X + \beta_2 A + \beta_3 X \times A$

Each parameter has a specified prior distribution, i.e. if the data is specified by D, then each parameter has posterior distribution:

$$p\left(\beta_i \mid D, y\right) \sim \frac{p\left(y \mid \beta_i, D\right) p\left(\beta_i \mid D\right)}{p\left(y \mid D\right)}$$

The mechanics of the regression is that each parameter value is drawn from its posterior distribution in order to approximate the posterior. This is implemented with some Monte Carlo techniques such as Markov Chain Monte Carlo. The outcome is a posterior distribution, not only for the response variable but also for each of the parameters. Whether or not a parameter should be included in the model can be assessed in different ways:

- Information criteria. Variants include Bayesian Information Criteria (BIC), Schwarz information criterion (SIC), Akaike information criterion (AIC) or the widely applicable information criterion (WAIC). These values balance the increased likelihood of the model achieved by adding parameters with a penalty for the number of parameters in the model. The optimal model choice is the one that minimised the information criteria. These are commonly used in Hierarchical Bayesian Modelling.
- Bayes factor. This is a likelihood ratio statistic that compares the posterior probability of two competing models. The larger the ratio, the more the support for the first model and the smaller the ratio the more support for the second model (and a Bayes factor of 1 indicating no support for one model over the other).
- Probability of direction (*p*-direction). This is probably the closest in feel to the frequentist *p*-value (although has a different meaning). For each parameter, the posterior distribution is examined to see whether that parameter has a value different from 0 (with a value of 0 meaning the parameter is adding nothing to the description of the observed data). The more of the posterior distribution that falls on one side of 0, the more the parameter has added to the model (and the larger the *p*-direction). The probability of direction can be converted to a *p*-value by $p - value_{one-sided} = 1 - p - direction/100$, or $p - value_{two-sided} = 2\left(1 - p - direction/100\right)$.

For a comprehensive description on a Bayesian workflow that can be used when approaching a task of modelling features of a dataset we recommend the work of Gelman et al [12].

7.5.2 MAKING MODELLING DECISIONS

Chapter 3 discussed how experiments could be used to assign probabilities (e.g. of transfer or persistence) or to investigate the mechanism of an event. The choice of which type of study is to be carried out will determine the sample makeup of the experiment. Will the experiment be to control for as many variables as possible and investigate one aspect of the event? This provides a greater insight into the mechanism but also may be less applicable to casework circumstances. Or, will the experiment be more uncontrolled, allowing multiple variables (such as the people participating) to vary in accordance with how they might be expected to in the real world? These types of experiments will be better suited for probability assignments in an evaluation but will not give as much insight into the mechanism as there will be many complicating, interacting and confounding factors not being directly modelled. Chapter 5 described this type of situation when assigning probabilities with respect to DNA transfer at a social gathering. Should the type of activity at the gathering be controlled and DNA transfer modelled for each activity separately? Or should any type of activity be allowed but then separately model how often these activities are expected to occur? The former is more onerous as it requires a much larger experiment to be carried out, but it would provide the information necessary to model a specific activity if information that that activity had occurred was available.

The other consideration is what the response variable will be. Again, this topic was touched on in Chapter 5, when considering what data to use in probability assignments. The earlier in the DNA profiling process that the measurements are taken, the more widely applicable the results will be, i.e. if the response variable is the number of major component alleles that could be interpreted from the donor, then this will incorporate the performance of the extraction, PCR, electrophoresis and interpretation that the laboratory uses. This will very much narrow down the pool of other laboratories that can use the data. If, however, the response variable is DNA amount, then everything downstream of that does not need to align with other laboratories. There is also the possibility that the model outcome will be binary in nature, e.g. the presence or absence of DNA on an item, or in various areas of an item (e.g. see [10, 11]). In these situations, a type of regression called a logistic regression will have been performed. Carrying out a regression in this way will limit the evaluations that the data that can be used on to those that consider observations in the same binary fashion.

Another option is that under certain circumstances a retrospective capture of casework data can be used to create models and inform probability assignments. Using casework can be difficult as it (by its very nature) is completely uncontrolled. Also, being casework, the truth of what has happened is not known (only someone's account of events). Therefore, the data can be affected by assumptions about what has occurred. For example, a study could be conducted into the level of semen present on vaginal swabs at various times after sexual intercourse has occurred. If sexual assault cases are used, then an assumption is being made that the assault took place when and how the complainant has described it and that ejaculation has occurred, which is known not to be true in a significant proportion of rape cases [13]. Or, if the amount of DNA transferred to a firearm from handling is being studied, and firearm cases are used, then there is an assumption that the defendants have indeed handled the firearms as they are accused of doing. However, there are other situations where data gathered from casework can be used with minor assumptions. For example, a study into the level of background DNA on firearms could be carried out which looks at the occurrences of DNA on firearms that cannot be accounted for by the defendant.

7.6   OBJECT-ORIENTED BAYESIAN NETWORK (OOBN)

7.6.1   THEORY OF OOBNs

Case circumstances can range from quite simple to very complex. Factors that add complexity to an evaluation involve the presence of multiple activities (or components of activities), multiple analyses being carried out, analyses of different types being carried out and the level of transfers being considered within the evaluation. When cases become complex, it is often the case that certain parts of the BNs are repeated. For example, it may be that part of a BN requires the consideration of a person (either the POI or an AO) as having contacted two similar items in a similar manner. The probabilities for DNA transfer in each of these events may be the same (or assumed to be the same given the available information about the case circumstances and available knowledge in the literature). That small part of the BN could be considered a discrete unit (or class network) that is repeated at multiple points within the larger BN. Various authors [14–16] refer to these as 'idioms'.

The BN constructed for case 1 was modified in several ways in Chapter 6, to demonstrate the flexibility of BNs to incorporate different aspects within the evaluations. Figure 7.23 shows one of the variants from the case 1 discussion, where the persistence of DNA on the tops of C and D was being considered.

Within the BN shown in Figure 7.23, there are various repeated structures. For example, nodes 2, 5, 13 and 17 represent the potential for DNA to transfer from a contact between a person and a

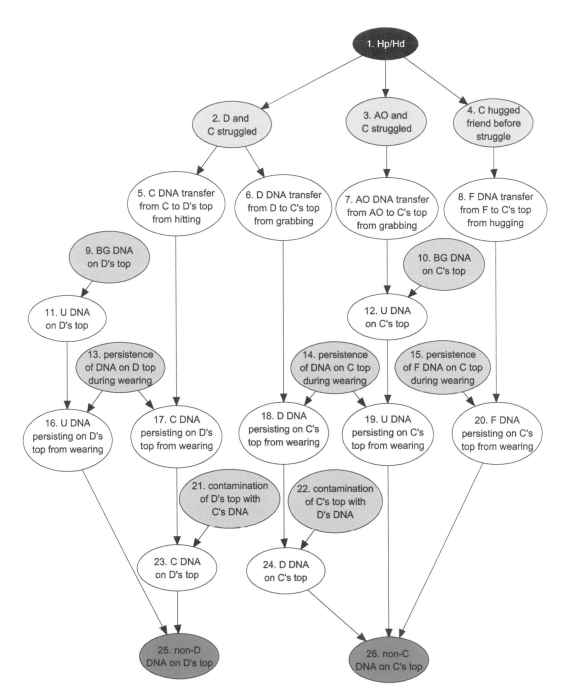

FIGURE 7.23   BN from case 1 showing BN with transfer and persistence considerations. This is a replication of Figure 6.28.

TABLE 7.5

Repeated Structure of Nodes 2, 5, 13 and 17 in the BN Seen in Figure 7.23

Description	Nodes in structure 1	Nodes in structure 2	Nodes in structure 3
Activity performed leading to contact with top	2	2	4
DNA transfer resulting from contact	5	6	8
Persistence of DNA on top	13	14	15
DNA persisting on top	17	18	20

top, and then that DNA potentially persisting, referred to as 'structure 1'. This same structure of nodes (an 'idiom'), using the same probabilities, is repeated three times in the BN seen in Figure 7.23, shown in Table 7.5.

If just the node structures from Table 7.23 were used to create a general mini-BN, then the architecture it would possess is shown in Figure 7.24. This type of structure is called a class network, named in Figure 7.23 as the 'TP class network' (T for transfer and P for persistence).

In Figure 7.24 the nodes are given particular types of borders to represent the way this small BN would fit into a larger BN. The nodes that have a thick dashed black border and a grey boundary are designated as input nodes. Both the 'activity performed leading to contact with top' and the 'persistence of DNA on top' nodes are inputs. Input nodes represent nodes that have parents coming from outside the class network. The nodes that have a thick solid black outline and a grey border are output nodes. The 'DNA persisting on top' is an output node in the class network in Figure 7.24. If a node is an output node, then it has a child node outside the class network. The 'DNA transfer to top' node in Figure 7.24 is neither an input nor an output node. It is referred to as a private node. It exists only within the class network and is d-separated from all nodes outside the class network.

Having set up the structure of the TP class network shown in Figure 7.24, it can now be used to replace the repeated structures it represents within the BN from Figure 7.23 (and specified in Table 7.5). The resulting BN, now an OOBN, is shown in Figure 7.25.

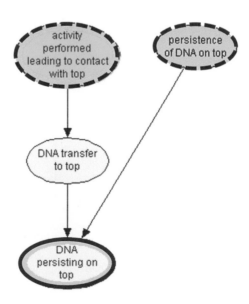

FIGURE 7.24   Small fragment of BN that is repeated three times in larger BN from Figure 7.23.

Notice that the OOBN in Figure 7.25 has been mildly simplified compared to the BN in Figure 7.23. The simplification has only reduced the complexity of the main BN by reducing the total number of nodes by three. This is because the repeated structure shown in Figure 7.24 was relatively small. The OOBN in Figure 7.25 works in exactly the same way as the BN in Figure 7.23. This is true of all OOBNs, i.e. there is always a fully expanded equivalent (i.e. with no class networks) that would carry out the identical evaluation.

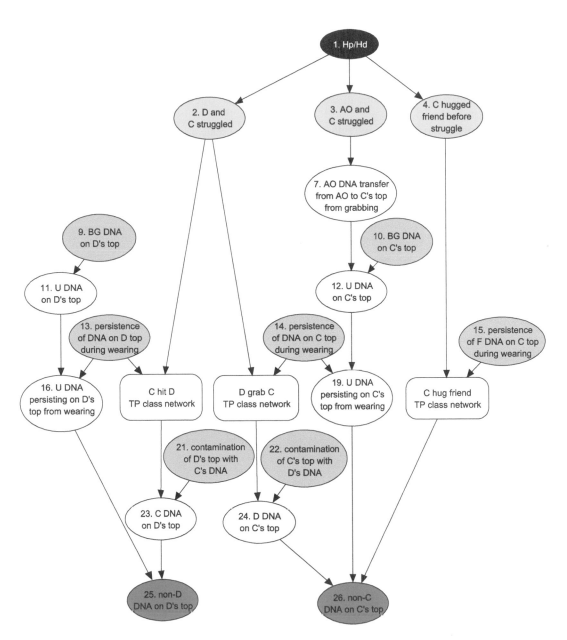

FIGURE 7.25   BN from Figure 7.23, updated to be an OOBN using the class network structure shown in Figure 7.24.

There are also several points that can be taken by examining the structure of the OOBN in Figure 7.25:

- The TP class network was able to remove the requirement for the final, 'DNA persisting on top' nodes in the BN from Figure 7.23 (i.e. nodes 17, 18 and 20) but did not remove the requirement for the activity nodes (2 and 4). To remove the activity nodes from the main BN, the TP class network would have required a proposition node input, which would have been linked to the proposition node in the main BN. Then the activity node would have been a child of the proposition node inside the TP class network. However, such a structure could not be set up because the activity nodes that are currently shown as inputs in the OOBN (nodes 2 and 4) have different conditional probabilities. If the activity node only existed within a single class network, then there would be only one way to set up the conditional probability table for it, and this would not work within the requirements of the main BN. It would be possible to set up two different class networks with two different activity node conditional probabilities; however, there would be little point as the class network set up for the C hugging friend route would not be repeated.
- The persistence nodes could not be removed from the main BN even though they were present in the TP class network, and they have the same conditional probabilities. The reason for the persistence nodes remaining on the main BN is that there are dependencies between nodes that exist outside the considerations of the TP class network. For example, the persistence nodes 13 and 14 in Figure 7.25 are parents of two nodes each. This is because they are being applied to the one item where the two sources of DNA being considered to have persisted over the same time period/conditions. In explanation, whether it was D's DNA or an AO's DNA present on the top of C at the point where she struggled with someone, they will have undergone the same environmental conditions leading up to the point of sampling for DNA profiling. The expectation is that persistence of these two sources of DNA is the same. The DNA transferred to C's top by hugging her friend has a separate persistence node to either D or AO's DNA on C's top, because the hug occurred at a different point in time and so has undergone different conditions leading up to sampling for DNA profiling. If the persistence nodes had been included in the TP class network, but not as an input (rather just a private node within the class network that did not require information from the main BN) and removed the persistence nodes from the main BN, then that dependency would no longer have been present i.e. it would have been equivalent to the BN in Figure 7.23 having a separate persistence node parent for nodes 16, 17, 18, 19 and 20.
- The transfer and persistence route from the activity node specifying whether AO and C struggled (node 3) was not replaced by the TP class network. The reason for this is that the BN architecture leading from node 3 to node 19 is different to the architecture identified as being in common with the other areas of the BN. Specifically the architecture from node 3 to node 19 includes a consideration of background DNA (node 10), which the others do not.
- If any of the three transfer probabilities from the activities had been different, then they could not be included in the TP class network as it has been defined. Due to the assumption that the probability of DNA transfer from hitting, grabbing and hugging to clothes was the same, it was possible to identify regions of the BN in Figure 7.23 that was exactly repeated. If any of the transfer probabilities had been different (for example if new data was found that suggested the probability of DNA transfer from hugging was much higher than for hitting or grabbing), then this repeated nature would no longer be the case. In this instance only two identical structures within the BN would be replaced. Alternatively, the activity node could have been removed from the class network and the DNA transfer node

made an input node and then the TP class network could be used to replace all three BN sub-structures again.

• If any of the transfer probabilities had been dependent on other transfer probabilities, then they could not be included in the TP class network as it was defined. This is the same concept as was described earlier for the reason that the persistence nodes needed to be input into the TP class network. Imagine that in the BN architecture a dependency between the transfer of DNA from C to D's top and the transfer of DNA from D to C's top was taken into account. This would have been equivalent to drawing an arc from node 5 to node 6 (or vice versa) in the BN shown in Figure 7.23. Perhaps if DNA has transferred in one direction, then this may suggest the struggle between C and D (under Hp) was intense and therefore it would raise the expectation of DNA transfer in the other direction. If this was the case, then the dependency would need to be modelled within the BN (with the arc between nodes 5 and 6) and the TP class network would no longer be applicable, as it treats the transfer probabilities as independent. In order to use a TP class network, the intensity of the struggle would need to be explicitly modelled with a node that was the parent of both nodes 5 and 6, and then this struggle intensity node would provide an input into the TP class network.

Given the added complexity of setting up an OOBN (evidenced by the many considerations of dependency just described), one may enquire as to the advantages of using OOBNs.

The first advantage, which was already mentioned, is that the use of OOBNs can visually simplify an otherwise cluttered BN. This is a minor advantage, as it does not affect the evaluation itself, but when BNs become highly complex and detailed, the effect can be dramatic. For some examples of OOBNs that dramatically simplify an otherwise complex BN, the reader is directed to [17–19], although note that these OOBNs are quite a step up in complexity from what has been discussed in this book. The visual simplification also has the advantage of making the evaluation more likely to be followed by a layperson as the general lines of thought and inference are still present but without the clutter of all the individual variables that are being considered for each event.

By modularising the repeated architecture, OOBNs are less prone to typographical errors. This same concept is used in most modern programming languages, which make use of object-oriented programming, i.e. a particular calculation that must be performed multiple times is placed into its own 'method', which can then be called by the program at any point required (rather than having multiple copies of the calculation code throughout the program). The programmer simply passes the information needed to carry out the calculation into the method when calling it and then is supplied back with the processed data. From this explanation, it should be apparent the similarities between object-oriented programming and OOBNs. Consider the example shown whereby the transfer and persistence have been encapsulated into the class network. When populating the conditional probabilities for the BN in Figure 7.25 with the transfer probabilities, the value need only to be entered twice (once within the class network and once for node 7) as opposed to four times in the original version of the BN from Figure 7.23 (nodes 5, 6, 7 and 8). With two less nodes to populate, there are two less opportunities for typographical error to occur and to less nodes for another analyst to review. When BNs become highly complex, and repeated structures are used multiple times, the effect of using OOBNs can be to reduce the required data entry by thousands of values. If another study was found to update the transfer probability, then again, the number of places which need to have a value changed is reduced if the value is within a class object. Such is the advantage of OOBNs in this way that BNs can be constructed specifically to use repeated units. For example, with some modification, the BN shown in Figure 7.23 could be altered so that all four transfers were able to be in a single class network. For example, consider that during the construction of the BN shown in Figure 7.25, the assumption was made that the background on C's top (node 10) was not affected by persistence (i.e. the modelling assumption being that background DNA on items is present at the time they are sampled and not as being deposited as a point in time and then lost). In this case then node 12 could be shifted to be between nodes 19 and 26, rather than where it sits now (between nodes 7 and 19). Doing this would then create a fourth instance of the repeated structure seen in Figure 7.24 and so a fourth instance of the class network could be substituted into the BN in Figure 7.25.

OOBNs can be nested. In the example provided there was the main BN and then in the main BN were instances of the TP class network. It is possible that within the TP class network another class network could have been added (had elements been repeated), creating a third 'layer' to the OOBN architecture. As more layers are added, then the reduction in the amount of data entry is exponential. However, care needs to be taken when considering dependencies, and at which point of the BN they sit. There can be instances when a factor is global to the entire evaluation (for example the match probability of a particular person's reference DNA profile) that may need to get passed through several layers of the OOBN before it is used in a calculation.

The modularity of OOBNs also has advantages outside of the immediate evaluation. When elements of the BN are constructed as class networks, then they are able to be used like building blocks in other BNs for cases that consider similar events. As more evaluations are carried out and more class networks are created, the easier it becomes to pull together class networks (which already have the probabilities populated from their previous use) to form an overall OOBN for the case.

When carrying out an assessment on whether the evaluation is robust (i.e. ensuring the *LR* is not very sensitive to probabilities for which there is little informative data), the effect of changes in the data on the evaluation must be considered. It is important to remember that when data is reused, then the test for the robustness of the *LR* to data must take into account its multiple uses. This point will be expanded on in the chapter on testing an evaluation for robustness. Another advantage is that when data is used in multiple places in the overall BN, but only through a single class network, then the analysis has already been set up to account for multiple data usages in a sensitivity analysis.

7.6.2 CREATING AN OOBN IN HUGIN

Each software has its own methods for creating OOBN class networks, and it is likely that software-specific assistance will be available through manuals, forums, online courses or tutorials. Some instructions are provided here for the creation of OOBNs within the software HUGIN, as it is the software used to generate the BN images that make up figures throughout this book.

Within the HUGIN software, multiple BNs can be opened at the same time. This is important as when a BN is open on the screen it can be imported (as an OOBN) into another network. Figure 7.26 shows a screen capture of HUGIN with the two BNs (each in their own separate window) that are required for the construction of the OOBN in Figure 7.25. The smaller BN is already set up in the screen capture in Figure 7.26 as a class network and so already has inputs and outputs specified. When setting up a class network for the first time, all nodes will be designated as 'private'. To designate nodes as inputs or outputs, right-click on the node, select the 'Set Input/Output' option and then choose either input or output as needed. In the screen capture in Figure 7.26, the 'persistence of DNA on top' node has been right-clicked, and as it is already set as an input node the only option to change its input/output designation is to private again (seen as the 'Set as private' option showing).

Once the BN that is destined to be a class network has been constructed and saved then it will be available to add to other networks through the 'nodes' toolbar option (i.e. the same option that would be chosen to add chance, utility, or function nodes, or decision tools). If multiple class networks have been constructed, then they will all be available for choosing to add as an 'instance tool' to the BN.

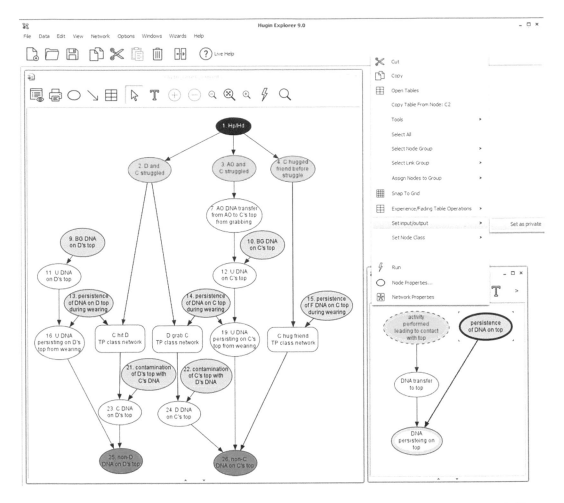

FIGURE 7.26   Screen capture of HUGIN showing the two BN structures used to construct the BN in Figure 7.25.

Once chosen, the class network can then be added by clicking on the screen (the same as for any other node addition).

Once a class network is added to a BN, it will appear as a small white box. By clicking on the corner of the box, it will expand to show the inputs and outputs of the class network. Figure 7.27 shows a section of the BN from Figure 7.25 with the 'C hug friend TP class network' expanded. Expanding this class network displays the inputs and outputs from that class network and also makes them available to be connected to other nodes in the BN. They can be connected in the same way as any other two nodes would be connected, by dragging an arc between them.

For any node in the BN to be connected to an input or an output node in the class network, it needs to have exactly the same states, i.e. it must be an exact copy of the outer node in order to translate the probabilities into and out of the class network.

A final note on the construction of OOBNs is that once a class network has been constructed and incorporated into a BN, then the types of changes that can be made are limited. Internal changes to the class network that only involve private nodes can be made without consequence. However, any changes to input or output nodes may either be unallowed, or they will cause all connections from any other BNs that have instances of that class network to be severed.

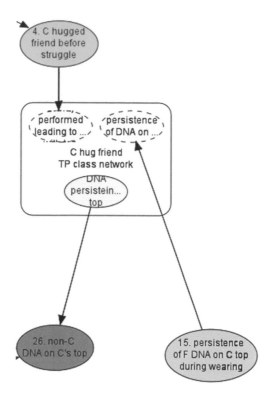

FIGURE 7.27   Section of BN from Figure 7.25 with the 'C hug friend TP class network' expanded to show inputs and outputs.

7.7   PRACTICE QUESTIONS

Q1)   Consider the BN shown in Figure 7.1 and the scenario that lead to its creation. Imagine now that the father also invited his friend, who is a very proficient fisherman along to join him and his son. The friend is expected to catch 5, 6, 7 or 8 fish with probabilities of 0.2, 0.3, 0.3 and 0.2 respectively. Construct a BN using numbered nodes to calculate the total number of fish expected to be caught from this new fishing party.

Q2)   Each time a person touches a drawer handle, it is expected that they will transfer some DNA (with the associated probabilities given in the table below).

DNA transferred (ng)	Probability of DNA transfer
0–1	0.50
1–2	0.30
2–3	0.15
3–4	0.05

When trying to open a locked cabinet, a person is expected to contact the handle between 1 and 5 times with equal probability. Construct a BN to show how much DNA is expected to be found on the handle (assume that no DNA is lost from subsequent touches).

Q3) Show, by derivation, how the BN is calculating the total amount of expected DNA in question 2.

Q4) The amount of DNA expected to be on a firearm is given in the table below:

DNA on firearm (ng)	Probability of DNA amount
0–1	0.50
1–2	0.25
2–3	0.15
3–4	0.10

Each time someone wipes down a firearm, they are expected to remove one-quarter of the DNA that is on it. Construct a BN that models the amount of DNA expected to be found on a firearm after 0–5 wipe-downs. If there was 1.6 ng of DNA detected on the firearm, how many wipe-downs is it most likely to have been subjected to?

Q5) Section 7.2.2.1 showed the derivation of:

Pr(3 ng < total DNA < 4 ng | transfer from object 1 = 2–4 ng, transfer from object 2 = 1–2 ng)

And obtained 0.25, which was one of the values seen in Table 7.2. Calculate the probabilities for:

Pr(4 ng < total DNA < 5 ng | transfer from object 1 = 2–4 ng, transfer from object 2 = 1–2 ng)

Pr(5 ng < total DNA < 6 ng | transfer from object 1 = 2–4 ng, transfer from object 2 = 1–2 ng)

Q6) Section 7.2.2.2 showed the derivation of:

Pr(2 ng < transferred DNA < 3 ng | DNA on object 1 = 4–6 ng, transfer proportion = 0.66–1)

And obtained 0.034, which was one of the values in Table 7.3 (the far-right column). Calculate the probabilities for:

Pr(3 ng < transferred DNA < 4 ng | DNA on object 1 = 4–6 ng, transfer proportion = 0.66–1)

Pr(4 ng < transferred DNA < 5 ng | DNA on object 1 = 4–6 ng, transfer proportion = 0.66–1)

Pr(5 ng < transferred DNA < 6 ng | DNA on object 1 = 4–6 ng, transfer proportion = 0.66–1)

Q7) An experiment to see how much DNA can be recovered from struck matches was performed. In ten experiments, the amounts of DNA obtained (in ng) were 0.094, 0.111, 759.455, 15.397, 602.348, 3.002, 0.727, 0.609, 0.002 and 875.931. Model these amounts with an appropriate distribution. How would you represent this distribution in a discrete node within a BN? Draw out a table with the states and probabilities you have decided on.

Q8) A person (D) is accused of attempting to break into someone's (C's) home. D went around and tried to open each of the three doors by the handle, but all were locked and so D left and is alleged to have spray-painted the wall as he did. C calls the police, who attend the next day. Each door handle is swabbed and submitted for DNA profiling. Overnight it has rained, and it is known from studies that there is a probability that this may wash away the DNA. It is also known that door handles have a high probability of possessing background DNA. Police recognise the spray-paint tag of D and arrest him. D denies ever having been near the house, and through discussions, you determine that there is no explainable route for the indirect transfer of D's DNA. In your evaluation you wish to consider that DNA matching D's profile could come to be on the door handles either through contamination of the door handle samples with D's reference, or because an alternate offender has a matching DNA profile.

Draw out a standard BN architecture (at first with just nodes and not any class networks). Note that this is a question only about the architecture, so the probabilities do not need to be assigned in any nodes. Identify repeated structures within the BN you have designed. Design an OOBN substructure that could replace these repeated idioms.

Q9) Construct the BN from Q8 with class network structure to create an OOBN.

Q10) Why must the match probability of the AO be considered outside the class network in the BN from Q9 and yet contamination is considered within the class network?

7.8 REFERENCES

1. G. Andrew, Prior distributions for variance parameters in hierarchical models (comment on article by Browne and Draper), *Bayesian Analysis* 1(3) (2006) 515–534.
2. A. Gelman, J.B. Carlin, H.S. Stern, D.B. Dunson, A. Vehtari, D.B. Rubin, *Bayesian Data Analysis* (3rd ed.), Chapman and Hall/CRC, 2013.
3. S. Carson, L. Volgin, D. Abarno, D. Taylor, The potential for investigator-mediated contamination to occur during routine search activities, *Forensic Science, Medicine and Pathology* 18 (2022) 299–310.
4. D. Lacerenza, S. Aneli, M. Omedei, S. Gino, S. Pasino, P. Berchialla, C. Robino, A molecular exploration of human DNA/RNA co-extracted from the palmar surface of the hands and fingers, *Forensic Science International: Genetics* 22 (2016) 44–53.
5. P. Manoli, A. Antoniou, E. Bashiardes, S. Xenophontos, M. Photiades, V. Stribley, M. Mylona, C. Demetriou, M.A. Cariolou, Sex-specific age association with primary DNA transfer, *International Journal of Legal Medicine* 130(1) (2016) 103–112.
6. R.W. Allen, J. Pogemiller, J. Joslin, M. Gulick, J. Pritchard, Identification through typing of DNA recovered from touch transfer evidence: Parameters affecting yield of recovered human DNA, *Journal of Forensic Identification* 58(1) (2008) 33–41.
7. M. Poetsch, T. Bajanowski, T. Kamphausen, Influence of an individual's age on the amount and interpretability of DNA left on touched items, *International Journal of Legal Medicine* 127 (2013) 1093–1096.
8. F. Dyson, A meeting with Enrico Fermi, *Nature* 427(6972) (2004) 297.
9. J. Mayer, K. Khairy, J. Howard, Drawing an elephant with four complex parameters, *Journal of Physics* 78(648) (2010).
10. M. Zuidberg, M. Bettman, L. Aarts, M. Sjerps, B. Kokshoorn, Targeting relevant sampling areas for human biological traces: Where to sample displaced bodies for offender DNA?, *Science & Justice* 59(2) (2019) 153–161.
11. P. Ramos, O. Handt, D. Taylor, Investigating the position and level of DNA transfer to undergarments during digital sexual assault, *Forensic Science International: Genetics* 47 (2020) 102316.
12. A. Gelman, A. Vehtari, D. Simpson, C.C. Margossian, B. Carpenter, Y. Yao, L. Kennedy, J. Gabry, P.-C. Bürkner, M. Modrák, Bayesian workflow (2020) arXiv preprint arXiv:2011.01808.
13. A.N. Groth, A.W. Burgess, Sexual dysfunction during rape, *The New England Journal of Medicine* 297(14) (1977) 764–766.
14. M. Neil, N. Fenton, L. Nielson, Building large-scale Bayesian networks, *The Knowledge Engineering Review* 15(3) (2000) 257–284.
15. K.B. Laskey, S.M. Mahoney, Network fragments: Representing knowledge for constructing probabilistic models, in: D. Geiger, P.P. Shenoy (Eds.) Proceedings of the Thirteenth Annual Conference on Uncertainty in Artificial Intelligence (UAI-97) (1997) 334–341.

16. D. Koller, A. Pfeffer, Object-oriented Bayesian networks, D. Geiger, P.P. Shenoy (Eds.) Proceedings of the Thirteenth Annual Conference on Uncertainty in Artificial Intelligence (UAI-97) (1997) 334–341.

17. D. Taylor, A. Biedermann, L. Samie, K.-M. Pun, T. Hicks, C. Champod, Helping to distinguish primary from secondary transfer events for trace DNA, *Forensic Science International: Genetics* 28 (2017) 155–177.

18. D. Taylor, L. Samie, C. Champod, Using Bayesian networks to track DNA movement through complex transfer scenarios, *Forensic Science International: Genetics* 42 (2019) 69–80.

19. D. Taylor, L. Volgin, B. Kokshoorn, C. Champod, The importance of considering common sources of unknown DNA when evaluating findings given activity level propositions, *Forensic Science International: Genetics* 53 (2021) 102518.

8 Testing Robustness of Evaluation

Duncan Taylor

CONTENTS

8.1 ROBUSTNESS VS SENSITIVITY VS ERROR RATES

Having heard the details of a case, extracted the relevant information, set propositions and carried out an evaluation in order to obtain an *LR*, the analyst may wish to test the robustness of their evaluation. In literature there are many different terms that are used to describe processes of quantifying one's uncertainty and these are often (incorrectly) used interchangeably. Such terms include

DOI: 10.4324/9781003273189-8

confidence interval, measurement error, margin of error, credible interval, sensitivity analysis, probability interval, measurement uncertainty and level of precision or accuracy. Each of these terms has their own specific meaning and are generally not interchangeable. This chapter is dedicated to investigating the robustness of an evaluation by testing how sensitive the *LR* is in relation to the data that has been used to assign probabilities, the assumptions made during the evaluation and the choice of evaluation architecture. Such an investigation is known as a sensitivity analysis, and there are different ways they can be carried out depending on the architecture and mechanics of the evaluation. However, before discussing sensitivity analyses, some time is dedicated to framing the concepts of uncertainty, error, probability and sensitivity analyses and hope to dispel common misconceptions.

8.1.1 ERROR RATES

Start first by considering the concept of applying an 'error rate' to an *LR*. The call for error rates to accompany forensic testimony has arisen in large part from cases where the normative opinion being provided is one of identity. For example, this call for error rates has been made in disciplines such as fingerprints where the testimony is that the source of a fingerprint is the defendant, to the exclusion of all other people. An identity statement is difficult to justify when not every person on the planet is fingerprinted. Therefore, such an opinion of identity has been described as a 'leap of faith' [1]. There is always a probability that an identical trace will be found (even if this probability is vanishingly small) and so the step from a small probability to zero probability is a leap of faith being made stating that something which is very unlikely to happen, has not happened. However, even if there were no two identical fingerprints in the world, there are other complicating factors. Fingerprints can be low quality, degraded, smudged, partial or generally otherwise compromised. Each trace type will have its own equivalents, which means traces collected in casework are not necessarily as clear and pristine as controlled samples. These complicating factors may compromise the analyst's ability to effectively declare the identity of the donor. Even if the fingerprints were in perfect, clear and pristine condition, the declaration of a match still requires a human interpretation step. Any process making use of a human interpretation step is subject to the frailties, biases and imperfections of the human mind. As such, no process of matching could claim with certainty the identity of a trace, without this leap of faith to say it is so.

To counter the perception of an infallible process that can assign identity, there have been calls to carry out blind testing of the ability of analysts to measure their abilities. In order for the court to have a complete understanding of the weight of the expert's testimony, it is suggested that their opinion is provided along with the proportion of the time that they are incorrect in blind trials. For example, rather than only stating that the fingerprint came from the defendant, the testimony would include a statement that in some percentage of such opinions, the stated identity is not actually correct. It is then left to the court to decide how much the testimony (if at all) shifts their beliefs. The call for error rates to accompany identity opinions has been extended by some to be a call to provide error rates to all forensic evaluations, including those that carry out the evaluation probabilistically (to provide an *LR*). The concept of providing an error rate for an *LR* is similar in nature to the concept of providing an interval on a probability (a topic discussed next). There are fundamental reasons why doing so is illogical. At the core is the fact that the *LR* provides all the information required to understand the significance of the observations in relation to competing propositions (this includes 'error rates'). Any additional numerical information provided could at best confuse the recipient of information (by providing an answer that is not relevant to the case or the observations) or at worst mislead them. Consider the following thought experiment:

A comparative science exists in which analysts can opine that two traces have the same source. Imagine that a blind validation is carried out on the ability of the analyst to make these conclusions and it finds that the ability to do so is not better than chance (to make the scenario simple, consider that in nature the two states 'same source' and 'different sources' occur equally often). When the analyst concludes that the two marks are from the same source, this could be reported along with the 50% error rate. Alternatively, the analyst could report $LR = 1$, reflecting no information obtained from the traces to distinguish the same from different sources. In other words, the probability of concluding the items are from the same source when they are in fact from the same source is the same as when they are from a different source.

This unrealistic example is used to illustrate that the LR incorporates the error rate, and there is no benefit, and possibly some confusion, from providing an error rate along with the LR. Indeed, the error rate refers to a conclusion of the same source, whereas the LR is agnostic about the conclusion, and therefore the error rate adds no information to the LR. The question would arise as to how a jury would even interpret an $LR = 1$ reported along with an error rate of 50%. Would they interpret this as meaning there is a 50% chance that the LR is not 1? This would be a gravely misleading line of thinking. There are various other logic arguments as to why an error rate should not be applied to an LR, but these are not further iterated here. The interested reader is directed to [2] for further discussion on this topic. This leads to the concept of providing an interval or a range to probability assignments.

8.1.2 CONFIDENCE INTERVALS

As mentioned in Chapter 1, probabilities are personal. They reflect a person's belief about the world around them and this belief is based on experience and knowledge, which in turn (particularly in the sciences) is informed by experimental observations, and information about the framework of circumstances that surround the case. As probabilities are personal, it is common to refer to assigning a probability, rather than talking about 'the' probability of an event because the latter implies there is some true numerical value for the probability that exists external to the mind of the person assigning it. This is fundamentally different to the idea of frequency, which is classically thought of as a real-world property, with a true value upon which a long-running series of experiments would eventually converge. To reflect this idea, the convention is to talk about 'estimating' a frequency (as opposed to assigning a probability), and there can be confidence in the estimate based on the size of the sample taken from the population (hence it is sensible to talk about confidence intervals).

De Finetti famously stated that 'probability does not exist' [3], reflecting the idea that there is no objective value for probability that exists independently from any individual. The extension of this idea is that if there are two numerically different probabilities assigned to the same event, it does not mean that at least one is wrong. Nor does it mean that it is appropriate to consider an interval or a range for the probability as might be implied by many terms such as 'confidence interval'. Remember that probability is an expression of our uncertainty, we are not uncertain about a probability.

This concept can be demonstrated with a thought experiment. It is known that people who smoke have a higher chance of developing lung cancer. The average 55-year-old woman non-smoker has a probability of 0.002 of developing lung cancer in the next 10 years. If that person is a smoker, then the probability jumps to 0.026. Approximately 12% of women in Australia are smokers. A person attends a doctor's office and tells them that they are concerned that their 55-year-old female friend might develop lung cancer. At this stage, the person does not know anything about their friends' smoking habits and the doctor tells the person that their friend has about a 0.49% chance of developing lung cancer in the next 10 years, obtained by:

$$\Pr(cancer) = \Pr(cancer \mid nonsmoker)\Pr(nonsmoker) + \Pr(cancer \mid smoker)\Pr(smoker)$$

$$= 0.002 \times 0.88 + 0.026 \times 0.12$$

$$= 0.00488$$

In a follow-up appointment, the person says their friend has admitted they are a regular smoker and so the doctor updates their evaluation and provides the person with a revised diagnosis of the friend having a 2.6% chance of developing lung cancer in the next 10 years. Note that neither of these probabilities were incorrect, they used the available information in order to provide the best probability assignment possible, i.e. the best representation of the doctor's uncertainty. Nor would it be appropriate to have initially provided some value and interval to the person, such as 1.4% ± 1.2%. This would not have been useful to the person at the time given the information they had. If anything, something akin to a case assessment and interpretation style analysis could have been carried out that advised the person on the prevalence of smoking in the community and the risks associated with smokers and non-smokers.

8.1.3   Measurement Error

Another concept that is often misunderstood is measurement error. This occurs when the property being measured is a physical quantity of some kind, such as a length or weight, and requires a device to measure that property. Any measurement taken is done so with a level of precision, i.e. the closeness of repeated measurements of the object under identical conditions. For example, the width of a pencil could be 'measured' by eye with a precision that is likely to be a few mm, with a ruler with a precision of 0.5 mm or by callipers with a precision of 0.005 mm. Depending on the amount of data available and the device being used, it may be important to take into account one or both sampling effects and measurement error. For properties like allele frequencies (derived from counts of observations in a DNA database), measurement error is generally not considered, as the performance of the profiling system, the interpretation guidelines and protocols of sample reworking are assumed to unambiguously identify the presence of all reference alleles.

8.1.4   The Probability of a Probability

The estimate of a frequency, along with appropriate prior beliefs, is used to assign a probability to an event occurring. The frequency estimate is potentially affected by sampling or measurement uncertainty. The question then arises as to why it is not appropriate to talk about uncertainty in the probability. There are several philosophical and logic arguments that explain this point, and many of these are outlined in a series of papers written as part of a special issue in Science and Justice on precision in *LR*s [4–11], preceded by a discussion in 'Law, Probability and Risk' [12, 13]. In essence probability is a state of mind that one has about the world around them, given all the knowledge and experience possessed at the time of assignment. To assign two different values (with the same knowledge) is to have two different states of mind (or two different sets of beliefs) which is not logical. Put another way, to assign an interval on a probability is to assign a probability to a probability and so why not a probability on a probability on a probability and so on into an infinite regress? It is similar to thinking about measurement error and noting that using a ruler, the width of a pencil might be measured to be 10 ± 0.5 mm. But then it might be noted that the estimate of the measurement error itself may be subject to some error (say that the confidence in the size of the measurement error is only approximately 0.1 mm), and so the measurement could be reported as 10 ± 0.5 ± 0.1 mm. Again, it might be recognised that the estimate of the error on the measurement error is itself subject to error, and so on. Through all this, if someone wanted the best estimate for the width of the pencil, then it would be the main measurement of 10 mm.

8.1.5   Conservatism

The next point to consider is what information is best to present to the court to allow them in their ultimate deliberations. When carrying out deliberations, the jury or judges must be satisfied beyond reasonable doubt in the guilt of a defendant before returning a guilty verdict. This process requires the decision makers to have heard all the evidence in the trial and not simply being satisfied that it is more likely for the allegations to be true than false (a burden of proof applied in many civil courts) but that by some conservative, personal and indescribable level that they are satisfied beyond reasonable doubt in the guilt of the defendant. One point of view is that it is not up to each individual witness to try and apply a conservative adjustment to their testimony in an effort to try and concede all reasonable doubt to the side of the defence. This is not suggesting that a conservative approach shouldn't be taken when assigning probabilities (taking into account differences between studies, differences between the study and the case and uncertain aspects of the case). Rather, once this has been done, and the probabilities assigned to the scientist's best judgment, there should not be an additional conservatism layered onto the reported value. If each witness applied some 'reasonable doubt' adjustment to their own testimony the theoretical question arises, then whether a conviction beyond reasonable doubt would then just be based on a balance of probabilities of what was heard during the trial, i.e. would it be appropriate to apply another layer of conservativism to reach 'beyond reasonable doubt' when the testimony heard had already done so? And how would such a deliberation be managed when some witnessed applied an adjustment (and whose personal definition of the amount of adjustment might differ from those on the jury) and others did not? It follows that the best course of action is for each witness to provide the most truthful testimony, accurate estimates or personal assignments and leave the fact of whether this combined information proves an allegation beyond reasonable doubt to the jury. For scientific evaluations, this concept is related to providing an *LR* without adjustment for the specific purpose of being conservative. As a simple example, imagine that a jury wished to know whether a POI had enough petrol in their vehicle to make a journey between two towns where robberies had been committed in close succession. The towns are 100 km apart as the crow flies. The first witness, an expert in distance and driving, knows that there are numerous different road routes that could have been taken between the towns that would range from 105 to 125 km of road-driven length. They decide to be conservative and state 125 km as the road distance. The second witness is a car expert and knows that the make of car the POI was seen driving can have fuel tanks that range between 50 and 70 L. They decide to be conservative and state 50 L. The third witness is an expert in fuel consumption, and based on the variables of car make, maintenance, care and fuel type the witness knows that the car may have consumed between 0.2 and 0.75 L/km of fuel. They decide to be conservative and state that the car used 0.75 L/km of fuel. The jury, having heard this evidence that the distance was 125 km, the car held 50 L of petrol and used 0.75 L/km realises that the car could only travel 67 km on a full tank, which is only half the distance between towns. In fact, if it is assumed that the ranges for distance, fuel tank size and fuel consumption rate are all uniform within the ranges given, then there is a probability of approximately 0.6 that the POI would have successfully made the journey (found by integrating across the prior probabilities for the three uncertain elements). Using this value of 0.6, which is the best representation of the state of knowledge, the jury can then choose to adopt a conservative stance during their deliberations.

The idea described above can also be considered on a smaller scale within a single evaluation. Imagine that proposition required a series of transfer events to have taken place in order for DNA to have travelled from an individual's hand to an object on which it was found. In an evaluation, probabilities are required for the transfer of DNA in the various steps in the chain of transfers. Each of these probabilities rely on data, and a 'conservative' adjustment could be applied to each probability prior to assigning it. There are several issues that arise. Firstly, it is not clear what adjustment would be considered 'conservative', i.e. it could not be to make all transfer probabilities universally higher, or universally lower as there is no guarantee that this would favour the defendant (and in many cases, it would not). Even if there was a clear side in which one would consider a biasing of a probability to be 'conservative', it is not at all clear how multiple probabilities would be combined. Recall the old statistical adage that the product of intervals is not the interval of a product. Simply using a biased value for each probability would not yield the desired bias in the final *LR*.

8.1.6   CALCULATING A BAYES FACTOR

One option for dealing with uncertainty in the underlying data is to integrate the prior probability. Doing so technically calculates a 'Bayes Factor' rather than an *LR*. In fact, it is often the case that a Bayes Factor is calculated in forensic evaluations, but the terminology of *LR* is so entrenched that we continue to use it in this book. An *LR* does not incorporate prior probabilities into its calculation, or they are set to equal so that they cancel (or can be considered by others). If there are any nuisance parameters in the calculation of an *LR*, they are simply given their maximum likelihood estimate value. A Bayes Factor does include prior probabilities and can integrate across nuisance parameters within its calculation. For example, if calculating the probability of a genotype set, the probabilities of certain alleles in the population could be assigned as their frequency within a database (this is their maximum likelihood estimate) and would lead to an *LR* being calculated. Alternatively, the allele frequency database could be used to construct a distribution that represents the potential allele frequency in the population (given that the database is only a subset of it). Then this distribution is used to integrate across the calculation of the Bayes Factor. As the description above might suggest, the calculation of a Bayes Factor is computationally more expensive and can be impractical on even modest-sized forensic problems [6] and so is often not pursued.

8.1.7   SENSITIVITY ANALYSES

Given all of the above, some justification is required to show that an evaluation is robust. In essence the *LR* can be sensitive to probabilities within the evaluation. If the *LR* is highly sensitive, then it means that a small change in the probability leads to a large change in the *LR*. If the *LR* is insensitive, then there will be little change in *LR* from a change in the probability. If there is no change in the *LR*, then the *LR* is completely insensitive to the probability assignment, indicating that the particular probability (or event that leads to it being included in the evaluation) is not required in the evaluation. If an *LR* is highly sensitive to a particular probability assignment, then ideally the probability assigned will have been based on a large dataset with experimental properties that are closely aligned with the circumstances of the case. Then the data being used is highly relevant and has the power to inform an assignment. However, when assigning a probability without much data, and using a prior as shown in Chapter 5 (i.e. a uniform Dirichlet distribution), then the prior probability will dominate the posterior probability. Still, both assignments (well informed by data, or poorly informed by data) may ultimately lead to the same value.

An example of the conundrum is given in section 4 of Taylor et al. [11]. In the section of Taylor et al., they describe a scenario by which an *LR* is assigned to some observations given two competing propositions. In one variant of the scenario, the *LR* is based on a large amount of data and the *LR* is considered to be robust. In the other variant there is little to no data available and the *LR* is not robust. In both variants the *LR* = 1 but have quite different meanings. In the variant where there is no data, the *LR* is reflective of our ignorance about the underlying factors that are potentially important to the evaluation. It tells us that the most informative way to proceed is to collect data relating to the probability of interest. In the variant with large amounts of data, the *LR* is robustly indicating a lack of ability to use the observations to differentiate between propositions. It tells us that the test being performed (or the factor being considered) is irrelevant to the evaluation.

But how can it be determined which probabilities the *LR* is sensitive to, or whether the evaluation is robust? This leads to sensitivity analysis. A sensitivity analysis considers what the *LR* assignment would have been, if the information (experimental data or other relevant information) underlying the assigned probabilities had been different. It characterises the system that generates the *LR* not the evidence in a particular case. The following sections describe the different forms of sensitivity analysis, what they show and how to carry them out either from an equation or within a BN.

8.2   TYPES OF SENSITIVITY ANALYSES

8.2.1   Sensitivity Analysis for a Factor That Has Two States

The simplest form of sensitivity analysis is performed when applied to data that informs a binary decision, say for example whether DNA transfer will occur. In these situations, only a single probability term needs to be considered. For example, the probability of transfer, and the alternative probability of no transfer, is simply the mutually exclusive and exhaustive complement (and can be written instead as one minus the probability of transfer). Within a derived *LR* formula, the sensitivity analysis can be carried out by varying the value assigned to the probability of focus and noting the subsequent change in *LR*. It is common for the full range of possible probability values to be trialled, i.e. 0 to 1, although the range can usually be restricted to sensible bounds (either based on related literature, knowledge, experience or common sense) that lie within the range [0, 1].

Recall from Chapter 5 that for case 1, the *LR* formula that was ultimately derived was:

$$LR = \Pr\left(\overline{T_{C \to D}} \mid {}^{Activity}H_p\right)$$

where the single term related to the probability of transfer not occurring from the complainant (*C*) to the defendant's top (*D*), given that *C* hit *D*'s top. The value assigned to this probability was based on a Daly et al. [14] and was 0.53, and consequently, the *LR* was 0.53. The *LR* formula for case 1 is the simplest example possible, as the *LR* is equal to the single probability of focus. In other words, if the probability value is varied, the *LR* will change by the same amount, and it is straightforward to understand the sensitivity of the evaluation to the data.

As explained earlier, the convention is to report an *LR* ≥ 1 in support of H1 compared to H2, and if the *LR* < 1, then it is inverted, and the value is reported as > 1 in support of H2 compared to H1. This is done as it is easier to understand numbers greater than one rather than as a fraction or decimal, when comparing the relative support for two opposing propositions. In the same way it is a convention to consider the probability of transfer, rather than the probability of no transfer within a sensitivity analysis, simply because it is intuitively simpler to think of the positive outcome of an event. These conventions are mentioned in the results of the sensitivity analysis in case 1 the sensitivity of the inverted *LR* to the probability of transfer is examined. For example:

- If $\Pr\left(\overline{T_{C \to D}} \mid {}^{Activity}H_p\right) = 0.1$, then $LR = 0.1$ and $LR^{-1} = 10$
- If $\Pr\left(\overline{T_{C \to D}} \mid {}^{Activity}H_p\right) = 0.2$, then $LR = 0.2$ and $LR^{-1} = 5$
- etc.

Figure 8.1 shows the value of the inverse *LR* across the range of values for $\Pr\left(T_{C \to D} \mid {}^{Activity}H_p\right)$.

The graph shown in Figure 8.1 has some limitations. Firstly, because the *LR* becomes so extreme at probability values close to 1 it appears as though the *LR*⁻¹ is zero for a probability of transfer up to approximately 0.8. This is not the case and only appears as such due to issues of the vertical scale. For these reasons, it is common to display the *LR* axis on a log₁₀ scale (or alternatively to plot the log₁₀(*LR*) instead of the *LR*). Figure 8.2 makes such an adjustment and shows the same data as Figure 8.1, simply with the *y*-axis displayed on a log₁₀ scale.

Figure 8.2 shows the important point that the inverse *LR* is always greater than 1 (and not 0 as it visually appeared in Figure 8.1). Note the rapid rise in the *LR* (seen in either Figure 8.1 or Figure 8.2) as the probability of transfer approaches 1. In fact, the inverse *LR* in Figures 8.1 and 8.2 is only plotted up to $\Pr\left(T_{C \to D} \mid {}^{Activity}H_p\right) = 0.999$. Assigning a value of 1 to this probability results in the inverse *LR* becoming infinite.

The intuition for the ascent of the inverse *LR* to infinity comes from the fact that if transfer (and subsequent persistence and recovery) was certain to occur whenever someone was hit then if Hp

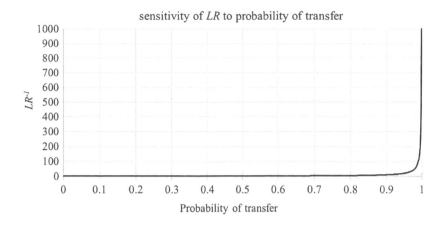

FIGURE 8.1   Sensitivity of *LR* to the probability of DNA transfer from hitting. Note that the graph shows the inverted *LR* (*LR⁻¹*).

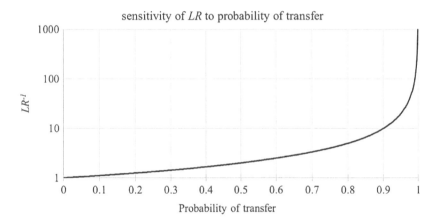

FIGURE 8.2   Sensitivity of *LR* to the probability of DNA transfer from hitting, with *LR* axis on a $\log_{10}$ scale. Note that the graph shows the inverted *LR* (*LR⁻¹*).

was true (which involves C hitting D's top) then, with certainty, the DNA of C would be present. An absence of C's DNA has zero probability of occurring if Hp has occurred and therefore $LR = 0$ and the inverse is infinite. At the other end of the spectrum, if the probability of transfer is 0 then the *LR* (and its inverse) is 1. The intuition for this result is that if DNA is expected to never transfer from hitting, and DNA transfer will not occur if no contact has occurred (i.e. the Hd scenario), then there is no power in this result that could discriminate between Hp and Hd (i.e. in either case no DNA will be transferred).

The question now comes as to whether the evaluation is robust. From Figure 8.1 and 8.2, it can be seen that the inverse *LR* can vary from 1 to infinity, which is quite a large range of possible values, and may initially seem too large for any opinion to be justifiably given. This line of thinking is a trap that can be fallen into when the entire possible range of a probability is included within a sensitivity analysis. Remember though, that the analyst is unlikely to be going into the sensitivity analysis without any knowledge whatsoever. The data from the Daly study gives us a value to be assigned to the probability of transfer of 0.47, which is at the centre of the *x*-axis in the plot in Figure 8.2. Some deviation might be expected from this value, but the prior probability of transfer being close to the extremes of 'never' or 'always' is extremely small. Taking a step back from the sensitivity analysis itself, recall that in the Daly study 47 instances of individuals grabbing material resulted in a transfer. Given this finding, it does not make sense to test the probability assignment which would have expected the Daly study to observe 0/100 transfers or 100/100 transfers.

A solution to this issue may be to limit the range over which the sensitivity analysis is conducted. This may be based on personal judgment, or it may be a range of values that encapsulates the findings from multiple different studies, or it may be based on one study but reflect some uncertainty around the effect of differences between the study and the case at hand. In case 1 quite generously wide bounds could be chosen to test, say 0.1–0.9 and even then the inverse LR across this range varies from just above 1–10. This entire range of LRs could be considered as providing slight support for the defence proposition compared to the prosecution proposition. Viewed in this manner, the sensitivity analysis shows that the evaluation would seem relatively robust. However, in practice a range of 0.1–0.9 represents a much greater level of uncertainty than is thought to exist, and the range would be limited to a much narrower band. As mentioned previously, this may be based on the spread of values seen in different published studies, or it may be a subjective interval that reflects a personal account for differences between the analytical methods in the published study and those in use at the forensic laboratory.

8.2.2 Sensitivity Analysis for Two Factors That Each Have Two States

Consider now the formula in case 1 that included the factor of persistence. The LR formula from Chapter 5 that incorporated both transfer and persistence terms was:

$$LR = \Pr\left(T_{C \to D} \mid {}^{Activity} H_p\right) \Pr\left(\overline{P_{C \to D}} \mid {}^{Activity} H_p\right) + \Pr\left(\overline{T_{C \to D}} \mid {}^{Activity} H_p\right)$$

Consider the sensitivity of the LR to the data underlying the persistence probability assignment, in a similar way that was done for transfer. In this instance the transfer probability is held constant at the data-derived assigned value of $\Pr\left(\overline{T_{C \to D}} \mid {}^{Activity} H_p\right) = 0.53$ and the value of persistence varied across the full possible range [0,1]. Figure 8.3 shows the result of a sensitivity analysis carried out in the manner just described.

Figure 8.3 shows that the LR is not very sensitive to the probability of persistence, when using the data-driven probability assignment of 0.47 for the probability of transfer. At the lowest probability value for persistence, the LR (and its inverse) is 1. The intuition behind this finding is that an

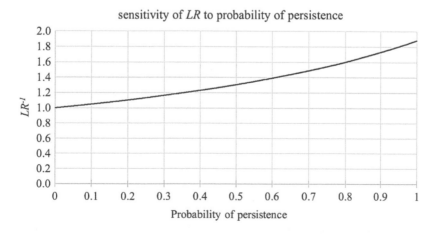

FIGURE 8.3 Sensitivity of LR to the probability of DNA persistence when the probability of transfer is 0.47.

assigned value of 0 means that if DNA is transferred to the top it is certain that it would not persist. Therefore, there is no difference in the expected outcome of either proposition and hence an $LR = 1$. At the other extreme (a probability assigned as 1 for persistence), the formula simplifies to the originally trialled version, $LR = \Pr\left(\overline{T_{C \to D}} \mid {}^{Activity}H_p\right)$, and so the $LR = 0.53$ (and its inverse approaches 2). The intuition is that if DNA transferred to the top is certain to exist, then the only way to explain an absence of material under Hp is that it was not transferred, to begin with. Note that the probability of recovering DNA will need to be considered here if the probability of recovery is not assumed to be 1 (or is not already accounted for by the experimental data being used to assign the probability). If recovery is already an implicit part of the transfer or persistence study results, then it should not be accounted for again separately (otherwise it will have been accounted for twice).

One final sensitivity analysis is shown for case 1, by stepping back to a version of the LR formula from Chapter 5 before some of the simplifications were made. The following formula has separate transfer probabilities for the defendant and the alternate offender:

$$\left[\Pr\left(T_{D \to C} \mid {}^{Activity}H_p\right)\Pr\left(\overline{P_{D \to C}} \mid {}^{Activity}H_p\right) + \Pr\left(\overline{T_{D \to C}} \mid {}^{Activity}H_p\right)\right] \times$$

$$LR = \frac{\left[\Pr\left(T_{C \to D} \mid {}^{Activity}H_p\right)\Pr\left(\overline{P_{C \to D}} \mid {}^{Activity}H_p\right) + \Pr\left(\overline{T_{C \to D}} \mid {}^{Activity}H_p\right)\right]}{\Pr\left(T_{AO \to C} \mid {}^{Activity}H_d\right)\Pr\left(\overline{P_{AO \to C}} \mid {}^{Activity}H_d\right) + \Pr\left(\overline{T_{AO \to C}} \mid {}^{Activity}H_d\right)}$$

Still make the assumption that the data from Daly et al. [14] is appropriate to apply to both $\Pr\left(T_{D \to C} \mid {}^{Activity}H_p\right)$ and $\Pr\left(T_{AO \to C} \mid {}^{Activity}H_d\right)$, and their complements of non-transfer. In Chapter 5 this was dealt with by cancelling components in the LR, but here both terms are retained so that a sensitivity analysis can be carried out on them. The sensitivity analysis could be carried out by holding all probabilities at their data-driven assigned values and varying only the terms $\Pr\left(T_{D \to C} \mid {}^{Activity}H_p\right)$ and $\Pr\left(\overline{T_{D \to C}} \mid {}^{Activity}H_p\right)$, plotting the LR over this range. Then the same could be done but vary the values of terms $\Pr\left(T_{AO \to C} \mid {}^{Activity}H_d\right)$ and $\Pr\left(\overline{T_{AO \to C}} \mid {}^{Activity}H_d\right)$, again plotting the LR over this range. However, doing this does not reflect the purpose of a sensitivity analysis. Recall that the purpose of a sensitivity analysis is to see how sensitive the LR is to the data informing the probabilities required in the evaluation. Therefore, if multiple probabilities use the same data in their assignment, then they all need to be tested together. In this case the sensitivity of the LR is being tested with respect to the data used to assign a probability of transfer (for a set value for persistence probability). Therefore, it is appropriate to:

- Assign $\Pr\left(T_{D \to C} \mid {}^{Activity}H_p\right) = 0.1$, $\Pr\left(\overline{T_{D \to C}} \mid {}^{Activity}H_p\right) = 0.9$, $\Pr\left(T_{C \to D} \mid {}^{Activity}H_p\right) = 0.1$, $\Pr\left(\overline{T_{C \to D}} \mid {}^{Activity}H_p\right) = 0.9$, $\Pr\left(\overline{T_{AO \to C}} \mid {}^{Activity}H_d\right) = 0.1$, $\Pr\left(\overline{T_{AO \to C}} \mid {}^{Activity}H_d\right) = 0.9$ and calculate the LR,
- Assign $\Pr\left(T_{D \to C} \mid {}^{Activity}H_p\right) = 0.2$, $\Pr\left(\overline{T_{D \to C}} \mid {}^{Activity}H_p\right) = 0.8$, $\Pr\left(T_{C \to D} \mid {}^{Activity}H_p\right) = 0.2$, $\Pr\left(\overline{T_{C \to D}} \mid {}^{Activity}H_p\right) = 0.8$, $\Pr\left(\overline{T_{AO \to C}} \mid {}^{Activity}H_d\right) = 0.2$, $\Pr\left(\overline{T_{AO \to C}} \mid {}^{Activity}H_d\right) = 0.8$ and calculate the LR, etc.

Doing this for a probability of persistence equal to 1 would yield exactly the same plot as seen in Figures 8.1 and 8.2. Normally such considerations are not present when using a derived LR formula in a sensitivity analysis as, if all simplifications have been carried out to the formula, terms that are assumed to be the same (and using the same underlying data) will be cancelled out. The consideration does become more important when carrying out a sensitivity analysis using a Bayesian network, as often the common probabilities are all present separately. This is discussed later.

Before continuing, imagine data is found to suggest that the probability of DNA being transferred through hitting, and the probability of DNA being transferred through grabbing are different. In this situation one set of data is used to inform the probabilities of $\Pr\left(T_{D\to C} \mid {}^{Activity}H_p\right)$, $\Pr\left(\overline{T_{D\to C}} \mid {}^{Activity}H_p\right)$, $\Pr\left(T_{AO\to C} \mid {}^{Activity}H_d\right)$ and $\Pr\left(\overline{T_{AO\to C}} \mid {}^{Activity}H_d\right)$ and a different set of data to inform probabilities of $\Pr\left(T_{C\to D} \mid {}^{Activity}H_p\right)$ and $\Pr\left(\overline{T_{C\to D}} \mid {}^{Activity}H_p\right)$. If the sensitivity of the LR was tested to the data used to inform the probability of transfer to C (i.e. the first set of four terms), and holding steady probabilities $\Pr\left(T_{C\to D} \mid {}^{Activity}H_p\right) = 0.47$, $\Pr\left(\overline{T_{C\to D}} \mid {}^{Activity}H_p\right) = 0.53$, and using a probability of persistence equal to 1, then the resulting LR graph is shown in Figure 8.4.

Figure 8.4 shows that the LR does not change, regardless of the value used for the probability of transfer to clothing from grabbing (the activity carried out that could have transferred DNA to C). Again, this sort of result from a sensitivity analysis for a derived LR formula is unusual, as formulaic simplification should remove terms that are equal in the numerator and denominator and thus do not impact the LR (as was done in Chapter 5). If a sensitivity analysis result such as this is obtained, then it suggests that perhaps some further simplification could be carried out in the derivation. But again, results such as this can occur more frequently when considering Bayesian networks as the BN is not architecturally simplified for such assumptions. For example, in Chapter 6, the BNs developed for case 1 had separate nodes that considered the transfer of DNA from D to C and from AO to C present.

Let's go back to the simplified formula:

$$LR = \Pr\left(T_{C\to D} \mid {}^{Activity}H_p\right)\Pr\left(\overline{P_{C\to D}} \mid {}^{Activity}H_p\right) + \Pr\left(\overline{T_{C\to D}} \mid {}^{Activity}H_p\right)$$

It is common practice, when conducting sensitivity analysis, to consider a single probability at a time and hold all others at their data-assigned values. You have seen this process so far in case 1 for the probability of transfer (Figures 8.1 and 8.2) and persistence (Figure 8.3). However, this does not necessarily give a feel for how the LR can be sensitive to both factors together. For example, given the results of the individual sensitivity tests, it may be expected that the inverse LR would become quite large for large values of transfer and persistence together, even though this combination of values has not been tested. One way to explore the sensitivity of the LR to both factors is to, in effect, carry out multiple sets of sensitivity analysis, traversing across one factor in larger steps. Figure 8.5

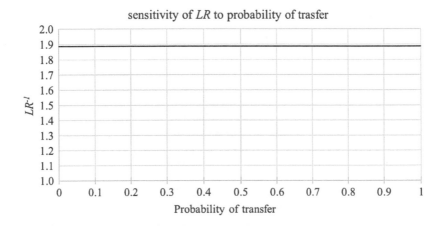

FIGURE 8.4   Sensitivity of LR to the data used to inform probability of DNA transfer to C.

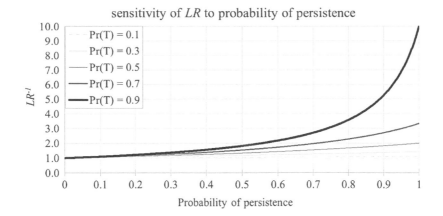

FIGURE 8.5   Sensitivity of *LR* to the probability of DNA persistence when the probability of transfer is varied.

shows the sensitivity of the *LR* to the data for persistence (i.e. as in Figure 8.3), but for five broad steps of values for the probability of transfer ranging from 0.1 to 0.9.

As an alternative to the plot shown in Figure 8.5, the data can be presented in a three-dimensional plot, as shown in Figure 8.6. Both Figures 8.5 and 8.6 show the same data, and it can be up to personal preference, or the availability of software and capability to produce 3D plots, as to which method of display is used. In fact, the five lines seen on the graph in Figure 8.5 come from slices across the transfer-plane in Figure 8.6 at values 0.1, 0.3, 0.5, 0.7 and 0.9. Most statistical packages, such as R, Stata, SPSS or MATLAB can produce 3D plots and there is some limited ability in HUGIN or Excel to do so also.

The results in Figures 8.5 and 8.6 show that if moderate bounds are set on the probability of persistence (again quite liberal bounds of 0.1–0.9 could be chosen as was done for the probability of transfer), then the *LR* does not obtain extreme value as when only the probability of transfer was

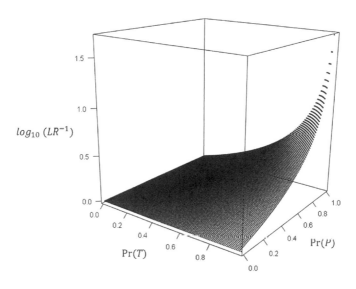

FIGURE 8.6   Sensitivity of *LR* to the probability of DNA persistence when the probability of transfer is varied, shown over a 3D surface.

considered (as in Figure 8.2). For example, if a probability of transfer was assigned a value of 1, it was previously shown that the value of the inverse *LR* became infinite. But if the model includes persistence, then even with a probability for the persistence of 0.9 a probability of transfer assigned as 1 will only result in an inverse *LR* of 10. Quite a bit less than infinite! This effect is often referred to as 'explaining away'.

Explaining away is the effect by which when there are multiple possible explanations for an outcome, if the probability of one of those explanations increases, then it will tend to decrease the probability of the other. In case 1 example in Figure 8.2, there was only one explanation for a lack of DNA under the prosecution proposition, and that was that DNA did not transfer. The consequence is that when it becomes very likely for DNA to have transferred the probability of the prosecution proposition becomes very low. Another possible explanation for a lack of DNA is added, perhaps it did transfer but did not persist. Now even for a high probability of transfer, the lack of DNA can be explained away by a non-persistence and so the probability of the prosecution proposition is not so low.

Another consequence of changes to models that explain away a result is that the sensitivity of a model to a single event will tend to decrease. This is the effect seen in Figures 8.5 and 8.6 compared to Figure 8.2.

8.2.3 Sensitivity Analysis for Factors That Have More Than Two States

So far, it has been shown how probability can be adjusted to see how sensitive the *LR* is to the data informing it. In the method shown for case 1, varying the probability is straightforward as the probabilities consider events that have only two states, i.e. transfer or no transfer, persistence or no persistence. The two states of these events are mutually exclusive and exhaustive so that when the probability of one of the states is varied, the probability of the complementary state is already determined.

The situation can become a little more complex when the event being considered has more than two states. For example, in case 3 from Chapter 5 (the family assault), the DNA can be transferred from cohabitation or biting, in high amounts, low amounts or none. In the derivation of the *LR* formula derived in Chapter 5 only consideration of transfer of low amounts or none was needed, as there were only low amounts of DNA detected from the D family on C's underwear. However, if high amounts of DNA had been detected, then consideration of high, low or no DNA transfer would be required. Also, in Chapter 6 when a BN was built for this case the states of high, low and none for the transfer nodes all had probabilities assigned. In these circumstances, a sensitivity analysis cannot simply be carried out by varying the probability of high levels of DNA transfer from 0 to 1, because for any value of high DNA transfer, there are multiple combinations of low and none that could be assigned. For example, if the probability of high levels of transfer was assigned a value of 0.1, then the following combinations of probability could be considered:

- A probability for low transfer of 0.9 and a probability of no transfer of 0
- A probability for low transfer of 0.8 and a probability of no transfer of 0.1
- A probability for low transfer of 0 and a probability of no transfer of 0.9

In fact, in this instance it would still be possible to create a plot, such as the one shown in Figure 8.5 or Figure 8.6, except that they would be for a single transfer event. However, any number of states beyond three could be considered, which would make even the 3D plotting impractical. An option in these situations is to consider a series of distributions for the probabilities of the different states that incrementally shift the mass of the probability from one extreme state to the other.

Consider the series of distributions shown in Figure 8.7.

The distributions shown in Figure 8.7 must only exist over the range from 0 to 1 (because the first law of probability tells us that probabilities can only exist over that range). The shape of the

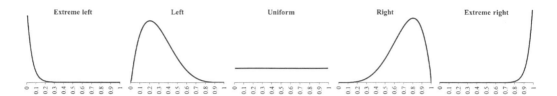

FIGURE 8.7   Series of distributions that could be used to vary probability values in a sensitivity analysis.

distribution must also be able to change. An ideal distribution that meets both these criteria is the beta distribution, which is what has been used to generate the distributions seen in Figure 8.7. The distributions in Figure 8.7 have been generated using the following distributions:

- Extreme left: $B(1, 20)$
- Left: $B(2, 5)$
- Uniform: $B(1, 1)$
- Right: $B(5, 2)$
- Extreme right: $B(20, 1)$

The number of distributions to create in the series, and the choices of parameters that define the distribution are arbitrary and need only reflect a shift in probability assignment from one state extreme to the other. The only exception to this is the uniform beta, which will always be a $B(1, 1)$. The distributions in Figure 8.7 are continuous and must be used to assign values to discrete states for the probability of an event. In order to do this, the distribution is broken up evenly across the range [0,1] and the area under the beta distribution curve is calculated across that range. For example, to assign probabilities to the three states: high, low and none, the range [0,1] is broken into three equal segments of [0, 1/3], [1/3, 2/3] and [2/3, 1] and then the area under the beta curve calculated for each curve and each bracket. Such a calculation can be done in a number of programs. For example, if access to Microsoft Excel is possible and the desired probability was for state 'low' using the 'Left' beta curve, then this could be obtained by the formula:

```
= BETA.DIST(2/3, 2, 5, TRUE) - BETA.DIST(1/3, 2, 5, TRUE)
```

which gives the value 0.333. Or if access to R is available the command:

```
pbeta(2/3, 2, 5) - pbeta(1/3, 2, 5)
```

will give the same result. Taking this approach across the range of distributions shown in Figure 8.6, for the states high, low and none, it would yield the probabilities shown in Table 8.1 for use in the sensitivity analysis.

TABLE 8.1
Probabilities for Use in a Sensitivity Analysis for States High, Low and None

	Extreme left	Left	Uniform	Right	Extreme right
High	0.9997	0.6488	0.3333	0.0178	0.0000
Low	0.0003	0.3333	0.3333	0.3333	0.0003
None	0.0000	0.0178	0.3333	0.6488	0.9997

The above method will provide some indication of the sensitivity of the model to the data for an event so long as two aspects are true:

1) The states have a natural order so that it is sensible to shift the probability from one extreme to the other; for example 'high/low/very low/none' would be an acceptable set of states to apply this method, whereas 'metal/glass/wood/plastic' would not.
2) It is sensible that the data is unimodal. This means that if (for example) the states high/low/none were being considered, then it makes sense that if one state has the highest probability, then the states stepping away are expected to have a decreasing probability. If an event had states high/low/none, and it was sensible that there might be probability distributed between the high and none states, with a relatively low probability assigned to the low state, then the method of sensitivity analysis described would also not be valid.

8.2.4   RESAMPLING THE RAW COUNT DATA

The sensitivity analyses up until this point were based on the premise of varying the value of some probability (or probabilities) across a, usually quite broad, range in order to determine whether the *LR* was particularly sensitive to any aspect of the evaluation. In a sense, carrying out the sensitivity analysis in this way, while informative, implies no knowledge, and that the probability could take any technically possible value. This is typified by the fact that the effects of persistence have been considered within case 1 but have never actually had a value assigned to its probability. Varying probabilities across the entire possible range caused issues when extreme values for probabilities were assigned and led to *LRs* that were infinite. This then required the introduction of some 'sensible' bounds on the sensitivity analysis so that it was informative and useful.

Of course, for most (and ideally all) probabilities being assigned, there will be experimental data that has been used to inform the analyst of the appropriate value. This experimental data could be a series of experiments, where the occurrences of a number of different types of outcomes are counted. Recall that in the probability assignment section of Chapter 5 the concept was introduced by which the raw count data was applied to a prior belief using a Dirichlet distribution, in order to obtain non-zero posterior probabilities for all states of an event. When the number of observations being used to inform a probability was low, then the prior probability tended to dominate the posterior. When there were many observations, then they would dominate the prior probability and the posterior would closely align with the raw experimental frequency. In terms of sensitivity analyses for the situation where there are low numbers of observations, little information is available to inform the probability assignment. If the experiments were carried out again, quite different results might be obtained, simply by the fact that frequencies of outcomes from small sample sizes are more prone to sampling variation. Consequently, it would be expected the *LR* would be more sensitive to this data than if the probability assignment were based on a large number of experiments. Note how talking about sensitivity analysis in this manner is quite different to how it has been considered up until now. This is the first time the source of information is being considered, and hence the robustness of the data is being considered, hence probability assignment, and hence evaluation. When carrying out a sensitivity analysis of the data that was used, the assumption is that the data is appropriate to the context in which it has been used and that the experiments have been carried out in such a way that the outcome has not been biased through experimental design. This assumption was not necessary for the previous sensitivity analyses, which had very little to do with the actual data used to assign the probability.

The same Dirichlet distribution that was used to obtain posterior probabilities to assign within an evaluation can be used to test the sensitivity of the *LR* to the data that was used to inform the probability assignment. The Dirichlet distribution can be used to resample the experimental counts under the assumption that the Dirichlet distribution correctly describes the data. For example, imagine that the distribution of the height of an average male adult can be described by a normal distribution with a mean of 177 cm and a standard deviation of 7.5 cm, but this knowledge wasn't available. Imagine the probability of a random male being over 185 cm tall was sought. With omniscient knowledge of the population distribution, the sought-after probability could be calculated relatively straightforwardly as the area under a normal distribution $N(177, 7.5)$ above the value of 185. This turns out to be approximately 0.14. However, without this knowledge, a decision is made to assign the probability empirically by taking a number of measurements of random males. Because time and resources are limited, only ten measurements can be taken. From an omniscient view, it is possible to simulate a number of possible samples of ten people, to see how dramatic the effects of sampling variation will be in this instance. One possible outcome from drawing ten numbers from an $N(177, 7.5)$ distribution is to obtain:

- 178.0252, 169.8589, 186.6270, 189.5946, 195.4958, 179.8478, 183.1064, 176.0407, 170.8141 and 182.7566

This can be achieved in Excel by populating ten cells with the formula:

```
=NORM.INV(RAND(),177, 7.5)
```

or in R with the command:

```
rnorm(n=10, mean=177, sd=7.5)
```

From the random sample above, there were three with heights greater than 185 and seven with values less than 185. Using the Dirichlet prior methodology would result in:

- $(3 + 1)/(10 + 2) \sim 0.33$ for the probability of choosing someone with a height above 185 cm
- $(7 + 1)/(10 + 2) \sim 0.67$ for the probability of choosing someone with a height below 185 cm

These values are somewhat different to the values of 0.14 and 0.86 which are known to be the population values. If another set of ten random measurements were taken, then slightly different probabilities would be assigned. If this process were repeated 10,000 times, then the probabilities assigned have a distribution as shown in Figure 8.8.

Figure 8.8 shows that there is a reasonable probability that from ten samples a value for the probability of being above 185 cm might be assigned as high as 0.5. The most common outcome is still in alignment with the population proportion of 0.14. It is expected that if a larger sample were taken, say 1000, then the observed variability in the probability value would be much less. The simulation that led to Figure 8.8 is repeated, except the number of measurements in each of the 10,000 simulations is increased from 10 to 1000. The resulting distribution of probability assignments for being above 185 cm is shown in Figure 8.9.

Comparing Figure 8.9 to Figure 8.8 shows that the larger the number of measurements taken, the less the variation expected in the probability assignment is caused by sampling uncertainty in the underlying dataset.

The same idea holds true for the data that is used to inform probability assignments used in evaluations. However, in the height example it is known that the data follows a normal distribution, whereas when assigning probabilities a Dirichlet distribution has been used. Therefore, some way of randomly drawing values from a Dirichlet distribution is needed in order to carry out resampling simulations. Luckily it turns out there is a relatively simple way to achieve this. If you have an N-dimensional Dirichlet distribution, where x_i is the posterior count (i.e. with prior added) of state

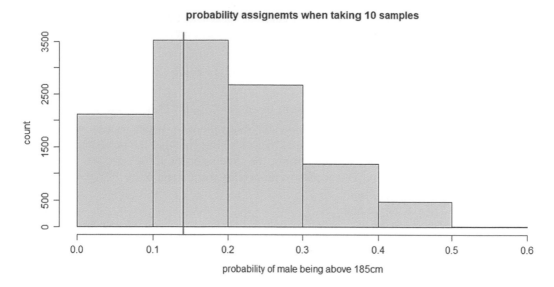

FIGURE 8.8 Probabilities assigned to a male being above 185 cm tall from 10,000 simulations of taking 10 random samples from the population. The red line shows the population level.

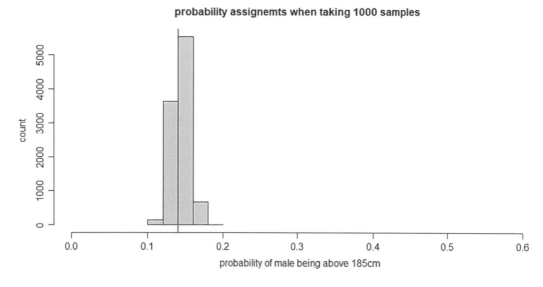

FIGURE 8.9 Probabilities assigned to a male being above 185 cm tall from 10,000 simulations of taking 1000 random samples from the population. The red line shows the population level.

i, a sample can be drawn from this distribution by drawing a sample for each count using a Gamma distribution:

$$f_1 = \Gamma(x_1, 1), f_2 = \Gamma(x_2, 1), ..., f_N = \Gamma(x_N, 1)$$

and then normalising to convert sampled counts (represented by f) to probability (represented by g):

$$g_i = \frac{f_i}{\displaystyle\sum_i f_i}$$

To produce a resampled $Dirichlet(g_1, g_2, ..., g_N)$ distribution, consider case 1, and the formula for the LR that included persistence and transfer terms:

$$LR = \Pr\left(T_{C\to D} \mid {}^{Activity}H_p\right) \Pr\left(\overline{P_{C\to D}} \mid {}^{Activity}H_p\right) + \Pr\left(\overline{T_{C\to D}} \mid {}^{Activity}H_p\right)$$

Data from Daly et al. [14] was used to inform us of $\Pr\left(T_{C\to D} \mid {}^{Activity}H_p\right)$, and hence then also $\Pr\left(\overline{T_{C\to D}} \mid {}^{Activity}H_p\right)$. The data was 100 experimental observations for which transfer was observed 47 times and no transfer was observed 53 times. Now also imagine that the persistence has come from four in-house experiments designed to mimic the circumstances of the case, and for which all four showed persistence of DNA. This leads to a probability assignment of:

- $(4 + 1)/(4 + 2) \sim 0.8$ for yes and
- $(0 + 1)/(4 + 2) \sim 0.2$ for no.

The LR, using these values, is:

$$LR = 0.47 \times 0.2 + 0.53 = 0.6083$$

And the inverse of the LR is 1.64. The sensitivity of the LR to the transfer data can be determined by resampling the counts from the Daly et al. study, i.e.:

1) Draw a value for the resampled transfer count, $f\left(T_i\right)$, from distribution $T_i = \Gamma(47+1,1)$
2) Draw a value for the resampled non-transfer count, $f\left(\overline{T_i}\right)$, from distribution $\overline{T_i} = \Gamma(53+1,1)$
3) Convert these to resampled transfer and non-transfer probabilities, g, by
$g\left(T_i\right) = f\left(T_i\right)/\left[f\left(T_i\right)+f\left(\overline{T_i}\right)\right]$ and $g\left(\overline{T_i}\right) = f\left(\overline{T_i}\right)/\left[f\left(T_i\right)+f\left(\overline{T_i}\right)\right]$
4) Use $g\left(T_i\right) = \Pr\left(T_{C\to D} \mid {}^{Activity}H_p\right)$ and $g\left(\overline{T_i}\right) = \Pr\left(\overline{T_{C\to D}} \mid {}^{Activity}H_p\right)$ in the LR calculation
5) Repeat steps 1–4 for $i = 1 \to 1000$, so that 1000 LRs are obtained

In a similar manner the sensitivity of the LR to the persistence data can be determined by resampling the four counts from the fictitious in-house experiments. Usefully, the sensitivity of the evaluation to all of the data used can be determined by carrying out resamplings of both sets of data and calculating the LR. Figure 8.10 shows the range of LR values obtained when all data is resampled (in blue), just the transfer data is resampled (in red) and when just the persistence data is resampled (in yellow).

It can be seen from Figure 8.10 that when the sampling uncertainty in all the data is taken into account, the inverse of the LR varies from approximately 1 to 2.4. The advantage of additionally carrying out the sensitivity analysis to the individual datasets is that it highlights what factors are contributing most to the spread of LRs. When only the transfer data is resampled, and the persistence probability is held at the value of 0.8, then Figure 8.10 shows the change in the LR value from approximately 1.3 to 2.2. This is less than the range for the LR when all data is considered, and that is as expected given that the transfer data only makes up part of the data used in the evaluation. When the transfer probability is held at its value of 0.47, and the persistence data is resampled, the range of inverse LRs obtained is approximately 1–1.9. There are two observations from the data in Figure 8.10:

1) The sensitivity of the LR to the persistence data is similar to the sensitivity of the LR to the transfer data, even though the persistence dataset is very small. This is the effect of 'explaining away' that was spoken of earlier. In fact, Figure 8.5 shows that using a fixed value for the transfer probability of approximately 0.5, the inverse LR can only vary

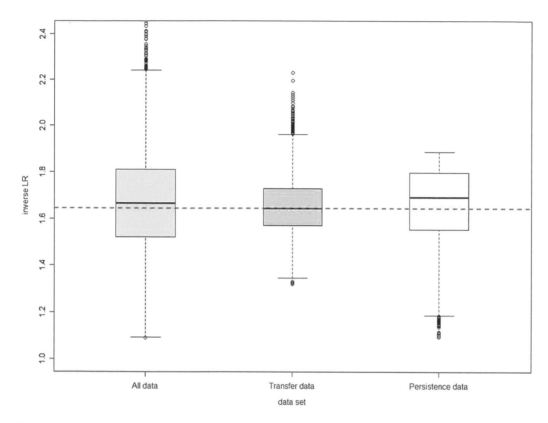

FIGURE 8.10 Sensitivity of *LR* to all data (blue), just the transfer data (red) and just the persistence data (yellow). The red dashed line is the value of the inverse *LR* calculated using the data-assigned values.

between 1 and approximately 1.9 (obtained by 1/0.53) across the full possible range of values for the probability of persistence.

  2) The sensitivity of the *LR* to both datasets is not dramatically increased from the sensitivity to the individual components. For example, the range of the all-data *LR* is not a multiple (or even an addition) of the ranges of the individual components. This fact comes from the *LR* formula construction.

With regards to the first point above, for a fixed probability of persistence (with a relatively high value, such as 0.9 shown in Figure 8.5), the inverse *LR* is bound by 1 (at the lower end) and the inverse of the probability of persistence (at the upper end). For a probability of persistence of 0.2, this bounds the range of the inverse *LR* in the sensitivity analysis to 5. However, the sensitivity analysis in Figure 8.10 only shows a range for the *LR* to transfer probability data up to approximately 2.2. The discrepancy between 5 and 2.2 comes from the fact that according to the sampling scheme that redraws data from a Dirichlet distribution, obtaining a transfer probability close to one is virtually impossible. This demonstrates that by using a resampling scheme to test the sensitivity of the *LR* to the underlying data, there is no need to set artificial bounds on the sensible range that probabilities can take. Limiting comes from the model itself.

Imagine now that in case 1 the transfer data was also based on a small dataset, say ten observations, with five observed transfers. Carrying out the same process as that which led to Figure 8.10 would give the sensitivity analysis results seen in Figure 8.11.

The results in Figure 8.11 show that the *LR* is most sensitive to the data used to assign the probability of transfer (given the much larger range of the *LR* to the transfer sensitivity analysis than

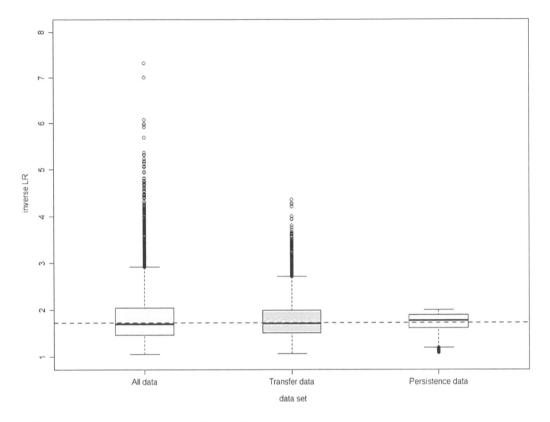

FIGURE 8.11 Sensitivity of *LR* to all data (blue), just the artificially low transfer data (red) and just the persistence data (yellow). The red dashed line is the value of the inverse *LR* calculated using the data-assigned values.

the persistence sensitivity analysis). Figure 8.11 also shows the upper bound of the *LR* sensitivity to transfer approaching the theoretical maximum of 5, as the low sample number used to generate the transfer probability is much more prone to sampling variation. If the variation in the *LR* was very high (high enough to question whether a robust evaluation could be provided), then the results shown in Figure 8.11 shows that the most impactful factor for which to invest resources to gather more data is transfer.

The ability to carry out a sensitivity analysis by resampling the underlying data is very versatile and can be applied to any event, regardless of the number of states the event is broken up into. The disadvantage is that generating these types of analyses requires some bespoke calculations in software such as Excel, or R (or any statistics or programming package from which random sampling from basic distributions is available).

8.2.5 SAMPLE THE DATA AND NOT THE TERM

Previous sections demonstrated the importance of considering the sensitivity of the *LR* to data rather than to a single factor. The importance of this was highlighted in Figure 8.4 using case 1 and the probabilities of DNA transfer from D to C or from AO to C (both being assigned probabilities from the same data). In that case the effect was seen as two probabilities should cancel out in the *LR*, and hence the *LR* is expected to be insensitive to the data used to assign those probabilities (which was observed as the horizontal line in Figure 8.4). Another effect of a correct sensitivity analysis, this time in relation to the resampling scheme, is that the more a dataset is reused in a chain of events, the more sensitive it is expected that the *LR* will be to it. This is demonstrated with an example that uses a BN to visualise the effect.

Consider the following example: The hands of a person accused of assault are swabbed, and DNA from only the accused and the victim is found. The prosecution stated that the accused touched the victim directly. The defence state that someone else touched the victim directly (call this person 1), then they shook hands with a second person (person 2), who shook hands with person 3, etc., until eventually someone in the chain (person $N - 1$) shook hands with the accused. Setting up the BN in the manner described in [15] yields the structure shown in Figure 8.12.

Within each of the 'C DNA transfer from Pn hands to Pn+1 hands from contact' nodes, only presence or absence states are considered. Imagine an in-house study is carried out to determine the rate of DNA transfer between hands and find that out of eight hand-shaking experiments, DNA was not detected. This would lead to the conditional probability table shown in Table 8.2.

Note that in the BN the simplification is made that there is no consideration given to the presence of any DNA other than C's. It is possible to consider the presence or absence of unknown DNA (and it would in fact exacerbate the effect being described); however, it is not required to demonstrate the point.

The BN has been set up so that the proposition node has two states (Hp and Hd) with probabilities of 0.5 for each. The 'DNA on C hands' represents whether or not C's DNA is on their own hands (and for the purposes of the example this node was set to have states yes and no with probabilities of 1 and 0, respectively). The 'C DNA on D hands' is a pure accumulation node. Then the *LR* is:

- 1 – if the defence proposition requires one transfer event (i.e. also a direct contact between C and D).
- 10 – if the defence proposition requires two transfer events in the chain (i.e. through one intermediate person, P1).
- 100 – for three transfer events in the Hd chain.
- 10^{N-1} – for N transfer events in the Hd chain.

Note that if the chain of transfers required by Hd is very complex the *LR* being reported could become extremely large (e.g. billions or more), but it is being informed on only ten empirical observations. An increase in the *LR* by a factor of 10 for each additional transfer step required under Hd is expected, and the correct outcome, given the assumptions that are used in the described evaluation.

The impact of reusing data in this way can be demonstrated by carrying out a sensitivity analysis in a manner as described in [11]. Consider the sensitivity of the LR to increasing numbers of transfer events being invoked by Hd, in two ways:

1) First, consider the effect if the data underlying each node was resampled separately. Note that this is not the correct manner to carry out a sensitivity analysis in this case as the data has all come from one source. Such a sampling scheme would be appropriate if all the data sources were separate.
2) Second, consider resampling the original observations once and then applying that to all appropriate nodes in the network.

Figure 8.13 shows the distribution of LR values obtained from resampling under the two schemes described above.

Note a few important points that can be taken from the results in Figure 8.13. Using either sampling scheme, the point estimate for the LR increases by a factor of 10 for each person added to the chain of transfers under Hd. Again, this is the expected, and desired, behaviour of the LR.

The important point here is that sensitivity analysis is carried out on the data and not the nodes of a BN. When the correct sampling scheme is used (sampling scheme 2 in Figure 8.13 right), the LR is highly sensitive to the underlying data. By the time there are five people in the transfer chain,

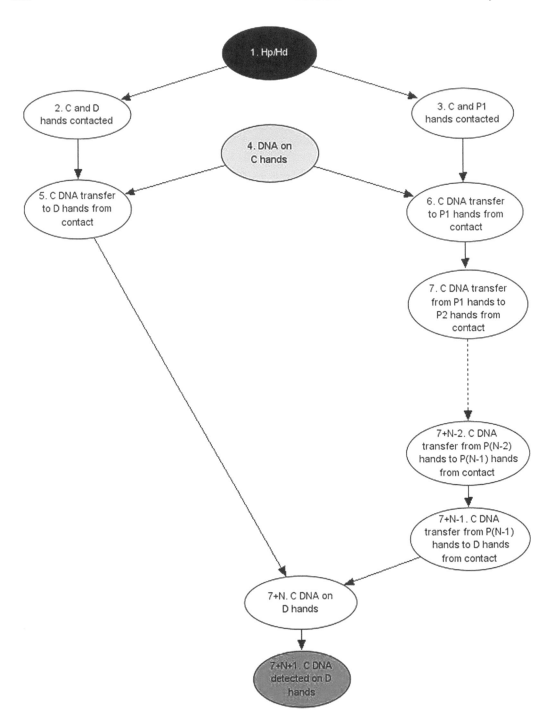

FIGURE 8.12 BN for the scenario which requires a primary transfer under the prosecution scenario and a complex N-step transfer under the defence scenario.

TABLE 8.2

Conditional Probability Table with Probability Assignments

Victim's DNA on Y hand from contact between X and Y		Yes	No
Victim's DNA on Z hand from contact between Y and Z	Yes	$(0 + 1)/(8 + 2)$ $= 0.1$	0
	No	$(8 + 1)/(8 + 2)$ $= 0.9$	1

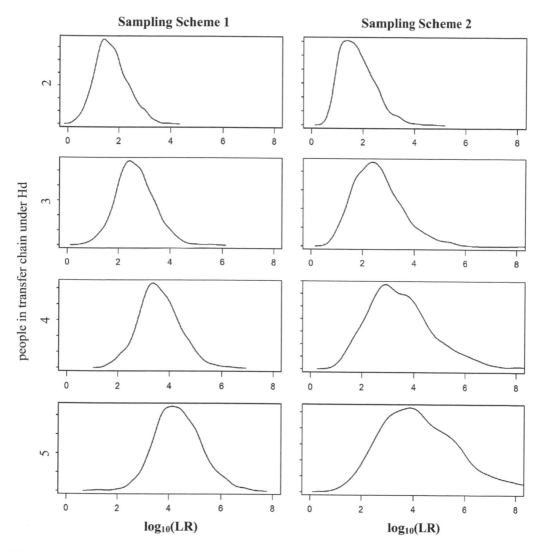

FIGURE 8.13   Sensitivity of *LR* to data underlying nodes when each node is resampled separately (scheme 1, left) and when the data is resampled once and applied to all nodes (scheme 2, right) for a range of transfers steps invoked under Hd.

99% of the distribution is within log10(LR) of 1.5 and 9.9, as opposed to 99% being between 0.7 and 4.0 when only two people are in the chain.

8.2.6 Sample the Term and Not the Data

Up to this point, the ethos behind carrying out sensitivity analyses has been described as an exploration of the effect of the underlying data (or uncertainty about that data) on the LR. In this ethos, if data is used in more than one way, then the sensitivity analysis is carried out by changing all probabilities assigned from that data source. Having made the importance of this fact through several examples, it is now time to reveal that there are situations where the sensitivity of the LR to individual probabilities (all informed from the same underlying data) may be examined. This is most often carried out in an investigative sense, to determine where further experimentation may be of greatest benefit. Take for example the formula used earlier in Section 8.2.2 for case 1:

$$LR = \frac{\left[\Pr\left(T_{D\to C} \mid {}^{Activity}H_p\right)\Pr\left(\overline{P_{D\to C}} \mid {}^{Activity}H_p\right)+\Pr\left(\overline{T_{D\to C}} \mid {}^{Activity}H_p\right)\right]\times}{\Pr\left(T_{AO\to C} \mid {}^{Activity}H_d\right)\Pr\left(\overline{P_{AO\to C}} \mid {}^{Activity}H_d\right)+\Pr\left(\overline{T_{AO\to C}} \mid {}^{Activity}H_d\right)}$$

$$\frac{\left[\Pr\left(T_{C\to D} \mid {}^{Activity}H_p\right)\Pr\left(\overline{P_{C\to D}} \mid {}^{Activity}H_p\right)+\Pr\left(\overline{T_{C\to D}} \mid {}^{Activity}H_p\right)\right]}{}$$

There are three different activities represented in this formula, which may have different effects on the evaluation:

- D grabbing C,
- C hitting D, and
- AO grabbing C.

Recall that during the evaluation of case 1, the assumption was made that the probability of DNA being transferred from all three of these actions was the same, and consequently the same data could be used to inform the transfer probability assignments. Imagine that the forensic laboratory handling this case was able to carry out experimentation to better inform the probability assignments. The following experimentation is possible:

- D grabbing C – getting D to grab cloth a number of times in the manner described by the victim as being done by the offender,
- C hitting D – getting C to hit cloth a number of times in the manner she did to the offender, and
- AO grabbing C – getting random people from the population to grab cloth a number of times in the manner described by the victim as being done by the offender.

The question arises as to which aspect of the evaluation the LR is most sensitive. In this instance, prior to conducting any experiments, sensitivity analyses may be carried out for each of the probabilities separately, i.e.:

- Varying $\Pr\left(T_{D\to C} \mid {}^{Activity}H_p\right)$, while holding $\Pr\left(T_{C\to D} \mid {}^{Activity}H_p\right)$ and $\Pr\left(T_{AO\to C} \mid {}^{Activity}H_d\right)$ at their data determined values
- Varying $\Pr\left(T_{C\to D} \mid {}^{Activity}H_p\right)$, while holding $\Pr\left(T_{D\to C} \mid {}^{Activity}H_p\right)$ and $\Pr\left(T_{AO\to C} \mid {}^{Activity}H_d\right)$ at their data determined values
- Varying $\Pr\left(T_{AO\to C} \mid {}^{Activity}H_d\right)$, while holding $\Pr\left(T_{D\to C} \mid {}^{Activity}H_p\right)$ and $\Pr\left(T_{C\to D} \mid {}^{Activity}H_p\right)$ at their data determined values

Similarly, there may be instances when (in an investigative capacity) the sensitivity of the *LR* to some specific aspect of the evaluation is of interest. In this instance, the value of only some of the probabilities informed by a single dataset may be varied for the sensitivity analysis; however, it must be understood why this is being done, and how to interpret the results of the sensitivity analysis.

8.2.7 RESAMPLING MODELLED DATA

Up until this point, the application of sensitivity analyses has been to data that comes in the form of counts being used directly to assign probabilities. Depending on the type of sensitivity analysis carried out, they can be applied to factors that have either binary states, or multiple states, and can be applied to all aspects of the data simultaneously to which the *LR* relies, or on individual components of the data. However, it is not clear how to apply any of these methods when probabilities are assigned by modelling data with a distribution.

In Chapter 2, a fourth case was introduced (the gun in the laundry), and in the previous chapter, it was used as an example which used data to inform us of distributions relating to DNA amounts or the proportion of DNA being transferred. This data, as any other, is subject to sampling uncertainty, and the fewer points on which the modelling is made, the more sampling variation will have an effect on the sensitivity analysis.

Return now to the example given previously, whereby the probability of a random male being above 185 cm in height was sought. From population data, it is known that the height distribution for males was modelled by $N(177, 7.5)$ and the proportion of this distribution above 185 is 0.14. In the example given previously it was explained that without the knowledge of the model for the heights of males in the population, it was decided to assign the probability of a male being above 185 cm by direct random sampling of individuals from the population. In each sampling, the proportion of males above 185 cm was measured and used to assign the probability directly. An example of a set of ten samples was given, for which three were above 185 cm, and seven were below. The probability assigned for being above 185 cm was 0.33. An alternative to directly counting the number of observations above 185 cm in each set would be to model the ten observations with their own normal distribution and then assign a probability of drawing a random number from that distribution above 185. For the ten example draws previously given, the average is 181.26 and the standard deviation is 8.1. The probability of an individual having a height above 185 cm if heights were modelled by $N(181.26, 8.1)$ is 0.32. Remarkably close in this instance to the direct counting result. Figure 8.8 showed the distribution of probabilities that would have been assigned from multiple resamplings of ten individuals from the population. Now consider carrying out a similar exercise, but instead of assigning a probability from the count directly, the ten observations are used to model a new normal distribution and then the value is chosen that corresponds to the area above 185. The result is shown in Figure 8.14, where the 1000 resampled distributions have been overlaid and drawn in grey, the true distribution is shown in blue and the point at 185 cm is marked in red.

In Figure 8.14 the 1000 normal distribution curves are overlaid, and opaque, but careful inspection around the 185 cm point shows that some curves have almost their entire mass to the left of it (with only the smallest fraction of the right tail past the 185 cm mark), whereas for others the 185 cm mark sits almost at the half-way point.

A similar exercise is carried out and a histogram plot of the probabilities that would be assigned to males above 185 cm, when using the fitted normal distributions. The results are shown in Figure 8.15. Note the similarity to Figure 8.8. This is a good thing, as it means there is a method for carrying out a sensitivity analysis on modelled data that is equivalent to the sensitivity of the direct counting method previously described.

Now turn to the evaluation of case 4. Recall from the previous chapters:

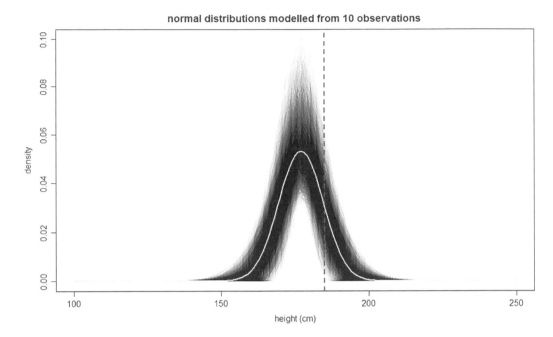

FIGURE 8.14   Overlaid normal distributions from 1000 resampling of 10 individuals.

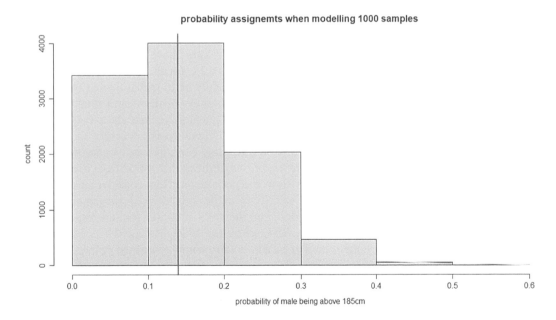

FIGURE 8.15   Probabilities assigned to a male being above 185 cm tall from 10,000 simulations of taking 10 random samples from the population and using those to model a normal distribution for heights. The red line shows the population level.

8.2.7.1 Propositions
- H1) D owned and handled the gun
- H2) An AO owned and handled the gun

8.2.7.2 Background
- The gun was found in the laundry bag of D

In this evaluation there were four sets of data used to model prior distributions for DNA amounts or proportion of DNA transfer. Specifically, these were:

- Six observations of DNA amounts from items of laundry that were tested in the forensic laboratory that were used to model a normal distribution for the amount of DNA available for transfer to the gun.
- Twelve observations of DNA amounts of owners of guns from the study by Gosch et al. [16] that were used to model a normal distribution for the amount of DNA of an owner on their own gun.
- Twelve observations of DNA amounts of unknown origin (or 'background DNA') on guns from the study by Gosch et al. [16] that were used to model a normal distribution for the amount of DNA of an unknown origin on a gun.
- Fourteen observations of DNA from pillowcases and 12 observations of DNA transferred from those pillowcases to hard objects from Carson et al. [17] that were used to model a beta distribution for the amount of DNA transferred from cloth to gun.

As with probabilities assigned using counts, the *LR* is also sensitive to probabilities assigned using modelled data. For example, some observations were modelled with a normal distribution that was then used to assign probabilities for the amount of DNA available for transfer from the laundry. Had these samples been different, then the normal distribution would have been different and then so too would the assigned probabilities have been different. If very few observations are used to fit the initial normal distribution, then the parameters of that distribution (i.e. the mean and the standard deviation) are expected to be affected by sampling variation to a larger extent than if many observations were used. This translates to greater variation in the assigned probabilities and ultimately the *LR* (as long as the *LR* is not completely insensitive to that data).

In the same way as just described on the height data, samples can be redrawn from the distributions that were used to fit the data, and then new distributions fit to the resampled data and the *LR* recalculated. This can be done for all datasets at once, or one at a time, and for each resampling, the *LR* can be calculated and stored in order to show the sensitivity of the *LR* to the data and modelling. Figure 8.16 shows the results of the sensitivity analysis.

A slight variation on this method would be to take draws randomly (with replacement) from the original set of observations and then model each set of drawn values with a distribution.

Figure 8.16 shows that the spread of values for the *LR* when considering its sensitivity to all data ranges from approximately 0 (neutral) into the millions ($\log_{10}(LR) = 6$) with regularity and up as high as 10^{10} in the simulation carried out. When considering the individual datasets, the component that is responsible for pushing the *LR* to lower values is the background DNA present on the gun. This makes sense as the background on the gun explains the presence of unknown DNA, if it has not come from an AO. If there is very little background DNA expected on the gun (much less than the amount of unknown DNA detected on the gun), the background becomes an improbable explanation and the alternative (direct handling by an AO) is favoured. There is very little sensitivity of the *LR* to the expected amount of DNA of the handler on the firearm or the transfer proportion from the

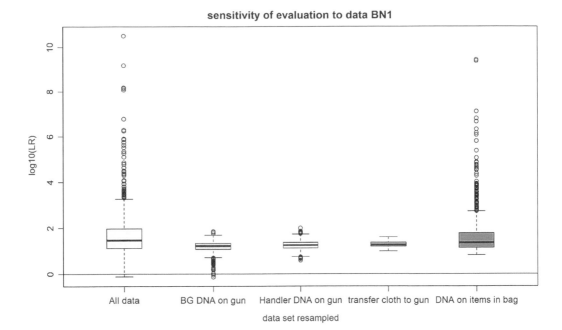

FIGURE 8.16   Sensitivity of *LR* to all data and individual components of the evaluation.

items of laundry to the firearm and this may partially be due to the fact that they each explain away the presence of D's DNA on the firearm under Hp. The aspect of the evaluation to which the *LR* is most sensitive is the amount of D's DNA on the items of laundry in the bag available for transfer. This is not unexpected given the low sample number that went into the fitted normal distribution. However, it is important to note two points:

1) The sensitivity of the *LR* to this dataset tends to be in the higher direction, i.e. resampling the data from these aspects of the evaluation can lead to quite high *LRs*, but not low *LRs* (and none that cause the *LR* to reach a neutral value).
2) The data being used to inform the amount of DNA available for transfer from the laundry items is different in nature to the other datasets used in the study. The difference lies in the fact that the distribution is being modelled from biological testing carried out as part of the case, rather than looking to model a transfer from a controlled study. As such, this provides data that is directly applicable to the case at hand.

To explain point 2 further, another architecture for the BN constructed for case 4 (see Chapter 7, Figure 7.13) would be to add nodes that consider the general amount of DNA expected to be found from people on their own laundry, informed by general studies that investigated the issue (i.e. not specific items from this case). These could be used to generate prior distributions for the mean and variation of DNA amount. Then results nodes would be added as child nodes to the mean and variance nodes and represent the amount of DNA on the specific items in case 4. Instantiating these items with the observed DNA amounts would update the distribution of their parent nodes (mean and variance for the general amount of DNA expected to be found from people on their own laundry), which would then be propagated throughout the rest of the network. During a sensitivity analysis, the literature-based mean and variance distribution could then be resampled, and not the distribution created from the case-specific observations. Due to a lack of literature information on the expected amounts of DNA from someone on their own laundry, in the evaluation for case 4, the case observations for DNA amounts on laundry were used for both purposes, i.e. the amount of DNA expected on laundry and also the case-specific observations. Or put another way the amount of DNA expected specifically from D on his own laundry was modelled.

However, the points discussed about aspect 2 above raise another important aspect of sensitivity analyses that can be explored. That is the sensitivity to the observed results in the case, which is discussed in the next section.

One final point to make in this section is that it may be the case that the probability has been assigned completely based on expert opinion, or partly based on data but adjusted based on a comparison of experimental design to case circumstances. In these instances, it is possible that there is not an easily identifiable data count or model to sample. When this occurs, the sensitivity analysis for that data will have to take the form of either varying the value of the state probabilities (such as described in Section 8.2.2 or 8.2.3) over a sensible range or considering an informed distribution for the potential values and integrating across the distribution (this type of calculation is discussed in more detail in Section 8.2.9).

8.2.8   SENSITIVITY TO THE RESULTS

There is much variation in the laboratory process that leads to a quantification result or DNA profile (see Bright et al. [18]). The results in case 4 will again be used as an example as they present the situation which best demonstrates the use of such an investigation into sensitivity analysis; however, the sensitivity of the LR to the observed result can be carried out in any evaluation. In essence this is done whenever a case assessment and interpretation is carried out during the triaging of a potential case (see Chapter 1), although the level of exploratory detail described in this chapter will not be carried out routinely in case evaluation or a CAI. An assessment of the sensitivity of the LR to case results is most easily carried out using a BN rather than a derived LR formula. The reason for this is that the derived LR formula will usually be derived out for a specific set of observed results, and testing multiple sets of potentially observed results will require multiple derivations.

In case 4, values for DNA amounts of D's DNA and U's DNA on the firearm are available which is based on the quantification of DNA extract from the swab of a gun, and the mixture proportions obtained from DNA profile analysis. The work by Bright et al. [18] demonstrated that there are stochastic effects present in all aspects of generating DNA profile data (and although not tested in their study, the same would be true of DNA extraction, and sampling a trace). It then seems reasonable that when an evaluative framework has been constructed that considers amounts of DNA (or other analogous observations on a continuous spectrum, such as the timing of a presumptive test, or the concentration of a chemical), an investigation into the sensitivity of the LR to small perturbations in those values is warranted. This may be particularly important if the laboratory values are close to the boundary that separates two discrete states in the discretised data (i.e. states in a BN). Put bluntly, if the evaluation was highly sensitive to the results obtained, and it is known that the results are stochastically affected to a large degree (and this fact isn't taken into account in the evaluation), then it may be felt that the evaluation is not robust.

There are a few different ways to address this issue in practice. If the number of observations is limited to one or two, then an equivalent practice can be carried out to the sensitivity analysis described at the beginning of this chapter, namely the results can be instantiated, starting at one extreme stated and varied through all states until the other extreme is reached, noting the LR at each step. Using case 4, the different potential results for the amount of D and U DNA obtained from the gun swab can be trialled to see how sensitive the LR is to the combination of results. The results of a sensitivity analysis carried out in this manner are shown in Figure 8.17.

Figure 8.17 shows a 3D surface created from all combinations of discretised states used in the BN for observed D or U DNA. The grey plane shows the plane of neutrality (i.e. where the LR does not provide support for either proposition) and the red point shows the case value. The results from Figure 8.17 show an expected trend given what was observed in the sensitivity analysis of

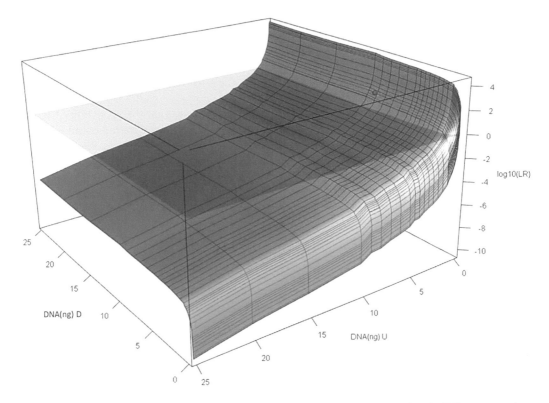

FIGURE 8.17 $\log_{10}(LR)$ obtained from different laboratory results for case 4 using the BN structure shown in Figure 7.13. The grey plane is drawn at $\log_{10}(LR) = 0$ and the red spot is the observed case values.

underlying data (which was given in Figure 8.16). When the unknown DNA amount is high, then the *LR* decreases (recall the discussion of the sensitivity of the *LR* to the BG DNA on guns factor in Figure 8.16). Also, the *LR* is less sensitive to the amount of D DNA (except when the levels of D DNA are very low). The only point at which the *LR* will change rapidly for a small change in the observed DNA amounts is when either the amount of D DNA or U DNA becomes very low, which is not the case in the observed results (although the level of U DNA is on the cusp of where the *LR* starts to become quite sensitive).

If there are many observations, then a similar analysis can be carried out to that shown in Figure 8.17, however varying one result at a time and producing a series of 2D plots. Carrying out sensitivity analyses in this manner has the same limitations as trialling all possible values for a probability within a sensitivity analysis, namely that much of the range will not sensibly apply. For example, in Figure 8.17 it is not reasonable to think that there is close to 0 ng of D's DNA in the sample, considering that the amount detected from the laboratory work was close to 9 ng. Note that the extension to this idea is that in some BN constructions, it does not even make sense to trial all possible outcome values as part of a sensitivity analysis, i.e. if the BN results nodes were discretised into presence or absence, or into categories such as 'D DNA', 'D + U DNA', etc. In these situations, the different results can be considered as diametrically opposing outcomes, so one may expect vastly different *LR*s from different instantiations, and this would not necessarily mean the evaluation was not robust.

An alternative to varying the results nodes over the entire range of states is to express the uncertainty in a value as a distribution. For example, it may be that the amount of DNA detected, on a log scale, can be described by a normal distribution, with the mean at the value detected, and the standard deviation set to some level that reflects the stochastic variability in quantitation within the laboratory. Note that unlike the previous sampling scheme, in which Dirichlet distributions were used, the results nodes will not necessarily be modelled by a particular distribution. Instead, the distribution that best reflects the analyst's uncertainty will need to be assigned, based on experience and data such as validation studies that investigate variability. In the example of case 4 the amount of D's DNA detected was 8.5 ng and for unknown DNA was 0.5 ng. Imagine that from repeated quantification validation work, it is known that a sample quantification can be modelled on a $\log_{10}$ scale (making the $\log_{10}$ DNA amount 0.93 for D's DNA) with a normal distribution and is expected to have a standard deviation of 0.1. A decision must be made on how to carry out the sensitivity analysis. Either:

1) Assume that there is complete dependence between the amount of DNA from D and the amount of DNA from U when sampling and generating a DNA profile. Therefore the amount of $\log_{10}$(ng) total DNA is modelled by an $N(0.95, 0.1)$. Then the amount of D's DNA is 93% of this sampled total, and the amount of unknown DNA is 7% of this sampled total. Knowing the total DNA distribution, random values can be drawn from it, and then the amounts of DNA from D and U determined and instantiated within the results nodes of the BN.

2) Assume that there is no dependence between the amount of DNA from D and the amount of DNA from U when sampling and generating a DNA profile. Therefore the amount of D's DNA in $\log_{10}$(ng) is modelled by an $N(0.93, 0.1)$ and the unknown DNA by an $N(-0.3, 0.1)$. Knowing these distributions, random values can be drawn from them and instantiated within the results nodes of the BN. Under these assumptions, the sensitivity of the LR can also be tested to the quantification result of both sources of DNA at once, or each one individually.

Reality is likely to fall somewhere between these two extremes, and therefore the resulting sensitivity analyses of both sets of assumptions is shown in Figure 8.18.

As expected, if the amounts of DNA from D and U are treated as completely independent from each other, then this maximises the sensitivity of the LR to the results. However, it is unlikely that these two values are independent, and it may be felt that the completely dependent scenario better reflects reality. The sensitivity of the LR to the results (taking into account the potential stochastic

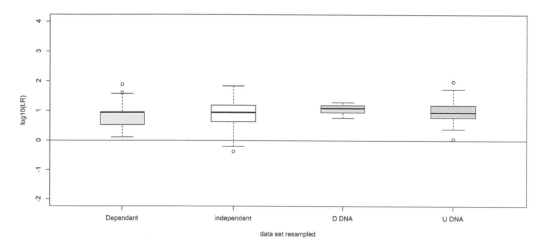

FIGURE 8.18 Sensitivity of the LR to results considering the two DNA amounts to have dependent mixture proportions (blue), or independent (yellow) and then if they are independent; the sensitivity of the LR to just the quantification of D's DNA (green) or just the U DNA (red).

variation expected from laboratory processes) is less than the sensitivity of the *LR* to the data underlying the probability assignments seen in Figure 8.16. Also, Figure 8.18 shows that when considered independently, the sensitivity of the *LR* is greater to unknown DNA than D DNA, which is reflective of the results seen in Figure 8.17 around the area of the case values.

The final point to make in this section is that, as an alternative to carrying out a sensitivity analysis of the *LR* to the results, the stochastic variability can be included within the evaluation and in effect, the variability is integrated to create a Bayes Factor rather than an *LR* (although common practice is still to call the value an *LR*). For case 4, for stochastic variability to be considered within the evaluation, the BN would include a node for the variability, which would be a parent of the DNA amount detected node. The BN construction is shown in Figure 8.19.

Note that the variance node in Figure 8.19 (node 14) could be a single value, or a distribution reflecting some laboratory calibration. If a single value is used, then node 14 could be emitted altogether and the variance value used in the expression terms within nodes 13 and 15. The other point to note is that the BN in Figure 8.19 has added two nodes (13 and 15) with DNA amounts expressed in a $\log_{10}$ scale. This is to apply the variance (which in this case is modelled for DNA on a $\log_{10}$ scale with a normal distribution). Also, if an independent variability between the amount of D and U DNA is assumed, then separate quantification variance nodes would need to be incorporated into the BN.

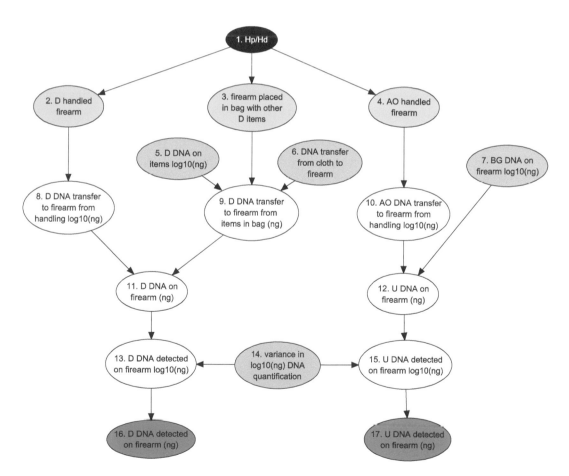

FIGURE 8.19 Updated BN structure from case 4 (original BN seen in Figure 7.13) showing the addition of structure for considering stochastic variability in the DNA amount detected.

A final note is that consideration will need to be given as to what practices will be most acceptable within the forensic laboratory, or to stakeholders. There are a number of factors of uncertainty that can be considered as distributions and integrated out of the evaluation, so as to produce a Bayes Factor. However, it may be of interest to the mandating authority how the evaluation is affected by different values for that factor, i.e. to see a graph of *LR* sensitivity rather than simply being told that all reasonable values (and their effect on the *LR*) have been taken into account in the reported *LR*. This latter practice (if taken too far) runs the risk of appearing as a mysterious black-box calculation.

8.2.9   SENSITIVITY TO THE UNKNOWN ASPECTS OF THE CASE CIRCUMSTANCES

This section follows on from the thoughts in the previous section but is subtly different to previous sensitivity analyses. The previous section considered the potential stochastic variability that might be present when observing the results, and how sensitive the *LR* was to that lab process variability. This is a type of uncertainty about the framework of circumstances, but one which is generated in the laboratory. The framework of circumstances however extends beyond the laboratory and applies to matters in the case. For example, a model could have been introduced for case 4 that relates not only the transfer proportion from cloth to metal but also how the amount of time the two have been in contact contributes to the transfer proportion. It may not be known exactly how long the firearm had been in the laundry, but the best estimates (from information about when washing was last done) might be 2 days but could be anywhere from 1 to 3 days.

Again, this can be treated in a couple of ways, which are similar to the methods previously described:

1) Develop a distribution for the uncertainty around the timeframe, which adequately portrays the level of uncertainty in this aspect of the framework of circumstances in the case. Use the best estimate from case information for the time, but carry out a sensitivity analysis by drawing times from the distribution.
2) Model the amount of time the firearm was in the laundry with its own node. In this node, the states that reflect the amount of time the gun was in contact with the laundry would have probabilities assigned in accordance with the distribution that portrays the level of uncertainty in this aspect of the framework of circumstances in the case. This uncertainty is then incorporated into the evaluation and technically this calculates a Bayes Factor rather than an *LR*.

It is often the case that for each aspect of the evaluation, there will be uncertainty in both the framework of circumstances in the case and also the sampling variation (or uncertainty) in the data from the literature. In the example just provided the uncertainty in the framework of circumstances is the amount of time the gun was in contact with the laundry; however, the data that was used to develop the model relating time to transfer amount is subject to sampling uncertainty or modelling uncertainty. Each of these two aspects can be investigated individually or together in the sensitivity analysis.

8.2.10   SENSITIVITY TO THE CHOICE OF DISCRETISATION

Another aspect of evaluations that has not been extensively explored is the sensitivity of the *LR* to the choice of discretisation of continuous variables. For example, if the amount of DNA is modelled, on a $\log_{10}$ scale, a range of values may be chosen to span from -3 to 3, stepping in levels of 1, so that the brackets of $\log_{10}(DNA)$ amount are:

- -3 to -2
- -2 to -1

- −1 to 0
- 0 to 1
- 1 to 2
- 2 to 3

A probability would then be assigned to each of these states that reflected the area under the normal distribution being used to describe the DNA amount between the extremes of each bracket. However, the choice could have equally applicably been made to break the range of values up into:

- Brackets of 0.1,
- Brackets of 0.001,
- A continuous distribution may have been used (if possible),
- Non-numerical categories of high/low/none which were then defined by bespoke ranges (for example, none = $[-\infty, -2]$, low = $[-2, 1]$ and high = $[1, \infty]$),
- Binary choice of present or absent around a low $\log_{10}(DNA)$ value such as −2.

Each of these different discretisation schemes will have an effect on the *LR*. Consider that any discretisation scheme is just a way to represent an underlying, continuous variable. Therefore, very small changes in fine-scale discretisation schemes are unlikely to have much effect on the *LR* (the *LR* would not be sensitive to such modelling choices). However, as the discretisation moves further from continuous towards binary, the changes in discretisation may have a larger effect on the *LR* (i.e. the *LR* would be sensitive to such modelling choices). This is particularly true if the observed case values approach the boundaries of a bracket discretisation value. In this case it would be expected that different discretisation would have their greatest effect on the sensitivity of the *LR*. For example, imagine that the amount of DNA observed from D was 2.25 ng and that the DNA amounts in the evaluation were broken into high, low or none. A choice between a categorisation for a high of greater than 20 ng or greater than 30 ng will not have much effect on the evaluation. However, a choice between a categorisation for a high of greater than 2 ng or greater than 2.5 ng could have a dramatic effect on the evaluation.

As a general rule of thumb (i.e. in general, but not for every case), the more coarse the discretisation of the factor, the less power there will be in the evaluation to distinguish between propositions. This comes from the fact that more coarse discretisation acts reduce the amount of information being used, which can only act to reduce discrimination capability. If a factor can be described in a continuous manner, which is true of most forensic observations (e.g. amounts of DNA, timing or strength of a presumptive body-fluid test result, level of degradation), then the most informative evaluation will occur when the variables are treated as continuous, i.e. not discretised at all. However, this treatment is often complex, and the paucity of available data to inform the models means that in reality, a quite coarse discretisation of factors will have almost the same level of power to discriminate between propositions.

8.2.11 SENSITIVITY TO THE BN ARCHITECTURE

The final aspect of sensitivity analysis that will be considered is the sensitivity of the evaluation to the assumptions made within the model. This could also be thought of as the sensitivity of the evaluation to the BN structure (with the different structures coming from making different assumptions). It could also be thought of as the sensitivity of the evaluation to different analysts carrying them out (again with different analysts making different assumptions), although the assumption would have to be made that the analysts had (and chose to make use of) the same sources of data to assign probabilities.

Again, take case 4 as an example of the sensitivity of the *LR* to assumptions. Recall from Chapter 7 that the simplification was made that there was no consideration of unknown DNA being transferred from the clothing in the laundry to the gun. The underlying assumption that would describe this simplification is that it is assumed there is no unknown DNA transferred to the gun from the laundry. This was done so as to simplify the evaluation, but it may be of interest what effect it would have (or if it would have dramatically changed the *LR*) had the decision been made to increase the complexity of the evaluation to include unknown DNA being transferred from the laundry to the gun.

The architecture of the BN from Chapter 7 (seen in Figure 7.13) is adjusted to include the possibility of unknown DNA being present on the clothing in the laundry bag and transferring to the firearm. Two nodes are added that act in the same way as the D DNA on items and transfer probability act to dictate how much of D's DNA is transferred to the gun. The new BN construction is shown in Figure 8.20. This is just one formulation of the BN, and in this architecture, the idea has enforced that transfer of D DNA and U DNA is dependent on each other. In other words, if high proportions of available D DNA are transferred from the laundry to the gun, then so too will high proportions of available U DNA. This can be seen by the shared 'DNA transfer from cloth to firearm' node for the two sources of DNA. A dependency could also have been enforced between the available amount of D and U DNA on the items by having a parent node to both of these that represented the total amount of DNA, and which then had child nodes 5 and 7 in the BN in Figure 8.20. This was not done given the different proportions of D and U DNA on the different samples in the case.

The same evaluation is now carried out into the sensitivity of the *LR* to different results (which previously led to Figure 8.17) with the result shown in Figure 8.21. The results in Figure 8.21 show that the addition of BN architecture to include unknown DNA transferring from clothing to the gun mainly has an effect on the evaluation when large amounts of unknown DNA are present. Therefore, the *LR* is sensitive to this assumption in the zone of U DNA > 5 ng, but relatively insensitive below that. Again, this comes from the fact that the unknown DNA now has multiple paths in which its presence can be explained away. In case 4 only 0.5 ng of U DNA was observed and so *LR* for the case is not sensitive to this modelling assumption. This brings into focus the concept of the sensitivity analysis being for a particular result, i.e. even if the same BN was used in a different

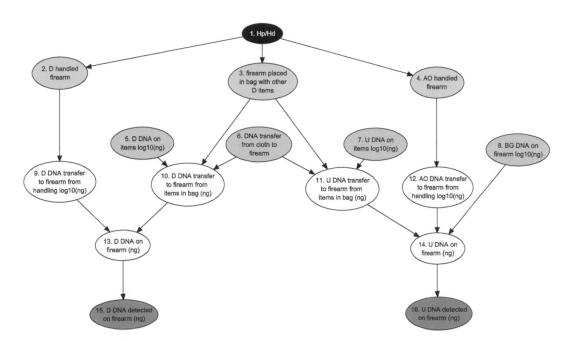

FIGURE 8.20 Updated BN structure from case 4 (original BN seen in Figure 7.13) showing the addition of structure for considering unknown DNA being transferred within the laundry bag.

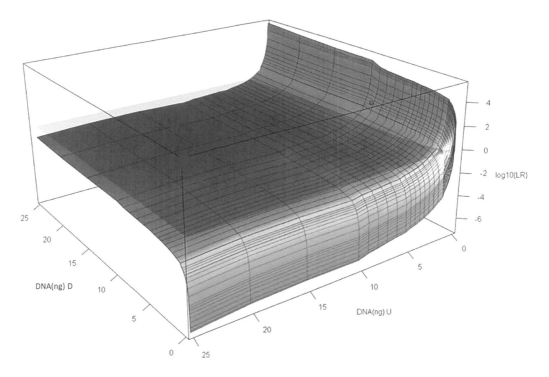

FIGURE 8.21 $Log_{10}(LR)$ obtained from different laboratory results for case 4 using the BN structure shown in Figure 7.13. The grey plane is drawn at $log_{10}(LR) = 0$ and the red spot is the observed case values.

case, the sensitivity analysis needs to be redone (sensitivity analyses are not translatable unless both evaluation and results are identical).

Having the new BN shown in Figure 8.20 and the range of *LR*s it would give for different results in Figure 8.21, it might be expected that the sensitivity of the *LR* to the background DNA on the gun will be less than for the simpler version of the BN. The reason for this is that an additional route is now present to explain away the presence of unknown DNA on the gun (other than from an AO or from the background). Carrying out the same sensitivity analysis as was done to produce Figure 8.16, but on the data that was used to inform probabilities in the BN shown in Figure 8.20, produces Figure 8.22.

As expected, the results in Figure 8.22 show us that the sensitivity of the *LR* to most of the components is less than it was in the simpler form of the BN. This is especially true for the background DNA, which now no longer crosses the *LR* line of neutrality. The only node for which there appears to be the same level of sensitivity (and which is the main driver of *LR* variation) is the amount of D DNA available for transfer from the items of clothing in the bag. As previously explained, this node is subtly different than others in the BN in that it is not built from models of controlled experiments but rather relates to casework observations.

Also note that there has been a slight increase in the *LR* by including the ability for unknown DNA to have transferred from the items of laundry to the gun. This is a relatively mild shift in the *LR*, which, as explained above, is not unexpected given the low levels of unknown DNA that were observed in the case.

There are numerous ways in which a typical evaluation can be extended or changed in order to avoid making assumptions or simplifications. In the same way as was mentioned in Chapter 6 (BN construction), the choice of how far to extend the evaluation will be a choice on data availability, and also some understanding of whether the *LR* will be sensitive to these potentially added factors. If the question exists as to whether an assumption is having a large effect on the *LR*, but data availability

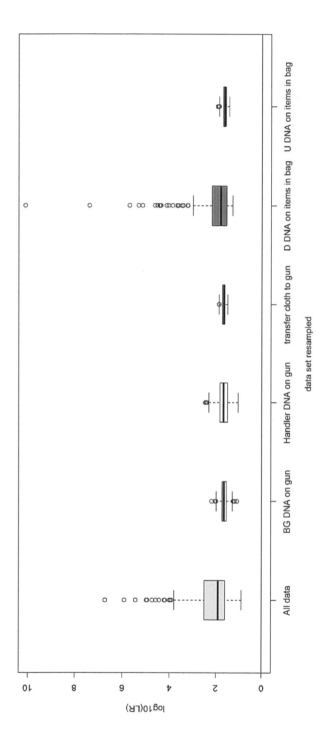

FIGURE 8.22   Sensitivity of *LR* to all data and individual components of the evaluation for the BN seen in Figure 8.20.

is an issue, and the sensitivity of the *LR* can be explored by trialling different potential distributions of probability on the new node.

A final note is that this type of sensitivity is not normally done in practice, but there are circumstances where the analyst may want to investigate the importance of their modelling choices.

8.3 SETTING UP SENSITIVITY ANALYSES IN A BAYESIAN NETWORK

The sections up to this point have been largely theoretical, dealing with the different conceptual ways in which sensitivity analyses can be carried out. The next section deals with how sensitivity analysis can be set up and carried out using a BN. The initial demonstration is carried out using case 1 to describe how to set up the sensitivity analysis. In the previous section it was seen how the sensitivity analysis could be carried out using the *LR* formula and by varying a probability over a range, or by resampling counts. When using a *BN* setup in a point-and-click interface-driven software, the easiest way to carry out a sensitivity analysis will be to vary the probabilities relating to different datasets across a reasonable range.

Table 8.3 shows the conditional probability table associated with the transfer of C's DNA from C to D's top that could occur from C hitting D. This is node 5 from the BN shown in Figure 6.22. There is one parent node with states Yes or No relating to whether D and C struggled. In Table 8.3 it can be seen that if D and C did not struggle then no DNA could be transferred, and the probabilities will not be varied when the parent state in No. However, data was used to inform us when assigning the probabilities when the parent state is Yes. One way to carry out the sensitivity analysis would be the manually change the values in this table with values for the node states:

- 0 for Yes and 1 for No
- 0.1 for Yes and 0.9 for No
- ... up to ...
- 1 for Yes and 0 for No

and for each conditional probability table set used, instantiate the case results and note the probabilities in the proposition node.

While this method would produce the desired result, it is a laborious and time-consuming way to carry out the evaluation and there are measures that can be taken in order to make the process easier. This section of the chapter describes those measures. Note that this may be different if setting up a sensitivity analysis for a BN constructed using a code-based software like the bnLearn package for R, as the changing of probability values within nodes can be done automatically within a code loop, but such code setups are beyond the scope of this book.

The general concept of setting up a BN for doing a sensitivity analysis involves:

1) Setting up the individual sensitivity nodes,
2) Setting up the case results node,

TABLE 8.3

The Original Conditional Probability Table for Node 5 from the BN Seen in Figure 6.22

D and C struggled		Yes	No
C DNA transfer from C to D's top from hitting	Yes	0.47	0
	No	0.53	1

3) Setting up the *LR* calculation node and

4) Carrying out the sensitivity analysis.

8.3.1 SETTING UP THE INDIVIDUAL SENSITIVITY NODES

The first stage of setting up a BN for an activity level evaluation is to identify all the nodes that have probability assigned which is based on a personal belief in an event occurring (hopefully also informed by some data). For each of these nodes, a sensitivity node is created that becomes a parent to the node with the conditional probability being explored. An example of this type of construction is shown in Figure 8.23, where a fragment of the BN is shown from case 1 (Figure 6.22). Because there are possibly going to be numerous sensitivity nodes within a BN setup in the manner described, each sensitivity node is labelled with the name 'sensitivity' and a prefix number that represents the node to which the probability being explored belongs.

The sensitivity node will possess a number of states that have values that span the range over which to vary the probability. The number of states used is up to personal preference, too few may fail to show the relationship between the *LR* and the data, but too many will simply cause more work than is required. An example of the node setup for the sensitivity is shown in Table 8.4, where 12 states have been designated. Eleven of the states in the sensitivity node in Table 8.4 are the values over which the probability will be varied: 0, 0.1, 0.2, 0.3, 0.4, 0.5, 0.6, 0.7, 0.8, 0.9 and 1.0. The final state is labelled 'case' and holds the value that was used in the evaluation of the case observations. The reason for including the 'case' state is to allow the easy investigation of multiple different nodes

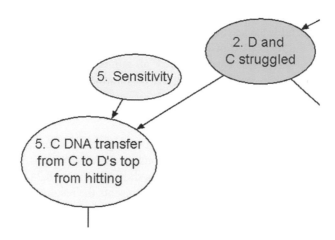

FIGURE 8.23 The BN structure for setting up a sensitivity node within a BN.

TABLE 8.4

The Conditional Probability Table for a General Sensitivity Node

Sensitivity		
	0	1/12
	0.1	1/12
	...	...
	0.9	1/12
	1	1/12
	case	1/12

TABLE 8.5

The Conditional Probability Table for Node 5 in the BN in Figure 6,22, Updated to Include the Sensitivity Node

D and C struggled		Yes					No			
Sensitivity		0	0.1	...	1	Case	0	...	1	Case
C DNA transfer from C to D's top	Yes	0	0.1	...	1	0.47	0	...	0	0
from hitting	No	1	0.9	...	0	0.53	1	...	1	1

within a BN, i.e. all but one of the sensitivity nodes can be instantiated with the 'case' category and the focus node instantiation can cycle through all values between 0 and 1. Then the focus node can change and the same process is repeated, until all nodes have had a chance to be the focus node.

Each state within the sensitivity node shown in Table 8.4 has an equal prior probability entered; however for the evaluation, this is not important, as at all times all sensitivity nodes will be instantiated with some value. The only prior probability value that may cause issues if entered into the conditional probability table is 0, as this will cause a number of exceptions if that state is instantiated. When looking at the conditional probability table for node 5 in Figure 8.23 with the sensitivity node parent, there are now transfer probability values required for each state of the sensitivity node (Table 8.5). If the event of D and C struggling has not occurred, then all probabilities are unchanged from the case values, i.e. transfer cannot have occurred. If the state of the parent node 'D and C struggled' is yes, then the value for the transfer state 'Yes' is assigned as the same value in the sensitivity parent node. For example, in table 8.5 see that when 'D and C struggled' is 'yes' and 'sensitivity' is 0.1, then the 'yes' state for the 'C DNA transfer from C to D's top from hitting' node is also 0.1. This manner of probability assignment is carried out for all numerical states of the parent 'sensitivity' node, and the final state 'case' is assigned the values used in the case evaluation.

8.3.2 SETTING UP THE CASE RESULTS NODE

Another time-saving measure that can be taken when setting up a BN for carrying out a sensitivity analysis is to add a single results node that represents all the observations in the case. This node becomes a child of all the individual results nodes, as shown in Figure 8.24.

The 'case results' node has states 'observed' that represents the results that were observed in the case and 'other' that represents all other possible combinations of results that were not observed (Table 8.6). The probability assignment for the 'case results' node is 1 assigned to the 'observed' category (and hence 0 for the 'other' category) for the combination of states that correspond to the observations in the case. Recall for case 1 the 'DNA on D's top' node result was that an unknown DNA was observed (state 'U'), and for the 'DNA on C's top' node, the case observations were that the friend's DNA was found (state 'F'). All other combinations of parent node states are assigned a probability of 0 for the 'observed' state of the 'case results' node and 1 for the 'other state'. When

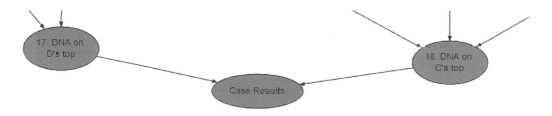

FIGURE 8.24   The BN structure for the combination of all results in a case into a single-case results node.

TABLE 8.6

The Conditional Probability Table for the Case Results Node

17. DNA on D's top	C + U		U	C	None
18. DNA on C's top	All states	F	All other states	All states	All states
Observed	0	1	0	0	0
Other	1	0	1	1	1

the 'observed' state of the 'case results' is instantiated, this is equivalent to instantiating each of the observed states in the results nodes individually.

Note that if only a single results node is present, then there is no need to create a case results node. Even when there are multiple results the time saving for creating the 'case results' node is minimal if the sensitivity analysis is carried out just once, and all at the same time. However, it is very common for small changes to be made during sensitivity analysis (either as changes that correct some component of modelling, or to fix data entry errors) and when the BN contains many result nodes, only having to instantiate a single node each time the sensitivity analysis is carried out can accumulate to save significant time.

8.3.3 SETTING UP THE *LR* CALCULATION NODE

The final alteration to the BN is one that automatically calculates an *LR* from the probabilities in the proposition node. The construction requires the addition of one node for each proposition and an *LR* node, as shown in Figure 8.25.

The individual proposition nodes are set up as numbered nodes so that the different states must be numerical in nature, and the numbers have meaning (as opposed to being treated as labels). The probabilities assigned to the states of the 'Hp' and 'Hd' nodes given the states of the proposition node are shown in Table 8.7.

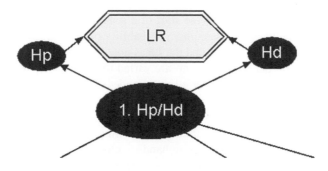

FIGURE 8.25 The BN structure for calculating the *LR*.

TABLE 8.7

The Conditional Probability Table for the Hp and Hd Nodes in the *LR* Calculation Structure

1. Hp/Hd	Hp	Hd	1. Hp/Hd	Hp	Hd
Hp	0	0	Hd	0	1	0
	1	1		1	0	1

The construction shown in Figure 8.25 and Table 8.7 will result in the 'Hp' node '1' state having a value equal to the 'Hp' state in the proposition node, and the 'Hd' node '1' state will take a value that is equal to the probability of the 'Hd' state in the proposition node. The final step is to add a function tool node, labelled '*LR*', as the child of the 'Hp' and 'Hd' nodes, and possessing the formula:

$$Hp/Hd$$

So that the value shown in the *LR* node is the *LR* calculated from the division of the probabilities of the 'Hp' state and 'Hd' state in the proposition node. It is also possible to instead get the *LR* node to calculate the inverse of the *LR*, if the formula:

$$Hd/Hp$$

was used. Note that both an *LR* and an inverse *LR* utility node could be added so that both values were automatically calculated. The advantage of having nodes that calculate these values automatically comes from the fact that it saves some time being able to note down the *LR* directly, rather than probabilities, which must then be divided. As with the other measures explained, the gain is minimal, but significant when hundreds of *LR*s need to be calculated.

8.3.4 CARRYING OUT THE SENSITIVITY ANALYSIS

Applying all the BN modifications as described in Sections 8.4.1–8.4.3 leads to case 1 BN (originally shown in Figure 6.22) shown in Figure 8.26. This BN is now ready for the sensitivity analysis to be carried out. To do so, the 'observed' states of the case results are instantiated, along with the 'case' state of all sensitivity nodes. Then, for whichever component of data to which the sensitivity of the *LR* is being tested, the probabilities are varied over the desired range, by instantiating the node states one increment at a time.

Remember from earlier that it is not a node to which the sensitivity of the *LR* is considered, but rather the data. Therefore, if the same data is used to assign multiple probabilities, then they should all be changed together. Figure 8.27 shows the BN for case 1, with selected conditional probability tables showing and nodes instantiated for the start of a sensitivity analysis on the Daly et al. [14] data, which was used as the basis for assigning a probability of DNA transfer from hands to cloth. This probability is varied from 0.1 to 0.9, and the *LR* is noted at each stage. Note that nodes 5, 6, 7 and 8 all assign the same probability and so all are varied together, so the BN seen in Figure 8.27 has probability values of 0.1 instantiated for all these nodes. The *LR* is 0.9, with the inverse being approximately 1.1.

As the probabilities are moved through from 0.1 to 0.9 in nodes 5–8 from the BN in Figure 8.27, the *LR* is noted at each value. Figure 8.28 shows the resulting inverse *LR* plotted against the value of the probability used for transfer of DNA from hands to clothing. The graph in Figure 8.27 shows the same values as when the sensitivity analysis was carried out using the formula (see Figure 8.2), which is the expected outcome, given that the BN is simply a graphical representation of an underlying formula.

8.3.5 THE SENSITIVITY NODE SETUP FOR THE FAMILY ASSAULT

The final aspect of setting up a BN for a sensitivity analysis being demonstrated is the application of distributed priors for high/low/none categories as were used in case 3. Figure 8.29 shows a small fragment of the BN shown in Figure 6.41, with the addition of a sensitivity node.

The sensitivity node shown in Figure 8.29 is now provided states such as 'extreme left', 'left', 'uniform', 'right' and 'extreme right', each with equal prior probabilities. The probability assignments for node 9 are then set up with values that represent these distributions of data. Examples of

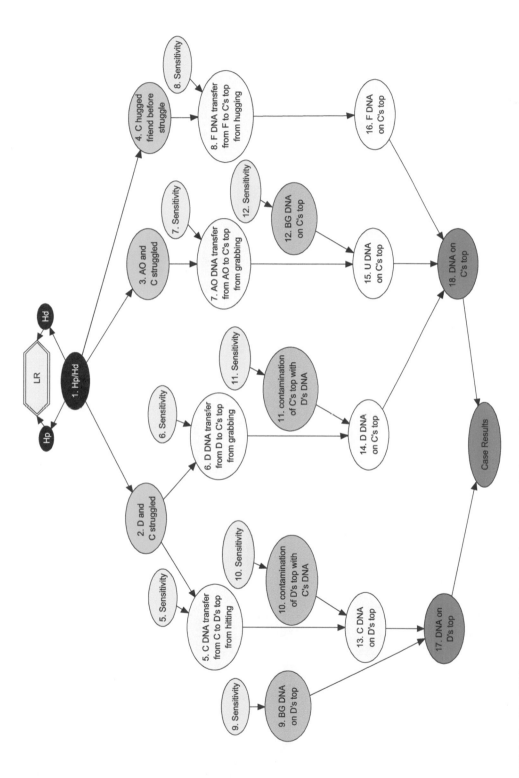

FIGURE 8.26   The BN for case 1 set up for carrying out a sensitivity analysis.

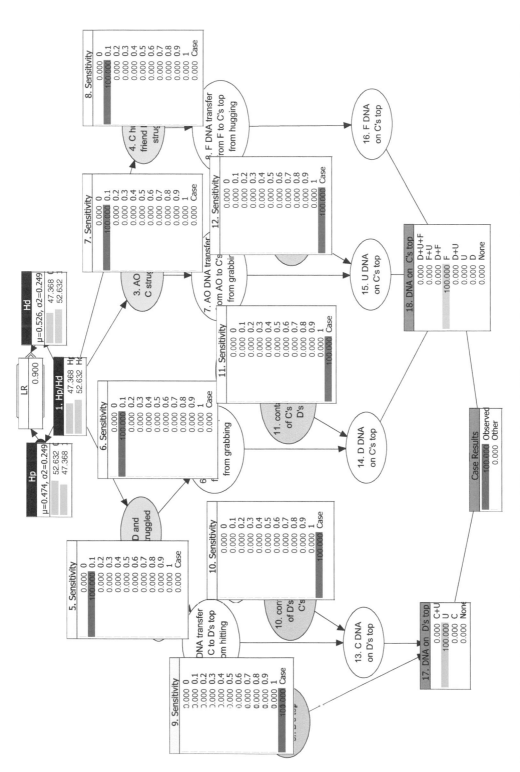

FIGURE 8.27 Instantiated form of sensitivity BN shown in 8.26, testing the probability of transfer from the hand to the cloth being tested (with the currently tested value being 0.1).

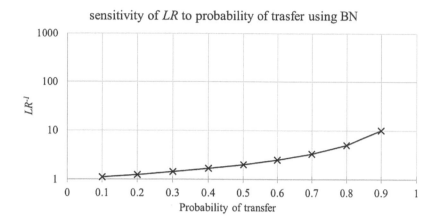

FIGURE 8.28 Resulting sensitivity graph produced from different values of probability of transfer.

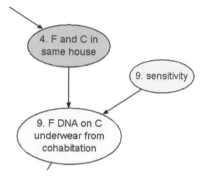

FIGURE 8.29 Fragment of BN shown in Figure 6.41 with a sensitivity node added.

TABLE 8.8

Conditional Probability Table for Node 9 Showing Assignments for Sensitivity Analysis (Only Values for State 'Yes' in Parent Node 4 Are Shown)

4. F and C in same house		Yes				
9. Sensitivity		**Extreme left**	**Left**	**Uniform**	**Right**	**Extreme right**
9. F DNA on C underwear	High	0.9997	0.6488	0.3333	0.0178	0.0000
from cohabitation	Low	0.0003	0.3333	0.3333	0.3333	0.0003
	None	0.0000	0.0178	0.3333	0.6488	0.9997

how distributions could be used to distribute probability were shown in Table 8.1, and those same values are used for states high/low/none in the transfer node. The probability assignments for node 9 are shown in Table 8.8 for the 'yes' state of the activity node 'F and C in the same house', noting that when the state of this parent is no, all probability is assigned to the 'none' state of node 9.

The sensitivity analysis is then carried out in the same way, with the different states of the sensitivity nodes iterated through and the *LR* recorded at each step. When carrying out a sensitivity analysis in this manner, there is no probability value to use for the *x*-axis and so a bar

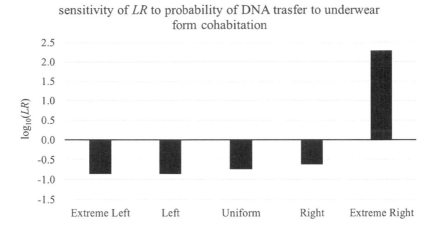

FIGURE 8.30   Sensitivity analysis for the *LR* in case 3 to the probability of DNA transfer to underwear from cohabitation.

chart is likely to be the more appropriate form of data display than a line graph (for example Figure 8.30).

The graph in Figure 8.30 shows the *LR* to be relatively insensitive to the probability of transfer of DNA to underwear from cohabitation, except when the 'extreme right'. Why could this be? The extreme right refers to a scenario where almost all the probability is assigned to 'none', i.e. no DNA transfer is expected 99.97% of the time. Therefore, within the evaluation, the much more reasonable explanation for low levels of the family YSTR profile on the underpants of the complaint is the Hp mechanism (biting) and so the *LR* dramatically increases. However, the extreme right probability assignments used in the evaluation are very extreme, and it is unlikely that people would reasonably expect such a high probability for no DNA to be transferred. This is another example of needing to consider sensible bounds within which to carry out a sensitivity analysis.

8.4   FINAL WORD ON SENSITIVITY ANALYSES

Carrying out a sensitivity analysis is an important task. It gives confidence that the evaluation is robust as well as knowledge to answer questions about how different factors affect the weight of evidence. Anecdotally, a sensitivity analysis is also a common time to find refinements within a BN or *LR* derivation. The sensitivity analysis can highlight when assumptions that have been made cause the *LR* to change in counter-intuitive ways. It also highlights when further investigation, or even bespoke experimentation, may be required to better inform a critical part of the evaluation.

But how far should an analyst go with sensitivity analyses in casework? This is an important question, and as this chapter has demonstrated, there are many different ways that a sensitivity analysis can be carried out, and many different aspects of sensitivity can be considered. Not all of these will be required, or practical, to carry out for every case. Some of the methods described are not possible without custom-written code to carry out the calculations required, and so will not be accessible to most forensic practitioners.

Ultimately the depth and breadth of sensitivity analysis carried out are similar to any other aspect of evaluation (i.e. the resolution of the data used, the assumptions made, the simplifications, the breadth of different transfer mechanisms). It will depend on the case and the assessment of the analyst as to what is required. It is likely that, at a minimum, a basic test of different probability values will be carried out whenever data is sparse. Questions that have arisen from prosecution or defence may also guide the analyst on the aspects which are likely to be explored in greater depth. Whatever testing is carried out beyond this is a personal and professional choice.

8.5 PRACTICE QUESTIONS

Q1) Imagine that a particular transfer event was experimentally found to occur in 3 of 100 instances. Using a Dirichlet distribution, resample the counts for this experiment and graph the distribution. As a hint to get started, you will sample the 3 observations 100 times from a Gamma(3, 1) distribution, and the 97 observations 100 times from a Gamma(97, 1) distribution. You will then add one to these drawn counts (to represent uniform prior belief) and normalise pairs of these counts to form probabilities. Note that 100 values can be drawn from a Gamma(N,1) distribution in Excel using the formula in 100 separate cells:

```
=GAMMA.INV(RAND(),N,1)
```

Or in R by using the command:

```
rgamma(100,N,1)
```

You will then create a histogram of the 100 values.

Q2) Now imagine that the transfer experiment in question 1 was able to carry out 1000 tests and found 30 transfers occur (note this is the same ratio of transfers to non-transfers as in experiment 1). Again, use a Dirichlet distribution to resample the counts for this new number of experiments and graph their distribution. How does it compare to the distribution from question 1? Why is it different?

Q3) In Chapter 6 a BN was developed for case 2: The three burglars (shown in Figure 6.33). Add sensitivity nodes (as discussed in Section 8.3) to the BN.

Q4) Using the sensitivity prepared BN in question 3, carry out a sensitivity analysis on the data underlying the different nodes in the BN.

Q5) Having completed the sensitivity analysis from question 4, what data are the *LR* the most sensitive to? Why do you think this is? With this information, do you think the *LR* is robust?

Q6) In Section 6.8 a BN was constructed for case 3, the family assault. Using the method described in Section 8.2.3, add sensitivity nodes to the BN.

Q7) Using the BN from question 6, explore the sensitivity of the *LR* to the amount of saliva transferred during biting and from cohabitation. What data are the *LR* the most sensitive to? Why do you think this is? With this information, do you think the *LR* is robust?

Q8) In Figures 7.24 and 7.25 a BN was shown that demonstrated an object-oriented architecture to evaluate the observations in case 1. Using this OOBN architecture, set up the BN for a sensitivity analysis of the data underlying the 'DNA transfer to top' node within the class network (shown in Figure 7.24). Using this sensitivity-enabled OOBN, carry out a sensitivity analysis and graph the *LR* for different values of the probability.

Q9) Using the BN for case 2 (in Figure 6.33, the three burglars), explore how sensitive is the *LR* to the presence of unknown DNA on the samples taken from the case. How would you explain this sensitivity to someone who was concerned with the robustness of the evaluation and the performance of forensic DNA profiling in general?

Q10) How would you expand the BN in Figure 6.33 (for case 2) to take into account the fact that unknown DNA might be present in the samples, but undetected by the DNA profiling system (through stochastic effects), thereby taking the concerns raised in question 9 into account.

8.6 REFERENCES

1. D.A. Stoney, What made us ever think we could individualize using statistics, *Journal of the Forensic Science Society* 31(2) (1991) 197–199.
2. D. Taylor, D. Balding, How can courts take into account the uncertainty in a likelihood ratio?, *Forensic Science International: Genetics* 48 (2020) 102361.
3. B.d. Finetti, *Theory of Probability: A Critical Introductory Treatment*, Wiley, Hoboken, 2017.
4. G. Morrison, Special issue on measuring and reporting the precision of forensic likelihood ratios: Introduction to the debate, *Science & Justice* 5 (2016) 371–373.
5. G. Morrison, E. Enzinger, What should a forensic practitioner's likelihood ratio be?, *Science & Justice* 5 (2016) 374–379.
6. J. Curran, Admitting to uncertainty in the LR, *Science & Justice* 5 (2016) 380–382.
7. D. Ommen, C. Saunders, C. Neumann, An argument against presenting interval quantifications as a surrogate for the value of evidence, *Science & Justice* 5 (2016) 383–387.
8. C. Berger, K. Slooten, The LR does not exist, *Science & Justice* 5 (2016) 388–391.
9. A. Biedermann, S. Bozza, F. Taroni, C. Aitken, Reframing the debate: A question of probability, not of likelihood ratio, *Science & Justice* 5 (2016) 392–396.
10. A. van den Hout, I. Alberink, Posterior distribution for likelihood ratios in forensic science, *Science & Justice* 5 (2016) 397–401.
11. D. Taylor, T. Hicks, C. Champod, Using sensitivity analyses in Bayesian networks to highlight the impact of data paucity and direct future analyses: A contribution to the debate on measuring and reporting the precision of likelihood ratios, *Science & Justice* 56(5) (2016) 402–410.
12. A. Nordgaard, Comment on 'Dismissal of the illusion of uncertainty on the assessment of a likelihood ratio' by Taroni F., Bozza S., Biedermann A. and Aitken C., *Law, Probability and Risk* 15 (2016) 17–22.
13. F. Taroni, S. Bozza, A. Biedermann, C. Aitken, Dismissal of the illusion of uncertainty in the assessment of a likelihood ratio, *Law, Probability and Risk* 15 (2016) 1–16.
14. D.J. Daly, C. Murphy, S.D. McDermott, The transfer of touch DNA from hands to glass, fabric and wood, *Forensic Science International: Genetics* 6 (2013) 41–46.
15. D. Taylor, A. Biedermann, T. Hicks, C. Champod, A template for constructing Bayesian networks in forensic biology cases when considering activity level propositions, *Forensic Science International: Genetics* 33 (2018) 136–146.
16. A. Gosch, J. Euteneuer, J. Preuß-Wössner, C. Courts, DNA transfer to firearms in alternative realistic handling scenarios, *Forensic Science International: Genetics* 48 (2020) 102355.
17. S. Carson, L. Volgin, D. Abarno, D. Taylor, The potential for investigator-mediated contamination to occur during routine search activities, *Forensic Science Medicine and Pathology* 18(3) (2022) 299–310.
18. J.-A. Bright, K. Stevenson, J. Curran, J. Buckleton, The variability in likelihood ratios due to different mechanisms, *Forensic Science International: Genetics* 14 (2015) 187–190.

9 Cell Type Testing

Bas Kokshoorn

CONTENTS

9.1 INTRODUCTION

Much of the focus of this book, and the field of forensic genetics in general, has been on the use of DNA testing to address case-relevant questions. The focus on this one particular type of molecule is understandable, as the power of DNA is in the rapid identification of individuals who may be associated with the crime under investigation. Forensic DNA testing is therefore a very powerful investigative tool, and the presence of DNA of an individual on a particular item or location will generally lead to questions on how and when those traces have been deposited.

However, DNA is only one of the many biological traces that may be deposited by an individual. DNA may be deposited as part of intact cells, or from extracellular sources (so-called 'Cell Free DNA') [1]. Such DNA sources generally are embedded in a matrix of other biomolecules. The non-DNA content, be it protein, mRNA, or other molecular markers, may be used to investigate the type

DOI: 10.4324/9781003273189-9

of cellular material that was deposited [2], or even the time of deposition – for instance by looking at the relative degradation of different mRNA transcripts [3]. Both body fluids and tissues may consist of multiple cell types, an example of which is human blood containing amongst others red and various types of white blood cells. However, the forensic tests that are applied to identify them generally target one biomarker only, which is commonly selected to identify one specific cell type within the body fluid or tissue. Throughout this chapter, we will refer to 'cell type testing' for testing for body fluids and human tissue types.

Information on the type of cellular material may be highly informative when addressing an issue at the activity level. Consider for example a case where the complainant and defendant are known to have regular social interaction. The presence of DNA of the complainant on the fingers of the defendant may (depending on the case circumstances) not be able to distinguish between them hugging and having sexual contact (such as in a case of sexual assault by digital penetration of the vagina). Examining such traces and finding evidence to support the presence of vaginal epithelial cells in the sample, however, may provide some level of support for sexual assault over social interaction only.

In this chapter we will focus on techniques to determine the nature of the human biological material to address source-level issues. Beyond that, we will look at how these types of tests can be incorporated in evaluations of both DNA and cell type testing results, given activity-level propositions.

There are two broad applications for cell type testing:

1. Screening for stains that may contain DNA of a person of interest, and
2. Determining the nature of cellular material deposited by an individual.

Let us consider a case where a woman has been sexually assaulted by a male. The clothing of the woman is submitted for DNA testing with the primary purpose of identifying her assailant. In such a case we may consider (depending on the alleged activities performed by the perpetrator) screening the items of clothing for body fluid stains left by the male. These could be for instance saliva or seminal fluid. Screening for these stains, using forensic light sources or immunological or biochemical tests, may reveal locations on the clothing where we may expect to find cellular sources that are rich in DNA, potentially that of the assailant. Such stains may be sampled and subsequently submitted for DNA testing. Here we are not necessarily interested in the nature of the biological material but rather use this information to detect (rich) sources of DNA relevant for identification purposes.

If the nature of the biological material itself is at issue, which it may be if the nature of the activities is questioned, we will work at identifying the biological source of the DNA found. Rather than screening for sources of DNA, we may apply different testing strategies to inform ourselves on whether the sample contains one or more of the body fluids that could be related to sexual activity. As such the results from such testing strategies may become important when addressing issues at the activity level. Different methods for both screening for human biological traces and cell type testing are discussed in Sections 9.1.1 and 9.1.2, respectively.

Evidently the two applications of testing are not mutually exclusive. Screening methods will provide information on the type of cellular material that may be present in a sample, while some of the cell type testing methods are used for screening purposes. We do maintain the distinction as their intended purpose also in part dictates the relative sensitivity and specificity needed for the test. We will discuss this aspect in relation to the interpretation of test results in more detail in Section 9.2.

When analysing a sample, the integrity of the cellular material is lost. This is the purpose and consequence of the DNA, RNA or protein extraction processes. After the extraction, we are left with, for instance, a DNA extract and a fraction containing protein or RNA. In this process the link is lost between these components, which results in challenges to the interpretation of the findings, particularly the association of a cell type to a particular donor of DNA. This can be illustrated by a fictitious sample that tested positive for the presence of blood (and let us assume for now that the sample indeed contains blood) and that resulted in a mixed DNA profile of three or more contributors. It will be challenging to determine based on these findings whose blood is in the sample, from one contributor only (and which one), or from more than one? We discuss cell type association in Section 9.3.

In Section 9.4 we will discuss particular aspects that need to be considered when including cell type testing in evaluations of these and other findings, given activity-level propositions.

9.2   CELL TYPE TESTING

The focus of forensic biology has been on anything but DNA for most of its history. The very early anecdotal example of the murder in 11th century China which was given in the Preface for instance focussed on identifying the presence of blood on a potential murder weapon through a 'fly test'. An earlier example of 'blood stain pattern analysis' has been described from 2nd century Rome, the case of 'The Wall of Handprints' ('Paries palmatus' in 'Declamationes Maiores', attributed to the Roman legal scholar Quintilian). In this case a man was found stabbed to death in his bed. A trail of bloody handprints covers the wall all the way to the bedroom of his son, who was put on trial for the murder of his father. Defence counsel pointed to the stepmother of the boy, who was alleged to have made the handprints to incriminate her stepson in the murder. The case description suggests that the inferences on the body fluid type and their attribution to the source (the deceased) could be made based on contextual observations only.

Such visual observations of body fluid stains were supplemented by chemical analysis techniques that were developed for forensic purposes in the late 19th and 20th centuries. In 1901 Ziemke [4], for instance, described the use of an antigen test to determine whether an observed bloodstain was from a human or animal source. In the first half of the 20th century, a test for the attribution of a bloodstain to an individual was described by Boyd & Boyd [5]. This method of blood grouping later adapted to include other blood factors and thereby increasing the resolution of the test remained the major endeavour of forensic biology until the advent of forensic DNA typing in the 1980s. However, finding potentially relevant traces of body fluids is the first step in the investigative process.

9.2.1   Screening Methods

9.2.1.1   Non-invasive Screening

By non-invasive testing, we refer to tests that do not physically alter the state of the trace while they are being performed, or at least to an extent that it does not interfere with subsequent testing or analysis of the trace. Non-invasive testing is generally applied for screening purposes. This starts by simply looking at an item in daylight or with artificial lighting. Based on the colour of a stain, we might for instance infer the presence of blood or semen. Specific forensic light sources may be used to enhance our observation using specific filters that filter out certain wavelengths. Body fluids have specific characteristics that may cause certain wavelengths to be absorbed or cause fluorescence effects. Using for instance blue light at a 450 nm wavelength and a yellow filter, we may observe the fluorescence of saliva stains [6]. Another example is the use of light sources in the near infrared to visualise bloodstains on dark surfaces [7].

More advanced non-invasive spectroscopic techniques are based on X-ray analyses of a substrate. Macroscopic X-ray Fluorescence (MA-XRF) has been shown to detect stains of blood, semen, saliva, sweat and urine on fabric based on their elemental signatures [8]. Raman spectroscopy [9] is another such application for the non-invasive screening of items for body fluid stains.

9.2.1.2   Invasive Screening

Screening techniques are also available that do interact with the substrate and the stains present there. As these techniques generally alter the physical state in which the stains are in, we deem them to be invasive. An example of such an invasive technique used to screen fabrics for saliva is the amylase pressure test. Amylase is an enzyme that is present in human saliva in high concentrations. A wet piece of paper containing a substrate is applied to the surface that is being examined.

By applying pressure to the paper, the substrate is brought into contact with any stains containing the enzyme amylase. The technique is considered invasive as part of the trace is likely to be lost to the testing paper, while the biological material on the item is also moisturised which increases the risk of biological degradation of the trace.

Amylase, re-activated by the moisture, will convert the substrate on the paper into a blue dye which is visible with the naked eye. The blue staining on the paper will thus reveal the presence of potential saliva stains on the corresponding area of the fabric. A similar approach with a different substrate may be used to detect acid phosphatase in seminal fluid stains.

Other applications are available, among which are techniques used to visualise latent finger-marks. Vacuum metal deposition (VMD) has for instance been applied to detect latent DNA stains on fabrics [10]. Cyano acrylate fuming, commonly used to visualise latent fingermarks on non-porous substrates, is also used to direct sampling for latent DNA traces (as the presence of ridge skin impressions indicates contact between bare skin and a surface).

An example of a test that uses biochemical luminescence to detect latent blood staining is Luminol [11]. The chemical is sprayed in an aqueous solution on a surface. The chemical binds to haemoglobin and emits fluorescence when doing so. In a darkened space, such luminescence can be detected and it can function as a screening for potential blood staining.

9.2.2   Cell Type Testing Methods

There are a number of recent review papers that discuss body fluid and tissue identification methods [2, 12–14]. We refer to these studies for in-depth discussion of the technological aspects of these methods. Here we will briefly discuss the types of tests and provide some examples.

9.2.2.1   Protein-Based Tests

Most classic forensic tests are based on the presence of proteins that are present in high amounts in specific bodily fluids (and thus in body fluid stains). Such tests generally either test the presence of activity of an enzyme or detect the presence of (partial) protein (e.g. peptides).

An example of these approaches can be found in the enzyme 'Amylase', which we mentioned before is present in high amounts in the saliva. An amylase pressure test (for example the Phadebas® forensic press test) works by having a substrate converted by the active enzyme into a visible colour staining [15, 16]. ELISA tests are available (like the RSID™ Saliva test) which work by attaching an antibody to a specific epitope of amylase on a lateral flow strip [17]. These types of antibody tests do not require the enzyme to be activated.

Mass spectrometry as an application to cell type testing is also available [18]. With this tech-nology, a number of relevant proteins are detected through their amino acid sequence in a single analysis (e.g. 'proteomics' type approaches).

9.2.2.2   Biochemical Tests

The Tetrabase test is an example of a highly sensitive biochemical test for the presence of blood [19]. The test works by applying Tetrabase to a sample of a stain, followed by the application of bari-umperoxide. Any haemoglobin present will act as a catalyst in the chemical reaction between the tetrabase and peroxide, causing a visible blue colouration. The two-step nature of the test eliminates many potential false positives. The test is not specific to human blood, as other animal blood will also result in a positive test result.

9.2.2.3   DNA-Based Testing

The fact that the DNA sequence in all nucleated cells in our bodies is the same is foundational to forensic DNA testing. This means that DNA tested from a buccal swab can be compared to DNA taken from, for instance, a blood or semen stain. Epigenetic characteristics of the DNA may vary between cells and cell types. A cell may silence certain genes by binding proteins to the backbone of the DNA or by methylating cytosines in CpG dinucleotides, both of which will prevent transcription of the DNA. As different cells, with different functions, will silence different genes, this epigenetic information may be used to determine the cell type, and DNA methylation is generally used for this purpose [20]. A sample will be subjected to bi-sulphite conversion, a chemical process which will cause non-methylated cytosines to be converted into uracil. Sequencing the DNA for body fluid-specific targets then substitutes those uracil bases with thymines. By comparing the untreated DNA sequence to the bi-sulphite-treated DNA sequence, a methylation profile can be deduced. This methylation profile can then be compared to a library to infer the cell types present in a sample.

When a sample may contain spermatozoa, differential lysis may be applied to enrich one fraction of the DNA extract with DNA from the spermatozoa relative to a fraction of the DNA extract which contains DNA from other cellular sources. While the purpose of the differential lysis is to increase DNA typing success rates, the method can also be used to infer the presence of spermatozoa in a sample when there is an enrichment of the sperm fraction with male DNA [21].

9.2.2.4 RNA-Based Testing

The most commonly used RNA targets for forensic cell typing are messenger RNAs (mRNAs) [22–24]. These are transcripts from protein-coding genes. Other RNA targets that have been described for forensic use are smaller, regulatory RNAs, specifically micro-RNAs (miRNA), circular RNAs (CircRNA) and piwi-interacting RNAs (piRNAs) [23, 25–32]. RNA can be co-extracted during the DNA extraction process, resulting in separate DNA and RNA extracts. RNA targets for body fluids are commonly combined in a multiplex analysis to target multiple body fluid and cell types in a single analysis. Multiplexes have been described for the analysis of a number of forensically relevant body fluids (saliva, semen, blood, menstrual secretion) or mucosa tissue lining bodily orifices (nasal mucosa, vaginal epithelial cells, rectal epithelial cells). A multiplex kit for organ tissue has also been developed which can be used to infer the presence of liver, lung, kidney, skeletal muscle, larger muscle and central nervous system (e.g. brain, spinal cord) tissue in a sample [33].

9.2.2.5 Testing Based on Microbiome

A substantial portion of the human body weight consists of associated micro-organisms like (gut) bacteria, archaea, fungi and yeasts. It is well documented that different parts of the skin and internal parts of the body host differently composed 'ecosystems' of these micro-organisms [34]. These differences may be used in a forensic setting to determine the presence of most forensically relevant body fluids and cell types. Microbiome analyses generally make use of microarrays, which are glass slides covered with large numbers of different probes [35] that bind to specific microbial cellular proteins or DNA sequences. Another method applied to microbial analyses is Massively Parallel Sequencing [36].

9.3 EVALUATION OF CELL TYPE TESTING RESULTS

Tests that are used to determine the cellular nature of stains are often referred to as being 'indicative', 'presumptive' or 'confirmatory'. The meaning of these terms is unclear as they are usually not properly defined when used. A 'confirmatory test' suggests that such a test will provide absolute certainty on the presence or absence of the body fluid in the stain that was tested. From this follows that 'indicative' and 'presumptive tests' do not provide that level of absolute certainty. The term 'indicative' would suggest that the presence of the body fluid that is tested for is one explanation for a positive test result (but there may be other explanations, like a false positive). The term 'presumptive' suggests that, although the test has no absolute specificity, it may be 'presumed' that the body fluid is present when the test is positive.

Let us take a step back from this terminology and consider the statistical interpretation of a test. The application of the Bayesian theorem to the interpretation of test results was explained in Chapter 1 (Section 1.2). A classic example of statistical treatment of test results is a test for the

Human Immunodeficiency Virus (HIV) which is administered to an individual. Let's say the test is positive and the doctor wants to inform the patient of the probability that they actually have contracted the virus. We can calculate this probability using the following formula:

$$\Pr(HIV \mid pos) = \frac{\Pr(HIV)\Pr(pos \mid HIV)}{\Pr(HIV)\Pr(pos \mid HIV) + \Pr(not\ HIV)\Pr(pos \mid not\ HIV)}$$

where Pr(*HIV*|*pos*) is the posterior probability of the patient actually having the virus, i.e. the probability that the doctor is interested in. Pr(*HIV*) is the incidence of the virus in the relevant population. It is the prior probability of anybody with the relevant characteristics of the patient carrying HIV. Depending on the availability of data, this probability could be informed by the general population, the male or female proportion of the population, the proportion of the population of people who donate blood, etc. Pr(*not HIV*) is the inverse of Pr(*HIV*), the prior probability of anybody with the relevant characteristics of the patient *not* having HIV. Pr(*pos*|*HIV*) is the probability of a true positive, while Pr(*pos*|*not HIV*) is the probability of the test returning a false positive result.

Let us assume that the test that was administered has a sensitivity (which is Pr(*pos*|*HIV*)) of 0.97 (which means that the test will return a positive outcome in 97 out of 100 cases if the test condition – here the HIV – is actually present). The specificity of the test is 0.94, which results in Pr(*pos*|*not HIV*) = 1 – 0.94 = 0.06 (the test will give a *positive* outcome in 6 out of 100 cases if the test condition is *not* present).

The HIV infection prevalence in the population of the Netherlands is estimated to be 0.2%. If we assume this to be the relevant population for our patient, Pr(*HIV*) = 0.002.

With these probabilities, we can now calculate the probability that our patient has HIV:

$$\Pr(HIV \mid pos) = \frac{0.002 \times 0.97}{0.002 \times 0.97 + 0.998 \times 0.06} = \frac{0.00194}{0.06182} = 0.03 \text{ or } 3\%.$$

Although the test for HIV was positive, our patient has only a 3% chance of actually carrying the virus. This may seem to be counter-intuitive, as one would expect a test with high specificity and sensitivity (here 0.97 and 0.94, respectively) to accurately predict the presence of the condition. The test does have strong predictive power, but as the incidence rate for HIV in the relevant population is very low, the chance becomes very high that the test returns a false positive. The crucial lesson here is that although the test itself may be very powerful, the *posterior probability* of the test condition being present is also crucially dependent on the *prior probability* of the test condition occurring in the tested sample.

We can further illustrate this if we would assume that the patient is a 30-year-old resident of Botswana. As the HIV infection prevalence in Botswana for that age category is 24.8%, Pr(*HIV*) is 0.248 (and Pr(*not HIV*) 0.752). Pr(*HIV*|*pos*) now becomes 0.24056/0.28568 = 0.84. This patient, based on the very same test result with the very same sensitivity and specificity of the test, has an 84% probability of carrying the virus.

This very same principle of including the prior probability when inferring if a test condition is present holds for the cell type testing that is used in forensic biology. Let us consider a test for the presence of Prostate Specific Antigen (PSA) in a sample which was taken from the vaginal cavity of a complainant. We are interested in whether or not the seminal fluid is present in the sample and apply the test, which returns a positive result.

We can calculate the probability of seminal fluid being present given a positive test result using the same formula as that which we used to calculate the probability of the patient having HIV:

$$\Pr\left(seminal\ fluid \mid positive\ test\right) = \frac{\Pr\left(seminal\ fluid\right)\Pr\left(positive\ test \mid seminal\ fluid\right)}{\begin{bmatrix} \Pr\left(seminal\ fluid\right)\Pr\left(positive\ test \mid seminal\ fluid\right) + \\ \Pr\left(no\ seminal\ fluid\right)\Pr\left(positive\ test \mid no\ seminal\ fluid\right) \end{bmatrix}}$$

Based on a literature review, De Zoete et al. [37] have assigned a probability of 0.974 to Pr(*positive test|seminal fluid*). The probability of the test giving a positive result if no seminal fluid is present can be taken from the same review and we assign a probability of 0.017 when only vaginal secretion is present (with the exclusion of the presence of all other cell types or sources for false positive test results for the purpose of this example).

This leaves us to assign Pr(*seminal fluid*) and Pr(*no seminal fluid*), the prior probability of our complainant (not) having seminal fluid in her vaginal cavity prior to testing. As the relevant population would be women who made a complaint of sexual assault, we could assign this probability based on retrospective case file studies. Ingemann-Hansen et al. [38] reported 35% of samples taken from complainants of sexual assault to be positive for semen in a Danish population. In samples taken from a population of complainants of sexual assault, McGregor et al. [39] have reported 'sperm positive forensic results' from 100 out of 262 samples tested (38.2%) in a Canadian study. A study on such a population from Kenya [40] reported 43% of samples returning a positive test result.

If we would assume these three studies are representative of the population that our complainant came from, the posterior probabilities of seminal fluid being present in the sample are:

- Denmark 0.97
- Canada 0.97
- Kenya 0.98

We notice here that, because the prior probabilities for the seminal fluid being present are very similar in the three populations, the posterior probability is also very similar as the sensitivity and the specificity of the test are the same (and both quite high). While this makes sense for these intimate samples where we expect limited variation in the prevalence of traces across populations, this may be very different for samples taken from different items or surfaces. Consider for instance the prior probability of seminal fluid being present in a sample from a knife taken from a dish washer, a balaclava sold in a hardware store or underwear that has been worn for a week by a sexually active male teenager. Where the prior probability of seminal fluid being present in a sample from the first two items might (hopefully) be considered very low, it might be high for the latter. If we would, for argument's sake, assign a prior probability for the seminal fluid being present in the sample of 0.001 to the knife and 0.5 to the sample from the underwear, the posterior probability for the seminal fluid being present in the sample given a positive test result with the same test as before would be 0.05 and 0.98, respectively. Note that this example is merely used to illustrate the concept of cell type test interpretation. The scientist may not be in the best position to assign a prior probability. Whose role it is to do so is discussed in more detail in Section 9.3.2.

What we can take from these examples is that the prior probability of a body fluid being present in a sample is an important factor to consider when determining, based on a test result, whether or not the body fluid is actually present in the sample. If we thus consider that contextual information will need to be considered, and consider that *every test* has false positive and false negative results (e.g. the specificity and sensitivity of a test are never 100%), we can never state that a *test* is confirmatory. At best the scientist may make an informed decision (or take a leap-of-faith) and decide, based on the posterior probability, that the test results, in combination with contextual information on the sample, and considering 'utilities' like the impact of their decision [41], justify the conclusion that a body fluid is present in a sample. The term 'confirmatory test' may therefore be misleading, as a positive test result reported from a 'confirmatory test' could be understood by others as providing absolute certainty on the presence of the body fluid. By extension, there is no logical basis to refer to individual tests as either presumptive or indicative, as there are no confirmatory tests. This terminology is therefore better abandoned. The merits of a test need to be determined by its diagnostic value, not by the conclusions we wish to draw based on that.

There may be instances when the 'leap of faith' by a scientist that a particular cell type is present in a sample is justified. An example could be where the body of a person is found with clear wounds to the head. A red fluid is seen coming from the lacerations and pooling on the floor beside the head. The head of a cotton swab is dipped in the pool of liquid and subsequently tested for the presence of blood. The test is positive.

Now in such an instance one would readily be willing to conclude that the red liquid is blood (more often than not the test for blood would not even be performed). Note that the observations on the body and pool of liquid are contextual, and not part of the test that is performed. This again illustrates that the prior probability of a body fluid being present in a sample is a crucial factor in the decision-making process. We will discuss examples of the interpretation of cell type testing in the next section.

9.3.1   EXAMPLES OF MODELS FOR INTERPRETATION OF TEST RESULTS

The interpretation of cell type testing may be quite complex, but statistical models have been published on the interpretation of different types of tests. We will discuss some of them in more detail.

An early probabilistic approach to the interpretation of a forensic cell type test was published by Hooft and Van de Voorde [42]. They set out to determine the predictive value of a modified Zinc test and the acid phosphatase (AP) test for the presence of spermatozoa in a sample. To this end, they collected vaginal samples from 261 and 195 female volunteers in two study cohorts. Based on these samples, they calculated the positive and negative predictive values for the two tests separately, and for the two tests performed in a series or parallel. For the combined analysis of both test results, the authors considered one of the tests positive in the series as a positive series test result, while both tests need to be positive for a parallel test result.

The positive predictive value of a test is the posterior probability of the test condition being present given a positive test result, or Pr(*condition*|*positive test*). The negative predictive value similarly represents the posterior probability of true negatives, or Pr(*not condition*|*negative test*). These values differ from the sensitivity and specificity of a test, as these are Pr(*positive test*|*condition*) and Pr(*negative test*|*not condition*), respectively.

Based on their analysis of the data, the authors conclude that the positive predictive value of the tests and test combinations for the presence of spermatozoa in the sample is 0.87 (Zinc), 0.80 (AP), 0.78 (parallel) and 0.90 (serial). In other words, if for example both tests would be applied to a sample in sequence, and either one of them is positive, there is a 90% probability of the sample containing spermatozoa. These posterior probabilities were calculated using a 'neutral' prior probability of semen being present in the sample of 0.5. Similarly, the negative predictive values were calculated as 0.98, 0.89, 0.99 and 0.90, respectively. Using this approach, they suggested the tests could be used in laboratory processes to screen samples for the potential of semen being present before committing to more labour-intensive microscopic analysis.

One could also decide to use these data to report the weight of the evidence of the test results, given propositions specifying the presence or absence of seminal fluid. However, one would need to decide whether it is reasonable to assume a 'neutral' prior (the decision to use a specific prior is not neutral!). Another aspect is whether the scientist is the right person to assign this prior probability. The question that always needs to be asked when a probability needs to be assigned is 'Who would be in the best information position to assign this probability?' It is a common practice in forensic science to leave the assignment of the prior probability of propositions (be it at sub-source, source or activity level) to the trier of fact. As they have access to the entire case file, they are generally in the best position to assign these probabilities.

A recent example of the statistical interpretation of body fluid tests comes from mRNA profiling. Modern multiplexes contain multiple markers that show high expression in body fluids of forensic interest [26, 43–47]. Ypma et al. [47] for instance describe an mRNA multiplex designed to infer the presence of 9 body fluids using 15 mRNA markers. They use a multinomial logistic regression model to calculate a likelihood ratio given propositions:

H1: The sample contains at least one body fluid of interest (and possibly other body fluids).
H2: The sample contains none of the body fluids of interest (but possibly other body fluids).

As the purpose of the mRNA multiplex is to assess the presence of multiple body fluids, the propositions reflect the possibility of a mixture in a sample. We will discuss proposition setting to address source-level issues in more detail in Section 9.4.

Bayesian networks, including a helpful user interface for their use, have recently been developed to ease the interpretation of commonly used cell type tests for blood, semen and saliva by Samie et al. [48]. Prior to this, De Wolff et al. [49] constructed a Bayesian network to interpret the results from two commonly used tests for the presence of amylase: The amylase pressure test and the RSID™ Saliva test (Figure 9.1).

We show this network here and discuss its structure in some detail, as we will touch upon several relevant issues that need to be considered when assessing cell type testing results.

9.3.2 WHOSE PRIOR?

De Wolff et al. [49] have argued that, when it comes to assigning a prior probability to the presence of a body fluid in a sample, there may be instances where the scientist is in an equal or better information position than the trier of fact may be. Their argument on when to report the posterior probability for the presence of a body fluid rather than an LR is illustrated by an example situation:

Consider e.g. a case where a balaclava has been examined, and signs of wear have been found (i.e. human hairs on the inside). The fact that balaclavas are commonly worn over the head will reduce the issue of the prior probability of saliva being present primarily to the probabilities of transfer and persistence of saliva. The latter two probabilities are typically the domain of the forensic practitioner. We, therefore, suggest that […] he or she could report a posterior probability rather than the likelihood ratio in [these] type of cases.

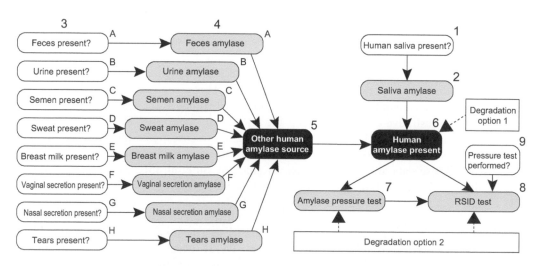

FIGURE 9.1 Bayesian network structure to address the issue of whether or not human saliva is present in a sample (reproduced from De Wolff et al. 2015).

While this example may intuitively make sense, one might overlook case-specific aspects that could be of relevance to the assignment of the prior probabilities. Information may, for instance, be available to the fact finder (and not to the scientist) that the balaclava was recovered from a barbershop and that it is uncertain whether it is connected to the incident under investigation because there are conflicting statements from witnesses. It is not the scientist who is in a position to assess this information, nor would they be expected to assign a prior probability based on this information (as this would require them to assess the merits of the witness statements). One could therefore argue that there will always be aspects to be considered in the prior probability that are outside the domain of the forensic scientist.

De Wolff et al. acknowledged this situation in a second example:

> However, in a sexual assault case concerning oral intercourse for instance […], the prior probability of the complainant's and suspect's allegations being true will have a major influence on the prior probability of saliva being present. When the alleged activities will strongly influence the prior probability, it may be best to restrict the conclusion to the likelihood ratio, and assist the court in the discussion concerning the prior probability.

(Note that the prior probability that needs to be assigned is on the presence or absence of human saliva and not on whose saliva it is.)

Aitken and Nordgaard [50] have discussed this issue in more detail and have demonstrated mathematically how different sets of background information can (and should) be used by the different actors in the decision-making process.

It is therefore generally advisable for the scientist to restrict their conclusion to providing a likelihood ratio (or Bayes factor) to the findings from cell type tests. They will, in doing so, need to guide the trier of fact on the relevant information they have incorporated in their assessment of the evidence. In Figure 9.1 prior probabilities would have to be assigned to the presence or absence of human saliva in the sample (node 1), but also to the presence or absence of all other body fluids that may contain amylase (nodes 3A to H).

9.3.3   What Are We Testing?

Nodes 7 and 8 in Figure 9.1 represent the amylase pressure test and RSID test results. They both have the same parent node 6, which defines the presence or absence of human amylase in the sample that is being tested. This structure should convey the message that both tests are tests for the presence of amylase, not of saliva. Although both tests may assist in determining whether or not saliva is present in a sample, they do not directly test for the presence of saliva. In that sense, the *amylase* pressure test is more aptly named than the RSID™ *Saliva* test. This aspect holds for any and all body fluid tests that are in use for forensic purposes. All tests are designed to detect one or more biological markers that may be present in a body fluid of interest but are not exclusive to that body fluid. Amylase for instance is found in saliva but also in other body fluids as illustrated by nodes 4A to H.

Another very illustrative example of this is the mRNA molecule STATH (Statherin). This biomarker is commonly used in forensic mRNA multiplexes to investigate the presence of saliva. Figure 9.2 shows the expression levels of the mRNA molecule in a panel of 79 human and 61 mouse tissues. These data are from a microarray study by Su et al. [51], which have been made available through BioGPS (http://ds.biogps.org/?dataset=GSE1133&gene=6779; accessed 8 May 2021). Figure 9.2A shows that high expression levels of STATH are observed in the human trachea, salivary gland, thyroid, prostate and thalamus. As the salivary gland produces saliva, high levels of STATH are observed in saliva. Needless to say, the other tissues are not directly related to human saliva, but may produce body fluids that also contain STATH (which is for instance also observed in nasal mucosa and vaginal secretion) [22, 23, 52, 53]. If we look at the same data on a logarithmic scale (Figure 9.2B), we see that all tissues (both human and mouse) express the biomarker to some

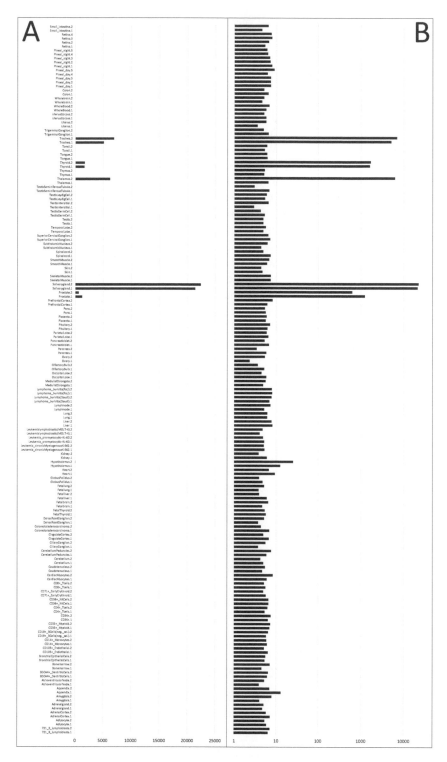

FIGURE 9.2 Expression levels of the mRNA molecule STATH in human and mouse tissues [51]. (A) Linear scale visualising the highest expression levels in human trachea, salivary gland, thyroid, prostate and thalamus. (B) Data shown on a Log10 scale, showing lower expression levels in all tissues tested.

degree. The marker itself is not body fluid specific nor is therefore the mRNA test based on this marker. The probability that a marker is expressed from a specific body fluid or tissue will depend on the sensitivity of the mRNA analysis, which will need to be considered in the statistical analysis of the data, given the relevant propositions.

9.3.4   Cross-Reactivity or False Positive Test Results?

Nodes 4A to H in Figure 9.1 define other sources of human amylase (coming from other body fluids). As the tests are defined by De Wolff et al. as tests to detect human amylase, these other human body fluid sources should not be considered causes for false positive test results. Note that this is a definition question. If we would have defined our tests as tests for human *saliva* rather than for human *amylase*, the other body fluid sources would need to be defined as potential causes for false positive test results. This means that the probability of a false positive test result is narrowly defined as coming from other, non-human sources. If we would define the tests as tests for the presence of amylase (human or non-human), causes for false positives that need to be considered would cross-react with other proteins or non-biological sources. While this could be a sensible definition for use of the amylase pressure test (which reacts to the presence of the active enzyme, regardless of its source), the RSID™ Saliva test is an antibody test that binds to a specific epitope of amylase that is found in humans and some primates, with the exclusion of other sources of amylase. It may therefore not be sensible to define the test as a generic test for amylase but rather for human amylase and consider potential primate amylase as a false positive.

9.3.5   Degradation of What?

As different tests act upon different biomarkers, we will need to consider the impact of degradation on the specific targets of the tests. Figure 9.1 shows two degradation options. Option 2 would assign different probabilities to the degradation of different aspects of the amylase. For the amylase pressure test, it would be defined as degradation of the active enzyme, while for the RSID™ saliva test, it would define degradation of the specific epitope to which the antibody binds. The level of degradation may be quite different, as a de-activated (damaged) enzyme which will be missed by the amylase pressure test may still have an intact epitope that is detected by the RSID test. Differential degradation specifically becomes important when multiple tests are used to infer the presence of body fluid.

The conditional probability tables under the nodes in the Bayesian network in Figure 9.1 require the assignment of probabilities based on data. As the study by De Wolff et al. has shown, there is limited data available from published sources. Extensive validation of the tests should remedy this, and interpretation models like these could provide guidance on the setup and scope of the validation studies to be performed.

9.4   CELL TYPE ATTRIBUTION TO DONOR

Testing for cell types in a sample and the interpretation of the test results is a crucial step to addressing source-level issues. However, source-level issues extend beyond the test for cell type. Cook et al. [54] provided two examples of source-level propositions in their paper on the hierarchy of propositions:

H1A: The semen came from Mr B
H2A: The semen came from some other man
H1B: The blood on Mr C's clothing came from Mr Z
H2B: The blood on Mr C's clothing came from an unknown person

In these examples the propositions address who the donor of a given body fluid (here semen or blood) is. Given these sets of propositions, the biological source (semen or blood) is not questioned, only the person who donated that body fluid is. These propositions would be relevant if the biological source is reasonably not in dispute, for example with a visible reddish-brown stain in an impact pattern, tested positive for blood, or a stain in underwear in which numerous intact human spermatozoa have been observed. However, as we have seen in the previous paragraph, such a decision is not always straightforward. Figure 9.3 shows graphically the three sources of uncertainty we need to consider when addressing source-level issues.

The attribution of DNA in a sample to an individual (issue 1 in Figure 9.3) is a core business in forensic genetics, and a multitude of statistical models [55–62] have been developed and implemented in daily casework practice globally. Issue 3, the inferences being made on the type of cellular material in a stain are supported by interpretation models as described in Section 9.2 of this chapter. Both of these issues need to be addressed probabilistically and for both the statistical approach and ground truth data are available to do so.

Issue 2, the attribution of the DNA in a sample to the same biological source as the target of the cell type test, has received less attention to date. The inference will generally need to be made by the expert based on their observations of a stain (clearly visible or latent, DNA amount as expected by the type of stain, single source or mixed sample, is masking of a minor contributor expected, biological gender of contributors for gender-specific cell types, etc.). As such, this issue is often the weak link in the inferential chain to address source-level issues. This attribution is not an issue with DNA methylation analysis to determine cell type as the DNA used for donor identification is also the target molecule for the cell type test. Some progress is also being made with mRNA typing. Sequencing of the mRNA targets allows for the analysis of single nucleotide polymorphisms (SNPs) which can be used to attribute the target to a specific contributor in a mixed stain [63]. These

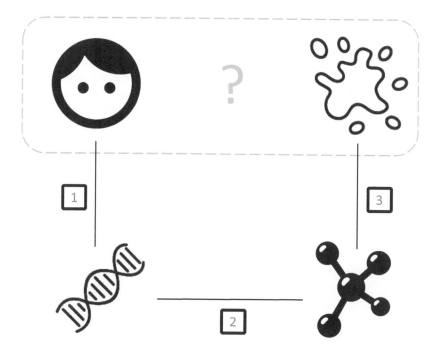

FIGURE 9.3 Three sources of uncertainty when addressing source-level issues. (1) The source of the DNA in a sample; can we attribute the DNA to a specific individual? (2) The association of the DNA and the target of the cell type test (be it a protein, mRNA, etc.) to the same biological source. (3) The nature of the cellular material; what does the target molecule tell us about the cell type present in a sample?

technical developments may in time assist in a more robust evaluation of forensic biology findings, given source-level propositions.

The example source-level propositions given by Cook et al. [54] consider the nature of the biological material (issue 3) as a given. The question remains whether the association to the DNA is also assumed to be a certainty (for instance when the result is a single source DNA profile and the amount of DNA fits the observed blood or semen stain and contamination with, or masking of, DNA is not considered), or whether the connection between DNA and cell type is uncertain (i.e. a mixed DNA profile of three individuals of the same biological gender).

In other instances the source of the DNA (1 in Figure 9.3) may not be in question, but the biological source of the DNA is. Table 9.1 shows different issues at the source level in which any or all of these three aspects are either being assumed or uncertain. The inference made on the source of a body fluid stain can thus be considered as a chain with three links. Any of these links may be uncertain or questioned, while others are not. The overall strength of the evidence will be determined by the weakest link in this chain.

Situation A in Table 9.1 is what was discussed in the previous paragraph. Assigning a weight of evidence to a cell type test result. This might come up as part of an attribution issue (i.e. evaluative situations E, F or H), or here as an investigative question on the nature of a stain.

Situation B is an issue that the example propositions by Cook et al. [54] might address. Neither the nature of the cellular material nor its attribution to the DNA found in the sample is at issue. This situation will most often be one where it is considered reasonable that the sample contains the biological material of a single individual, and the nature of the material is not in question because of contextual information and findings supporting this. An example could be a visible pattern of staining on a bed sheet on which the complainant stated that her assailant has ejaculated. A sample is taken, and upon microscopic examination, a large number of human spermatozoa are observed. The sample from the stain further results in a single source DNA profile of a male, with Y-STR markers not indicating the presence of DNA of more than one male. The DNA quantification fits what would be expected for a semen stain of the size sampled. Here the biological source of the stain may be considered certain (semen), and the attribution of the DNA to that biological source may not be at issue. As in this example, attribution of DNA to a biological source will be helped if the cell type is gender specific (semen, vaginal mucosa, menstrual secretion) and only one individual of that gender can be assumed to have contributed biological material to the sample.

Situation B is the only one where the carry-over of the likelihood ratio from sub-source to source level is valid (given the assumptions that are made), as the DNA association to the donor is the weakest link in the chain (the other two being considered an absolute certainty). In all other situations such a carry-over must be considered an association fallacy, specifically a 'source level fallacy' [64]. Such a fallacy will generally result in overstating the weight of the evidence, since the uncertainty on the biological source or the DNA attribution is overlooked.

The example propositions as given by Cook et al. [53] may also refer to situation C. Here too the source of the biological material is not in question. However, the attribution of DNA to the source is. This may happen if the sample contains a mixture of the DNA of multiple individuals, who cannot be associated with the observed biological source. Consider for example a sample resulting in a mixed DNA profile of three or more contributors. Given that the sample contains blood, attribution of that blood to any one of the contributors (or multiple contributors) may not be possible based on DNA profiling and cell typing information only. Here the relative amounts of DNA, size of the stain and particularly contextual information (who was wounded?) may assist in addressing this issue.

Intuitively one might also consider taking control samples from the area around the bloodstain to infer the presence of background DNA. This information might be considered in the assessment of what component of the DNA may have been derived from the blood and what component from the background DNA. While this is true, considerations of background DNA or prevalence of DNA are aspects that are best considered, given activity-level propositions. Or in a more general sense, if

TABLE 9.1

Different issues at the source level exemplified for a potential bloodstain

Issue	Which aspects are *not* questioned?	Relevant propositions
Does the sample contain blood? *Investigative issue*	A	H1: The sample contains blood H2: The sample does not contain blood (but possibly any other cell type) *Here we are only concerned with the diagnostic value of the cell type test.*
Whose blood is it? *Evaluative issue: Focus is on the stain*	B	H1: The blood is from X H2: The blood is from someone other than X *LR at the source level will be the same as LR for DNA at the sub-source level, as this is the only source of uncertainty.*
	C	H1: The blood is from X H2: The blood is from someone other than X *As the association between cell type and DNA is uncertain, LR at the source level will be equal to the LR for the weakest link.*
	D	*All three sources of uncertainty need to be accounted for. The question ('whose blood is it?') is not proper, given the uncertainty regarding the type of cellular material in the stain. Depending on the case circumstances, we are likely to either address A or G.*

(Continued)

TABLE 9.1 (CONTINUED)
Different issues at the source level exemplified for a potential bloodstain

Issue	Which aspects are *not* questioned?	Relevant propositions
Did X contribute blood? *Evaluative issue: Focus is on the PoI (X)*	E	H1: X contributed blood H2: X contributed any other cell type than blood *LR at the source level is the same as that for the cell type test result since the source of the biological material is not in question.*
	F	H1: X contributed blood H2: X contributed any other cell type than blood *LR at the source level will be equal to the LR for the weakest link.*
	G	H1: X contributed blood H2: X contributed any other cell type than blood *LR at the source level will be equal to the LR for the attribution.*
	H	H1: X contributed blood H2: X contributed any other cell type than blood H3: X did not contribute any biological material *A third proposition is required as X being a source of the material is also questioned.*

information from multiple samples is combined, this is best done, given propositions at the activity level.

Where the focus of the evaluation is on the source of the stain in situations A to C (D being an improper way of dealing with the evidence), situations E to H focus on the person of interest rather than the source of the stain. In situations E, F and G the person is assumed to have contributed DNA to the sample. The issue in these situations is what the type of cellular material is that the PoI has contributed. Situation E is one where the association of DNA and cell type test target molecule is not questioned (only the nature of the cellular material is questioned), while in situation F both the nature of the cellular material and its association to the DNA donor are at issue. In situation G it is only the attribution of the cell type to the known donor that is at issue. An example report of an evaluation of findings given source-level propositions as in situation E is provided in Box 9.1.

We have seen that three issues need to be addressed when making inferences on who is the source of a particular body fluid stain. These three issues can be considered serial evidence, with the elements in the series (the person of interest, the DNA, the cell type test target and the body fluid) being connected by 'links' to form a chain. The link with the lowest weight of evidence determines the strength of the entire chain.

In contrast to serial evidence, parallel evidence refers to multiple 'chains' connecting two elements. An example of both serial and parallel evidence with regard to inferences being made given source-level propositions is shown in Figure 9.4. Here we see the familiar 'chain' connecting the individual to the body fluid stain (serial evidence), with two cell type tests (or rather two different targets) being used to infer cell type. These two chains (2a + 3a vs 2b + 3b) connecting the DNA to the body fluid stain can be considered parallel evidence. The strength of chain 'A' is determined by the weakest of the links 2a or 3a, and the strength of chain 'B' by the weakest of 2b or 3b. The strength of the chain between the DNA and body fluid stain elements is determined by both chains A and B. If 2a is considered conditionally independent of 2b, and 3a of 3b, the weight of the evidence of both chains can be multiplied to calculate the overall strength of the evidence connecting the DNA to the body fluid stain. If one or more of these links are not conditionally independent, the overall strength can be calculated by taking this conditional dependence into account. This is generally very complex as data to support such calculations are often not available, as the conditional dependence between different cell type tests is not a common issue to investigate.

If data to support such inferences are lacking, a pragmatic approach would be to report the strongest of the two chains as a 'lower bound' for the combined LR. This concept of evidence schemes using serial and parallel evidence is developed (and the statistical background worked through) in De Koeijer et al. [65]. We will come back to this topic in Chapter 12 when we discuss combining evidence from multiple forensic disciplines.

We have seen and discussed some statistical models that have been developed to address cell type testing results (e.g. situation E in Table 9.1). Models have also been published that address the

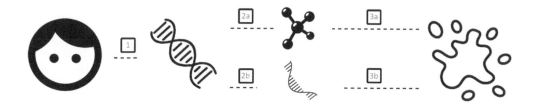

FIGURE 9.4 Interpretation of findings given source-level propositions when multiple tests for a body fluid are used. In this example two tests are used to infer the presence of one body fluid type.

issues of cell typing evidence and its association to a DNA donor for situations F, G and H, which we will briefly refer to here.

Taylor [66] addressed the DNA attribution to spermatozoa in a situation where differential extraction is performed. This is an example of the treatment of issue G in Table 9.1. A Bayesian network was constructed which was used to infer whether a known male person of interest has contributed to the assumed presence of semen in a sample. To this end, the results from the macroscopic analysis of slides made from the sample of interest, DNA typing results and DNA quantities from both fractions of the differential lysis were included in the evaluation. Based on this study, it is concluded that particularly the aspects of activities proposed in the context of the case impact strongly the weight of the evidence that is calculated. This is due to the importance of considering other sources of DNA being present in the sample from the person of interest (e.g. prevalent DNA).

A further study by Taylor et al. [67] addressed the issue of blood attribution to a contributor in a mixed sample as in situation F in Table 9.1. In their Bayesian network construction, they included visual observations on the stain and two tests (HemaStix and HemaTrace) as results for cell type testing. DNA amount and DNA typing results were further included to address whether or not the person of interest had contributed blood to a sample.

De Zoete et al. [37] took a similar approach but looked at the issue of attribution of multiple cell types in a sample. Specifically, they included results from four cell type tests in their Bayesian network construction, one for blood, two for seminal fluid and one for saliva. The evaluative framework fits situation H as shown in Table 9.1 but is limited to assumed two-person mixtures, with each individual contributing only one cell type. While limited in scope, they worked out a mock case example of an oral sexual assault in which the victim spat out semen on the street. A sample from this stain fitted the scope of the framework for interpretation. Examples of sensitivity analyses on data sources for probability assignments (as discussed in Chapter 8 of this book) are also given.

BOX 9.1   EXAMPLE REPORT ON EVALUATION, GIVEN SOURCE-LEVEL PROPOSITIONS

A woman was assaulted in a park. She is hit over the head and loses consciousness. After waking up, she files a report with the police. The case is treated as a robbery, and no forensic examination of the victim takes place. Her clothing is seized for forensic examination. A suspect is arrested a few days after the incident and samples are taken from his fingernails. At this point, a suspicion has arisen that the assault might have had a sexual component, as semen stains have been found on the outside of her clothing. The issue at this point becomes whether or not penetration of the complainant's body has taken place. The suspect concedes that he did rob the victim but denies any sexual activity. The samples from his fingernails were submitted to both DNA and mRNA profiling. Samples from two fingers resulted in a mixed DNA profile of the suspect, and possibly the complainant (LR over 10 billion supporting inclusion of complainant). For the purpose of this example, the other samples and their results are left out of scope. The interpretation of the DNA and mRNA typing results for the two samples were reported as follows:

ASSUMPTIONS

From samples A (under nails left-hand suspect) and B (upper side nails left-hand suspect), DNA has been recovered that can be attributed to the suspect, as well as DNA that may be from the victim [*Note that the findings from the DNA analyses and associated evaluation will have been discussed in a separate section of the report that is not given as part of Box 9.1.*] Based on the DNA typing results, I have no reason to assume that the samples contain DNA of more than two individuals.

- Based on the high weight of the evidence supporting the proposition that the victim is a contributor, and because the suspect has stated that he was in contact with the victim, I assume that both samples from his left hand contain her DNA.
- I also assume that both samples contain DNA from the suspect and victim only.

This means that both samples contain the DNA of one male (the suspect) and one female (the victim). This also means that, if the samples contain vaginal epithelial cells and/or menstrual secretion, the biological material can only be derived from the victim.

[*Note that these assumptions address issues 1 and 2 as shown in Figure 9.3. They link the victim to the DNA and the DNA to the same biological source as the mRNA molecules used for the cell type test, resulting in situation E as shown in Table 9.1.*]

PROPOSITIONS

Vaginal epithelial cells and menstrual secretion are both linked to the same body cavity (vagina) and are therefore frequently encountered together. In the statistical model that is used to calculate the weight of the evidence, the presence or absence of these cell types is evaluated together. This aspect is included in the formulation of the propositions.

The results from the mRNA typing from both samples A and B are evaluated, given the following propositions:

Proposition 1: The sample contains vaginal cells and/or menstrual secretion (either with or without other biological material) from the victim

Proposition 2: The sample does not contain vaginal cells and/or menstrual secretion (but only other biological material) from the victim

INTERPRETATION

The weight of the evidence is calculated, given the propositions and assumptions, using a validated statistical model. For more information on this model, I refer to: 'Ypma, R.J.F., Maaskant-van Wijk, P.A., Gill, R., Sjerps, M. & van den Berge, M. (2021). Calculating LRs for presence of body fluids from mRNA assay data in mixtures. Forensic Science International: Genetics, 52, 102455'.

CONCLUSION

For both samples A and B, the results from the mRNA analysis are over 1000 times more probable if proposition 1 is true rather than if proposition 2 is true. This means that the results are much more probable if the samples contain vaginal epithelial cells and/or menstrual secretion from the victim than if they don't.

[*Note that the report will contain a reference to a scale of verbal equivalents used to express the weight of the evidence, which includes numerical boundaries to those verbal qualifications.*]

9.5   INCORPORATING CELL TYPE TESTING TO ADDRESS ACTIVITY-LEVEL ISSUES

Any findings from forensic biological examinations may be relevant to help address issues at the activity level. As we discussed before, such findings may consist of relative or absolute quantities of DNA contributed to a sample by persons of interest, the number or distribution of traces found on an item or scene of the crime and the nature of the biological material. In any type of case, but particularly in sexual assault cases, the type of cellular material may be highly informative to distinguish between scenarios proposed by the prosecution and defence. One could imagine that the presence of semen on the body of the complainant might better fit a scenario of sexual assault rather than legitimate social interaction only. Similarly, the presence of vaginal epithelial cells under the fingernails of the accused could support a prosecution scenario of digital penetration rather than, again, superficial social interaction. In both these cases, latent DNA traces only might provide no or poor distinction between the proposed scenarios if the legitimate interaction between the persons involved is not in dispute.

The way in which results from cell type testing can be included in an evaluation, given activity-level propositions require very similar considerations as DNA testing results.

9.5.1   How Is the Result Defined?

While it is possible (and sometimes necessary) to include the uncertainty on the source of the DNA, more often than not the LR at the sub-source level is extremely high, and in the context of the case, the presence of DNA of an individual is often not contested. Throughout this book, we have thus seen that we usually assume the presence of DNA of persons of interest. With cell type testing, we will also need to consider whether or not we assume the presence of the cell type or make the weight of evidence at the source-level part of our assessment, given the activity-level propositions.

If the source of the biological material is not in question (and we have seen examples of this in previous paragraphs), we will assume the presence of a cell type. However, if the biological source of the material is uncertain, we are left with two options: (1) We include the weight of the evidence we have calculated for our cell type tests in our evaluation, or (2) we do not address the issue of cell type but rather assign probabilities to a positive or negative test result rather than the probability of a cell type being present.

We might for instance be interested in the presence of saliva in a sample from underwear after an alleged oral intercourse. Let us assume that the RSID™ Saliva test was positive. We could (1) assign a probability to the presence of saliva in the sample, where the LR we calculated for the positive test result given propositions on the presence or absence of saliva in the sample is part of this assessment. Or (2) we assign a probability to a positive test result from the RSID™ Saliva test given both propositions at the activity level. Note that in the latter instance we do not need to worry ourselves with the diagnostic value of the test, since we are not addressing the issue of whether or not *saliva* is present.

By choosing the second approach, we may lose resolution in our assessment as we omit information that has the potential to distinguish between the propositions. In the example this is because the presence of *saliva* may be much more informative than the 'presence' of a positive RSID™ Saliva test. The latter may of course also be explained by the presence of other body fluids, increasing the probability of the observation relative to observing saliva under a scenario with no oral sexual intercourse. The reason we will nonetheless often be forced to choose the second option is the lack of data on TPPR of *body fluids* on relevant items and substrates, while data are available on TPPR of *testing results*. We will therefore more often be able to provide an empirically supported opinion under option 2 rather than option 1.

9.5.2   Discretisation of the Results

As discussed in Chapter 5, we might choose discrete states for our DNA testing results. These discrete states could be very broad (e.g. presence or absence of DNA) or very narrowly defined (e.g. small intervals for relative or absolute DNA amounts). We could also use continuous distributions on such observations for the amount of DNA in a sample. The decision on how to define our states will be made based on the availability of empirical data to assign probabilities to the states, as well as the case context and results obtained from our analysis.

The same decision needs to be made for cell type testing results. We might for instance decide to use broad categories (cell type present/absent; test positive/negative/inconclusive) or define the states more narrowly. This could include information on the amount of biological material or the strength of the test result. Examples of this could be the number of spermatozoa observed in a microscope slide, the strength of the immunological reaction in a lateral flow strip, the strength of fluorescence of probes in a microarray or the expression levels of individual mRNA markers in a multiplex. Here too we will need to be mindful of the trade-off between the level of discretisation and the availability of data. It is generally easier to find sufficient empirical data to support the assignment of probabilities for broader categories than a greater number of narrow categories or to fit a continuous distribution.

9.5.3 Combining Cell Type Test Results with DNA Analysis

If we have cell type testing results and consider including them in our assessment, we will have to combine them with the results from our DNA analyses. At this point, we will need to consider whether or not there is a conditional dependency between the findings.

Conditional dependence in brief means asking the question; given the propositions, assumptions and case circumstances, will the result of one test change my belief in the outcome of another test? Or:

Is Pr(Outcome test A|H,I) equal to Pr(Outcome test A|H,I, Outcome test B)?
Let us consider the following example.

A sample is taken from the inside mouth area of a balaclava. The sample is subjected to the RSID™ Saliva test and subsequently to DNA profiling. The RSID test was positive, and the DNA analysis resulted in a single source DNA profile of X. The context is that the balaclava was worn by a perpetrator of a crime and secured shortly after. The prosecution proposes that X is the perpetrator, while the accused states that it may be his balaclava, but that someone else has borrowed it from him (and wore it during the robbery). We could be presented with the following propositions:

H1: X wore the balaclava during the robbery
H2: Someone other than X wore the balaclava during the robbery

Given this set of case circumstances, we construct the Bayesian network shown in Figure 9.5.

We will define the outcome of the RSID test as either positive or negative, while we define the outcome of the DNA analysis in four broad categories as shown in Figure 9.5.

The question that needs to be answered is whether the outcome of the saliva test would change our assignment of the probability to the outcome of the DNA profiling, *given the propositions, assumptions and case circumstances*. Under H1, it is given that X wore the balaclava right before it was secured and sampled. We might assign a high probability to finding a positive RSID test as well as assign a high probability to finding only his DNA in the sample, *given the RSID test was positive* (Pr(ss X|H1, RSID p)). Would our probability assignment to finding only X DNA in the sample change if we know the RSID test is negative (Pr(ss X|H1, RSID n))? One may argue that a positive test might increase the probability of finding a single source DNA profile over a mixture. But not necessarily increase our belief in finding the DNA of the accused only, as we already know (given the proposition) that he was the last person to wear the balaclava. Hence there is conditional

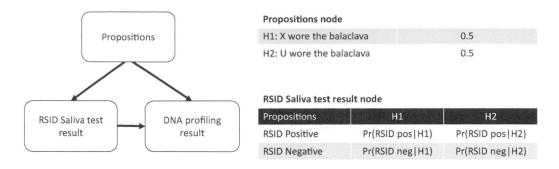

Propositions node

H1: X wore the balaclava	0.5
H2: U wore the balaclava	0.5

RSID Saliva test result node

Propositions	H1	H2
RSID Positive	Pr(RSID pos|H1)	Pr(RSID pos|H2)
RSID Negative	Pr(RSID neg|H1)	Pr(RSID neg|H2)

DNA profiling result node

Propositions	H1		H2	
RSID test result	Positive	Negative	Positive	Negative
Single source X	Pr(ss X|H1, **RSID p**)	Pr(ss X|H1, **RSID n**)		
Single source U				
Mix X+U				
Other				

FIGURE 9.5 Bayesian network to evaluate the results from the RSID™ Saliva test and the DNA profiling result in the fictitious case example.

dependence between the outcomes of the two tests. Experimental data may provide guidance when assigning these conditional probabilities.

9.6 ADDRESSING THE ISSUE IN THE CASE AT THE SOURCE OR ACTIVITY LEVEL?

In those instances where the findings are relatively straightforward (think about a clearly visible stain, high template DNA, single source DNA profile, etc.), the findings may be addressed at the source level with a quite high likelihood ratio supporting the presence of a specific cell type of a particular individual. However, when dealing with low template samples or mixtures, non-gender-specific cell types and cell type tests with low diagnostic value, the attribution of the cell type (if it is present!) to a particular donor (or multiple?) may be very complex and the findings may provide limited support for one proposition over the other.

One may ask how relevant the cell type determination and attribution issue is in the context of a case. An illustrative example comes from Taylor et al. [67]. As was briefly discussed in the previous paragraph, the paper described a model to make inferences on the presence of blood and its association to a contributor, given source-level propositions. The study is motivated by a case where a victim was assaulted, and the top of a suspect was examined for the presence of her blood. The findings were a positive test for blood and a mixed DNA profile matching both suspect and victim. While the issue may seem to address whose blood is present on the top of the suspect, the underlying issue was formulated as: 'Could the DNA you detected be from saliva [rather than blood] which happens to have been deposited by the victim on the suspect's top when they were talking a few days ago?' This is clearly an issue at the activity level. Discussing the case with the mandating authority is a crucial step to getting clarity on the issue that really needs to be addressed.

The good news is that we don't need to go through the source attribution issue before evaluating findings given activity-level propositions. Depending on the decisions we make in the

TABLE 9.2

Evaluation of the findings from the study by Breathnach and Moore (after their Table 9.5).

Outcomes	P(E|H1) (Fellatio)	P(E|H2) (No fellatio)	LR
No amylase	0.0100	0.9420	0.0106
Amylase	0.9900	0.0580	17.0680
Amylase no DNA	0.0099	0.0145	0.6827
Amylase DNA wearer only	0.1980	0.0145	13.6550
Amylase DNA other than wearer	0.7920	0.0290	27.3100

discretisation in the states of our network, and on the definition of our 'findings', we may very well evaluate these given activity-level propositions even when the cell type test result, its association as well as the donor are uncertain. An example of this is provided by Breathnach and Moore [68]. They studied the transfer to and prevalence of saliva on underwear in scenarios with and without oral sexual intercourse. They tested the underwear for possible saliva stains using a two-step approach. First a screening for amylase was performed using the Phadebas pressure test, followed by the application of the RSID saliva test on stains that gave a positive Phadebas result. Samples that tested positive for both were subsequently subjected to DNA analysis. They used the study's results in an evaluation given propositions at the activity level stating that fellatio either occurred (H1) or did not occur (H2). They then calculated the likelihood ratio for the findings from the amylase tests only, and for the combination of these tests with the DNA typing results. This resulted in Table 9.2. This shows that the presence or absence of a positive test for amylase (not necessarily saliva) may provide moderately strong to strong support for propositions H1 or H2, respectively. The combination of a positive amylase test and the presence of non-wearer DNA further strengthens support for H1.

Note that neither a conclusion is drawn on the presence or absence of saliva in the samples nor is the association between the amylase test and the DNA results discussed. Either test (but consider their conditional dependence) may provide evidence supporting one proposition at the activity level over another. We do potentially lose resolution in our assessment, as the presence of saliva that can be attributed to a source other than the wearer in the underwear is likely to provide even more support for H1 over H2. Data to support probability assignments given source-level propositions are scarce, and evaluation of such findings given activity-level propositions may be the most sensible way to evaluate the findings and communicate their meaning in the context of the case to a court.

9.7 EXAMPLE CASE 3 – 'THE FAMILY ASSAULT'

In Chapter 2 the family assault case was introduced. An evaluation of the findings (faecal staining in the underwear, a positive RSID saliva test and a (low) Y-STR profile corresponding to the male family lineage) was performed in Chapter 5 using hand derivation of the appropriate formula. In Chapter 6 the case was revisited, and a Bayesian network was constructed for this case, the structure of which can be found in Figure 6.41.

Let us now consider the 'three sources of uncertainty' when attributing a body fluid to a specific donor (see Figure 9.3). As discussed in the previous paragraph, we can avoid these issues when addressing issues at the activity level. We can, as was done in Chapter 6, consider the probability of a positive test result for amylase as well as consider the presence of a familial Y-STR profile separately. There may however be added diagnostic value to the test results if we consider the context of the trace more closely. As in Chapter 6, we will assume that the source of the DNA is not in dispute.

First, we address the nature of the biological material in more detail. We extend the network by including the activity of wearing the underwear by C (which is true under both propositions), which may give rise to the transfer, persistence and recovery of a number of relevant body fluids from C to C's underwear. Which body fluids are considered relevant will be determined by the case context, such as the type of item (here underwear), location of sampling and activities performed with it (here wearing). Other aspects may be the duration of wear, and whether or not C was menstruating at the time (we will assume for now she was not). As the amount of DNA from C that is recovered may be informative as to the nature of the biological material (we may for instance expect higher amounts of DNA of C with high amounts of saliva from C present compared to latent DNA from wearing only), we include findings from autosomal DNA analysis for C as well. Figure 9.6 shows the additional activity node and extended network configuration coming from that with respect to the activity of wearing the underwear by C when considering the potential presence of saliva, faeces, vaginal secretion, urine and latent DNA.

Note that some of the background nodes from Figure 6.41 have now been assimilated in the 'wearing by C' activity rather than being defined as separate root nodes. A single background node remains which specifies the probability of recovering amylase from an unknown source (other than C, D or F) from the underwear. A similar structure for TPR of amylase is used in the network for the interpretation of saliva testing results (Figure 9.1). The transfer and persistence of the body fluid is considered in a separate node (i.e. 'C saliva on C underwear from wearing') from the probability of recovery of a *detectable* amount of amylase (i.e. node 'C saliva amylase on C underwear'). These nodes connect to 'summary nodes' (or binary nodes) that subsequently connect to the test result node. This principle is extended to the other relevant activities and actors (biting, co-habitation). The resulting network structure is shown in Figure 9.7. Here too a distinction is made between the transfer of saliva and the transfer of latent DNA (that is not/cannot be attributed to a specific cellular source). Only saliva is considered here as a potential source of amylase other than C. Depending on the case circumstances, one may decide to include other potential sources (e.g. male urine or semen for instance). Amylase from different donors (C, F, D and background unknown – U) combines in node 'Amylase present on C underwear'.

The network structure presented here allows for a more detailed exploration of the diagnostic value of the RSID saliva test. Multiple potential sources for amylase in the sample can be explored. This requires probability assignments for TPR of the individual body fluids, as well as the cumulative probability of a positive test result, given the presence of these body fluids. Data to assign such probabilities should be derived from validation studies of the tests used (and the need for such data may also direct the design of such studies) as well as from published sources that provide relevant information on this topic (for instance [69, 70]).

The attribution of a cellular source to a donor is one final aspect that may be considered. Recent developments have shown that such attribution may be informed by testing, for instance through peptide sequencing to detect genetically variant peptides or through mRNA sequencing [63]. If such testing would be available to us in the family assault case, one could incorporate this as a test result node in the network structure as shown in Figure 9.8. The states for this node 36 would be all potential sources of amylase and the combination thereof (D, F, C, U, D and F, D and C, … , Other, None).

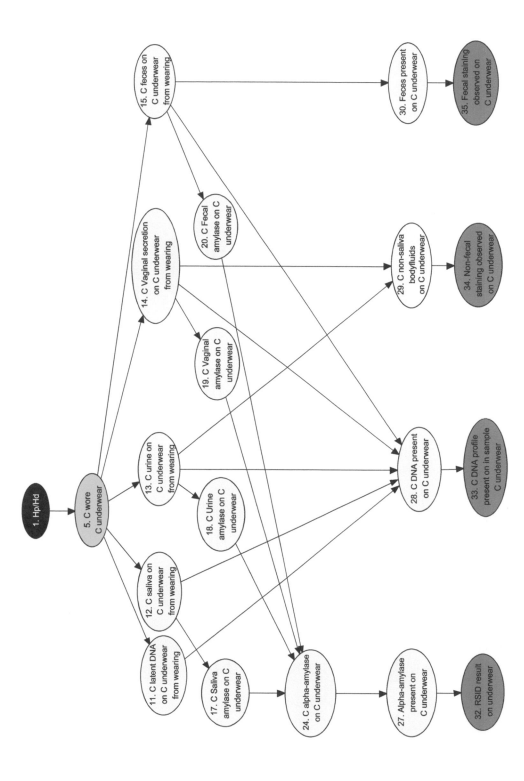

FIGURE 9.6 Extension to the network presented in Chapter 6, Figure 6.41. The activity 'C wore C underwear' is added, resulting in consideration of TPR of DNA and body fluids to the underwear.

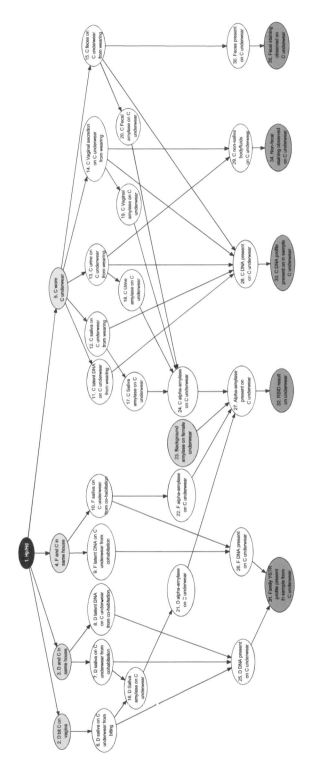

FIGURE 9.7   Bayesian network structure for the family assault case with added emphasis on the interpretation of the body fluid testing result.

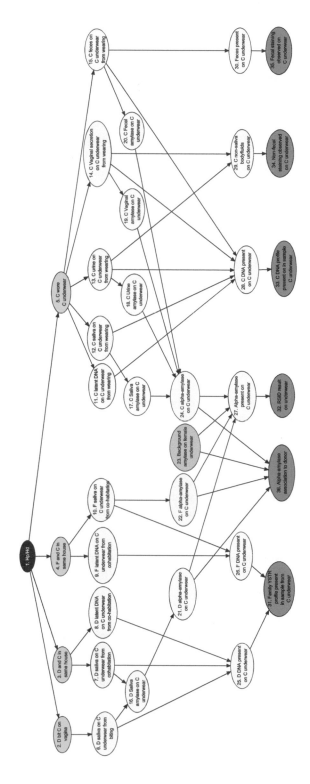

FIGURE 9.8 Network structure including an additional test result node 'Alpha-amylase association to donor' (node 36).

9.8   PRACTICE QUESTIONS

Q1) Would you consider the examination of a microscope slide and the reported observation of 12 spermatozoa on that slide a confirmatory test for the presence of human semen? Why (not)?

Q2) The issue is whether a sample contains seminal fluid. You perform a PSA test on the sample. The test is positive. What would you report?

Q3) A forensic report that you have been asked to review contains the statement: 'The sample was tested for the presence of saliva. The test was positive. This is an indication for the presence of saliva in the sample'. Would you comment on this statement in your review? And if so, what would that comment be?

Q4) You apply 'ABAcard® HemaTrace® for the Forensic Identification of Human Blood' to a sample. The test is positive. Which assumptions do you need to make to *identify human blood* in the sample? Under what circumstances would you make these assumptions?

Q5) Following from Q4, assume (for argument's sake) that the sample contains human blood and is taken from the hands of the defendant. The sample also contains DNA from the defendant and a female individual. What assumptions would you need to make to attribute any human blood to the female contributor?

Q6) You perform differential DNA extraction on a sample and find that there is no enrichment of male DNA in the stringent lysis fraction compared to the mild lysis fraction. What do you conclude with regard to the presence or absence of spermatozoa in the sample?

Q7) Consider the Bayesian network structure shown in Figure 9.8. Node 'C faeces on C underwear from wearing' has three states:

- Yes (High)
- Yes (Low)
- None
- Draw out the conditional probability table for this node. What probabilities would you assign to this table? What information would you rely on?

Q8) Consider the Bayesian network structure shown in Figure 9.8. Would you consider the results nodes 'RSID result on underwear' and 'C DNA profile present […]' or 'Family Y-STR profile present […]' conditionally dependent? Why (not)?

Q9) Consider example case 3 'The family assault' for which the Bayesian network shown in Figure 9.8 was constructed. From the case information provided, we understand that the complainant was picked up by a friend who took her to the police station where her underwear was secured for forensic examination. Now let us assume that the victim used the toilet in the facility prior to the underwear being secured and that the police officer who took the underwear was talking to the complainant at the time while not wearing a face mask.
- How would you include this information in the model shown in Figure 9.8?
- Would different modelling decisions be made if a reference DNA sample was available from the police officer who secured the item?

Q10) Following Q9, what other sources of contamination may be relevant to consider?

9.9 REFERENCES

1. I. Quinones, B. Daniel, Cell free DNA as a component of forensic evidence recovered from touched surfaces, *Forensic Science International: Genetics* 6(1) (2012) 26–30.
2. T. Sijen, Molecular approaches for forensic cell type identification: On mRNA, miRNA, DNA methylation and microbial markers, *Forensic Science International: Genetics* 18 (2015) 21–32.
3. A.P. Salzmann, et al., Degradation of human mRNA transcripts over time as an indicator of the time since deposition (TsD) in biological crime scene traces, *Forensic Science International: Genetics* 53(2021) 102524.
4. E. Ziemke, Weitere Mittheilungen über die Unterscheidung von Menschen- und Thierblut mit Hilfe eines spezifischen Serums, *Dtsch Med Wochenschr* 27(42) (1901) 731–733.
5. W. Boyd and L. Boyd, Blood grouping in forensic medicine, *The Journal of Immunology* 33(2) (1937) 159–172.
6. W. Lee, B. Khoo. Forensic Light Sources for Detection of Biological Evidences in Crime Scene Investigation: A Review, *Malaysian Journal of Forensic Sciences* 1(1) (2010) 17–28.
7. T. Schotman, et al., Understanding the visibility of blood on dark surfaces: A practical evaluation of visible light, NIR, and SWIR imaging, *Forensic Science International* 257 (2015) 214–219.
8. K. Langstraat, et al., Large area imaging of forensic evidence with MA-XRF, *Scientific Reports* 7(1) (2017) 15056.
9. C.K. Muro, et al., Forensic body fluid identification and differentiation by Raman spectroscopy, *Forensic Chemistry* 1 (2016) 31–38.
10. J. Fraser, et al., Visualisation of fingermarks and grab impressions on fabrics. Part 1: gold/zinc vacuum metal deposition, *Forensic Science International: Genetics* 208(1–3) (2011) 74–78.
11. W. Specht, Die Chemiluminescenz des Hämins, ein Hilfsmittel zur Auffindung und Erkennung forensisch wichtiger Blutspuren, *Deutsche Zeitschrift für die gesamte gerichtliche Medizin* 28(1) (1937) 225–234.
12. F. Kader, M. Ghai, A.O. Olaniran, Characterization of DNA methylation-based markers for human body fluid identification in forensics: A critical review, *International Journal of Legal Medicine* 134(1) (2020) 1–20.
13. J.M. Robinson, et al., Forensic applications of microbiomics: A review, *Frontiers in Microbiology* 11(3455) (2021). https://doi.org/10.3389/fmicb.2020.608101
14. T. Sijen, S. Harbison, On the identification of body fluids and tissues: A crucial link in the investigation and solution of crime, *Genes* 12(11) (2021). https://doi.org/10.3390/genes12111728
15. G.M. Willott, An improved test for the detection of salivary amylase in stains, *Journal of the Forensic Science Society* 14(4) (1974) 341–344.
16. G.M. Willott, M. Griffiths, A new method for locating saliva stains — Spotty paper for spotting spit, *Forensic Science International: Genetics* 15(1) (1980) 79–83.
17. L. Quarino, et al., An ELISA method for the identification of salivary amylase, *Journal of Forensic Sciences* 50(4) (2005) 873–876.
18. K. Van Steendam, et al., Mass spectrometry-based proteomics as a tool to identify biological matrices in forensic science, *International Journal of Legal Medicine* 127(2) (2013) 287–298.
19. B. Lomholt, N. Keiding, Tetrabase, an alternative to benzidine and orthotolidine for detection of haemoglobin in urine, *Lancet* 1(8011) (1977) 608–609.
20. H.Y. Lee, et al., Potential forensic application of DNA methylation profiling to body fluid identification, *International Journal of Legal Medicine* 126(1) (2012) 55–62.
21. G. Alderson, G., et al., Inferring the presence of spermatozoa in forensic samples based on male DNA fractionation following differential extraction, *Forensic Science International: Genetics* 36 (2018) 225–232.
22. S-M. Park, et al., Genome-wide mRNA profiling and multiplex quantitative RT-PCR for forensic body fluid identification, *Forensic Science International: Genetics* 7(1) (2013) 143–150.
23. A. Lindenbergh, A., et al., A multiplex (m)RNA-profiling system for the forensic identification of body fluids and contact traces, *Forensic Science International: Genetics* 6(5) (2012) 565–577.
24. C. Haas, et al., mRNA profiling for body fluid identification by reverse transcription endpoint PCR and realtime PCR, *Forensic Science International: Genetics* 3(2) (2009) 80–88.
25. D. Zubakov, et al., Stable RNA markers for identification of blood and saliva stains revealed from whole genome expression analysis of time-wise degraded samples, *International Journal of Legal Medicine* 122(2) (2008) 135–142.

26. G. Dørum, et al., Predicting the origin of stains from next generation sequencing mRNA data, *Forensic Science International: Genetics* 34 (2018) 37–48.
27. C. Haas, et al., RNA/DNA co-analysis from human menstrual blood and vaginal secretion stains: Results of a fourth and fifth collaborative EDNAP exercise, *Forensic Science International: Genetics* 8(1) (2014) 203–212.
28. E.K. Hanson, H. Lubenow, and J. Ballantyne, Identification of forensically relevant body fluids using a panel of differentially expressed microRNAs, *Analytical biochemistry* 387(2) (2009) 303–314.
29. D. Zubakov, et al., MicroRNA markers for forensic body fluid identification obtained from microarray screening and quantitative RT-PCR confirmation, *International Journal of Legal Medicine* 124(3) (2010) 217–226.
30. Y. Zhang, et al., Evaluation of the inclusion of circular RNAs in mRNA profiling in forensic body fluid identification, *International Journal of Legal Medicine* 132(1) (2018) 43–52.
31. F. Song, et al., Microarray expression profile of circular RNAs in human body fluids, *Forensic Science International: Genetics Supplement Series* 6 (2017) e55–e56.
32. S. Wang, et al., The potential use of Piwi-interacting RNA biomarkers in forensic body fluid identification: A proof-of-principle study, *Forensic Science International: Genetics* 39 (2019) 129–135.
33. A. Lindenbergh, et al., Development of a mRNA profiling multiplex for the inference of organ tissues, *International Journal of Legal Medicine* 127(5) (2013) 891–900.
34. H.E. Blum, The human microbiome, *Advances in Medical Sciences* 62(2) (2017) 414–420.
35. F.C.A. Quaak, et al., Human-associated microbial populations as evidence in forensic casework, *Forensic Science International: Genetics* 36 (2018) 176–185.
36. C.C. Benschop, et al., Vaginal microbial flora analysis by next generation sequencing and microarrays; can microbes indicate vaginal origin in a forensic context?, *International Journal of Legal Medicine* 126(2) (2012) 303–310.
37. J. de Zoete, et al., Cell type determination and association with the DNA donor, *Forensic Science International: Genetics* 25 (2016) 97–111.
38. O. Ingemann-Hansen, et al., Legal aspects of sexual violence: Does forensic evidence make a difference?, *Forensic Science International: Genetics* 180(2–3) (2008) 98–104.
39. M.J. McGregor, J.D. Mont, and T.L. Myhr, Sexual assault forensic medical examination: Is evidence related to successful prosecution?, *Annals of Emergency Medicine* 39(6) (2002) 639–647.
40. M. Miheso, M. Mutugi, Perpetrator DNA profiling in samples from rape victims in Kenya, *Journal of Forensic Research Crime Studies* 2 (2015) 1–9.
41. A. Biedermann, S. Bozza, F. Taroni, Decision theoretic properties of forensic identification: underlying logic and argumentative implications, *Forensic Science International* 177(2–3) (2008) 120–132.
42. P.J. Hooft, H.P. van de Voorde, Bayesian evaluation of the modified zinc test and the acid phosphatase spot test for forensic semen investigation, *The American journal of forensic medicine and pathology* 18(1) (1997) 45–49.
43. J. de Zoete, J. Curran, and M. Sjerps, A probabilistic approach for the interpretation of RNA profiles as cell type evidence, *Forensic Science International: Genetics* 20 (2016) 30–44.
44. D. Iacob, A. Fürst, and T. Hadrys, A machine learning model to predict the origin of forensically relevant body fluids, *Forensic Science International: Genetics Supplement Series* 7(1) (2019) 392–394.
45. G. Dørum, et al., Predicting the origin of stains from whole miRNome massively parallel sequencing data, *Forensic Science International: Genetics* 40 (2019) 131–139.
46. S. Fujimoto, S., et al., Distinct spectrum of microRNA expression in forensically relevant body fluids and probabilistic discriminant approach, *Scientific Reports* 9(1) (2019) 14332.
47. R.J.F. Ypma, et al., Calculating LRs for presence of body fluids from mRNA assay data in mixtures, *Forensic Science International: Genetics* 52 (2021) 102455.
48. L. Samie, L., et al., Use of Bayesian networks for the investigation of the nature of biological material in casework, *Forensic Science International* 331 (2022) 111174.
49. T.R.D. Wolff, et al., A probabilistic approach to body fluid typing interpretation: An exploratory study on forensic saliva testing, *Law, Probability and Risk* 14(4) (2015) 323–339.
50. C. Aitken, A. Nordgaard, The roles of participants' differing background information in the evaluation of evidence, *Journal of Forensic Sciences* 63(2) (2018) 648–649.
51. I. Su Andrew, et al., A gene atlas of the mouse and human protein-encoding transcriptomes, *Proceedings of the National Academy of Sciences* 101(16) (2004) 6062–6067.

52. A.D. Roeder, C. Haas, mRNA profiling using a minimum of five mRNA markers per body fluid and a novel scoring method for body fluid identification, *International Journal of Legal Medicine* 127(4) (2013) 707–721.

53. R.L. Fleming, S. Harbison, The development of a mRNA multiplex RT-PCR assay for the definitive identification of body fluids, *Forensic Science International: Genetics* 4(4) (2010) 244–256.

54. R. Cook, et al., A hierarchy of propositions: Deciding which level to address in casework, *Science & Justice* 38(4) (1998) 231–240.

55. R. Puch-Solis, et al., Evaluating forensic DNA profiles using peak heights, allowing for multiple donors, allelic dropout and stutters, *Forensic Science International: Genetics* 7(5) (2013) 555–563.

56. R. Puch-Solis, R., A dropin peak height model, *Forensic Science International: Genetics* 11(0) (2014) 80–84.

57. M.W. Perlin, et al., Validating TrueAllele® DNA mixture interpretation, *Journal of Forensic Sciences* 56 (2011) 1430–1447.

58. D. Taylor, J.-A. Bright, and J. Buckleton, The interpretation of single source and mixed DNA profiles, *Forensic Science International: Genetics* 7(5) (2013) 516–528.

59. G.C. Robert, Validation of an STR peak area model, *Forensic Science International: Genetics* 3(3) (2009) 193–199.

60. Ø. Bleka, G. Storvik, and P. Gill, EuroForMix: An open source software based on a continuous model to evaluate STR DNA profiles from a mixture of contributors with artefacts, *Forensic Science International: Genetics* 21 (2016) 35–44.

61. R.G. Cowell, S.L. Lauritzen, and J. Mortera, Probabilistic expert systems for handling artifacts in complex DNA mixtures, *Forensic Science International: Genetics* 5(3) (2011) 202–209.

62. H. Haned, Forensim: An open-source initiative for the evaluation of statistical methods in forensic genetics, *Forensic Science International: Genetics* 5(4) (2011) 265–268.

63. S. Ingold, et al., Assigning forensic body fluids to donors in mixed body fluids by targeted RNA/DNA deep sequencing of coding region SNPs, *International Journal of Legal Medicine* 134(2) (2020) 473–485.

64. G.E. Meakin, et al., Evaluating forensic DNA evidence: Connecting the dots, *WIREs Forensic Science* 3(4) (2021) e1404.

65. J.A. de Koeijer, et al., Combining evidence in complex cases: A practical approach to interdisciplinary casework, *Science & Justice* 60(1) (2020) 20–29.

66. D. Taylor, Probabilistically determining the cellular source of DNA derived from differential extractions in sexual assault scenarios, *Forensic Science International: Genetics* 24 (2016) 124–135.

67. D. Taylor, et al., Evaluating forensic biology results given source level propositions, *Forensic Science International: Genetics* 21 (2016) 54–67.

68. M. Breathnach, E. Moore, Oral intercourse or secondary transfer? A Bayesian approach of salivary amylase and foreign DNA findings, *Forensic Science International* 229 (2013) 52–59.

69. D. Sari, et al., Amylase testing on intimate samples from pre-pubescent, post-pubescent and post-menopausal females: Implications for forensic casework in sexual assault allegations, *Australian Journal of Forensic Sciences* 52(6) (2020) 618–625.

70. V. Pham-Nguyen, et al., Reactivity of the RSID™-Saliva test to α-amylase present in vaginal secretions, *Australian Journal of Forensic Sciences* (2021). 1–9. https://doi.org/10.1080/00450618.2021.1973099

71. G.J. Parker, et al., Forensic proteomics, *Forensic Science International: Genetics* 54 (2021) 102529.

10 Reporting

Bas Kokshoorn

CONTENTS

10.1 INTRODUCTION

In previous chapters of this book, the different aspects of evaluation of findings given activity-level propositions have been discussed. The crucial next step is to communicate the findings of the evaluation to those requesting this information (e.g. the police, prosecution, defence counsel and the court). This chapter discusses the communication of the interpretation of the findings and the conclusions drawn from those. The focus will primarily be on the formal report, but other forms of communication will briefly be discussed as well.

There are three broad categories of forensic science expert opinions. These are:

- Technical (in some sources also referred to as 'factual')
- Investigative opinions
- Evaluative opinions

The first category does not contain any (or only very limited) interpretation of results (and as such does not contain an opinion formulated by the expert). It is a report of a measurement of the property of an object (like the size of a bullet, the colour of a fibre, the length of a DNA fragment, or the concentration of a chemical or biological compound). An example of a technical report is the measurement of the alcohol content of a blood sample taken from the driver of a vehicle. Reporting the blood alcohol value, possibly with an error margin based on the validation studies of the system, would usually be all that is required by the criminal justice system, as legislation generally defines a threshold value above which driving is not permitted.

The second category contains investigative opinions, which will require some level of interpretation of the data by the forensic scientist. As was discussed in Chapter 1 (Section 1.6), the scientist will provide these types of opinions when they are in an investigative role. An example from forensic biology could be the reporting of saliva testing results. When a sample is tested using an amylase test and the test is found to be positive, the scientist may decide to provide an investigative opinion as: 'The test provided an indication for the presence of saliva in the sample'. Such an opinion reflects the interpretation of the positive amylase test by the scientist, as apparently the presence of saliva is found to be the best *explanation* for the positive test result. To get to this opinion, the scientist would have considered the following:

- Amylase was detected (and the possibility of a false positive test result is considered very low).
- Amylase is found in high concentrations in saliva, but in lower concentrations in other body fluids.
- The (prior) probability that such other body fluids are present in the sample is considered very low.
- The (prior) probability that saliva is present in the sample is considered (reasonably) high. Therefore, the most likely explanation for the positive test result is that saliva is present in the sample.

From this, it is clear that reporting 'an indication for the presence of saliva' is an opinion that is at least in part based on the case context, and something quite different from providing a technical report that the test for amylase was positive.

The third category is evaluative opinions provided by the scientist, which are different from investigative opinions as they are a complete evaluation of the findings given case-relevant propositions. Such evaluative opinions could be provided at any level in the hierarchy of propositions. They are quite common at sub-sub-source level and sub-source level in forensic genetics, when a weight of the evidence is calculated for a major component or full-mixed DNA profile, respectively, given propositions addressing the presence or absence of DNA of a person of interest. Occasionally such opinions are provided at the source level when there is certainty about the source of DNA in a sample. Propositions may be formulated as 'The semen in this sample is from mr. X vs. The semen in this sample is from a random, unrelated male individual'. Chapter 9 deals more extensively with the evaluation of cell type testing results given propositions at the source level.

While these three categories of forensic science opinions are generally recognised, a forensic report may be found to contain all three different types of reported findings. For example, a single report in a case may contain the following three statements:

- The concentration of DNA in sample A is 0.0003 ng/µl, which is below our laboratory threshold for further analysis. Therefore, no DNA profile was generated for this sample.
- Sample B was tested for the presence of saliva. An indication was found for the presence of saliva in this sample.
- Sample C contains DNA of at least three individuals. I have calculated the weight of the evidence for this sample given the following propositions:

    H1: The sample contains DNA of the complainant and two random, unrelated individuals.
    H2: The sample contains DNA of three random, unrelated individuals.
       The results are more than a billion times more probable if H1 is true rather than if H2 is true.

The first statement on the quantity of DNA in the sample is a technical statement. The result from the DNA quantification is taken as below a determined threshold without further interpretation.

The second statement is an investigative opinion on the presence of saliva in the sample. As no suspect or defendant is yet identified, the scientist is operating in investigative mode. Hence, such an investigative statement on the presence of saliva may be opportune.

The third statement is an evaluative opinion as the findings are evaluated given case-relevant propositions. Again, in an investigative mode, it may be relevant to express the weight of the findings as this may provide further support for, or falsification of, scenarios being considered by the police in their investigation.

At the time of writing, there are no globally accepted *standards* for forensic science reports. There are arguments for and against setting generic standards, which are discussed in some detail in an opinion paper by Biedermann et al. [1]. The International Standards Organisation (ISO) is currently working on drafting standards for forensic science (ISO 21043) which will also cover reporting (ISO 21043-4). Although standards are currently not available, there are several documents providing guidance and recommendations on evaluative reporting.

10.2 GUIDANCE ON EVALUATIVE REPORTING

Most of the guidance on evaluative reporting focuses on evaluation of findings given activity-level propositions. A number of documents that have been drafted by professional bodies are listed and discussed chronologically below. Similarities and differences between them are highlighted.

10.2.1 THE ASSOCIATION OF FORENSIC SCIENCE PROVIDERS (AFSP) – 2009

The AFSP is an independent body that represents their member organisations in the United Kingdom and Ireland. Their mission, as quoted from their website (March 2021):

> To represent the common views of the providers of independent forensic science within the United Kingdom and Ireland with regard to the maintenance and development of quality and best practice in forensic science and expert opinion in support of the justice system, from scene to court.

As early as 2009 the AFSP has published 'Standards for the formulation of evaluative forensic science expert opinion' [2]. The scope covers all forensic disciplines and was formulated as: 'Forensic Expert Opinion formulated in the Evaluative or Evidential mode across all scientific disciplines'. Although considered standards for the AFSP member organisations, they can be considered guidelines for non-member organisations. The standards cover the entire process, but the focus is on documentation and reporting here. Taken from the standards with regard to documentation:

> 4.13 The following will be documented clearly on the case file:
>
> - The key issue(s) and propositions addressed.
> - Potential outcomes and assigned probabilities.
> - Sources of data used in probability assignments.
> - Examination strategy.
> - All discussions with clients and other experts including those working for the defence.

With the next point covering the content of the report:

> 4.14 Statements or reports will include:
>
> - Background information used in the assessment/interpretation.
> - The propositions addressed.
> - Relevant items received.

- Items examined.
- Significant findings.
- Conclusion(s).

The AFSP standards consider that the statement or report provided by the scientist does not necessarily cover the interpretation of findings (whichever findings they may be). This appears to be at odds with one of the four guiding principles formulated in the standards, which is *transparency*: 'The expert will be able to demonstrate how he came to his conclusion' (the other three being balance, logic and robustness). Although the document continues with the position that the expert 'will be able, if required, to provide the data he has used and its provenance'.

10.2.2   Guidance from Workshop 'Legal Expressionism' – 2011

Found and Edmond [3] have published guidance on the reporting of the comparison and interpretation of pattern evidence. This guidance comes from a workshop 'Legal Expressionism' which was hosted by the Program in Expertise, Evidence and Law, University of New South Wales, in 2011. The authors provide a report flow diagram, which lists the following contents of a report:

- Title
- Executive summary
- List of exhibits submitted for examination
- Statement on which exhibits are questioned source and which have a known source
- Statement of chain of custody of exhibits
- Statement of the examination request
- Statement of domain irrelevant information that was received
- Statement as to the qualifications and experience of the examiner
- Methods and materials
- Procedures
- Results (observations and discussion)
- Conclusions

With each point the authors provide guidance, with particular emphasis on the qualification and experience of the examiner (about which the authors state that 'since the majority of the pattern evidence sciences rely heavily on the interpretive skills of the examiner, detailed information should be provided regarding the background, training and experience of the individual evaluating the evidence') and the methods and materials. About the latter, the authors state:

> The method employed when making the comparison judgment should be described along with reference to relevant published works. This section should include a description of the reporting scale employed and the manner in which an interpretation is translated into a conclusion. Additional Issues which this section should address include:
>
> - Whether the method used has been described in detail in a publication and whether this method is widely accepted as the standard method. Where relevant, references should be provided.
> - Whether the method has been subjected to expertise testing for the purpose of showing whether expertise is required to conduct the analysis. If so, references should be provided.
> - Whether the method has been subjected to validation testing. If so, references should be provided. It should be noted whether the individual, whose findings are expressed in the report, and the peer review examiner, participated personally in the validation of the method.
> - Any likely limitations of the method, including limitations imposed by the quality of the evidentiary samples, along with their relevance to the current case. The established error rates for the procedure (both false positive and false negative) should be provided, or if these are not established this should be disclosed.
> - Any population database used to estimate the frequency of the properties used in the comparison process should be identified and described.

- A statement as to how relevant the population described is to the case under examination.
- Whether the individual, whose findings are expressed in the report, and the peer review examiner have participated in proficiency testing. If so, the proficiency test providers should be stated (e.g. Collaborative Testing Services (CTS)).

Evaluative opinions on forensic biology evidence given activity-level propositions, based on statistical modelling decisions and probability assignments, rely heavily on the training, experience and expertise of the scientist. As such, the guidance provided in this chapter should be considered relevant to the reporting of such opinions.

10.2.3 Australian Standard 5388.4 – 2013

Standards Australia (which is the Australian representative organisation of the International Standards Organisation (ISO)) has published standard 'Forensic analysis Part 4: Reporting' (accessed 23-12-2021, https://www.standards.org.au/standards-catalogue/sa-snz/manufacturing/ch-041/as--5388-dot-4-2013). The scope of the standard is formulated as:

> This standard sets out requirements and recommendations for the reporting of results and opinions, derived from scientific data and observation in the analysis and examination of physical material, for investigative purposes and which may be used in legal proceedings.

While a general standard for forensic reporting, it provides some guidance on the contents of an evaluative report. Section 9.1 of the standard describes the content of reports in general, as well as those intended for use in court:

> All written reports shall include –
>
> (a) the date of issue;
> (b) the name of the forensic facility;
> (c) a unique case identifier;
> (d) the name of the person responsible for the report;
> (e) a means of ensuring that each page is a part of the report; and
> (f) a means of signifying the end of the report.
>
> Reports prepared for legal proceedings should include the following (whether in the body of the report or an appendix):
>
> (i) Collection and continuity of forensic material.
> (ii) Analysis and comparison of material.
> (iii) Results.
> (iv) Limitations.
> (v) Opinions.
> (vi) Qualifications and experience of the author.
> (vii) Analysis and comparison of material.
> (viii) Definitions or explanations for technical terms used.

Of specific interest to reporting findings given activity-level propositions are requirements relating to opinions provided by the scientist. Section 9.5 of the standard provides guidance on the reporting of opinions, which is echoed in recommendations from other guidance documentation discussed below.

10.2.4 The European Network of Forensic Science Institutes (ENFSI) – 2015

'The purpose of ENFSI as a network of experts is to share knowledge, exchange experiences and come to mutual agreements in the field of forensic science' (from their website, March 2021). ENFSI adopted guidelines on evaluative reporting which were drafted as part of a project funded by the

European Commission [4]. On the content of a report that expresses an evaluative opinion, the guidelines state:

> 3.12 Reports should include (not exhaustive list):
>
> - Conditioning information used
> - Mandate and questions asked, if required
> - The propositions of interest
> - Relevant items collected/received
> - Items examined
> - Significant findings
> - **Discussion and evaluation**
> - Conclusion(s)
> - **A caveat that any change in conditioning information may require assessments, conclusions and/ or propositions to be reviewed**.

Two aspects are highlighted here (**bold**) that differ from the standards drafted by the AFSP. The ENFSI guideline states that the report should include a discussion and evaluation of the findings (which enhances transparency). An important second addition is a caveat *in the report* that the weight of the evidence is conditionally dependent on the propositions, assumptions, results and other task-relevant information (as we have discussed in earlier chapters in this book).

With the addition of the discussion and evaluation of the findings, the report will enhance transparency as to how the scientist arrived at their conclusions. The guideline however does not specify the level of detail in which the evaluation should be documented in the report.

10.2.5 ROYAL STATISTICAL SOCIETY (RSS) PRACTITIONER GUIDES – 2010–2015

The UK's Royal Statistical Society has drafted four law practitioner guides. Their purpose: 'The guides look at communicating and interpreting statistical evidence in the administration of criminal justice. They are intended to assist judges, lawyers, forensic scientists and other expert witnesses in coping with the demands of modern criminal litigation' (from the RSS website, March 2021).

The fourth part of these practitioner guides 'Case Assessment and Interpretation of Expert Evidence' [5] deals with reporting evaluative expert opinions. The authors state: 'Forensic reports can be structured around a set of generic headings. The following paragraphs summarise key headings that might – and ideally, should – be included, mapping onto the sequential stages of the CAI process'. After which the authors elaborate on the following four paragraphs:

A. 'Case circumstances' (or 'Background information')
B. 'Purpose'
C. 'Interpretation'
D. 'Conclusion'

For each paragraph, they describe what the content of the report should ideally be.

10.2.6 THE NATIONAL INSTITUTE OF FORENSIC SCIENCE (NIFS) – 2017

NIFS is a directorate under the Australia New Zealand Policing Advisory Agency (ANZPAA). ANZPAA is a collaborative agency connecting the Police organisations of Australia and New Zealand. As part of their mission, they promote the quality and consistency of police practice. In 2017 NIFS published 'An introductory guide to evaluative reporting' [6].

The guide provides an overview of the process that leads to an evaluative opinion including the report that follows. On reporting the guide echoes what earlier guidance documents have said:

> The explanation of the whole process for the fact-finder, including the information and method used to form propositions, the available case items, the analyses carried out on them, the resulting observations, the method of evaluation and the conclusions drawn from it. Assumptions made during the evaluative processes should be made clear to the fact-finder.

The guide continues with a more detailed explanation of how evaluative reports differ from other types of reports (like technical reporting and investigative reporting). More importantly, the guide also provides guidance for the implementation of evaluative reporting in a forensic science institute. Implementation of 'activity-level reporting' is the topic of the next chapter of this book.

10.2.7   THE SUPREME COURT OF VICTORIA (AUSTRALIA) – 2017

The Supreme Court of Victoria has published practice note SC CR 3 in January 2017 (accessed 23-12-2021, https://www.supremecourt.vic.gov.au/law-and-practice/practice-notes/sc-cr-3-expert-evidence-in-criminal-trials). It is a re-issue of a note published in 2014 and has been unchanged. This note has been written by lawyers, legal scholars and forensic scientists and has been approved by the supreme court judges. The scope of this note, which can be used in criminal trials to obtain relevant information from the forensic scientist, has been defined as:

> The purposes of this Practice Note are:
>
> - to enhance the quality and reliability of expert evidence relied on by the prosecution and the accused in criminal trials and proceedings under the Crimes (Mental Impairment and Unfitness to be Tried) Act 1997.
> - to encourage the early identification of issues in dispute that will be the subject of expert evidence.
> - to improve the utility of expert evidence by ensuring that it is focused on the issues genuinely in dispute.
> - to make use of existing pre-trial and trial processes at the earliest practicable opportunity to advance these purposes.

To attain these goals, the practice note specifies what information a forensic science report should contain. Section 6 of the note reads:

> 6.   *CONTENT OF ALL EXPERT REPORTS*
>
> 6.1   All expert reports to which this Practice Note applies (including primary expert reports and responding expert reports) shall state the opinion or opinions of the expert and shall state, specify or provide –
>
> (a)   the expert's name and place of employment;
> (b)   an acknowledgement that the expert has read this Practice Note and agrees to be bound by it;
> (c)   **whether and to what extent the opinion(s) in the report are based on the expert's specialised knowledge, and the training, study experience on which that specialised knowledge is based;**
> (d)   **the material, observed facts, reported facts, assumed facts and other assumptions on which each opinion expressed in the report is based (a letter of instructions may be annexed);**
> (e)   **(i) the reasons for, (ii) any literature, research or other materials or processes relied on in support of, (iii) a summary of – each such opinion;**
> (f)   **(if applicable) that a particular question, issue or matter falls outside the expert's specialised knowledge;**

(g) any examinations, tests or other investigations on which the expert has relied, identifying the responsible laboratory by which, and the 4 relevant accreditation standard under which, the examination, test or other investigation was performed;

(h) a declaration that the expert has made all the inquiries and considered all the issues which the expert believes are desirable and appropriate, and that no matters of significance which the expert regards as relevant have, to the knowledge of the expert, been withheld;

(i) any qualification of an opinion expressed in the report, without which the report would or might be incomplete or misleading;

(j) any limitation or uncertainty affecting the reliability of (i) the methods or techniques used; or (ii) the data relied on – to arrive at the opinion(s) in the report; and

(k) any limitation or uncertainty affecting the reliability of the opinion(s) in the report as a result of – (i) insufficient research; or (ii) insufficient data.

6.2 Where an expert is aware of any significant and recognised disagreement or controversy within the relevant field of specialised knowledge, which is directly relevant to the expert's ability, technique or opinion, the expert must disclose the existence of that disagreement or controversy.

This note goes beyond the guidance and recommendations that other sources provide. Much attention is given to the argumentation to support opinions and assumptions that are being made, stressing that these should be made transparent. Evidently, this note pertains to criminal trials in the state of Victoria and not beyond, but recommendations highlighted (bold print) are likely to be applicable to any jurisdiction. Specific legal requirements however may differ between jurisdictions.

10.2.8 THE INTERNATIONAL SOCIETY FOR FORENSIC GENETICS (ISFG) – 2020

The ISFG is an international professional organisation for forensic biology. DNA commissions of the ISFG address current topics in forensic genetics and provide recommendations for best practices. In 2018 and 2020, a DNA commission of the ISFG published a two-part paper on 'Assessing the value of forensic biological evidence – Guidelines highlighting the importance of propositions' [7, 8]. The second part focused on the interpretation of forensic biology findings given activity-level propositions. In this chapter, the DNA commission made a recommendation based on the ENFSI guidelines:

Recommendation 9: In the statement, the source of the data that are used to provide the probabilities should be disclosed, as advised in the "ENFSI guidelines for evaluative reporting".

This recommendation again emphasises the need for transparency in the interpretation of the findings by the scientist and to disclose the sources of scientific data used to support their opinion.

10.2.9 SUMMARY OF RECOMMENDATIONS FOR REPORTING GIVEN ACTIVITY-LEVEL PROPOSITIONS

The practitioner guide from the Royal Statistical Society [5] states about interpretation in an evaluative report:

The expert should lead the reader through the process of assigning probabilities for the scientific findings, describing and explaining whatever data have been relied on from (specified) other sources, as well as any information derived from the scientist's own personal experience. It should be possible for an informed reader to retrace the logic of the scientist's approach, recheck calculations or reconsider the scientist's findings in the light of changed assumptions. It should be crystal clear to anybody relying on the report how the expert arrived at her opinion, and what assumptions have been made.

This quote summarises quite clearly what the purpose of the written statement should be, and what all guidance documentation aspires to achieve. The recipient of the information should be able to reconstruct how the scientist came to their conclusion and on what scientific data they base their opinion.

In addition to this, it should be considered crucial to make two aspects of Bayesian reasoning in forensic reports very clear:

1. The scientist restricts themselves to expressing an opinion on the weight of the findings *given* the propositions, and not on the propositions themselves.
2. The weight of the evidence that is reported is crucially dependent on the conditioning information, e.g. the propositions, contextual information, assumptions and findings from the forensic analyses.

Example texts for such caveats in evaluative reports can be found in table 8 of [9]

10.3   VERBAL EQUIVALENTS

When attempting to communicate the results of examinations, and the subsequent interpretation of those findings and the conclusion that are being drawn from this, this is done in (mostly) plain language. The highly technical methodology that was used, as well as the statistical modelling and calculations performed, need to be translated into verbal statements that properly reflect those aspects. As words may have different meanings in different contexts, there is always a risk of mis-interpretation if that context differs between the sender and receiver of the information. This is especially so with terms that relate to quantities. One person may for instance say that they consider something big or high, while it is entirely unclear to another person what the actual size is of the item that is being referred to.

In the context of forensic reports, this is an issue when communicating probabilities verbally, or the likelihood ratio itself. A scientist may say that the probability of a certain event occurring is low. Without a frame of reference, or some numerical boundaries to such a term, it will be unclear to the reader what is meant by 'low'. Evidently, this could lead to the reader not receiving the intended meaning of the message conveyed by the scientist.

Gill [10] has provided a table for probability terms. Table 10.1 shows an adaptation of the table by Gill, as it is currently in use in reports provided by the Netherlands Forensic Institute (NFI).

This table is included in evaluative reports in which verbal equivalents for probabilities are used. In that way the table provides a shared frame of reference for the scientist and the recipient of the report.

TABLE 10.1

Verbal Equivalents Used for Probabilities at the NFI (Translated from Dutch)

The Probability Is ...	Means That the Probability Assigned Is ...
Very low	Under 0.05
Low	Between 0.05 and 0.25
Reasonably high	Between 0.25 and 0.75
High	Between 0.75 and 0.95
Very high	Over 0.95

Note: Probabilities cover a range between zero and one. A probability of 0.05 can be read as a 5% chance, a 0.95 probability as a 95% chance.

There has been some debate on whether verbal equivalents should be used to express the weight of the findings in a report and, if they are, how their meaning should be conveyed. A good overview of the arguments can be found in a paper by Marquis et al. [11]. The reasons for using a verbal equivalent are listed by the authors as:

- Numerical data are not available.
- The expert feels more comfortable providing a verbal statement on the order of magnitude of the weight of evidence.
- The intended recipient has poor understanding of probabilistic reasoning and numbers may put them off.
- The legal community prefers words over numbers.
- Using the same verbal terminology will enhance harmonisation in forensic reports.

Whether or not these are considered valid arguments, there are several studies that point to the risks inherent in using verbal equivalents for the likelihood ratio.

Studies by Martire and co-authors [12, 13] have shown that verbal equivalents are less effective in communicating the intended strength of the findings than numerical expressions. They also describe a 'weak evidence effect' when findings with low weight of evidence are verbally expressed. This weak evidence effect means that when a low weight of evidence supporting proposition A is reported, this lowers the belief in proposition A rather than strengthening it. In other words, it is perceived as support for an alternative proposition rather than as support (although weak) for proposition A. Such an effect was not found when the likelihood ratio was expressed numerically.

On the other end, Garett et al. [14] as well as Van Straalen et al. [15] have shown that verbal terms like 'extremely strong support' may be overvalued. Van Straalen et al. studied the perception of different types of forensic conclusions commonly used in fingermarks comparisons. They found a tendency of different groups (jury, students, legal professionals, police) to overstate the meaning of the likelihood ratio when it was presented verbally.

These studies clearly demonstrate that there is a risk that the intended weight of the evidence communicated by the scientist is under- or overvalued by the recipient of the information. This has led an ISFG DNA commission [8] to make a recommendation on the use of verbal equivalents:

Recommendation 11: The verbal scale is optional but cannot be used by itself. If it is used, then the numeric equivalents must also be available/provided. In practice, one would provide first one's likelihood ratio, then the verbal equivalent is applied afterwards.

An example would be: 'My LR is in the order of 60. This means that – in my opinion – the results are much more probable (in the order of 60 times) if the proposition that 'X' is true than if the alternative 'Y' is true. According to the verbal scale used in our laboratory, these results provide moderate support for the first proposition rather than the alternative'.

These 'verbal equivalents' are by nature subjective and a number of verbal equivalence tables have been published. So, it is above all a convention.

This recommendation summarises a number of key issues in the use of verbal equivalents for the likelihood ratio. It also guides in the current best practice:

- Verbal equivalents are optional. Based on our current knowledge, numerical expressions are the best way to communicate the weight of evidence.
- If a verbal equivalent is used, it should be accompanied by the context in which it is used. This context should be a list with all verbal equivalents used by the institute, each accompanied by the range of numerical values that they cover.
- When the weight of the evidence is expressed verbally, ideally a numerical LR is provided first, followed by the verbal equivalent.

Several example verbal scales have been published. An overview of some of these is given in Table 10.2.

TABLE 10.2

Examples of Verbal Equivalent Scales for the LR

AFSP		ENFSI		NFI		NFC	
Verbal	Numerical	Verbal	Numerical	Verbal	Numerical	Verbal*	Numerical
No support for either proposition	~1	No support for either proposition	1	No support for either proposition	1	Level 0	1–6
Weak support	>1–10	Slightly more probable	2–10	Slightly more probable	2–10		
Moderate support	10–100	More probable	10–100	More probable	10–100	Level +1	6–100
Moderately strong support	100–1,000	Appreciably more probable	100–1,000	Appreciably more probable	100–10,000	Level +2	100–6,000
Strong support	1,000–10,000	Much more probable	1,000–10,000				
Very strong support	10,000–1,000,000	Far more probable	10,000–1,000,000	Far more probable	10,000–1,000,000	Level +3	6,000–1,000,000
Extremely strong support	1,000,000 and up	Exceedingly more probable	1,000,000 and up	Exceedingly more probable	1,000,000 and up	Level +4	1,000,000 and up

*Note that the scale in use by the National Forensic Center (formerly SKL, Sweden) also has verbal equivalents of level −1 to −4 if the findings support the alternative proposition. The ranges for the level of support mirror those of the positive scale.

The verbal equivalent should be read as: 'The findings are [verbal equivalent] if proposition A is true, rather than if proposition B is true', or as: 'The findings provide [verbal equivalent] for proposition A over proposition B'. Sources are ASFP [2], ENFSI [4], NFI [16] and NFC [17].

While the examples shown in Table 10.2 are not exhaustive (more have been published [18, 19]), there is generally some overlap in the terminology used. The terms are however not necessarily used for the same numerical LR. Note for instance that 'much more probable' may refer to an LR of 100 to 10,000. While verbal equivalents are thus subjective terms there are some general 'rules' for the use of verbal equivalents:

- Findings resulting in an LR = 1 of one will (and should) in all instances be considered 'neutral' or providing 'no support' for one proposition over another.
- The same verbal equivalents should be used for all evaluations regardless of the evidence type that generated the LR. From this it follows that, if verbal equivalents are used by an institute, a standardised list of verbal equivalents should be adopted by that institute. The numerical range that verbal equivalents cover should be the same for all disciplines and all experts. This follows from the logical concept that an LR of 100 in fibre evidence provides the exact same level of support for a proposition as does an LR of 100 in DNA evidence for the same proposition.
- There are limits to the use of words expressing extremely high or low values of the likelihood ratio. In other words, there is a lack of sufficient superlatives to express orders of magnitude that may be encountered with statistical calculations of the weight of the evidence. How would one for instance distinguish between LRs with orders of magnitude of 10^6, 10^9 or 10^{26}? The scale of verbal expressions will thus necessarily be capped at an arbitrary high value, and anything above will be referred to with the same verbal equivalent.
- It is ok to go from numbers to words but not the other way around. Hence, a verbal expression should follow from a calculation of the LR based on probabilities assigned (whether based on hard experimental data or entirely subjective). It is difficult to conceive how a personal verbal expression (I feel that these findings are much more probable if H1 is true rather than H2) could ever result in the conclusion that the findings are thus 1,000–10,000 times more probable under H1 than under H2.

10.4   OTHER FORMS OF COMMUNICATION

In the course of a criminal investigation, multiple moments of contact may occur between the scientist and other parties in the criminal justice system. Chronologically, these could for example be:

- a case intake in the investigative phase of the case,
- questions by police on the meaning of findings reported on at source level,
- a request for a case pre-assessment by the prosecution or defence counsel,
- court appearance and testimony.

At each of these occasions, issues at the activity level may be part of, or central to, the discussions. In this section we share some insight from personal experience as well as those of others on these matters.

10.4.1   PROVIDING AN INVESTIGATIVE OPINION

At any point during the investigative phase of a case, the scientist may be asked to place their findings in the context of the case. This could for instance be when a report has been submitted which addresses examination of samples and items from a crime scene, in which the findings are evaluated at the source level. Questions may arise on their meaning of such findings given specific scenarios that police are considering at that time on activities that have taken place. At this point the scientist may, based on their knowledge and experience, provide *explanations* for their findings that they consider reasonable. One could imagine providing the following opinion in a case meeting or email to the police officer in charge of the case:

> We found DNA in three samples taken from locations at the crime scene. They are single source DNA profiles, and we can establish that they correspond to each other and may thus be derived from the same individual. The DNA is found in samples taken from the inside windowsill of the broken window, from the handle of the cabinet that has been opened and from which items have been removed, as well as from the inside door handle of the back door through which the burglar left the scene.
>
> Given that we would expect to find DNA of the inhabitant of the house on such locations, and given that we don't often find single source DNA profiles from trace DNA samples (let alone three samples), it is my opinion that the most likely explanation for these results is that the DNA has been left by the inhabitant rather than the burglar. We would need a reference sample of the inhabitant to verify or falsify this.

Here we have provided an explanation for the findings about the way in which the traces were deposited. Such investigative opinions may assist the police team in the direction of their investigation. In this instance, by questioning the inhabitant about their cleaning routine and possibly requesting a voluntary reference DNA sample, and possibly submitting other items for testing.

10.4.2   SHARING RESULTS FROM A CASE PRE-ASSESSMENT

Depending on the jurisdiction, the scientist may be asked to perform a case pre-assessment and share this with a mandating authority (be it police, prosecution or an investigative judge). Based on the outcome of such a pre-assessment, the mandating authority may be better informed to decide whether an examination of items of evidence is able to provide relevant information. A form of case pre-assessment before the start of any examinations is commonly performed. Whether this is an explicit pre-assessment (following a formal procedure) or implicit reasoning, the principle will be applied to nearly any case intake. Informally this is commonly called an 'exhibit triage' step.

A pre-assessment may also be requested after initial examinations have been performed, and the question is whether any additional examinations may provide additional relevant information. Additionally, a requesting authority may want to know if a formal evaluation of the findings given activity-level propositions has potential to lead to a conclusion that is of assistance to the court's deliberations.

Whether or not such assessments of existing findings in combination with a pre-assessment of additional work occurs may depend on the jurisdiction. In the Dutch inquisitorial system, evaluation of findings given activity-level propositions is recognised as a specific field of expertise that builds on but differs from source-level expertise. This can for instance be seen from the separate registration areas in the Netherlands Register for Court Experts (see Chapter 11, Section 11.1.4). Due to this separation, it is not up to the scientist to decide whether they report their findings given propositions at the (sub)source or activity level, such a decision needs to be made by a mandating authority (either a prosecutor or an investigative judge). As a formal evaluation given activity-level propositions tends to take quite some time and resources, it is not uncommon for the prosecutor or investigative judge to request a pre-assessment. Based on such a pre-assessment they may decide on whether or not to proceed with a formal evaluation of the findings, and with any additional examinations if required. Such a pre-assessment would be performed with probabilities assigned based on knowledge and expertise, but generally without explicit reference to any published data. Hence, it is a 'quick and dirty' assessment intended to give a rough indication of the potential of the findings to discriminate between case-relevant scenarios. Such a pre-assessment would generally contain:

• The issue as it is understood at that point,
• Suggested propositions for the mandating authority to decide on,

- Any assumptions that would need to be made,
- Other task-relevant information for the evaluation, including any additional task-relevant information being requested from both parties,
- Current findings,
- Expected level of discrimination of the findings given the propositions,
- Potential for additional examinations and possible outcome,
- Cost and delivery time of the report.

The case pre-assessment, particularly in situations where some initial findings are already available, is a form of communication about the relevance of those DNA examination results in the context of a case. Any opinion on these matters by the scientist may be taken by the users of such information in their decision-making process. It is therefore crucial that a case pre-assessment (be it communicated in a report, email or verbally) is properly based in logic, and that the expressed opinions are balanced (address at least two competing propositions), robust and that it is transparent to the users how the opinion is reached. With these 'quick and dirty' pre-assessments particularly this is crucial, and a disclaimer highlighting that the pre-assessment does not represent a formal evaluation of the findings based on scientific data may be required.

10.4.3 Court Testimony

We must first recognise that a courtroom is not the place to provide investigative opinions. The scientist can and should provide relevant explanations for their findings when the investigation team or the findings call for it. However, in the courtroom we are by its very definition at the end of the evaluative phase of the case. The scientist will therefore refrain from providing explanations for their findings like 'In my opinion these findings are more likely to be caused by primary than by secondary transfer' or 'It is possible that these traces were deposited when the defendant tied up the victim'. When questioned in court about these issues, it is crucial that the scientist explains the scientific framework they operate in, explain the hierarchy of propositions and explain what an evaluative opinion entails. They should refrain from answering leading questions on the likelihood of any scenarios or propositions given their findings (as was discussed in Section 1.6).

We must recognise that it is not possible to perform a formal evaluation of the case findings on the stand. Acquiring the task-relevant information, setting propositions and assigning a probability to the findings given the set of task-relevant information, while basing yourself on relevant scientific studies, are simply not something that can be done on the fly. Such an evaluation requires a thorough study of the relevant literature, assigning probabilities based on that data and performing sensitivity analyses to make sure that the evaluation (and the resulting LR) is robust. It is therefore important that all parties in a jurisdiction understand that such issues are best put to the scientist before the trial.

When being questioned on a report that was submitted on the evaluation of findings given activity-level propositions it is the duty of the scientist to explain the statistical framework they used, the basis for their probability assignments, and how that led to their conclusion. Communicating complex statistical evaluations, including those that are part of an evaluation of findings given activity-level propositions, to a lay audience is an area of ongoing research. There have been several studies into the efficacy of communication styles and formats, whether that be numerical or verbal. In a review of studies into juror comprehension, Eldridge [20] found that there have been numerous suggestions for communication styles, some numerical, some verbal and some visual. While highlighting many deficiencies in our knowledge of the topic, their three conclusions from the literature were:

- Examiners should avoid the term 'match' as it is perceived more strongly by lay people than intended by scientists,

- The scientist cannot assume the jury have an understanding of any key terms or concepts and that more explanation (particularly of scientific terms) is always better, and
- Visual aids are effective in representing probabilistic concepts.

There have been various studies that show jurors do not understand probability statements (even on topics they deal with every day, such as weather forecasts [21]) or do not assign the same weight to the expressions of evidential value as intended by the scientist [12, 14]. Therefore, there is clearly a need for additional research into the most effective manner of communication of probabilistic findings, and this is likely to encompass the fields of both forensic science and cognitive psychology.

10.5   EXAMPLE REPORTS

An early description of evaluative reports as drafted by the former Forensic Science Service in the United Kingdom [22] provides some insight into how the setup of the report can – and possibly should – follow the flow of the case assessment and interpretation process. In their paper they discuss the chapters or paragraphs that make up the FSS evaluative reports:

- Mandatory requirements (which may be dictated by the jurisdiction)
- Framework of circumstances (task-relevant information received)
- Purpose (the purpose of the examination as discussed with the mandating authority)
- Technical issues (including explanation of methodologies applied)
- Examination and results (all *relevant* observations)
- Interpretation (assessment of the probability of the findings given the propositions)
- Conclusion
- Appendix

With the latter explained as: 'The purpose of the Appendix is to isolate the more esoteric technical information and results from the main body of the statement'.

This leads us to consider the communication of the statistical model used in our evaluation of the findings, be it a Bayesian network or hand-derived formula. There has been some debate on whether or not a Bayesian network should be part of the report submitted by the scientist. Sjerps and Berger [23] have discussed this point in some detail in their paper 'How clear is transparent?'. They see a trade-off between clarity and transparency, where full disclosure of all information may lead to reduced understanding by the recipient of the information. The recipient needs to make a decision based on the information provided to them. To do this, they need to be able to follow the scientist's reasoning and see how they arrived at their conclusions. And while this may be reasonable with relatively simple models, it may not be reasonable to assume that a jury or judge is able to grasp the intricacies of Bayesian networks without careful guidance by the scientist. Sjerps and Berger [23] use the analogy of a pilot who needs to understand how to operate a plane to be able to fly it, but does not need to know all the inner workings of the mechanics to do so. They conclude with the following recommendations with respect to reporting an evaluative opinion:

We summarise our conclusions and recommendations below:

- We have defined expert reasoning as the argumentation showing how the conclusion follows from the results. It is the output of a thought process that involves many aspects (concerning deriving questions, selecting and structuring information, formulating and checking a conclusion).
- Transparency of reasoning concerns two decisions: which aspects to report and at what level of detail.
- Transparency is required because the legal practitioner needs to understand the expert's reasoning to check whether the results support the conclusion. However, in complex cases, the legal practitioner can only understand the reasoning up to a certain level.
- There can be a trade-off between clarity and transparency.

- The scientist should only report those aspects of his reasoning that have significant added value to the argumentation. Other aspects should be documented in the case file but omitted from the report for sake of clarity.
- The expert should aim for a high level of detail only for crucial parts of the reasoning. Less important arguments can be presented at a more global level. The report should make clear that only the main arguments are presented and that further details are available on request.
- Legal practitioners and scientists should not tell each other how to do their job in the setting of legal cases, but discussion between the two groups in other settings is to be promoted.
- There is currently room for improvement in the expert's reasoning itself, and sometimes in reporting it. The likelihood ratio framework and Bayesian networks promote transparency and logic.
- Transparency requires making clear whether a conclusion is a consensus and reporting diverging opinions on request.
- We recommend that reporting guidelines explicitly address transparency of expert reasoning.

In textboxes 10.1 and 10.2, we have reproduced the reports that were submitted in two of the cases that we have discussed throughout this book. These are example case 1 (the attempted kidnap) in textbox 10.1 and example case 2 (the three burglars) in textbox 10.2. The reports have been de-identified and rewritten to current best practice at the respective laboratories. The general setup of the report and, more importantly, the conclusions have not been changed compared to the formal case reports that were submitted. We encourage readers to compare the reports to the evaluations performed in Chapters 5 and 6 of this book to reflect on the way a formal evaluation of findings given activity-level propositions can be reported.

As can be seen when reading through these reports, different decisions have been made on the level of detail that was used to describe the interpretation of the findings.

Example report 1 is quite brief with the statistical model and the formal calculations provided in an annexe to the report. The report itself discusses the conditioning information and propositions as well as the probability assignments and the literature sources used. While maximally transparent, it may also, given the arguments raised by Sjerps and Berger [23], raise questions and detract from the arguments presented in the report.

Example report 2 is slightly longer but lacks an appendix. The documentation on the statistical model used is part of the forensic laboratory case file and is provided on request (as is made explicit in the report).

Another difference in the style of the reports is the way in which the sources of data are presented that support the probability assignments. In example report 1 the probabilities are assigned numerically, and the sources are discussed in the main body of the text. Example report 2 discusses verbal probabilities (e.g. a high probability or a low probability) in the main body of the text with reference to footnotes that contain details on the sources of the data that have been used. The numerical data are part of the documentation of the Bayesian network, which is not part of example report 2.

Both reports restrict the discussion of probabilities to those that are a parameter in the calculation of the LR. Irrelevant probabilities are not discussed as that may detract from the main message. An example of such an irrelevant probability would be the probability of finding background DNA on the hair of the victim in example case 2, where no DNA of unknown individuals was found.

Both reports, in their own way, follow the general guidance provided in Sections 10.2 and 10.3 of this chapter. Crucially,

- Both reports provide a numerical LR followed by a verbal equivalent (which in both instances is linked to the scale of verbal statements and associated numerical LRs).
- Both reports also flag important caveats with their conclusion; that the weight of the evidence relates to the findings and not the propositions themselves, and that the weight of the evidence is conditional on the information that was provided.

The format and content of the reports containing opinions on findings given propositions at the activity level that are provided by the authors have changed over the years and are likely to continue to do so. Best practices will continue to develop and as such the example reports provided here in Sections 10.5.1 and 10.5.2 reflect the current state of the art. They are intended to assist and inspire those who start reporting given activity-level propositions. Those who do are encouraged to develop their own reporting style, following guidance provided by relevant bodies and adapting to specific requirements in their own jurisdictions.

BOX 10.1   REPORT FOR EXAMPLE CASE 1 – THE ATTEMPTED KIDNAP

Forensic Science Laboratory

EVALUATIVE REPORT ON THE EXAMINATION OF ITEMS IN THE CASE INVOLVING:

Mr D

(Defendant)

and

Ms C

(Complainant)

by

The Scientist

Forensic Scientist

Case Number: 123456

Incident Report Number(s):   111111

Report dated DD/MM/YYYY

This report should be read in conjunction with the report dated DD/MM/YYYY identified by The Laboratory case number 123456, written by The Scientist

Contents of this report:

1.0   – CASE INFORMATION

1.1   Request for Work

I was asked by the Ms Prosecutor on DD/MM/YYY to carry out an evaluation of the findings in this matter considering activity-level propositions. The case has undergone an appeal, in which both the prosecution and defence version of events were provided, and the court has requested any further information available to assist with their decision.

1.2   Case Scenario

The prosecution case stated that the victim in this matter (C) was walking along Prospect Road on 24/11/2010 wearing a singlet top. It is alleged that a car pulled over and a single male (D), exited, grabbed C by the right upper arm and attempted to pull her into the vehicle. C struggled with D, which included her hitting him in the chest and he abandoned his attempt to kidnap C and drove off.

I will refer to this as the prosecution scenario.

C called police within an hour stating that she had been attacked. Shortly after police attended the home of D and seized his clothing that he had been wearing at the time in question. The following day police seized as evidence the clothes worn by CF.

D stated that he was driving home along Prospect Road, but never stopped his vehicle. He also claims to have had no previous contact with C either directly or indirectly as far as he can recall. It is being suggested by the defence Barrister acting for D that someone other than D must have attempted to kidnap C.

I will refer to this as the defence scenario.

1.3   Access to information

While carrying out my evaluation and writing my report I had access to:

- South Australian Police submission forms relating to items submitted under case number 123456
- Transcripts of evidence given by all expert witnesses during the trial
- Copies of all expert reports tendered as evidence during the trial
- A copy of the Forensic Laboratory casefile 123456

2.0   – DESCRIPTION OF THE ITEMS RECEIVED
AND RESULTS OF EXAMINATION

Detailed descriptions of the items received, and the DNA results can be found in the report dated DD/MM/YYYY, with case number 123456, written by The Scientist (hereafter called the DNA report). The results that I will use in the evaluation given activity-level propositions are a summary of the findings presented in the DNA report and are:

1) *A top taken from C.* This was examined at the laboratory and a tape lift taken of the right shoulder strap. From a tape lift of the singlet top of C, a mixed DNA profile was obtained that could be explained by the presence of DNA from three individuals. C's alleles corresponded to the major component of the mixture and D was excluded as a minor contributor by the forensic scientist. A male friend of C, who had previously hugged her that day, could account for the alleles of one of the minor contributors. Y-STR profiling was conducted and a single-sourced profile containing the same alleles as C's male friend was obtained. No sign of a second male was detected from Y-STR analyses and so the conclusion was that the third component of the mixture from C's top was female. Therefore, it has been assumed that there is also no DNA detected of an alternate offender.

2) *A top taken from D.* This was examined at the laboratory and a tape lift of the chest area. From a tape lift of the top of D, a mixed DNA profile was obtained that could be explained by the presence of DNA from two individuals. D's alleles corresponded to the major component of the mixture and C was excluded as the minor contributor by the forensic scientist.

3.0 – EVALUATION

I explain my evaluation with regard to the points in Section 2.0. In my assessment, I have used Bayesian networks (graphical models for reasoning under uncertainty), as well as the literature available and my knowledge. I summarise below my reasoning and refer the reader to the Appendix for a full description.

The findings have been evaluated by considering two competing propositions:

H1) D struggled with C.
H2) D did not struggle with C.

The background information I am using is:

• D and C did not have any prior contact.

The assumptions I am making are:

• If D did not struggle with C, then some other male struggled with C (i.e. there is no dispute that a struggle occurred).

Exhibit 1 (Tapelift of the Top from C)

I have assessed the probability of the findings (i.e., as described in Section 2.0) given both prosecution and defence's propositions.

There are four mechanisms I have considered by which DNA matching the defendant could be present on C's top:

a) DNA was transferred to C's top during a struggle with D.
b) C's top (or the tapelift or DNA extract from the tapelift) was contaminated with the DNA of D during forensic analysis.

Under the prosecution case, both mechanisms are considered.

Under the defence case, there is one mechanism by which the D's DNA could have come to be on C's top, mechanism 'c' from above.

The absence of D's DNA on C's top under the prosecution proposition would occur if during the struggle there was no transfer of DNA from D to C's top and there was no contamination.

The absence of D's DNA on C's top under the defence proposition would occur if there was no contamination within the laboratory.

For the probability of transfer to fabric, I use the result of Daly et al. [24]. Because there is no distinction between holding and hitting in the study of Daly et al., I will assume the probability of no transfer from grabbing is the same as the probability of no transfer during hitting, which is 0.53. I also use the same value for the probability of DNA transfer due to hugging.

For the probability of contamination, I will use the results from internal laboratory Quality Assurance monitoring. Using these monitoring systems, I have assigned a probability of contamination of the kind being considered for this evaluation as 0.0001.

For the presence of background DNA on C's top, I use the work of Breathnatch et al. [25] and Berge et al. [26] who found that approximately 90% of tops tested possessed background DNA.

The results, i.e.

• no DNA from D
• no DNA from an unknown male
• DNA from C's friend

detected on C's top, are approximately equally probable given defence's proposition to prosecution's proposition.

Exhibit 2 (Tapelift of the Top from D)

I have assessed the probability of the findings (i.e., as described in Section 2.0) given both prosecution and defence's propositions.

There are three mechanisms I have considered by which DNA matching the C could be present on D's top:

a) DNA was transferred to D's top during a struggle with C.
b) D's top (or the tapelift or DNA extract from the tapelift) was contaminated with the DNA of C during forensic analysis.

Under the prosecution case, mechanisms 'a' and 'b' are considered.

Under the defence case, there is one mechanism by which C's DNA could have come to be on D's top, mechanism 'b' from above.

The absence of C's DNA on D's top under the prosecution proposition would be because during the struggle there was no transfer of DNA from C to D's top and there was no contamination.

The absence of C's DNA on D's top under the defence proposition would be because he did not have any contact with C and there was no contamination within the laboratory.

For the probability of transfer to fabric, I use the result of Daly et al. [24]. Because there is no distinction between holding and hitting in the study of Daly et al., I will assume the probability of no transfer from grabbing is the same as the probability of no transfer during hitting, which is 0.53.

For the probability of contamination, I will use the results from internal laboratory Quality Assurance monitoring. Using these monitoring systems, I have assigned a probability of contamination of the kind being considered for this evaluation as 0.0001.

For the presence of background DNA on D's top, I use the work of Breathnatch et al. [25] and Berge et al. [26] who found that approximately 90% of tops tested possessed background DNA.

The results (i.e., no DNA of C on D's top) are approximately twice as probable given defence's proposition than prosecution's proposition.

4.0 – CONCLUSION

The consideration of the factors of transfer and the data upon which I have relied in forming my opinion are provided in the Appendix. The findings have been evaluated by considering two competing propositions from prosecution (H1) and defence (H2):

H1) D struggled with C.
H2) D did not struggle with C.

The background information I am using is:

- D and C did not have any prior contact.

The assumptions I am making are:

- If D did not struggle with C, then some other male struggled with C.

Value of the Results

After evaluating the findings in light of the two scenarios given above, I conclude that it is approximately two times more probable to have obtained the observed results if the course of events occurred in the manner described by the defence scenario rather than in the manner described by the prosecution scenario.

This conclusion should be seen as a reinforcing factor to the perception of the propositions that existed before the technical evidence was taken into account. I do **not** give an assessment of how probable it is that the prosecution or defence proposition is true.

The results therefore slightly support the proposition that D and C did not struggle, rather than the proposition that they did struggle.

I remain at your disposal for any questions and/or additional information.

Signature
DD/MM/YYYY
The Scientist | Forensic Scientist
The Laboratory
Laboratory address
P: phone numbers| F: fax number

APPENDIX TO THE REPORT

Formal Calculations That Are Relevant to the Opinions Provided in the Main Report

In order to help address the propositions that are relevant to this matter, I have considered the findings of the DNA report. By considering the mechanisms of transfer, I have constructed the Bayesian network (BN) shown in Figure 10.1. BN is ideally suited to deal with the complexity of such a case. For further information see the review article:

Taylor, D., B. Kokshoorn, and A. Biedermann, Evaluation of forensic genetics findings given activity level propositions: A review. *Forensic Science International: Genetics*, 2018. 36: p. 34–49.

Background Information Considered During Evaluation

- D and C had no contact prior to the alleged attempted kidnapping.
- Someone attempted to kidnap C (note that whether any attempted kidnapping occurred is a matter that was later raised in the media, and we will investigate the effects of a disputed activity later).
- C hugged her friend within an hour of the attempted kidnapping.
- The offender was male.
- The contact between the hands of C and D's top was made at the area tapelifted.
- The contact between the hands of D and C's top was made at the area tapelifted.

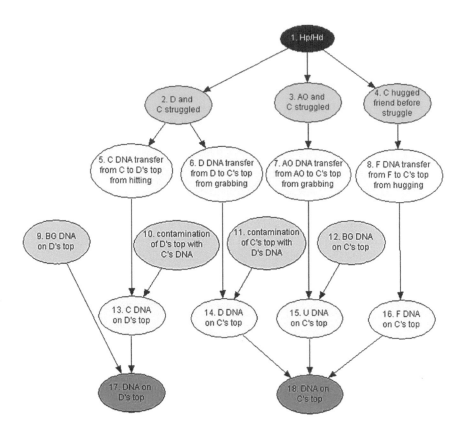

FIGURE 10.1 Bayesian network considering the case scenarios. Nodes are coloured so that black is the propositional node, blue are the activity nodes, grey are the prior decision nodes, yellow are the calculation nodes and pink are the findings nodes.

Assumptions Made During Evaluation

- The DNA of C's friend (F) was present on C's top.
- C's DNA was present on her own top.
- There was no DNA from an alternate offender found on C's top.
- Neither C, D and/or the offender were wearing gloves or had washed their hands within a few minutes of the offence.
- Neither C, D and/or the offender had any skin conditions that meant that would be particularly high shedders of DNA.
- Neither C nor D had washed the clothing that was seized by police between the time of the alleged offence and the seizure.
- The clothing supplied by C and D were the items worn at the time of the alleged offence.
- The items of clothing were appropriately stored and packaged after seizure and prior to submission to the forensic laboratory.

Node 1 (Hp/Hd): The proposition node with possible values of 'Hp' or 'Hd'. The 'Hp' option is the prosecution scenario as given in Section 4 of the main report and 'Hd' considers the defence scenario as given in Section 4 of the main body of the report. I assign these two options with equal prior probabilities. Note that this does not mean the prior odds in this case are equal, I apply equal prior probabilities for the propositions so that the values obtained by the BN inform me of the likelihood ratio.

1. Hp/Hd	
Hp	0.5
Hd	0.5

Node 2 (D and C struggled): Considers whether D struggled with C. Under Hp this has occurred and under Hd it has not.

1. Hp/Hd		Hp	Hd
2. D and C	Yes	1	0
struggled	No	0	1

Node 3 (AO and C struggled): Considers whether an alternate offender struggled with C. Under Hp there was no alternate offender. Hd specifies that D did not struggle with C but instead there was an alternate offender (AO) who struggled with C.

1. Hp/Hd		Hp	Hd
3. AO struggled	Yes	0	1
with C	No	1	0

Node 4 (C hugged friend before struggle): Considers the act of C hugging her friend prior to the alleged offence. This has occurred under both propositions.

1. Hp/Hd		Hp	Hd
4. C hugged friend	Yes	1	1
before struggle	No	0	0

Node 5 (C DNA transfer from C to D's top from hitting): This node considers the probability of C's DNA being transferred to D's top if she hit him. For the probability of transfer of DNA to fabric, I use the result of Daly et al. [24] where cloth samples were held by volunteers for 60 seconds before sampling using tapelifts (the same sampling technique as using in the samples taken for this matter) and extracted using Qiagen® QIAamp DNA mini kit. Table 1 of the Daly study shows that 53 out of 100 samples produced less than 0.01 ng/µl of DNA and I will use this as an approximate cut-off for obtaining DNA profile information for Profiler Plus™. Because there is no distinction between holding and hitting, I will assume the probability of no transfer from grabbing is the same as the probability of no transfer during hitting, which is 0.53. Another study into the rate of DNA transfer to clothing in simulated 'struggle' scenarios was carried out by Sethi et al. [27]. In this study 'assailant' volunteers were asked to grab the arms, elbows and wrists of 'victim' volunteers for 15 seconds, while they struggled to get free. Three types of cloth were grabbed (cotton, polyester and a cotton/polyester blend) and sampled after 12 hours and 7 days. DNA extraction was carried out on 0.5 cm² swatches of cut cloth using the QIAamp DNA micro kit. From the 27 samples tested at 12 hours, only three yielded a detectable amount of DNA. This is approximately 11% of samples. I will use

a transfer rate of 0.47 from the first study for two reasons. The sample size in the Daly et al. (2013) study is much larger than the Sethi et al. (2013) study and the sampling technique is equivalent to that used in this matter (i.e. tapelifts).

The Daly paper also goes on to produce DNA profiling results but uses a different DNA profiling system to that in this matter, and hence those results are less applicable than the quantification result. Only fabric samples that yielded 0.03 ng/µl were profiled in the Daly et al. [24] study due to the low probability of obtaining a useable (defined by the Daly et al. [24] authors as obtaining six alleles they could attribute to a single contributor) DNA profile. The quantification value of 0.01 ng/µl was used as a cut-off in this work because (a) it was the lowest quantification category recorded in the study, and (b) it is the author's experience with using Profiler Plus™ that DNA profiles would regularly not be obtained from less than this concentration of DNA.

2. D and C struggled		Yes	No
5. C DNA transfer from C to D's top from hitting	Yes	(47+1)/102 ~0.47	0
	No	(53+1)/102 ~0.53	1

Node 6 (D DNA transfer from D to C's top from grabbing): This node considers the probability of D's DNA transferring to C's top. This node is set up in the same manner as node 5 and uses the same probabilities.

2. D and C struggled		Yes	No
6. D DNA transfer from D to C's top from grabbing	Yes	0.47	0
	No	0.53	1

Node 7 (AO DNA transfer from AO to C's top from grabbing): This node considers the probability of AO's DNA transferring to C's top. This node is set up in the same manner as node 5 and uses the same probabilities.

3. AO and C struggled		Yes	No
7. AO DNA transfer from AO to C's top from grabbing	Yes	0.47	0
	No	0.53	1

Node 8 (F DNA transfer from F to C's top from hugging): This node considers the probability of F's DNA transferring to C's top. This node is set up in the same manner as node 5 and uses the same probabilities.

4. C hugged fried before struggle		Yes	No
8. F DNA transfer from F to C's top from hugging	Yes	0.47	0
	No	0.53	1

Node 9 (BG DNA on D's top): Considers the probability of background levels of DNA being present on D's top. To inform this node, I use the work of Breathnatch et al. [25], who found that 87.3% of garments (in that case underwear) demonstrated detectable background DNA. While the study used underwear and the case involves tops, a secondary study by Berge et al. [26] also found similarly high levels of background DNA on gloves (90%, $n = 20$), trouser ankles (100%, $n = 48$) and armpit of shirts (100%, $n = 16$).

9. Background DNA on D's top		
	Yes	0.9
	No	0.1

Node 10 (contamination of D's top with C's DNA): This node considers the possibility that the DNA of C could, through laboratory process, have come to be on D's top. In this instance, the reference of D was processed at all stages on different batches to the sample taken from C's top. Additionally, the tapelift taken from the top of D was processed at different times than the tapelift taken from the top of C. The Laboratory has never observed a contamination event such as this in approximately 10,000 extraction batches; therefore, the values of 0.0001 and 0.9999 have been assigned for yes and no, respectively.

10. Contamination of D's top with C's DNA		
	Yes	0.0001
	No	0.9999

Node 11 (contamination of C's top with D's DNA): This node considers the possibility that the DNA of D could, through laboratory process, have come to be on C's top. It has the same properties as Node 8 and so the same probabilities have been assigned.

11. Contamination of C's top with D's DNA		
	Yes	0.0001
	No	0.9999

Node 12 (BG DNA on C's top): Considers the probability of background levels of DNA being present on C's top. This node is set up in the same manner and with the same probabilities as I node 9.

12. Background DNA on C's top		
	Yes	0.9
	No	0.1

Node 13 (C's DNA on D's top): Is an accumulation node that considers the different pathways for C's DNA to be deposited on D's top?

5. C DNA transfer from C to D's top from hitting		Yes		No	
10. Contamination of D's top with C's DNA		Yes	No	Yes	No
13. C's DNA on D's top	Yes	1	1	1	0
	No	0	0	0	1

Node 14 (D's DNA on C's top): Is an accumulation node that considers the different pathways for D's DNA to be deposited on C's top?

6. D DNA transfer from D to C's top from grabbing		Yes		No	
11. Contamination of C's top with D's DNA		Yes	No	Yes	No
14. D's DNA on C's top	Yes	1	1	1	0
	No	0	0	0	1

Node 15 (U's DNA on C's top): Is an accumulation node that considers the different pathways for unknown DNA to be deposited on C's top?

7. AO DNA transfer from AO to C's top from grabbing		Yes		No	
12. Background DNA on C's top		Yes	No	Yes	No
15. U DNA on C's top	Yes	1	1	1	0
	No	0	0	0	1

Node 16 (F's DNA on C's top): Is an accumulation node that considers the different pathways for F's DNA to be deposited on C's top?

8. F DNA transfer from F to C's top from hugging		Yes	No
16. F's DNA on C's top	Yes	1	0
	No	0	1

Node 17 (DNA on D's top): Is an accumulation node that compiles the different combinations of sources that could have donated DNA to D's top?

9. BG DNA on D's top		Yes		No	
13. C DNA on D's top		Yes	No	Yes	No
17. DNA on D's top	C+U	1	0	0	0
	U	0	1	0	0
	C	0	0	1	0
	None	0	0	0	1

Node 18 (DNA on C's top): Is an accumulation node that compiles the different combinations of sources that could have donated DNA to C's top?

16. F DNA on C's top		Yes				No			
15. U DNA on C's top		Yes		No		Yes		No	
14. D DNA on C's top		Yes	No	Yes	No	Yes	No	Yes	No
18. DNA on C's top	D+F+U	1	0	0	0	0	0	0	0
	F+U	0	1	0	0	0	0	0	0
	D+F	0	0	1	0	0	0	0	0
	F	0	0	0	1	0	0	0	0
	D+U	0	0	0	0	1	0	0	0
	U	0	0	0	0	0	1	0	0
	D	0	0	0	0	0	0	1	0
	None	0	0	0	0	0	0	0	1

Having populated the probability tables with values, I use the following results in the BN:

- No DNA from D or an alternate offender was detected on C's top, only the DNA from C's friend, F.
- No DNA from C was found on D's top, only unknown DNA was found.

When the above information is instantiated in the BN shown in Figure 10.1, the result is propagated through the BN to give the results seen in Figure 10.2.

Note that the probabilities shown in the 'Hp/Hd' node are posterior probabilities as they are a product of the prior information provided to the states within this node (seen in tables) and the information provided by me at other nodes. In the *BN* I have constructed, the two states that the 'Hp/Hd' node can take are given equal prior probabilities. The *LR* produced is from a division of the posterior probabilities $\dfrac{\Pr(H_p \mid E)}{\Pr(H_d \mid E)}$; however, as the priors are equal, the likelihood ratio $\dfrac{\Pr(E \mid H_p)}{\Pr(E \mid H_d)}$ (without taking into account prior probabilities) is the same, i.e.

$$\frac{\Pr(H_p \mid E)}{\Pr(H_d \mid E)} = \frac{\Pr(E \mid H_p)}{\Pr(E \mid H_d)} \text{ when } \Pr(H_p) = \Pr(H_d)$$

In this matter $\dfrac{\Pr(H_p \mid E)}{\Pr(H_d \mid E)} = \dfrac{\Pr(E \mid H_p)}{\Pr(E \mid H_d)} \approx 0.53 \approx (1.9)^{-1}$

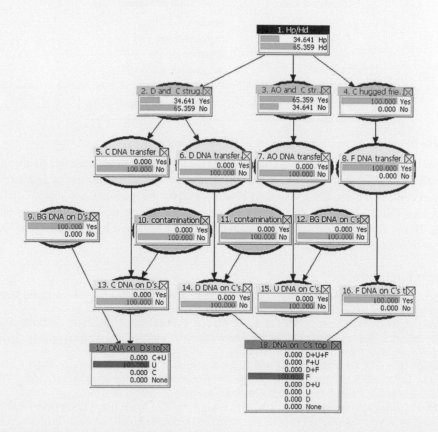

FIGURE 10.2 BN from Figure 10.1 with prior probabilities entered and information instantiated.

Background Explanation on the Hierarchy of Propositions

The classification in these different levels has been named the hierarchy of propositions: The higher the level, the more information and the more expertise (on behalf of the scientist) are needed, thus enabling more value to be added.

Examples of the different levels in the hierarchy of propositions are given hereafter:

Sub-level 1 (sub-source): The DNA originated from Mr A versus the DNA originated from an unknown person.

Level 1 (source level): The semen originated from Mr A versus the semen originated from an unknown person.

Level 2 (activity level): Mr A has sex with Ms B versus Mr A only socially contacted Ms B.

Level 3 (offence level): Mr A raped Ms B versus Mr A had consensual sex with Ms B.

The Verbal Equivalence Being Used

Likelihood ratios are numerical by nature. Words can be assigned to these numerical values that act as verbal descriptors of the level of support that likelihood ratio assigns to one proposition over the other. Verbal descriptions of numerical values are by their nature subjective, and a number of different scales exist within the forensic community. The Laboratory uses verbal descriptors based on those provided by Evett and Weir [19], which we give below.

Likelihood Ratio	Level of Support
1	Neutral
1–10	Limited support
10–100	Moderate support
100–1000	Strong support
1000–1,000,000	Very strong support
Greater than 1,000,000	Extremely strong support

Aspects Regarding Quality (Assurance/Management)

All aspects of relevant Quality Assurance and control have been maintained throughout the analysis of DNA samples in this case. This involves the minimisation of and screen for potential contamination. The Laboratory is accredited by the National Association of Testing Authorities (NATA) and all members of the scientific group undergo yearly proficiency testing. If any further information is required, it can be obtained by contacting the Quality Assurance Manager at The Laboratory.

Relevant Literature

D.J. Daly, C. Murphy, S.D. McDermott, The transfer of touch DNA from hands to glass, fabric and wood, *Forensic Science International: Genetics* 6 (2013) 41–46.

M. Breathnach, L. Williams, L. McKenna, E. Moore, Probability of detection of DNA deposited by habitual wearer and/or the second individual who touched the garment, *Forensic Science International: Genetics* 20 (2016) 53–60.

M. van den Berge, G. Ozcanhan, S. Zijlstra, A. Lindenbergh, T. Sijen, Prevalence of human cell material: DNA and RNA profiling of public and private objects after activity scenarios, *Forensic Science International: Genetics* 21 (2016) 81–89.

V. Sethi, E. Panacek, W. Green, J. NG, S. Kanthaswamy, Yield of Male Touch DNA from Fabrics in an Assault Model, *Journal of Forensic Research* (2013) doi: 10.4172/2157-7145. T1-001.

I. Evett, B. Weir, *Interpreting DNA Evidence: Statistical Genetics for Forensic Scientists*, Sinauer Associates, Sunderland, MA, 1998.

STATEMENT OF WITNESS

Statement of: The Scientist
Age: Over 21
Occupation: Forensic Scientist
Address: The Laboratory address
Telephone: Telephone number

This statement, consisting of **19** pages signed by me, is true to the best of my knowledge and belief.

Relevant qualifications of the scientist

Items have been examined relating to the matter of **C** (complainant) and **D** (Defendant).

An Evaluative Report, prepared by me, and relating to the examination of the items is attached. The report, signed by me, is identified by The Laboratory Case Number **123456**. I hereby incorporate that report as part of my statement.

Signed:

Signature
Date:
DD/MM/YYYY

BOX 10.2   REPORT FOR EXAMPLE CASE 2 – THE THREE BURGLARS

> Return address: PO Box 12345, Forensic Laboratory
Court
Addressed to examining magistrate
Address

EVALUATIVE REPORT

Evaluation of the results of an examination for biological traces given propositions at the activity level following a robbery committed in [Location] on [Date]

Case number   12345 (request number 5)
Date of request   date
Examining Magistrate's number   12/345
Public Prosecutor's Office number   12/345-67
Public Prosecutor   Prosecutor
Defendant   Defendant

1.   QUESTION

The examining magistrate requested that the findings of the examination for biological traces and DNA analysis in this case be evaluated given the following pairs of propositions:

With regard to the DNA found on the knife:

Proposition 1:   The defendant's DNA was found in the sample taken from the knife because he had held the knife.

Proposition 2:   The defendant's DNA was found in the sample taken from the victim's hair because he had grabbed her hair.

With regard to the DNA found in the victim's hair

Proposition 1:   The defendant's DNA was found in the sample taken from the knife because it was transferred via the hands of the person who had held the knife. This person was wearing the defendant's gloves, took them off, then held the knife with bare hands.

Proposition 2:   The defendant's DNA was found in the sample taken from the victim's hair because it was transferred via the hands of the person who grabbed her hair. The perpetrator was wearing the defendant's gloves, took them off, then grabbed the victim's hair with bare hands.

From these propositions and from further information received (see paragraph 3 of this report), it appears that:

1. The robber threatened the victim with a knife. The robber pushed the victim's head downwards with his left hand while pressing the blade of the knife against the victim's cheek with his right hand.
2. The defendant has declared that he was not involved in the robbery. At some point, the (unknown) perpetrator had worn gloves belonging to the defendant. After taking off the gloves, this unknown perpetrator pushed the head of the victim in a downwards direction and pressed the knife against her cheek.

Based on this information, and to place the focus on the activities that could have led to the set of traces, the propositions, with the written permission of the examining magistrate, were formulated as follows:

Proposition A: During the robbery the defendant pushed the head of the victim in a downwards direction, and pressed the knife against her cheek.

Proposition B: During the robbery somebody other than the defendant pushed the head of the victim in a downwards direction, and pressed the knife against her cheek. Before doing so, this person wore gloves belonging to the defendant.

2.   METHODS

In this report, the results of the examination for biological traces and DNA analysis conducted as part of this case were evaluated given propositions at the activity level.[1] The method[2] used by the expert is based on guidelines from the European Network of Forensic Science Institutes (ENFSI).[3]

The Issue and the Propositions

The propositions were formulated in consultation with the commissioning party. In drafting the propositions, the expert ensured they were consistent with the applicable guidelines and recommendations.[4,5,6]

Assumptions

For the purpose of the evaluation, assumptions have to be made. These assumptions relate to information that is missing or about which there is uncertainty. If this information is relevant to the evaluation of the findings, the expert must make assumptions about it. For each assumption, the expert explains the basis on which the assumption was made. If one or more assumptions are disputed, this could affect the conclusion presented in this report.

Assignment of Probabilities

As part of the evaluation, the expert has assigned probabilities. These include the probability that biological material could be transferred, the probability of it persisting, and the probability of it being detected in the forensic examination. Probabilities of finding biological material that is not related to specific criminal acts, such as 'background DNA', have also been assigned. Experts assign these probabilities on the basis of scientific publications or from studying the files of comparable cases. The circumstances will never be identical to those of a specific case. The probability estimate is therefore always partly based on an expert's knowledge of and experience with forensic examinations for biological traces and DNA analyses.

Since these probabilities cannot be determined with precision, experts in the field of human biological traces and DNA use verbal terms for these probabilities in their reports. These verbal terms are defined as follows[7]:

The Probability Is ...	Means That the Probability Assigned Is ...
Very low	Lower than 0.05
Low	Between 0.05 and 0.25
Relatively high	Between 0.25 and 0.75
High	Between 0.75 and 0.95
Very high	Higher than 0.95

Explanatory note: Probabilities cover a range from zero to one. A probability of 0.05 can also be read as a 5% probability, while a probability of 0.95 can also be read as a 95% probability.

Probability Theory and Evidential Value

In the evaluation, the probability of the findings of the examination for biological traces and the DNA analysis given each proposition was calculated. These calculations were performed with a Bayesian network.[8] Bayesian networks are statistical models that allow for calculations with conditional probabilities. In the evaluation, they are used to calculate the weight of the evidence based on the conditional probabilities that were assigned to DNA transfer, persistence, prevalence and recovery events. The Bayesian network and associated documentation can be provided upon request.

The evidential value of the findings that were calculated crucially depends on the propositions, the contextual information, the assumptions and the analysis results. If any of these elements change, it could affect the evidential value. A new evaluation can be performed under a different set of conditioning information if so required.

3. INFORMATION OBTAINED

The information used to evaluate the findings was obtained from:

- Letter concerning the appointment of an expert by the Examining Magistrate, dated [date];
- Official records of the reports made by the victim on [dates] to Police;
- Official records of the interrogations of the defendant on [dates] by Police;
- Email from Examining Magistrate, of [date] giving permission for the propositions to be reformulated as described in Section 1;
- The forensic laboratory case file, which a.o. contains the results from the forensic examinations.

The following information from the above documents was used to evaluate the findings of the examinations for biological traces and DNA analysis:

- The defendant is suspected of threatening the victim with a knife, pushing her head in a downwards direction and pulling her hair with his left hand, in her home, on [date] at approximately 8:25 pm.
- The defendant has declared that he was not involved in the robbery at the victim's home. He claims that an unknown perpetrator used the knife, pushed down the victim's head and pulled her hair. He also claims that at some point this person had worn gloves belonging to him.
- The defendant has never been in the victim's home (prior to the disputed incident).
- According to the victim's statement, the perpetrator was not wearing gloves while threatening her, at least not on his right hand.
- The knife originated in the victim's home.
- A male witness lives with the victim in the house where the robbery took place.

4. ASSUMPTIONS

For the purpose of evaluating the analysis results, the following assumptions were made:

- The knife that was examined is the knife that was used during the robbery and to threaten the victim.
- The perpetrator did not wear gloves during the incident.
- The defendant had no direct (physical) contact with the victim prior to the time of the incident, and had not been inside the victim's home on any previous occasion.
- The result of the DNA analysis of the sample taken from the handle of the knife is 100,000 times more probable if the sample contains DNA from the defendant than if this is not the case (see Section 5, analysis results). Based on this result, it is assumed, for the purpose of the evaluation given propositions at the activity level, that this sample contains DNA from the defendant.
- The result of the DNA analysis of the sample taken from the victim's hair is 10 million times more probable if the sample contains DNA from the defendant than if this is not the case (see Section 5, analysis results). Based on this result it is assumed, for the purpose of the evaluation given propositions at the activity level, that this sample contains DNA from the defendant.
- The items that have been examined for DNA analysis were packaged, stored and transported in accordance with prevailing forensic guidelines.

5. SUMMARY OF THE FINDINGS

For a complete overview of the analysis results, please refer to previous reports in this case dated [dates].

In the sample taken from the victim's hair, DNA originating from the defendant was found, in addition to DNA from the victim herself. There are no clear indications that DNA of unknown individuals was present in this sample.

In the evaluation we consider the findings from the sample of the hair of the victim to be the presence of DNA of the victim and the defendant.

In the sample taken from the knife handle, DNA originating from the defendant was found, in addition to DNA from the victim herself and DNA from the cohabiting witness. Based on the assumption that these three individuals have contributed their DNA to the sample, there were no clear indications that the DNA of unknown individuals was present in this sample.

In the evaluation we consider the findings from the sample of the handle of the knife to be the presence of DNA of the victim, the defendant and the witness.

6. EVALUATION OF THE FINDINGS

Given the hypotheses set out above, many factors play a role in the interpretation of the analysis results. In the evaluation, I restricted myself to a discussion of the circumstances that play an important role in determining the weight of the analysis results obtained.

6.1 Interpretation

With Regard to the Sample from the Handle of the Knife

According to the case information, the offender pressed the knife against the victim's cheek. Although it is unclear from the information received how long the offender handled and held the knife, in the evaluation it was assumed to be a relatively short period of time (one to a few minutes), during which time the handle was gripped with considerable force with bare hands. When a knife is held in this way, without gloves being worn, there is a high[9] probability of finding DNA on the knife handle from the person who held the knife.

This means that under proposition A, the probability is high that DNA from the defendant is recovered from the handle of the knife. Conversely, under proposition B the probability is low for *not* recovering DNA from the unknown offender.

With Regard to the Sample from the Hair of the Victim

According to the case information, the offender pushed the victim's head in a downwards direction and pulled her hair. In doing so, his bare hand came into contact with the victim's hair. Although it is unclear from the information received how long the offender pushed the victim's head downwards, in the evaluation it was assumed to be a relatively short period of time (a few seconds to one minute), during which time the head was held down with considerable force. Because of this, a high[10] probability is assigned to recovering DNA from the person who pushed the victim's head downwards and pulled her hair.

This means that under proposition A the probability is high that DNA from the defendant is recovered from the hair of the victim. Conversely, under proposition B the probability is low for *not* recovering DNA from the unknown offender.

With Regard to Transfer of DNA through Gloves

Given proposition B, DNA from the defendant may have been transferred to the knife handle and the hair of the victim via the hands of this unknown offender, after this person wore gloves belonging to the defendant.

In this route for indirect transfer of DNA from the defendant to the handle of the knife and the hair of the victim, we need to consider the probability that DNA from the regular wearer of the gloves (in this case the defendant) was present on the gloves when they were worn by the unknown offender, and subsequently, the probability that DNA from these gloves transferred to the hands of the offender while wearing it and persisting before handling the knife and the hair.

Studies have shown that the probability of recovering DNA of a wearer from gloves is very high.[11] There is no specific data available about DNA transfer from gloves to the hands of the wearer. Under the most favourable conditions (the gloves contain a high amount of DNA, the gloves are not worn in the intervening period, the unknown perpetrator wore the gloves for a long time prior to the disputed incident and removed them shortly before using the knife and grasping the victim's hair), I expect that the probability of transfer of DNA from the defendant from the gloves to the hands of the offender and subsequently persisting on these hands before handling the knife and hair would be comparable to the probability of such transfer via the shaking of hands. If the conditions were less favourable, this likelihood would decrease.

A reasonably high probability is assigned to DNA[12] from the defendant transferring to the hands of the unknown offender through wearing of gloves.

6.2   Conclusion

The findings have been evaluated given the propositions:

Proposition A: During the robbery the defendant pushed the head of the victim in a downwards direction and pressed the knife against her cheek.

Proposition B: During the robbery somebody other than the defendant pushed the head of the victim in a downwards direction and pressed the knife against her cheek. Before doing so, this person wore gloves belonging to the defendant.

Based on the considerations set out in Section 6.1 above, given the assumptions made and given the information on the case circumstances that I received, I consider the findings approximately 140 times more probable if proposition A is true rather than if proposition 2 is true.

This means that I consider the findings to be *appreciably more probable*[13] if during the robbery the defendant pushed the head of the victim in a downwards direction, and pressed the knife against her cheek, than if somebody other than the defendant did so.

Note that no opinion is given concerning the probability of the propositions.

Instead, an opinion is given on the probability of the analysis results obtained if the proposed propositions are true.

The weight of forensic analysis results depends on the propositions, the contextual information, the assumptions and the analysis results. If these change, this could affect the conclusion.

In that case a new evaluation can be performed, if necessary.

Signature

I declare that I have drawn up this report truthfully, fully and to the best of my knowledge as an expert in forensic examinations for biological traces and DNA analyses, registered as a court expert in the Netherlands Register of Court Experts for the 'DNA analysis and interpretation – activity level' area of expertise.

Place
Date
Author of the report

NOTES

1. For general information about the scientific framework within which the scientist works, see: Kokshoorn, B., Aarts, B., De Blaeij, T., Maaskant-van Wijk, P. and Blankers, B., 2014. *Bewijskracht van onderzoek naar biologische sporen en DNA. Deel 1: Theoretisch kader en aandachtspunten bij conclusies in het deskundigenrappor*t [Evidential value of examinations for biological traces and DNA, Part 1: Theoretical framework and key points in expert report conclusions]. Expertise en Recht 2014–6: 197–203.
2. This method is described in Kokshoorn, B., Aarts, B., De Blaeij, T., Maaskant-van Wijk, P. and Blankers, B., 2014. *Bewijskracht van onderzoek naar biologische sporen en DNA. Deel 3: Activiteitniveau* [Evidential value of examinations for biological traces and DNA, Part 3: Activity level]. Expertise en Recht, pp. 213–219.
3. Willis, S.M., McKenna, L., McDermott, S., O'Donell, G., Barrett, A., Rasmusson, B., Nordgaard, A., Berger, C.E.H., Sjerps, M.J., Lucena-Molina, J.J. and Zadora, G., 2015. ENFSI guideline for evaluative reporting in forensic science. European Network of Forensic Science Institutes. Investigation question and hypotheses.
4. Gill, P., Hicks, T., Butler, J.M., Connolly, E., Gusmão, L., Kokshoorn, B., Morling, N., van Oorschot, R.A., Parson, W., Prinz, M. and Schneider, P.M., 2020. NA commission of the International society for forensic genetics: Assessing the value of forensic biological evidence – Guidelines highlighting the importance of propositions. Part II: Evaluation of biological traces considering activity level propositions. *Forensic Science International: Genetics*, 44, 102–186.
5. Kokshoorn et al., (2017) Activity level DNA evidence evaluation: On propositions addressing the actor or the activity. *Forensic Science International* 278: 115–124.
6. Taylor, D., Kokshoorn, B. and Hicks, T., 2020. Structuring cases into propositions, assumptions, and undisputed case information. *Forensic Science International: Genetics*, 44, 102–199.
7. The verbal terms used and the associated range of probabilities are based on Gill, P. (2014): *Misleading DNA Evidence: Reasons for Miscarriages of Justice*, Academic Press, London: Elsevier.
8. Sjerps, M. and Kloosterman, A. (2010). Het gebruik van Bayesiaanse netwerken in de forensische (DNA-) statistiek. Ars Aequi, 59(7), 502–508.
9.
- Samie et al. (2016). Stabbing simulations and DNA transfer. *Forensic Science International: Genetics* 22: 73–80.
  In this study, it was found that DNA from the person who used a knife to stab an object was recovered in 95% of 64 samples taken from knife handles.
- Meakin et al. (2017). Trace DNA evidence dynamics: An investigation into the deposition and persistence of directly – and indirectly – transferred DNA on regularly used knives. *Forensic Science International: Genetics* 29: 38–47. In this study, DNA from the regular user of knives was recovered in 100% of 12 samples taken from used knives.
  Both studies used knives that were cleaned prior to the experiments being performed. This increases the probability of recovery of the handlers of the knife compared to the case-relevant situation where the offender used a knife from the home of the victim. The duration of use (in the Meakin study) or the intensity of use (in the Samie study) are higher than likely under the given set of case circumstances. Therefore, a slightly lower probability is assigned to the recovery of DNA from the offender from the handle of the knife.
10. Fonneløp, A. E., Ramse, M., Egeland, T., and Gill, P. (2017). The implications of shedder status and background DNA on direct and secondary transfer in an attack scenario. *Forensic Science International: Genetics*, 29, 48–60.
    In this study attackers grabbed the shirts of victims in simulated attack scenarios. DNA from the assailant was recovered from 16 out of 17 shirts that were grabbed. While no data are available on DNA transfer to hair, it is assumed that hair behaves similarly to smooth fabrics. Therefore, a high probability is assigned to the probability of recovery of DNA from the offender from the hair of the victim.
11. van den Berge, M., Ozcanhan, G., Zijlstra, S., Lindenbergh, A., & Sijen, T. (2016). Prevalence of human cell material: DNA and RNA profiling of public and private objects and after activity scenarios. *Forensic Science International: Genetics*, 21, 81–89.

In this study, it was found that from all 20 samples taken from regularly worn winter gloves, DNA from the wearer was recovered. The probability of DNA from a regular wearer being present on winter gloves is therefore very high.

12. In this study 24 samples were taken from the right hand of volunteers who shook hands with somebody else. DNA from that other person was recovered from 11 out of these 24 samples. As these samples were taken after a brief handshake and after pressing the hand on a smooth glass surface five times, it is my expectation that the probability of DNA transferring to the hand from wearing the glove for an extended period of time is higher. I therefore consider this probability reasonably high.

13. This term is part of a standard verbal scale (the left column in the table below). This scale is used when the scientist has no or insufficient numerical data to explicitly substantiate a numerical conclusion. The selection of the specific verbal term is based on expert knowledge, experience in research and casework, etc. To promote the transparency for the reader and the uniformity among the different experts, the forensic laboratory has defined the verbal terms numerically. These definitions are expressed in orders of magnitude and are listed in the right column in the table below. For example, the term 'slightly more probable' means that the probability of observing the results of the investigation is considered 2 to 10 times larger when one hypothesis is true than when the other hypothesis is true.

Verbal Equivalent	Order of Magnitude of Evidential Strength
Approximately equally probable	1–2
Slightly more probable	2–10
More probable	10–100
Appreciably more probable	100–10,000
Far more probable	10,000–1,000,000
Extremely more probable	>1,000,000

The conclusion expresses the evidential strength of the results regarding the propositions. The conclusion does not represent the probability that a particular hypothesis is true. That probability depends on other evidence and information outside the domain of forensic expertise and falls outside the scope of this report. More information about this way of concluding is available in the professional annex (reference to laboratory website).

10.6   PRACTICE QUESTIONS

When you have started reporting findings given activity-level propositions it is likely that at some point you will be required to testify in court on these matters. Below are ten questions that have come up in court when the authors have been examined about their testimony. We share these with you for the purpose of preparation. We encourage you to formulate an appropriate response to each question. In Chapter 14 we do not provide full answers to these questions but hint at aspects that may be considered in your answer.

Q1) You have made the assumption that no gloves were worn by the perpetrator. Neither the victim nor the witnesses have provided any testimony on this matter. Why did you make this assumption and how does it impact the weight of the evidence?

Q2) You have assumed that the DNA is from my client, the defendant. Why did you assume this, as the report states that the scientific conclusion is on the findings only and not about the propositions themselves? In other words, I understand that you make a leap of faith that is for the jury/judge to make.

Q3) What is a Bayesian network?

Q4) Is this statistical model that you use validated?

Q5) You have built a very simple model that you claim reflects the complexities of the crime we are considering. How can you be certain that this model is accurate and appropriate?

Q6) You have evaluated the presence of DNA of the defendant on the clothing of the victim. Why did you not consider the amount of DNA? Wouldn't that be very informative?

Q7) You have used this scientific study from another country in which they used only four volunteers in an experiment. The results, from 12 samples only, you use to assign a probability to the presence of DNA of my client on the shirt of the victim. You cannot possibly be convinced that these data are representative for this case?

Q8) How many reports have you written that address findings given activity-level propositions?

Q9) Can you explain to the court what your expertise in this matter is? You have experience with DNA analysis and reporting on source-level issues, but how does this make you an expert in activity-level reporting?

Q10) The report you provided makes claims on the likelihood of my client having committed the robbery. You must be aware that such deliberations are up to the court and not to you! Why do you think it is appropriate for a DNA scientist to discuss these matters while they are obviously outside your area of expertise?

10.7 REFERENCES

1. A. Biedermann, J. Vuille, F. Taroni, C. Champod, The need for reporting standards in forensic science, *Law, Probability and Risk* 14(2) (2015) 169–173.
2. Association of Forensic Science Providers, Standards for the formulation of evaluative forensic science expert opinion, *Science & Justice* 49(3) (2009) 161–164.
3. B. Found, G. Edmond, Reporting on the comparison and interpretation of pattern evidence: Recommendations for forensic specialists, *Australian Journal of Forensic Sciences* 44(2) (2012) 193–196.
4. S.M. Willis, L. McKenna, S. McDermott, G. O'Donell, A. Barrett, B. Rasmusson, A. Nordgaard, C.E.H. Berger, M.J. Sjerps, J.-J. Lucena-Molina, G. Zadora, C. Aitken, T. Lovelock, L. Lunt, C. Champod, A. Biedermann, T.N. Hicks, F. Taroni, *ENFSI Guideline for Evaluative Reporting in Forensic Science*, European Network of Forensic Science Institutes. http://enfsi.eu/sites/default/files/documents/external _publications/m1_guideline.pdf, 2015.
5. G. Jackson, C. Aitken, P. Roberts, *Case Assessment and Interpretation of Expert Evidence: Guidance for Judges, Lawyers, Forensic Scientists and Expert Witnesses*, Royal Statistical Society, 2015.
6. K. Ballantyne, J. Bunford, B. Found, D. Neville, D. Taylor, G. Wevers, D. Catoggio, An introductory guide to evaluative reporting, in: T.N.I.o.F.S.A.N. Zealand (Ed.) www.nifs.org.au, 2017.
7. P. Gill, T. Hicks, J.M. Butler, E. Connolly, L. Gusmão, B. Kokshoorn, N. Morling, R. van Oorschot, W. Parson, M. Prinz, P.M. Schneider, T. Sijen, D. Taylor, DNA commission of the international society for forensic genetics: Assessing the value of forensic biological evidence: Guidelines highlighting the importance of propositions. Part I: Evaluation of DNA profiling comparisons given (sub)source propositions, *Forensic Science International: Genetics* 36 (2018) 189–202.
8. P. Gill, T. Hicks, J.M. Butler, E. Connolly, L. Gusmão, B. Kokshoorn, N. Morling, R.A.H. van Oorschot, W. Parson, M. Prinz, P.M. Schneider, T. Sijen, D. Taylor, DNA commission of the International society for forensic genetics: Assessing the value of forensic biological evidence: Guidelines highlighting the importance of propositions. Part II: Evaluation of biological traces considering activity level propositions, *Forensic Science International: Genetics* 44 (2020) 102186.

9. T. Hicks, J. Buckleton, V. Castella, I. Evett, G. Jackson, A logical framework for forensic DNA interpretation, *Genes* 13 (2022) 957

10. P. Gill, *Misleading DNA Evidence: Reasons for Miscarriages of Justice*, Elsevier, 2014.

11. R. Marquis, A. Biedermann, L. Cadola, C. Champod, L. Gueissaz, G. Massonnet, W.D. Mazzella, F. Taroni, T. Hicks, Discussion on how to implement a verbal scale in a forensic laboratory: Benefits, pitfalls and suggestions to avoid misunderstandings, *Science & Justice: Journal of the Forensic Science Society* 56(5) (2016) 364–370.

12. K.A. Martire, I. Watkins, Perception problems of the verbal scale: A reanalysis and application of a membership function approach, *Science & Justice: Journal of the Forensic Science Society* 55(4) (2015) 264–73.

13. K.A. Martire, R.I. Kemp, M. Sayle, B.R. Newell, On the interpretation of likelihood ratios in forensic science evidence: Presentation formats and the weak evidence effect, *Forensic Science International* 240 (2014) 61–68.

14. B. Garrett, G. Mitchell, N. Scurich, Comparing categorical and probabilistic fingerprint evidence, *Journal of Forensic Science* 63(6) (2018) 1712–1717.

15. E.K. van Straalen, C.J. de Poot, M. Malsch, H. Elffers, The interpretation of forensic conclusions by criminal justice professionals: The same evidence interpreted differently, *Forensic Science International* 313 (2020) 110331.

16. N.F. Instituut, *Vakbijlage Waarschijnlijkheidstermen*. <https://www.forensischinstituut.nl/over-het-nfi/publicaties/publicaties/2017/10/18/vakbijlage-waarschijnlijkheidstermen> (accessed 23/21/2021).

17. A. Nordgaard, R. Ansell, W. Drotz, L. Jaeger, Scale of conclusions for the value of evidence, *Law, Probability and Risk* 11 (2012) 1–24.

18. J. Buckleton, J.-A. Bright, D. Taylor, *Forensic DNA Evidence Interpretation* (2nd ed.), CRC Press, Boca Raton, 2016.

19. I. Evett, B. Weir, *Interpreting DNA Evidence: Statistical Genetics for Forensic Scientists*, Sinauer Associates, Sunderland, 1998.

20. H. Eldridge, Juror comprehension of forensic expert testimony: A literature review and gap analysis, *Forensic Science International: Synergy* 1 (2019) 24–34.

21. G. Gigerenzer, R. Hertwig, E. Van Den Broek, B. Fasolo, K.V. Katsikopoulos, "A 30% chance of rain tomorrow": How Does the Public Understand Probabilistic Weather Forecasts?, *Risk Analysis* 25(3) (2005) 623–629.

22. I.W. Evett, G. Jackson, J.A. Lambert, S. McCrossan, The impact of the principles of evidence interpretation on the structure and content of statements, *Science & Justice* 40(4) (2000) 233–239.

23. M. Sjerps, C. Berger, How clear is transparent? Reporting expert reasoning in legal cases, *Law, Probability and Risk* 11(4) (2012) 317–329.

24. D.J. Daly, C. Murphy, S.D. McDermott, The transfer of touch DNA from hands to glass, fabric and wood, *Forensic Science International: Genetics* 6 (2013) 41–46.

25. M. Breathnach, L. Williams, L. McKenna, E. Moore, Probability of detection of DNA deposited by habitual wearer and/or the second individual who touched the garment, *Forensic Science International: Genetics* 20 (2016) 53–60.

26. M. van den Berge, G. Ozcanhan, S. Zijlstra, A. Lindenbergh, T. Sijen, Prevalence of human cell material: DNA and RNA profiling of public and private objects after activity scenarios, *Forensic Science International: Genetics* 21 (2016) 81–89.

27. V. Sethi, E. Panacek, W. Green, J. NG, S. Kanthaswamy, Yield of male touch DNA from fabrics in an assault model, *Journal of Forensic Research* (2013). https://doi.org/10.4172/2157-7145. T1–001.

11 Implementation

Duncan Taylor and Bas Kokshoorn

CONTENTS

DOI: 10.4324/9781003273189-11

11.1 CONSIDERATIONS FOR THE LABORATORY AND THE CRIMINAL JUSTICE SYSTEM

The decision by a laboratory that they wish to carry out evaluations given activity-level propositions is a big and important step, but only the first of many. Once this decision is made, comes a series of questions that the laboratory will face, the answers of which will determine everything from who receives the reports, when they will be done, whether the process will be accredited and how many analysts will be trained. This chapter provides some points to consider and some guidance on how to approach them.

11.1.1 WHO WILL BE REQUESTING THE WORK, I.E. JUDGES/LAWYERS/POLICE/SCIENTISTS?

The answer to this question will depend on the legal system in which the laboratory operates. In an adversarial system, the request for work could come from a number of different stakeholders. The most common structure of government or private forensic laboratories is that work is mainly commissioned by investigating or prosecuting parties, i.e. police, or prosecutors. A fully operational forensic laboratory system of reporting activity-level evaluations can proceed in the same manner as the DNA work carried out considering sub-source-level propositions. It is likely that cases which are received will be evaluated given propositions at both positions in the hierarchy. In this framework it becomes possible to carry out meaningful case pre-assessments using case assessment and interpretation as a means to triage cases that are accepted. Ideally in a mature system there are also open avenues of conversation with the defence community, in both an openness to explain the evaluations but also for the defence to provide information that assists in forming their propositions. This will not always be the case, even in the most progressive of forensic institutions, as the defence have a right to silence, and at times this will be exercised by providing a 'no comment' response to requests for information that would be used in evaluations. It is hoped that the greater the understanding of the evaluation process, the more engaged both prosecution and defence would be. It is also possible that evaluations and reports could be commissioned by the defence, or by private parties directly. This situation requires different considerations as there are privacy issues with respect to the prosecution. For example, it may be that the DNA work has been completed in a forensic laboratory and evaluated using sub-source-level propositions. Then an evaluation is requested privately by the defence considering activity-level propositions that they have nominated. In these situations, even the fact that a report has been requested by the defence should be initially considered private, and so not only should the evaluation at the activity level be conducted by a different scientist as the evaluation at the sub-source level, but there should be limited (or no) discussion between the analysts about the report (at least until the report has been openly shared between prosecution and defence).

In a forensic laboratory that is part of an inquisitorial system, the general situation will be much the same as just described for an adversarial system. Specific legal aspects may mean that requests to evaluate findings given activity-level propositions come only from an investigative judge or public prosecutor. Nonetheless, in both an adversarial and inquisitorial system there is a need for both prosecution and defence to be involved and define their scenarios.

11.1.2 What Type of Cases Will It Be Applied To?

The question of which cases to apply activity-level evaluation depends on the level of maturity of activity-level evaluations within the forensic laboratory and judiciary. Ideally the answer to which case it should be applied is 'all of them'. In reality there are many cases which will not benefit:

- Cases where questions being asked are purely investigative in an identity sense, i.e. cases referred to as 'volume', 'database' or 'no-suspect' where the purpose of the work is to generate some information that can be used to assist in identifying a potential suspect (such as a DNA profile that can be searched against a database). While there are instances of no-suspect cases where a fuller, activity-level, consideration can assist (for example see [1]), this is not standard practice and would likely be reserved for select cases. But note that the concepts of evidence evaluation are also applied in case assessments.
- Cases where the work to carry out an evaluation would not benefit given admissions, hence the activities of interest are not disputed. For example, in a sexual assault case there may be some initial DNA work carried out that yields a DNA profile matching the defendant in a sperm fraction of a vaginal swab from the victim. At this point, the defendant may claim that the sexual intercourse was consensual. The issues in this scenario are now offence-level issues (as described in Chapter 1) and cannot be assisted by the scientist. The lack of ability for evaluation given activity-level proposition to assist the court becomes evident when the activity-level propositions are being set, which in this case would be 'the complainant and defendant had sexual intercourse' in both instances.
- When no admissions have been made, the case circumstances are such that an activity-level evaluation is unlikely to be helpful. For example, consider a case where items at an illicit drug manufacturing laboratory, including a drink can, were sampled. Imagine that a DNA profile is generated from the can which links to a person via a database search. No further information is provided by the defence. It is difficult to see in this instance how to proceed with an evaluation, given activity-level propositions. It seems reasonable that the arguments may lie with whether the defendant had any involvement with manufacturing the drugs, and not over whether they drank from the can.

When a laboratory first starts to implement activity-level evaluations, it will likely do so having carried out sub-source-level evaluations for some time. The transition into carrying out activity-level evaluations will most likely move through an era where activity-level evaluations are only considered after the cases have already been triaged, accepted, evaluated using sub-source-level propositions and reported. In such a workflow there are issues that can arise as there are samples which may have been accepted (or rejected) for which the triage decision would have been different had the case been viewed at the activity level from the start. For example, a case where an illegal weapon, alleged to belong to the defendant, is found wrapped in some clothing may accept the weapon only for sub-source-level evaluation. The classic thinking in this case is to test the item that most directly identifies the gun owner/handler. However, if the activity-level issue was that the defence stated the weapon did not belong to the defendant, but the clothing it was wrapped in did, then triage carried out with these activity-level issues in mind may also have accepted the clothing (as the amount/presence or absence of the defendant's DNA on the clothing becomes informative within the activity-level evaluation). In these instances, a laboratory policy needs to be made as to whether the additional items would be requested to assist with the activity-level evaluation (if they are still available), or whether the analyst will have to proceed with only the information they already have from the sub-source-level evaluation. Implementation of reporting given activity-level propositions would therefore benefit from implementing a case pre-assessment procedure in which current findings and potential additional examinations are being considered given the scenarios provided by prosecution and defence.

Another example of an issue that arises when carrying out activity-level evaluations after the case has been completed at the sub-source consideration level is that it becomes apparent the observations have very little power to distinguish between activity-level propositions. In fact, had the case been triaged with activity-level issues in mind, then it is possible the case would have been rejected. An example of this type of case could be one where an assault had occurred between two individuals that cohabitate. Consider that the items that have been targeted are clothing, in an effort to detect trace DNA from contact transfers. Any DNA found on the items of clothing will have a limited ability to distinguish between contact and cohabitation. While there is little information in the findings to distinguish between propositions and the case may have been rejected had it been triaged considering the disputed activities, this type of case is actually very important on which to carry out an activity-level evaluation. The reason for this is that the DNA results are now generated, and the sub-source-level evaluations have been reported. There is a clear risk that the results will be interpreted as being more informative towards activities than they are. A jury is likely to hear that the forensic analysis targeted areas of an item for contact DNA and confuse the findings of any DNA as therefore being from contact. An evaluation that highlights the relatively equal probability of finding DNA given either activity is an essential assistance to the court in understanding the significance of the findings now that they have been produced. There is a tendency with legal professionals as well as scientists to consider likelihood ratios close to 1 as uninformative and therefore not useful. Deciding on which case would benefit from an evaluation given activity-level propositions, given the example given before, should not be guided by whether or not the findings are able to distinguish between the propositions!

A good strategy when a laboratory is just beginning to issue activity-level evaluation reports is to pick the right cases. These will be the cases where the two propositions are clear, the DNA work is limited in amount but has provided clearly discriminating results and the issues of transfer, persistence, prevalence and recovery that relate to the case are well documented in the literature. Initially building a repertoire of small, relatively simple cases:

- Build the confidence of those that are carrying out the evaluation
- Shows stakeholders what these evaluations achieve, in circumstances that are simple enough that the logic is intuitive
- Finds points in the workflow that may need to be adjusted, or further refined in cases where the analyst will have to deal with evaluation complexities at the same time as new process complexities
- Will likely represent the real-world cases that most closely represent the idealistic cases that are used in training
- Will represent the most easily explained and defended cases if challenged in court.

11.1.3 How Many People Will Be Trained?

The number of people who are trained in activity-level evaluations will vary depending on the laboratory, the section of the laboratory and the maturity of activity-level evaluations in that laboratory. At some level, all evaluation/reporting/interpretation scientists will (or should have) an understanding of the basic ideas or transfer, persistence, prevalence and recovery of biological material. This may not necessarily come from targeted training in this area, but rather through reading, casework experience (i.e. success rates of different types of items/samples), triaging decisions, knowledge from other scientists or common sense. Such base-level knowledge is used to make standard casework decisions on what to accept and how to test items. Individuals (and the laboratory as a whole) benefit from all employees receiving some basic training on these fundamental concepts. The benefits extend to being able to choose the best items for testing and guiding stakeholders in what they should consider. For analysts who wish to carry out activity-level evaluations, then this basic training can be extended to cover the deeper understanding required.

However, when first implementing a system of reporting evaluations carried out using activity-level propositions, the likely situation will be that there is limited understanding or comprehension of them from the legal community. The consequence of this is that evaluations at the activity level will need to be driven by the analysts in the forensic laboratory. Also, in this scenario it is likely that there will be only a few people (maybe only one person) trained in carrying out these types of evaluations and so the number of cases they are applied to will be limited. The choice of which case will need to be made by the small team of individuals carrying out the evaluation and will be best based on cases for which the greatest risk of misinterpretation exists if the sub-source-level evaluation results alone were provided.

When a laboratory first extends into activity-level evaluations, the normal practice is for one or two individuals to undergo extensive and external training on the topic and then champion the implementation within the laboratory. There are numerous benefits of the initial training consisting of at least two people. One is that training is often more effective, and better absorbed when done in groups. A sole individual studying the topic, and not having colleagues to discuss the content can make for a more difficult time. Additionally, when both individuals have completed training, there is the capacity for evaluations to be reviewed. The alternative is that evaluations are reported unreviewed, or a review by an individual from an external agency is required (which can have its own logistical issues or cost). Importantly when training these first people is the choice of person. The process of implementing activity-level evaluation will initially require the individuals to drive it through various aspects of bureaucracy, administration and some resistance (from within the organisation and external to it). This is true of any change in any organisation. These first individuals will therefore need to be enthusiastic believers in the cause. Choose those who are able to develop solutions for the complexities presented by applying theory to the real world and those who can confidently present their work to stakeholders and in court. They are likely to be the first in the organisation to face courtroom challenges to the process and will pave the way for other scientists to come.

The initial 'team' or champion group will only have the capacity for a limited number of cases (which they will have to drive) and therefore activity-level evaluations will only slowly gain coverage among stakeholders. As time progresses and stakeholders become more aware of activity-level evaluations, they will begin to request evaluations and there will be a natural increase in the required capacity to carry out such evaluations. At that point, the biggest implementation hurdle has been cleared and the implementation process has progressed to its next phase, growth. From here, the decision can be made to grow the number of individuals slowly (i.e. a limited number each year), or a one-off large training event can be conducted. Both systems have advantages and disadvantages, and the intentions of the laboratory in how they wish to progress into the future will dictate what path is best.

The third phase of implementation occurs when a laboratory has a mature activity-level reporting program and there are a large number of employees trained and carrying out evaluations. In this third phase individuals begin to branch into specialities. With different types of cases being addressed at the activity level, natural occurrences of interdisciplinary evaluations may pop up. Some examples are:

- The combination of DNA and fingermarks on an item to address the order of handling of an item,
- The location of traces on duct tape [2], which may require physical end matching of the pieces of tape or
- The combination of findings from a forensic medical examination and DNA analysis in a sexual assault case.

Some individuals may specialise to perform these multidisciplinary evaluations [3]. Interdisciplinary evaluations are discussed in more detail in Chapter 12. Another speciality may be to concentrate on how evaluations of data can assist in the investigative phase of an investigation. Another may be to specialise in the more complex aspects of testing evaluations for robustness using bespoke computer coding. It is likely that with some of these specialities either further specialised training in areas such as statistics or computer programming will be beneficial (or it may be that someone already trained in these areas is specifically targeted for hiring).

11.1.4   How Will People Be Trained/Deemed Competent?

When the decision has been made as to how many and which individuals will be trained in activity-level evaluations, then the mechanism of how those people will be trained must be determined. Just as the decision on how many people will be trained depends on the maturity of the laboratory with respect to activity-level evaluation, so too will the mechanism of training, and deeming the individuals competent at the conclusion of their training. These aspects may be partially determined by the requirements of the local institution, accrediting body or expert registers.

The decision may also depend on a cooperative approach between multiple laboratories. For example, multiple laboratories within a country or region may decide to employ a 'centre of excellence', whereby a single laboratory is chosen to have individuals trained in activity-level evaluations, and all requests for such reports in the region would go to that centre (with other laboratories specialising in different areas, such as mRNA body fluid typing, or massively parallel sequencing, etc.). This approach has the advantage that because the training and implementation of different specialities are distributed across a laboratory network, the combined abilities exceed what a single laboratory could ever achieve (simply due to resource limitations).

When a laboratory is training its first individuals who will champion the process of activity-level evaluation, it is common for the training to be conducted external to the laboratory. This of course is not a requirement, and it may be that the initial training is conducted by one or a few individuals through their own study and understanding, possibly assisted by experts from other forensic disciplines within the institute (e.g. a forensic statistician or general criminalist). There is a difficulty in overcoming the circularity of individuals conducting their self-training, and who then would have to be the ones (as the only 'trained' individuals in the laboratory) to deem their own competence. The process is not impossible but will require additional checks and measures to mitigate against the (at least perceived) self-serving benefits. If external training is conducted for the first individuals carrying out activity-level evaluations, then there is an independent determining of a curriculum and assessment of ability. There are limited courses currently available that specialise in the area of activity-level evaluations, with the most well-known (at the time of this book being written) being the Statistical Evaluation of Forensic Evidence course run out of Lausanne University (https://www.formation-continue-unil-epfl.ch/formation/statistics-evaluation-forensic-evidence-cas/), which includes a competency test. It may also be possible (depending on the country) to organise on-site training by having external experts recognised in the field come to a host organisation to provide training. This model has the advantages that the training material can be delivered by external experts and that it can be delivered in a compact timeframe (i.e. over 1 or 2 weeks, perhaps at several points across a year). In such a model the trainees will need to be self-driven to maintain momentum in training between sessions. Lighter options that are available (but which may help structure and highlight training requirements) are one- or two-day workshops connected to major forensic science conferences, or incidental workshops organised by individual laboratories.

Once the process of activity-level evaluations has been implemented in the laboratory, then the next phase of implementation begins, in which additional individuals are trained in order to meet rising demand. At this stage, the laboratory will have at least one individual who has been trained (likely from external sources) and who has been producing activity-level evaluations in casework for some time. The laboratory may then decide that they have enough accumulated knowledge to develop and provide in-house training and assessment. Providing in-house training is almost another arm to the implementation of activity-level evaluations in a laboratory as it requires a series of new processes and protocols to be developed. Typically, there will be internal requirements for a curriculum, divided into separate training modules. Additionally, there will be a requirement for training materials to be developed, and this is likely going to require a series of lectures, presentations and tutorials to be run, and assignments to be set and marked. Finally, there will be an organisational need to define the method of assessment for competence and what constitutes a mark that would allow the individual to be deemed fit to start carrying out evaluations on active casework. When considering if in-house training will be attempted, there is also a human factor to recognise, i.e. are the people who are currently deemed competent going to be able to, and feel comfortable with, providing training to their peers? Regardless of whether the training of additional employees has occurred through an in-house training program, or an external program, there will likely be some period of mentorship during which the in-training analyst is mentored by an analyst who is already deemed competent. This may be for a period of time, a number of cases or both.

Some countries or regions also have an external body that recognises and registers expertise in different areas. One example of such a body is the Netherlands Register for Court Experts (NRGD; https://english.nrgd.nl/). The NRGD is an independent organisation with statutory powers in the Netherlands criminal justice system. They set standards for different fields of forensic science and provide competency testing for individual experts based on those standards. 'DNA analysis and interpretation – source level' was the first field to be demarcated by the NRGD in 2010. In 2021 the NRGD has, based on several rounds of consultation of an international group of forensic biologists and legal professionals, extended and updated the demarcation of the field. Currently they recognise four different fields to which experts can apply:

- 'DNA analysis and interpretation – sub-source level'
- 'DNA analysis and interpretation – source level'
- 'DNA analysis and interpretation – activity level'
- 'DNA analysis and interpretation – kinship'

The 'sub-source level' register is restricted to autosomal STR DNA profile interpretation (including statistical evaluation). The 'source level' register extends this expertise to haplotype marker sets, cell type testing and case assessment. The 'kinship' and 'activity level' registers are more self explanatory.

The 'core activities' of experts in the field of 'DNA analysis and interpretation – activity level' is described in the NRGD standard (version 4.1) as follows:

'Experts concerned with activity level activities mainly focus on the following tasks:

- Autosomal and YSTR DNA interpretation. This includes both high and low-template samples, single source and mixed DNA profiles.
- Probabilistic genotyping to assess the weight of evidence. The expert is able to set appropriate propositions and apply a proper statistical model to the data. The expert should be able to defend the decisions made, and should be able to explain the workings of the model in general terms. The expert is not expected to be able to have access to, or to have knowledge on the workings of, the source code of the software used. This falls within the domain of a forensic statistician.
- Probabilistic assessment of associations with uncertain cell type evidence and/or non-gender specific markers like saliva or blood that require thorough understanding of Bayesian inference.
- Case assessment. The expert is able to discuss the relevant issue with the mandating authority and is able to structure case information in relevant propositions, assumptions and undisputed case information. The expert is able to communicate limitations, for example if no reasonable assessment is possible due to lack of information. The expert is able to translate a case assessment based on the case information into an examination strategy.
- Statistical modelling. The expert is able to set up a proper model, e.g. a Bayesian Network, to calculate a likelihood ratio based on probabilities of transfer, persistence, prevalence, background and recovery of biological materials given the relevant propositions and findings in the case. As

part of this the expert is able to construct a Bayesian network if needed (e.g. with multiple find-ings that are conditionally dependent or when cell type attribution is at issue). The expert must be able to express the limitations of modelling decisions.
- Assigning probabilities. The expert is able to assign probabilities to the transfer, persistence, prevalence, background and recovery of biological materials given the propositions, assump-tions and case information. Experts must be able to make the boundaries of their knowledge explicit. The source on which the assignment of probabilities is based should be transparent (e.g. expert elicitation, case file data, published studies, case specific experiments).
- The expert is able to write a report on the evaluation of the findings given activity level proposi-tions following generally accepted guidelines from professional bodies (e.g. ENFSI, ISFG)'

For experts who wish to register for the 'activity' field, the requirements are formulated as:

'Basic requirements:

- work at the level of someone who has completed an academic Master's Degree, and must have a proven level of education, training and expertise;
- be familiar with the summary of concepts (see annex A) and keep abreast of state-of-the-art developments;
- have sufficient knowledge of the pros and cons of various techniques, specialisations and sci-entific methods used in the field, be aware and capable of explaining the possibilities and lim-itations of these techniques, specialisations and methods, and keep abreast of related recent developments;
- have spent an average of 40 hours a year over the past 5 years on forensically relevant profes-sional development (e.g. publications, attending conferences, running or attending courses).

Specific requirements Activity Level:

- have interpreted and reported on Source Level on at least 25 single source and/or complex and/or mixed DNA profiles divided over a minimum of 5 case requests in the past 5 years that have been subjected to collegial review and/or supervision or be registered as a Source Level or Kinship Analysis expert;
- have interpreted and reported on at least 5 cases containing propositions on activity level in the past 5 years that have been subjected to collegial review; In case the applicant is also acting as a supervisor, at least 3 cases containing propositions on activity level should be independently interpreted and reported on'.

The expert will have to submit a file containing a number of relevant cases, as well as documenta-tion demonstrating the competencies listed above. An independent panel of scientists as well as legal professionals will assess the competency of the experts based on the file and live questioning if deemed necessary by the competency testing advisory board.

If deemed competent to the set standard, registration will follow for a period of 5 years, after which the expert may apply for re-registration (which again includes competency testing following the procedure described).

The NRGD example has been described at some length to demonstrate that demarcation of 'reporting given activity level propositions' from the broader field of forensic biology is possible, that it requires specific expertise and that periodical competency testing of individuals to at least some generally agreed minimum standards is possible.

The final point to consider, and this may come at a point when activity-level evaluations are relatively mature within an organisation, is whether there is any benefit to providing individuals with specialised training. For example, if a laboratory has chosen to implement activity-level evalu-ations by deriving the LR formula by hand, then they may wish to consider the transition to using Bayesian networks. In order to achieve this then they may wish to provide an individual within the organisation with specialist training on the use and functionality of Bayesian networks in a general sense, i.e. not specifically for activity-level evaluations in forensic science, but rather their general use in logical evaluation or decision theory. It may be that the training is for a specific software implementation of Bayesian networks, or it may be training in programming related to evaluation techniques. Unlike other training, this specialist training is likely to be confined to a small group or even a single person as the need for this in-depth understanding is limited and mainly required to expand current capabilities into new areas or conduct research. Another example of specialist training may be in the area of experimental design and analysis for DNA TPPR studies, which would be used if the decision in a laboratory was to start carrying out bespoke experiments to collect data to inform probability assignments in their cases.

11.1.5 How Will Information Be Disseminated to Stakeholders?

The laboratory cannot decide to proceed with evaluations given activity-level propositions without consultation with other actors in their jurisdiction. All parties (police, prosecution, judiciary, defence) need to be informed and educated on the topic. This is essential as parties need to recognise in a case that the issue may be at the activity level and that specific expertise is required to address this issue. The experts or champion group performing these evaluations, therefore, have a duty to inform stakeholders. There are different ways to approach this and any or all of them may be considered and implemented, depending on the specific cultural and legal aspects of jurisdiction.

One avenue may be lectures or presentations to the relevant parties. These lectures could be part of periodic training that the laboratory is already involved in or be organised as seminars or workshops specifically addressing reporting given activity-level propositions. Lectures could cover:

- General principles of evidence interpretation,
- The hierarchy of propositions,
- The current knowledgebase of DNA TPPR,
- Reporting standards and
- Procedures.

Each of these topics could be covered separately to build understanding with the relevant parties. When starting with evaluations given activity-level propositions, attention may need to be focused on explaining what it is and why it is relevant. There may be considerable hesitation among the parties involved as it may be perceived as the expert taking on the role of the jury or judges (which requires explaining the concept of the Bayesian framework), parties may question the availability of data (which would require explaining the current knowledgebase but also the fact that probabilities can be assigned even without ample relevant data). Many of these recurrent concerns were addressed in Chapter 2. Case intake discussions or triaging with police may also be excellent opportunities to educate on the hierarchy of propositions and discuss the merits of evaluation findings given activity-level propositions.

Another avenue could be to set up a working group with representatives from stakeholders such as police, prosecution and defence. Such a working group could set out to develop practical aspects of the work like deciding on what types of cases may initially be covered, who decides on whether a case will be suitable (Should it be the expert? Or the mandating authority?) and flesh out any other legal or practical obstacles. Such a working group may pave the way for acceptance of the practice within the jurisdiction. Not in the least because champions with the police or prosecution services may contribute towards the acceptance of formal evaluations, given activity-level propositions in casework practice.

Further investments need to be made in practitioner guides and documentation on procedures that are specific to the jurisdiction. While quite some documentation on evaluative reporting is available in English (like the RSS practitioner guides, or the ENFSI and the Australia New Zealand Policing Advisory Agency (ANZPAA) guides to evaluative reporting), it may be needed to write such guidance in the national language. Such guidance could be generic or go into the more practical aspects that are relevant in the specific jurisdiction. Examples are several documents and papers that the NFI have published, on the need for extensive task-relevant information when addressing issues at the activity level (https://www.forensischinstituut.nl/publicaties/publicaties/2019/12/20/document-contextinformatie

-nfi-lay-out [last accessed 10 January 2023]. This document (in Dutch) describes the general concept of task relevant information and lists what type of information generally is needed (with examples). This document assists all parties involved to collect and share only the task relevant information with the scientist to reduce the risk of contextual bias. An associated infographic has also been developed: https://www.forensischinstituut.nl/publicaties/publicaties/2022/08/10/infographic-contextinformatie [last accessed 10 January 2023]), or decision support for prosecution and judiciary on when to ask for a formal evaluation given activity-level propositions in a case (https://www.forensischinstituut.nl/publi-caties/publicaties/2022/10/17/beslisschema-voor-een-evaluatie-op-activiteitniveau-door-een-deskun-dige [last accessed 10 January 2023]). The latter document has been developed in a collaboration between representatives of the judiciary, prosecution, and NFI. The document lists four subsequent check questions. When all are answered by 'yes' evaluation of findings, given activity-level propositions may be relevant and possible in the case. These questions are:

- Is there more than one case-relevant scenario that may explain the case findings?
- Are the scenarios practically feasible and is there supporting information in the case file?
- Can forensic findings distinguish between the scenarios?
- Are the scenarios substantial and concrete?

Each question is supported by a textbox with background information on the topic and references to further reading. If the answer to any of these questions is 'no', an evaluation of the findings given activity-level propositions by a scientist is likely to be difficult, if not impossible. Such an evaluation is then of little benefit as the expert cannot add any scientific knowledge, other than in very general terms. It should be stressed that if this is the case, this also holds true for the legal professionals discussing the findings in the context of the case!

11.1.6   How Will Evaluations Be Carried Out (Will the Software Be Required)?

As demonstrated in the chapters of this book, there are various ways that the mechanics of calculating an LR can be achieved. It will be the preference of a laboratory as to which method they wish to initially implement:

- Hand derivation – In this category the LR formula is derived by hand for each case. Examples of hand derivation were given in Chapter 5. This model of implementation is arguably the easiest to implement as it does not require any purchase or validation of software, it does not require any training in the use of speciality software and can be worked out with pen and paper. Relatively simple cases in which the observations relate to a single (meta)trace may also be addressed using readily available published formulae [4, 5]. Many laboratories have conducted, and continue to conduct, evaluations given activity-level propositions using hand-derived LRs. This is true in biology, and also in other forensic disciplines, as discussed in Chapter 12. If implementing a hand-derivation scheme, there will be a limitation to the complexity and scale of evaluation that can reasonably be handled. This was demonstrated in Chapter 5, where even a relatively simple evaluation becomes derivationally complex when considering DNA amounts (and even then, only in three brackets of high, low and none). Therefore, such an implementation may require bounds to be developed on either the types of cases (initially excluding major crimes for instance) that are evaluated at the activity level and/or bounds on the level of complexity within the evaluation.
- Point-and-click software – This implementation makes use of software tools that graphically develop BNs. It may require the laboratory to test/validate the software. It also requires training for individuals wishing to use the software; however, in reality the training required for the use of the software is not likely to be much more onerous than the training required for an individual to be able to derive complex LR formulae by hand. Depending on the software chosen, there may be an associated licence cost, and the decision will need

to be made whether this cost outweighs the associated professional support and updates that come with it, to the point that freeware is chosen instead. Point-and-click software that generates graphical BNs also have the advantage that they are more easily explained [6] to stakeholders (rather than trying to explain formulae or computer code).

- Code-driven evaluations – In this implementation scheme the laboratory chooses to use free software (such as R with BN add-ons) to carry out evaluations. It has the most flexibility to accommodate very complex BN constructions (as seen in Chapters 7 and 8 where quite detailed investigations into the robustness of the evaluation could be carried out). It is also the method that would be the most restrictive with regards to individuals who could apply the techniques, as it would not only require an understanding of activity-level evaluation but also an understanding of computer coding. It is likely that a laboratory that considered a computer code-driven evaluation implementation would do so alongside one of the other evaluation methods (i.e. hand derivation or point-and-click software use). It could be used only in circumstances where the level of flexibility and detail was required, and only by a subset of the activity-level evaluation team that specialised in this area. There would also be a requirement for the code to be validated in some way. Ideally this would be another individual who is familiar with the coding language used and could review the code itself. However, code validation would also require a demonstration that the code was carrying out the functions it is intended to be doing. This high validation burden on coding makes it attractive to produce code in an object-oriented way, i.e. coding as much into a function that is validated and then used within larger programs. Even more ideal is that if certain coding tasks become standardised, and validated functions (or modules) exist, then a graphical user interface can be constructed to allow greater access to the features the code provides, without the need for in-depth programming knowledge.

11.1.7 WILL THE PROCESS BE ACCREDITED?

The answer to this question is yes. At least to some extent. As most jurisdictions require laboratories to be accredited by a national accreditation body for forensic DNA analysis to ISO/IEC standards 17020/17025 and 21043, laboratories will have a quality assurance system (QAS) in place. Essential to this QAS is the documentation of standard operating procedures (SOPs) in the laboratory. Examples of such SOPs are provided in Section 11.5. The procedures that the laboratory adopts for evaluations of findings given activity-level propositions, therefore, need to be documented and will be part of the accreditation of the laboratory. This may include the standards for training and competency testing of the experts who perform such evaluations.

Having said this, as reporting given activity-level propositions relies more heavily on expert opinion than reporting given (sub) source-level propositions, individual competency may be considered more important than the accreditation of the procedures. Refer to Section 11.1.4 for competency testing.

11.1.8 WHAT WILL THE POLICY BE ON DATA COLLECTION?

Data to support probability assignment to DNA TPPR events may not be readily available. While the scientific literature provides a solid knowledge base, the data are likely to mismatch with case circumstances to some extent. If no data are available, one option (besides expert elicitation) would be to perform dedicated experiments based on the case circumstances. Within the laboratory, there needs to be a clear policy on whether or not such bespoke experiments will be performed and under which circumstances. A requirement could for instance be that there is consensus between the parties involved on crucial aspects of the case circumstances. Reality will be that due to constraints on time (the report needs to be filed soon) and resources (who is to perform these experiments, and who will analyse the data?) such experiments will not, or not often, be possible. Such policy may also need to be discussed at the senior management level and in consultation with police and prosecution, particularly if the laboratory works with service level agreements (SLA) with those parties. Any bespoke experiments may impact those agreements.

Further considerations are the sharing of the data that results from such bespoke experiments. Ideally data from all studies on TPPR of biological (and other traces) are published in the public domain as they may benefit others working on cases with a similar issue. As many laboratories are operated using public funds, the resulting data may also be considered public property. Publication and peer review may therefore be considered a requirement. For this purpose, it may also be advisable to reference the study in currently available databases like DNATRaC (https://bit.ly/2R4bFgL) or TTADB (http://www.uqtr.ca/LRC/BDATT-TTADB).

Peer review may also be considered crucial in the design of the experiment. As the evaluation will be subjected to collegial review, it may be considered good practice to review the experimental design prior to the execution of the experiments. This is to confirm that the experimental boundaries conform to the known case circumstances and that the experiment will deliver the data needed to assign the relevant probabilities. Such a review may involve a (forensic) statistician who may advise on the number of repeat experiments required to reach the required power to further analysis of the data.

An ISFG DNA commission has formulated the following considerations with regard to the design of experiments on DNA TPPR:

Consideration 4:

There are some important points to be made about experimental designs used to determine probabilities:

a) Be clear about the statistical model that will be used to assign the likelihood ratio beforehand, as this will dictate the factors to take into account and the data that will be needed.

b) Data within the same experimental design should ideally be used because recovery is dependent upon the sensitivity of the technique used in the analysis, i.e. the probability of recovering DNA is increased given both prosecution and defence propositions with a more sensitive method. It could be misleading to take data from different experimental designs without any consideration of the difference in methods.

c) Alternatively, it has also been proposed that laboratories could determine efficiency factors of commonly used collection devices and methods and extraction methods etc. If invariable, this may enable corrections to be applied across different techniques employed.

d) Experimental designs should at the same time, provide knowledge to assign for example the probability of DNA recovery given the activities alleged respectively by defence and prosecution.

e) Laboratories are encouraged to carry out comparative studies to parallel their results with others.

While bespoke experiments may be difficult to realise for each case that comes through the lab, some (major) cases may require such experiments, possibly by demand of the requesting parties. Nonetheless, the casework workflow may show common issues in multiple cases for which relevant and up-to-date data to assign probabilities to DNA TPPR are unavailable. The casework workflow may in this way direct research efforts in this area. A defendant may for instance claim that someone else wore their clothes while committing the crime. Such a scenario may come up more frequently in cases (at least in authors' experience) and may direct research in the area of persistence of DNA of the original wearer after subsequent wear by another person on clothing, as data on these specific issues are at the time of writing scarcely. As these experiments may be quite a resource and time intensive (particularly if different durations of wear and different items of clothing and type of fabric are considered), it may be advisable for laboratories to 'team up' in such efforts. The ISFG (in [7]) alludes to this, also to compare methodologies and study the robustness of the outcome of such studies:

Laboratories are encouraged to carry out comparative studies to parallel their results with others.

Within a case, there should be a level playing field for all parties. The data generated from bespoke experiments should therefore be made available to all if requested. One way to do this is by publishing the data and having it peer-reviewed. Another option is an appendix to the report with a detailed description of the experimental design, methods and results.

11.2   TRAINING RESOURCES

As reporting on findings given activity-level propositions requires different (additional) expertise compared to reporting findings given levels lower in the hierarchy of proposition, training of experts is required at some point. A number of sources are available that may assist in such training.

Quite a number of publications have addressed the evaluation of findings given activity-level propositions. This ranges from studies on DNA TPPR to build a knowledge base, to books setting out the general principles of evidence evaluation, and papers addressing specific (statistical) issues with aspects of modelling. While not exhaustive, many of these sources have been referenced throughout this book. This book itself is written as a training resource for students and scientists, as well as the legal community.

Specific and in-depth training is also available from online courses. Some excellent resources are provided by the University of Lausanne (UNIL). They provide online courses open to forensic practitioners on the use of Bayesian networks, Statistics and the Evaluation of Forensic Evidence (SEFE), as well as a dedicated course on reporting forensic biology findings given activity-level propositions. These courses are all followed by a competency test, which is of benefit for those wishing to test themselves as well as those wanting to make their competency on the topic demonstrable (for instance for court-going purposes). They have furthermore provided a Massive Open Online Course (MOOC) on 'Challenging Forensic Science: How Science Should Speak to Court (Coursera)' in which they discuss the common issues of evidence interpretation.

Other training resources can frequently be found through incidental workshops on the topic (often associated with one of the major conferences in the field) or seminars provided by one or more organisations. Such opportunities differ in length, frequency and depth in which topics are covered but collectively provide a good source for training and continuing professional development.

Laboratories may also decide to organise in-house training specifically tailored to the topic and the specifics of their jurisdiction. There are a number of organisations and laboratories that provide peer-to-peer training. Such in-house training may be a good opportunity to train a larger group of experts on the topic, as well as calibrate experts in their probability assignments.

11.3   CASE INFORMATION MANAGEMENT

Information received may change the assessment of an observation. This is a good thing, as observations must be considered in a context to make sense of what they mean. An example would be the observation of a photograph of an elephant. In and of itself, seeing a photograph of an elephant is not remarkable as it is known that these animals exist. If information is obtained that reveals the photograph is taken in Thailand, this will not change the assessment, as it is known that Asian elephants occur in the country and are quite common. However, if the information is obtained that reveals the photograph was taken in Antarctica, it is likely to change the opinion as it is common knowledge that elephants don't occur on that frozen continent. The photograph may either be considered as showing a very remarkable observation, or the truthfulness of the person who shared the information may start to be questioned.

Similar processes take place when observations of forensic examinations are interpreted. The base rate expectations (knowledge of where elephants commonly occur), as well as contextual information (the location where the photo was taken) and the source of the information (the person revealing that the photo was taken in Antarctica), may consciously or unconsciously impact on the assessment of the weight of the findings in the context of the case.

These cognitive processes have been well studied, including in the forensic sciences [8–11]. Some studies on the impact of contextual information on the interpretation of results from DNA analyses are also available [12, 13]. While not all aspects of these cognitive processes have been studied in detail in daily forensic practice, it is reasonable to assume that these processes do impact the interpretation and communication of the weight of the findings. Stoel et al. [14] have therefore stated, 'if there is at least a chance that contextual bias exists, it would be prudent to perform forensic examinations in such a way that the risk of contextual bias is minimized'.

The impact of cognitive biases on the interpretation of forensic findings is also addressed in a guidance document of the Forensic Science Regulator in the UK [15]. They recognise seven different types of cognitive bias that may impact the work of the forensic scientist:

- Confirmation bias
- Expectation bias
- Anchoring effects
- Contextual bias
- Role effects
- Motivational bias
- Reconstructive effects

While all may have a crucial impact on the work, *contextual bias* is discussed in some detail as the risk of it is something that can and should mitigate in work processed when addressing findings given activity-level propositions. Important here is the concept of *task-relevant information*. This was discussed in Chapter 4. To reduce the risk of contextual bias and make sure that the expert performing the evaluation only receives task-relevant information, some measures can be taken:

- Differentiate tasks between experts. Any evaluation given activity-level propositions will require consultation with the parties involved. It is crucial that the issue is clear and that relevant propositions are formulated. In this process it is key that all parties can express their positions, reference relevant information and findings, e.g. discuss freely. During this process, there is a high risk that the expert is exposed to task-irrelevant information, for instance results from examinations from other disciplines that support either propositions or statements from witnesses that mention having seen the defendant at the scene. This 'contamination' of the expert cannot be avoided and it may therefore be prudent to have one expert perform this case intake and have another perform the evaluation without exposure to this task-irrelevant information.
- Appoint a case information manager (CIM). To filter any potentially biasing information from the case information provided, a CIM may be appointed. This person, who has knowledge of the evaluative process and can recognise task-relevant information (commonly another expert), can make sure that only task-relevant information is provided to the expert performing the evaluation. This CIM can be the same individual who consulted with the parties involved.
- Separate source from activity evaluation. The expert that reported on examination earlier in the case, for instance when the case (and expert) was in an investigative mode, may have been contaminated with case findings, build a connection with the police officers working the case or have knowledge on specifics of the defendant. This information may lead to (unconscious) bias in an evaluation, given propositions at the activity level. It may therefore be advisable to separate tasks and have another expert perform the evaluation given activity-level propositions. Again, the expert who worked on the case on (sub)source level might be in a good position to discuss the issue in the case with the mandating authority and act as CIM to those performing the evaluation. Another benefit to this is that the discussion in the courtroom on issues pertaining to the (sub)source level evaluation and that at the activity level can be clearly separated as two different experts will testify on these issues.

Further to this point is that when there are already findings from the case based on earlier work, these findings could (unconsciously) bias the expert when assigning probabilities to potential observations, generally increasing their belief in the outcome they are already aware of.

To make these processes transparent, it is crucial that the report lists the documents received and the task-relevant information available to the expert at the time of the evaluation.

11.4 CASEWORK WORKFLOW

In practice a proper evaluation of findings given activity-level propositions is required by a mandating authority when those findings are already available. More often than not forensic DNA analyses have already been performed in the course of the investigation (with the purpose of database searching or addressing reconstruction-type questions). Only at a late stage in the case, be it in the courtroom or just before that, issues at the activity level are put to the scientist. This may appear at odds with the formal model of case assessment and interpretation as in this stepwise process the examination of the items and the DNA analyses are only performed when the issue is clear and the propositions at the activity level have been formulated. As discussed in the previous paragraph, the fact that these findings are known introduces the risk of contextual bias if the scientist is aware of them. However, not only the scientist can be biased by these findings, but the parties involved may also be influenced in their formulation of the case-relevant scenarios. This increases the risk of *explanations* for the findings being presented as *propositions* given that the probability of the findings needs to be assessed. While this is not an issue in itself, the risk is that it is not recognised that the a priori probability of such scenarios should be considered *without* considering the findings that lead to them being formulated. This might lead to either over- or underestimation of the posterior probabilities of the propositions as the findings may not discriminate between both *explanations*.

There are some practical aspects relating to these issues that may impact on the workflow in the laboratory. If no examinations have yet been performed, the formal model for case assessment and interpretation (CA&I) (which has been discussed in Chapter 10) can be adopted and implemented in a laboratory procedure. This model will work well at the onset of the case. If, however, some findings are already available from prior examinations, there are some aspects that need to be considered. This may require a second, different workflow.

The CA&I model describes a case pre-assessment based on propositions that are formulated based on the pertinent issue in the case. This pre-assessment will explore the potential outcomes of any examinations that can be performed, assign a probability to these expected outcomes given the propositions and other case circumstances and assess whether or not the examinations have the potential to provide results that can distinguish between the propositions (and the probability of obtaining such an outcome). The assignment of the probabilities at this stage is commonly 'guesstimated' based on experience and without extensive reference to scientific literature.

This information is then communicated to the mandating authority (or 'client' as it is defined in the model) together with the cost of the possible examinations. It is then up to the client to decide if and which examinations are to be performed. After the examinations, the scientist will then evaluate the outcome given the propositions and provide weight to the evidence. This evaluation will involve an assignment of probabilities to TPPR of traces for which the scientist will use data from literature or experiments.

When some findings are already available from prior examinations, the case pre-assessment becomes a mix of pre-assessment and formal evaluation of findings, as was discussed in Chapter 10 (Section 10.4). The power of the current findings to distinguish between the propositions needs to be communicated, together with the potential for further examinations. It is the mandating authority that needs to decide on whether or not additional examinations are required. To make this decision, they will need information on the power of the current results.

There are two ways to deal with this in the casework workflow. One could perform a formal evaluation of the current findings and report this, together with the results of the pre-assessment that detail which additional examinations may be performed and how they may impact on the outcome of the evaluation. If it is decided that additional examinations are required, a new evaluation will need to be performed, which makes the initial evaluation redundant. While this may seem inefficient, it is the most risk-free way to communicate with the parties in the case.

Another option is to include the current findings in the case pre-assessment and describe their power in general terms based on a guesstimate of the probabilities. This is more efficient, but it introduces the risk that the recipient of the report takes the rough assessment of the weight of the findings given the propositions for granted and introduces it in the courtroom. As such this pre-assessment may be misinterpreted as an evaluation based on scientific data.

The way in which such a pre-assessment which includes already available findings is communicated is therefore crucial. Such communication may include for instance a disclaimer that its purpose is a triage of further examinations and should not be perceived as an expert opinion for use in court.

11.5   EXAMPLES OF STANDARD OPERATING PROCEDURES (SOPS)

Much of the material on how an evaluation should be carried out within a laboratory can be taken from literature provided on evaluation, or indeed this book. Some examples of standard operating procedures (SOPs) are provided to assist in the formation of a laboratory's own SOPs but note that they will need to be tailored to the specific framework in which a laboratory operates. The SOPs provided can be used as a base when working through the practice questions provided at the end of the chapter in order to develop the paperwork required for a laboratory to implement activity-level evaluations. A typical SOP that will make up part of a laboratory's repertoire is a template report. Such an SOP is not provided here as example reports were provided in Chapter 10.

Section 11.5.1 documents the workflow for case intake. This document specifies what type of cases will be accepted for an evaluation given activity-level propositions. The initiative for such an evaluation is primarily with the mandating authority, but the scientist may initiate this after consultation with the client.

Section 11.5.2 provides an example of a review checklist that can be completed by a peer scientist while reviewing an activity-level evaluation case. This style of the checklist is designed to prompt the reviewer in what they should be checking during a review, and so has some administrative components and some technical components. Often review checklists are used as part of a laboratory's quality assurance system.

Section 11.5.3 is an example of an SOP that details the process of carrying out an evaluation. This SOP provides overview details on when a case should be accepted, how an evaluation should be carried out, what information should be provided in the case file and how the case should be reviewed. This SOP gives the overarching structure of carrying out activity-level evaluations, pointing the reader to other SOPs to give more specific details on one aspect of the process.

Section 11.5.4 documents the requirements for training and assessing staff to become competent in activity-level evaluation. The example given is brief but can be expanded on by laboratories, based on their own specific implementation of the activity-level evaluation.

Section 11.5.5 gives an example SOP document that describes Bayesian network construction for an activity-level evaluation. The SOP describes the

- Network architecture,
- Node naming conventions,
- Probability assignment and
- Conditional probability table layout

in the form that has been described in various chapters throughout this book.

Together these five SOPs (along with a report SOP) provide a laboratory with the foundational documentation that is required for implementing activity-level evaluation in an accredited forensic institution.

11.5.1 Case Acceptance/Initiation

11.5.1.1 Preface and Background

The group can perform an evaluation of forensic biology findings in light of competing proposed activities. This type of evaluation is undertaken after biological screening tests and DNA testing has been completed in a case.

Opportunities for activity-level evaluation (ALE) may originate directly from the police, prosecuting or defence lawyers or the Biology Reporting Scientist.

Defence requests are to be administered using the process described in the private case initiation standard operating procedure.

11.5.1.2 Scope

This procedure applies to all activity-level evaluation work accepted by the laboratory.

Suitable cases must have the following characteristics:

- ALE will be useful for prosecution or defence. This is to be established through direct communication with the relevant external party.
- Results available are amenable to ALE.
- Case circumstances allow the formation of reasonable competing propositions.

11.5.1.3 Procedure

ALE should only be undertaken when:

- Requested by police, prosecution or defence lawyers, or
- Confirmation from police, prosecution or defence lawyers who the work is needed when an enquiry is initiated by the Reporting Scientist.

11.5.1.3.1 Case Initiated by External Request

Enquiries should be directed to the Activity-Level Evaluation Team Leader or Manager to support that the work will be undertaken.

Requests, related correspondence and reasons for case rejection when work is not undertaken are to be recorded/stored in the case file.

11.5.1.3.2 Case Initiated by the Reporting Scientist

When the scientist identifies an opportunity for ALE in their own casework, they are to seek scientific advice from the Activity-Level Evaluation Team Leader or Manager to support the case and to help it meet acceptance criteria.

When work is supported, the Reporting Scientist contacts police and/or prosecution or defence lawyers and provides scientific advice on the activity-level evaluations that could be undertaken and how this may assist in the case.

Following confirmation that the work is needed from police, prosecution or defence lawyers , the request for work and related correspondence is to be recorded in the case file.

11.5.2   EXAMPLE OF A REVIEW CHECKLIST

(A copy of this is to be retained in the case file.)

Case file contents	Report
All pages numbered?	Information available to the analyst is detailed?
All pages have case file number?	Requesting authority mentioned?
No post-it notes or pencil?	All pages numbered correctly?
Non-contemporaneous errors crossed out, signed and dated?	All pages have case file number?
Each page identifies who did the work?	No spelling or significant grammatical errors?
Copies of referenced literature present?	Conclusions and expressions justified and correct?
External request for work documented?	Limitations and assumptions stated?
Case file cover	All literature used referenced?
Number of pages recorded?	All other court reports referred to?
Person/group(s) report to be sent to listed?	Report signed by RO?
Date and number of pages in report recorded?	Evidence Evaluation
Requesting authority recorded	Propositions appropriate?
Evidence recovery	Appropriate factors considered?
All exhibits and samples relevant to calculation mentioned in report?	Factors combined in a logical manner?
Results of presumptive tests mentioned?	Relevant data used to populate probability tables?
Justification for deeming sample irrelevant given?	Have appropriate priors been used?
Types of tests and results recorded?	Correct information entered into network?

Reviewer: Dated:

11.5.3   CARRYING OUT ACTIVITY-LEVEL EVALUATIONS

11.5.3.1   Preface and Foreword
- Each case being considered for Activity-Level Evaluation (ALE) will have different circumstances and each statement will require unique considerations. There are however generally agreeable and logical elements of the evaluation process which can be described and followed.
- This document provides an overview of the elements and considerations when evaluating activity-level propositions. This should be read in conjunction with 'ALE Bayesian Network Construction' and Biology training resources SOPs.
- The aim of this document is to describe activity-level evaluations within Biology at the laboratory and demonstrate they conform to guiding principles of evidence interpretation.

11.5.3.2   Scope
- This document applies to all cases where the evaluation of scientific observations is required to discriminate between mutually exclusive activity-level propositions.
- Evaluative reporting may be undertaken for prosecution or defence, in order to assist the court as much as possible.

11.5.3.3   Definitions and Abbreviations
- BN = Bayesian network
- LR = Likelihood ratio

- No-comment interview = Situation where the defence does not provide further information or does not suggest an alternate proposition in order to evaluate the evidence.
- RO = Reporting Officer

11.5.3.4 Principle

- The hierarchy of propositions is an accepted concept in forensic evidence. This concept describes the levels at which questions can be posed in evidential matters leading up to the ultimate issue of the court. The levels of the hierarchy and their propositions are:
  - Offence: Propositions that address issues which are legal in nature.
  - Activity: Propositions that specifically deal with disputed activities.
  - Source: Propositions that consider the biological material of a sample.
  - Sub-source: Propositions that consider the results of DNA analysis.
  - Sub-sub-source: Propositions that consider only a single component of a mixed DNA profile.
- Proposition setting is an important part of evaluating evidence at the activity level. These are two mutually exclusive events that consider the alleged and disputed activities. This should not include other aspects of the case circumstances that will be important to evaluate the evidence, as this will be considered separately as background information (I). Activity-level propositions should always be considered when the source of DNA is not being disputed, but rather the mechanisms that DNA transferred. Using sub-source-level propositions to make an inference on alleged activities can be a dangerous practice that is misleading to the court. It is always preferred for a scientist to consider different mechanisms of DNA transfer using a logical framework of evidence interpretation, rather than providing opinions on complex questions without preparation. If a scientist does not have an opportunity to explain the evidence in court, then the risk is that the court will be left open to unguided speculation.
- There are three principles of interpretation that should always be at the forefront of the scientists' decision-making process when evaluating evidence. They have been referred to as 'The single most important lesson for evaluative forensic science'. These are:
  - The findings are evaluated within a framework of circumstances (I)
  - The findings are evaluated given two competing and mutually exclusive propositions (Hp and Hd)
  - The role of the scientist is to only consider the probability of the evidence (E) given the propositions, and never the probability of the propositions directly
- Activity-level propositions are the highest that any scientist should ever address. The highest (offence) level is recognised as the domain of the court and the court alone. The scientist only ever evaluates the results given at least two competing and mutually exclusive propositions that will help the court to evaluate the propositions directly, with knowledge of the evidence in combination with all of the non-scientific evidence.
- The value of scientific evidence should always be presented in the form of a likelihood ratio (LR) in order to convey the strength of the findings. There are a number of different ways that this can be presented in order to express the magnitude of the LR, which are generally either qualitative or quantitative. Depending on the evaluation, the RO may choose to report the LR as a single value, a range of values or using a verbal equivalent. In order to calculate a numerical value for the LR, refer to the LR deviation SOP or the Bayesian network construction SOP.

11.5.3.5 Procedure

11.5.3.5.1 Case Creation

Cases appropriate for activity-level reporting are identified by the process described in the 'Acceptance of Activity-Level Reporting Cases' SOP.

11.5.3.5.2   When to Consider Cases for ALR

It is important to note that activity-level propositions should be considered when there is a potential risk that the evidence could be misinterpreted within the greater case context. If there is minimal risk of this misunderstanding, then sub-source-level propositions are adequate. This generally depends on the issues in court regarding what facts are disputed and what is agreed between both parties.

There are two situations where cases that are suitable for ALE can be identified. These are either recognised by the RO through routine criminal casework or when directly requested by the prosecution or defence. Refer to the 'Acceptance of Activity-Level Evaluation Cases' SOP for the minimum requirements that a case must possess in order to perform this type of evaluation.

There is no limitation for the type of case that is suitable for ALE with the exception that it should be a criminal case with a nominated suspect or person of interest. Criminal cases suitable for ALE can therefore be major indictable offences or minor indictable offences.

11.5.3.5.3   Case File Contents

- The case file and/or evaluative report must contain information that describes how and/or why a proposition was framed, including why a position of 'No comment interview' was reached or considered appropriate.
- The case file must contain all of the published material or sources of data that were used in the assignment of probabilities for the evaluation. If there are multiple studies that investigate a particular factor, then only the study relevant to the evaluation should be included, even if other papers have been referred to in the evaluative report. If multiple studies were used in the assignment, then they both must be included.
- The case file should contain a sensitivity analysis for all factors that rely on experimental data.
- The case file should also include the reviewer's comments and corrections made on the draft evaluation report.
- The case file must also contain a review checklist and case communications and requests.

11.5.3.6   Setting Activity-Level Propositions

11.5.3.6.1   Communicating with Prosecution and Defence

When evaluating the evidence, the RO needs to frame at least two alternate propositions. Generally, these will be designated as the prosecution hypothesis (Hp) and defence hypothesis (Hd), with the additional aspects surrounding the evaluation considered as background information and framework of circumstances.

To assist this process, the RO is required to obtain information about 'the framework of circumstances', i.e. relevant aspects of alleged events that will influence the interpretation. This is achieved through dialogue with the prosecution or defence and using any written information on the case circumstances and allegation(s) provided to the laboratory.

Even if the defence put forward a proposition, it may be possible and appropriate to use other information (such as eyewitness statements if made available) to provide a further reasonable proposition, which can be evaluated in addition to the proposition put forward by the defence.

Much of this information will be available on the police submission forms and/or medical examination notes however further discussion with stakeholders will be required for transparency and to encourage further input.

The two mutually exclusive propositions that will be considered should be provided to opposing parties to ensure their standpoints are captured accurately. If the work is being requested by the police, prosecution or defence, it is critical to seek approval from the requesting agency to disclose the propositions with the other party.

The non-requesting agency is under no obligation to provide their version of events, but the opportunity must be given for both parties to review the two scenarios with which the evidence is intended to be evaluated (as long as they provide their consent to contact the other party). As part of this process, it should be made clear that the prosecution and defence can only provide input on the proposition relating to their position.

11.5.3.6.2   No-Comment Interviews

It is critical the RO has an understanding of the different versions of events being proposed to explain the scientific evidence. If there is a no-comment interview, the scientist may:

- Use a reasonable alternative.
- It should be noted that alternatives such as 'An unknown person carried out the alleged activity and the POI was not involved' usually maximises the likelihood ratio and is not in favour of the defendant. This should be avoided where possible.
- Refrain from undertaking evaluative reporting.
- Evaluate results in light of source-level propositions, if relevant.

It is important to note that if no reasonable alternative is provided for the evaluation, then one will need to be developed on their behalf in order for the evaluation to be balanced.

11.5.3.6.3   Guide to Forming Propositions

The following can be applied to assist in the framing of propositions and to avoid making explanations. Propositions:

- Are in pairs; evaluative reporting needs to address (at least) two competing propositions.
- Need background information; specify all that is relevant in the case circumstances as they are provided. Define the propositions as simply as possible within that context.
- Are formal and relate logically to the framework of circumstances.
- Are mutually exclusive and exhaustive, they do not interrelate and are not open ended.
- Can be tested in a logical sense. It is helpful to ask 'Can I address the probability of the observations given this proposition?'
- Propositions should also capture the issue at hand regarding whether or not the activity is being disputed, the person involved is being disputed, or both. For example, propositions will be based on:
  - Whether or not the activity took place (either the accused or no perpetrator at all)
  - The activity was carried out by someone else (AO)
  - The accused carried out some alternate activity that led to the findings

Note that there may be occasions where both situations are being disputed, i.e. the accused performed some activity that led to the observed results; however, an alternate offender was responsible for the alleged criminal activity.

11.5.3.6.4   Assigning a Numerical Value to the LR

Similar to the assessment of DNA evidence at the sub-source level, ALR requires consideration of the probability of the evidence given the propositions. The RO needs to be conscious not to state they are considering the probability of the proposition.

Different situations may call for different LR assignment methods:

- Simple LR constructions – Relevant formulae can be derived by hand.
- Complex LR constructions – Use HUGIN software to construct Bayesian networks. Refer to the Bayesian network construction SOP.

11.5.3.6.5   Assignment of Probabilities

The most common source of material that will be used for evaluating evidence given activity-level propositions is peer-reviewed scientific publications. Probabilities can be assigned, but are not limited to, studies that investigate a particular factor through the number of times it has been observed in relation to the total number of replicates. An alternative method to using data counts when assigning probabilities is to model the data from the study and extrapolate/interpolate the model to the case-specific conditions to assign the probability.

There is no minimum number of observations or replicates in an experimental design for it to be regarded as insufficient to assign a probability. This is because prior counts are used which act to correct studies with small sample sizes and therefore have greater uncertainty (e.g. for the rate of transfer of a particular mechanism).

- The absence of appropriate data

    Options are available when there is absolutely no scientific data for a particular factor(s) relating to a scenario. These are to either:

    - Perform a sensitivity analysis to determine which probability assignments the LR is most sensitive to
    - Perform in-house experiments that closely replicate the case circumstances
    - Assign probabilities based on expert knowledge and experience
    - Only evaluate the evidence given sub-source-level propositions and comment on the current literature for the general mechanisms by which DNA can transfer
    - Refrain from conducting an evaluation at all

    Data that does not align with the case scenario – Some reasons why published studies do not align with the case circumstance include:

    - The study uses different laboratory protocols and analytical techniques
    - The data is not presented in the appropriate way or supplementary material is not available
    - The study uses a different amplification system with different sensitivity
    - The transfer mechanism has not been investigated but similar ones exist
    - Persistence is based on similar conditions but different amounts of time
    - Extraction efficiency is accounted for with different methods or protocols
    - DNA is recovered with different techniques

In these circumstances, it does not exclude using the data, as there will always be some aspects of published studies that do not use the exact same conditions. These can be accounted for partially with the inclusion of other factors or changes to the BN modelling structure. After an effort is made to account for this, the data may be used following a sensitivity analysis. The limitations and assumptions associated with the data need to be made clear in the report.

11.5.3.7   Sensitivity Analyses

These assess the level of sensitivity the LR has to individual probability assignments. Sensitivity analyses can be carried out in Excel, or if multiple probabilities are varied simultaneously the analysis can be carried out using software R.

There are two types of sensitivity analyses that may be performed in a case. First is a general sensitivity analysis for all of the factor(s) considered. Second is when there is a lack of data and the sensitivity analysis is being used as a justification for a particular assignment. The latter example must be included in the report.

11.5.3.8   Preparation of Reports

The report should make reference to any photographs, calculations, analyses, measurements and opinions. An ALE template is provided in the ALE report template SOP.

If a sensitivity analysis has been carried out due to a lack of data, then this should be made clear in the report. It should also be clear why the Reporting Scientist has come to the opinion that they have regarding the robustness of the opinion and the sensitivity of the LR to parameters within the model.

All sources of data or knowledge used to assign probabilities, and the construction of the BN (and hence assumptions of independence) should be clearly stated in the report.

11.5.3.9 Case Review

11.5.3.9.1 Technical Review

- The technical review is conducted by an RO who is competent to conduct ALE by completing the necessary training modules (refer to ALE training SOP) and has received authority for independent investigations to evaluate biology evidence at the activity level.
- Technical reviews may also be conducted externally. If done externally, then the external review checklist and review agreement forms must be completed by the external organisation. The external group must also be recognised as being proficient in ALE.
- The reporting officer will arrange the technical review. The final report must specify what aspects of the technical analysis have not undergone review (if any). The external reviewers will complete the *activity-level evaluation – external reviewer checklist*
- Points relevant to ALE casework the technical reviewer needs to consider include:
  - Alternate propositions are clearly stated, and there is clear reasoning for their use
  - Have appropriate factors been considered and is there appropriate documentation in the case file in relation to the framework of case circumstances?
  - Bayesian networks are constructed with a logical evaluative pathway.
  - Appropriate data was sourced for Bayesian network analysis and probabilities have been accurately entered into probability tables within HUGIN software.
  - Justification is provided for the use of conditional probabilities and prior probabilities.
  - Correct information has been entered into network.
  - Assumptions and limitations of the analysis are stated in the report.

11.5.3.9.2 Administrative Review (Applicable If External Technical Review Conducted)

- The reporting officer will arrange for an administrative review of the client report and case file, which can be performed by a member of Biology who has a general awareness of activity-level reporting.
- Reviewers are to complete the 'Activity Level Reporting Review Checklist'.

11.5.3.10 Case Cancellation

Work may be cancelled under the following scenarios:

- A client request – Work undertaken to date must be retained in the case file. For private cases, an invoice is to be issued for the number of hours of work performed to the time of notification.
- The RO is unable to base an opinion or formulate a conclusion, e.g. relevant data is not available to populate probability tables at the node(s) of Bayesian networks.

The scientist will contact the client advising them of this fact as early as possible and ask whether a report is required to this effect.

11.5.3.11 Validation

- A document has been prepared to describe the concept of general acceptance of this methodology in the forensic science community.
- A paper authored by the HUGIN software developers describes the use of HUGIN in forensic identification problems and DNA evidence.

11.5.4 Training Staff to Become Activity-Level Evaluators

11.5.4.1 Objectives
At the completion of this unit, the trainee will be able to:

- Understand and apply the hierarchy of propositions.
- Identify cases where activity-level propositions will have a significant impact of evidentiary evaluations.
- Understand the evaluation of evidence using Bayesian reasoning, including the formation of reasonable alternate activity propositions to explain biological evidence.
- Demonstrate an ability to create Bayesian networks using HUGIN software for the evaluation of evidence given specific case circumstances.
- Demonstrate an understanding of the types of data required in the various components of the evaluation and the ability to locate that data either from literature or in-house sources.

11.5.4.2 Method of Instruction
- Undertaking advanced training courses in evidence evaluation
- Tutorial sessions with experienced Reporting Scientist
- Discussions with Mentor
- Reading of relevant literature

11.5.4.3 Method of Evaluation
- Panel review covering objectives.
- Assessment of casework undertaken during the mentorship.

TRAINER: **TRAINEE:**

DATE COMMENCED: **DATE COMPLETED:**
ASSESSOR:
COMMENTS:

11.5.5 BN Construction or LR Derivation

11.5.5.1 Preface and Foreword
Bayesian networks (BNs) are graphical models which show the ROs logical reasoning with uncertainty and can be developed using specialised software. The laboratory uses the software package HUGIN. BNs are a useful tool to visually and logically build up complex evaluation frameworks (without the need to formally derive equations). This document provides an overview of the elements and considerations when constructing a BN. The aim of this document is to describe the process of constructing a BN in the HUGIN software within Biology, to demonstrate how to calculate a likelihood ratio (LR) using the HUGIN software and test the sensitivity of the LR to the underlying probability assignments.

11.5.5.2 Scope
This document applies to all cases that require the construction of a BN in order to assign a LR given mutually exclusive activity-level propositions. After constructing a BN and informing the factors with appropriate data, the impact that the choice of the data has on the overall LR can be tested with the use of sensitivity analyses.

11.5.5.3 Definitions and Abbreviations
AO = Alternate offender
BG = Background

BN = Bayesian network
C = Complainant
CPT = Conditional probability table
D = Defendant
LR = Likelihood ratio
OOBN = Object-oriented Bayesian network
RO = Reporting officer
TPPR = Transfer, persistence, prevalence, recovery
U = Unknown

11.5.5.4 Principle

- There are three distinct modelling steps for constructing a BN, these are:
  - Identifying the different factors to be considered in the evaluation
  - Identifying the dependencies between the different factors
  - Assigning probabilities to states in the conditional probability tables
- BNs are comprised of nodes and arcs, while sitting within the nodes are tables with a varying number of states (or rows) which is chosen by the RO based on how they would like the data to be categorised.
- Simplifying assumptions may be required during network construction, in order to appropriately model the complexities of reality. These should be stated and explained in the report.
- There may be instances where the data obtained does not completely align with the circumstances of a case. In the absence of other more appropriate data, the limitations should be clearly stated in the report. There may also be instances where there are multiple different sources of data for the same factor being considered for the scenario. In this case, a justifying statement should be included in the report as to why a certain study was chosen over (or combined with) another.

11.5.5.5 Procedure

11.5.5.5.1 *Standard Convention for the Network Structure*

BN development at the laboratory is performed using HUGIN software and by using the methods as described in:

Taylor D., Biedermann, A., Hicks, T., Champod, C. (2018). 'A template for constructing Bayesian networks in forensic biology cases when considering activity level propositions'. *Forensic Science International: Genetics* **33**:136–146

To use HUGIN refer to the 'Help' tab of the HUGIN software which will direct the user to an online manual. All BNs constructed for casework should apply the following standardised criteria:

- The BN should be constructed in such a way that each probability assignment is designated to its own node. This means that the structure will always have one node per factor, which is combined with the use of accumulation nodes. It is not recommended to perform calculations outside of HUGIN in order to assign probabilities that combine multiple factors.
- Factors within a network structure are categorised into proposition nodes which are coloured black, activity nodes which are coloured blue, prior decision nodes which are coloured grey, calculation nodes which are coloured yellow and result nodes which are coloured red.
- All results within a case that are relevant to the evaluation must be included in the BN, regardless of the outcome of the result, i.e. positive or negative results for a presumptive test, the presence or absence of an individual's DNA or indeed whether any DNA is found at all.

- Proposition nodes do not have entering arcs and will have exiting arc(s) connecting to all the activities under consideration. The standard naming for the proposition node is 'Hp/Hd' for prosecution and defence.
- Activity nodes have entering and exiting arcs and consider the different activities, whether disputed or agreed, that can lead to different mechanisms of transfer (i.e. DNA or biological material) to produce a given result.
- Prior decision nodes do not have any entering arcs and consider factors such as background DNA, persistence of material, contamination, case-specific events (e.g. whether or not ejaculation occurred or if a condom was used) or the probability with which someone will have the same DNA profile as a person of interest.
- Calculation nodes will have entering arc(s) and exiting arc(s) and consider factors such as transfer, accumulation of DNA or biological material from different mechanisms or accumulation of unknown sources of DNA.
- Result nodes will not have exiting arcs and will depend on the calculation nodes or prior decision nodes that may have led to the result being obtained.
- Nodes will always be numbered and naming should follow the following format:
  - Activity nodes should be called: '[PERSON] [ACTIVITY] [PERSON] [LOCATION]' or '[PERSON] and [PERSON] [ACTIVITY]'. For example D and C social interaction.
  - For transfer mechanisms, the node should be called: '[PERSON] [MATERIAL] transfer from [LOCATION] to [LOCATION] from [ACTIVITY]'. For example D DNA transfer from D hands to C underwear from digital penetration.
  - For persistence of DNA, the node should be called: '[PERSON] [MATERIAL] persisted on [LOCATION] through/for [EVENT]'. For example D DNA persisted on shirt through washing.
  - Accumulation nodes combine transfer events from different activities and can be called: '[PERSON] [MATERIAL] on [PERSON] [LOCATION]'. For example D DNA on C underwear.
  - Background nodes should be named: 'Background [MATERIAL] on [LOCATION]'. For example Background DNA on underwear.
  - Result nodes should describe the observation, DNA profile or presumptive test that is being used in the evaluation, in the format: 'Results of [TEST] on [ITEM]'.
- Each node label must be numbered starting from the proposition node, typically in a left to right, top to bottom format. Using this naming and numbering convention will avoid any two nodes having the same name that might be describing a similar mechanism for a different event that may or may not have occurred.
  - The use of OOBNs is permitted for complex networks that would benefit from repeating factors. OOBNs are a group of nodes known as a class network and contain multiple factors that can be repeated through the entire BN. These should be coloured white and follow the same naming/numbering format. OOBNs are generally only necessary when groups of factors are repeated multiple times throughout a network. These require the design of a separate BN which should be saved as: '[CASE NUMBER] _OOBN_ [NAME OF CLASS NETWORK]'.
  - BNs should be saved as '[CASE NUMBER] _BN'. Refer to Figure 11.1 for an example of a simple BN constructed in HUGIN which shows a proposition node, activity nodes, transfer nodes, accumulation nodes, background nodes and a result node.

11.5.5.5.2   Entering Probability Values into Node Tables

Values entered into CPTs depend on the number of states within a node, as all columns must sum to 1. Alternatively, leaving a value of 1 throughout the entire column will automatically normalise values based on the number of states. Probability assignments based on data from experiments or

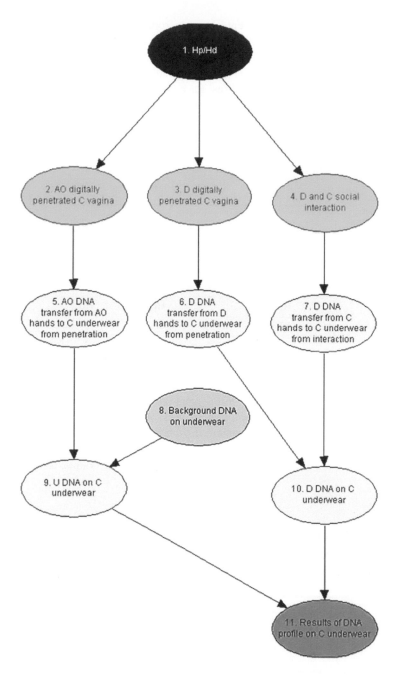

FIGURE 11.1 Simple BN of a sexual assault scenario with one result. This example shows both a disputed 'actor' and a disputed 'activity'.

peer-reviewed literature are entered into these tables. The number of columns will depend on the number of entering arcs and each additional arc will increase the complexity of the table.

Proposition nodes will generally have two states 'Hp' and 'Hd' where equal priors are given (the values are assigned as 0.5 for both states).

1. Hp/Hd		
	Hp	0.5
	Hd	0.5

Activity nodes will generally have one entering arc from the proposition node (with two states) and therefore two columns. The title of each column will depend on the name of the states from the previous node, e.g. 'Hp' and 'Hd' from the entering proposition node. Generally, there are two states for an activity node, either 'Yes' or 'No'. If a particular activity is agreed between prosecution and defence, then a 1 is entered into state 'Yes' for both columns and a 0 is entered into state 'No' for both columns.

Hp/Hd (1)		Hp	Hd
4. [ACTIVITY]	Yes	1	1
	No	0	0

If a particular activity is disputed between prosecution and defence, then a 1 is entered into state 'Yes' for column 'Hp' and a 1 is entered into state 'No' for columns 'Hd' and vice versa if it is said to have occurred for defence but not prosecution.

Hp/Hd (1)		Hp	Hd
[ACTIVITY]	Yes	1	0
	No	0	1

Transfer nodes can have one entering arc (two-column CPT), two entering arcs (four-column CPT), and so on. Again, the title of the column will depend on the node of the entering arc, e.g. 'Yes' or 'No' from an activity node. The states of a transfer node will generally be 'Yes' or 'No' indicating that a particular transfer has or has not occurred. A transfer rate is based on a number of observations and the total number of samples taken, which dictates how often a particular transfer is expected to occur. Probabilities are calculated with the use of prior counts, using the formula:

$$p_{i,k} = \frac{n_{i,k} + 1}{I + \sum_i n_{i,k}}$$

If a mechanism has not occurred, then it cannot transfer, and a 0 and 1 is entered into column 'No' for states 'Yes' and 'No', respectively.

[ACTIVITY]		Yes	No
5. DNA transfer from	Yes	$[p_{i,k}]$	0
Mechanism 1	No	$[1 - p_{i,k}]$	1

Prior decision nodes will have no entering arcs and therefore only one column with the two states, generally a 'Yes' or 'No'.

8. Background DNA on [LOCATION]		
	Yes	$[p_{i,k}]$
	No	$[1 - p_{i,k}]$

Accumulation nodes can have multiple entering arcs and combine the presence of DNA (defendants, complainants or unknown) from multiple different pathways (e.g. mechanisms of transfer). If there are two entering arcs, the table will therefore have four columns with a title that depend on the nodes with the entering arcs.

DNA transfer mechanism 1		Yes		No	
DNA transfer mechanism 2		Yes	No	Yes	No
10. D DNA on C [LOCATION]	Yes	1	1	1	0
	No	0	0	0	1

The example shows that if there is state 'Yes' from any of the previous nodes (both of which are different mechanisms of transfer), then a 1 is entered into state 'Yes', i.e. DNA will be present on the underwear if any of the previous mechanisms of transfer have occurred. Naturally, if DNA didn't transfer, then it cannot be present and a 1 is entered in state 'No' under columns 'No' and 'No'.

Result nodes can have multiple entering arcs where the states consider all possible outcomes that the result could show after doing the test. For this example, the complainant is an assumed contributor to a DNA profile obtained from a [SURFACE]. The entering arcs are the accumulation nodes of defendant's DNA or unknown sources of DNA and will therefore have four columns. If there is a 'Yes' in the table for U DNA and a 'Yes' in the table for D DNA, then a 1 must be entered into the state with both D and U present. Again, if there is no U DNA and no D DNA, then the state 'C' must contain a 1 (columns 'No' and 'No'). By convention, the states in the node table should be prepared in such a way that allows the values of 1 to be entered diagonally from top left to bottom right.

[Accumulation of U DNA]		Yes		No	
[Accumulation of D DNA]		Yes	No	Yes	No
	C+D+U	1	0	0	0
11. DNA on [SURFACE]	C+U	0	1	0	0
	C+D	0	0	1	0
	C	0	0	0	1

Using 'Run Mode' to calculate an LR:

- The run mode will only function correctly once all CPTs for each node have been filled out in a coherent manner, according to the laws of probability.
- Once in the run mode, all node tables should be displayed by highlighting all nodes and selecting 'show monitor window(s)'. This will display each table with the prior probabilities of each state in each node (as shown in Figure 11.2).
- To calculate an LR, all outcomes of analytical tests that were obtained in a case must be entered into the result nodes (see Figure 11.3). This is done by instantiating (setting a state so that it is known with certainty) a node depending on the result that was obtained. This is done by double-clicking the desired state in the result node monitor window. Once completed for all results that were obtained, the now-known information (following the analytical testing) that was entered will propagate through the network to update the posterior belief of the proposition node (see Figure 11.3).
- The LR is calculated by dividing the value obtained for the probability of 'Hp' by the value obtained for the probability of 'Hd'. Note that this is the posterior probability, however,

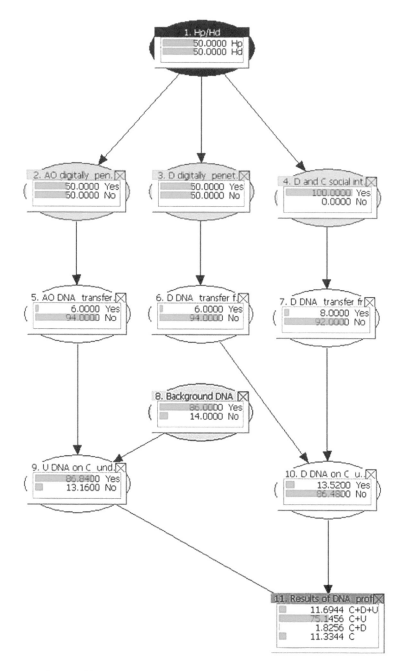

FIGURE 11.2 BN displayed in the 'Run Mode' after highlighting all nodes and selecting 'Show monitor window(s)', prior to instantiating result node(s).

calculating the LR in this way is only appropriate because equal prior probabilities for the proposition node were specified. Following Bayes Theorem, if the prior odds are equal, then the posterior probability is equal to the LR.

• Alternatively, the LR can be calculated by specifying a case findings node and then instantiating 'Hp' and 'Hd', respectively, in the proposition node, to calculate the LR.

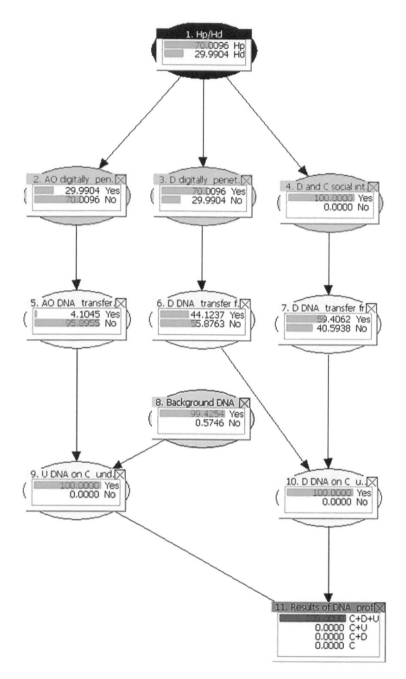

FIGURE 11.3 BN displayed in the 'Run Mode' showing the result node instantiated and the probability of 'Hp' and 'Hd' in the proposition node which is used to calculate the LR.

11.5.5.6 Sensitivity Analyses

A sensitivity analysis is a tool used to show how the choice of the data to inform nodes relating to TPPR may impact the evaluation. This will show how different probability assignments will vary the LR, one node at a time, and therefore which nodes the evaluation is most sensitive to.

- There are two ways to perform a sensitivity analysis, the first being that probability assignments of one node are varied, while all remaining nodes remain constant. This will show the relationship between the different possible values for a particular mechanism and the variability and trend with the LR. The second way is to vary multiple nodes at once and determine how it affects the LR. This method requires computer simulation and is outside the scope of this SOP.

- In order to perform a sensitivity analysis, a separate node is created within the BN, coloured green and labelled as '[NUMBER] sensitivity', with one arc entering the node of interest. This is done for every node which needs to be tested (see Figure 11.4). This version must be saved as a separate file in the format: '[CASE NUMBER] Sensitivity'.
- States within a sensitivity node should have discrete intervals of 0, 0.1, 0.2, ..., 1 followed by another state showing the case-specific value obtained from data (for example, a state labelled 'Case 0.86', where 0.86 would be the value obtained when prior counts were used with experimental data).
- Since there are no entering arcs for a sensitivity node, there will always be one column, which should have a 1 entered for each probability assignment and will therefore be normalised by the software. This is because states within a sensitivity node will be instantiated along with the results when calculating LRs.
- Since the node being tested now has an entering arc, the CPT will increase in complexity. Values should be entered in state 'yes' from 0 to 1 in discrete intervals of 0.1 and the final value is the case-specific value as shown in the table below. Note that this only shows half

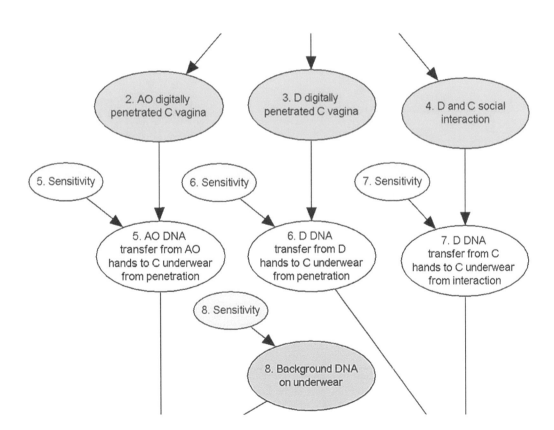

FIGURE 11.4   BN from Figure 11.1 with the addition of sensitivity nodes and arcs entering the variable of interest to be tested.

of the table for column 'yes' of an activity node. For column 'no' of the activity node, a 1 is entered for state 'no' for the remaining assignments.

[ACTIVITY]							Yes						
[SENSITIVITY]		0	0.1	0.2	0.3	0.4	0.5	0.6	0.7	0.8	0.9	1	Case 0.86
Mechanism 1	Yes	0	0.1	0.2	0.3	0.4	0.5	0.6	0.7	0.8	0.9	1	0.86
	No	1	0.9	0.8	0.7	0.6	0.5	0.4	0.3	0.2	0.1	0	0.14

- It is recommended that a 'case findings' node be included for large networks when performing this analysis, which will allow the results to be easily instantiated from a single node (Figure 11.4).
- The BN can also be structured to show the LR without having to calculate it from the numerator and denominator probabilities. This may be the desired approach for the casework or when performing a sensitivity analysis. This requires the creation of three additional nodes as shown in Figure 11.5.
- There are two states within the 'Hp' and 'Hd' nodes in a *numbered* format (the node type must be changed from labelled to numbered) which is 0 and 1. The tables of the 'Hp' and 'Hd' nodes are set out as shown below. Figure 11.5 shows the nodes labelled as 'Hp' and 'Hd'; however, it is important for the node names to also be changed to 'Hp' and 'Hd' in order for the calculation to function correctly.

Hp/Hd (1)		Hp	Hd
Hp	0	0	1
	1	1	0

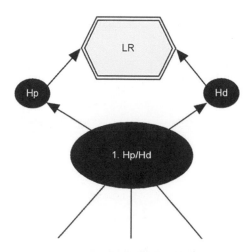

FIGURE 11.5 BN extension to the proposition node 'Hp/Hd'.

Hp/Hd (1)		Hp	Hd
Hd	0	1	0
	1	0	1

- The LR node requires an expression to function, which will then allow the calculation to be performed in the run mode. The expression is Hp/Hd. If the expression has not been entered correctly, then the text in the field will be in red font. The functioning of this BN extension can then be checked in the run mode by instantiated result nodes.
- The fully completed sensitivity BN is shown in Figure 11.6. Once this has been done, the sensitivity of the BN can then be tested for each node individually. This is done by instantiating the 'case' values of each sensitivity node and instantiating the results obtained in

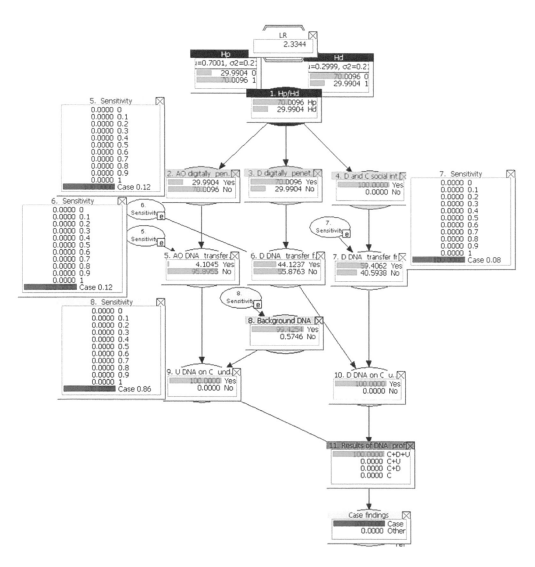

FIGURE 11.6   Sensitivity analysis as shown in the run mode, with case-specific values instantiated and the results node instantiated. The value of the evidence is then shown in the LR node at the top.

the case (Figure 11.6). The sensitivity of a node is tested by selecting a discrete value one at a time from 0, 0.1, …, 1, and recording the LR. This is then plotted to visually display the relationship and therefore the impact that a particular probability assignment will have on the overall LR within a case. This is then done for each sensitivity node which is then included in the case file.

- Sensitivity analyses are generally performed after all of the data is obtained for a BN; however, it may be desired to test the impact to the LR when data for a particular mechanism is unavailable.

11.6   PRACTICE QUESTIONS

Unlike other chapters, the practice questions in this chapter will not have answers provided at the end of the book. The questions given below provide more as a list of points to work through in order for a forensic institution/individual to prepare for the implementation of activity-level evaluation and reporting.

Q1) Identify the need and positioning of activity-level evaluation within your institution or for your stakeholders. Find some cases where activity-level evaluation could have assisted with understanding the significance of the findings that were reported. This can even be done by looking back at cases in your area that were considered miscarriages of justice and identifying whether activity-level evaluation would have assisted to prevent the miscarriage. Speak to managers, stakeholders or colleagues and explain the concept to bring them on board. Identify what other laboratories similar to you are doing … or thinking of doing.

Q2) By speaking with colleagues, identify individuals who are passionate and interested in learning about activity-level evaluation and reporting. This may range from a large group to just one person. Identify training opportunities, these may be online courses, conference workshops, targeted training courses for your institution or a fellow forensic institution that already has an established training plan. Determine how training can commence, who will undertake the training, how long it is expected to take to complete, what resources are required (cost for training, software required, etc.), is it possible to join with other laboratories and carry out training as a group.

Q3) Identify how activity-level evaluations will sit within your institution's workflow. How will cases be identified? How will they be authorised? How many cases will you be able to do, to begin with? Will the implementation start at case assessment and interpretation, or will it occur after a sub-source report has already been issued? Using the information you have gathered from the first three points write a business plan for introducing activity-level evaluation in your institution.

Q4) Write a standard operating procedure (SOP) that details how a case will be identified as one requiring an activity-level evaluation, what routes requests for activity-level evaluation work can come through, what authorisation is required to undertake the work and how the work will be allocated.

Q5) Explore the features of some Bayesian network construction software. Many have free or trial versions. Choose which software is going to be used by your laboratory. Carry out an in-house verification of the software by showing concordance with calculations done by hand (you can even use some examples that are already set up in this book). Write up the verification in a report that can be used as supporting material for your trust in the product.

Q6) BN construction method SOP. If you are not going to use BNs to carry out your evaluations, then write an SOP that describes the manner in which you will derive LR formulae and give several examples as your verification.

Q7) Determine how you will test and show the robustness of evaluations you carry out in your institution. When will a sensitivity analysis be done and how (or will) the results be conveyed to stakeholders? What aspects will be included in the sensitivity analysis? Create an SOP that describes all of these points.

Q8) Determine how results will be reported. Design a report template that gives examples of different types of cases and reports. Show this template to stakeholders to gain their feedback and adjust if necessary.

Q9) Make a plan for how you will train the next round of people to carry out activity-level evaluations in your institution. Decide what information is important and how it can be learnt. Will external training be sought or will those already trained then provide in-house training going forward? How will the competency of those being trained be assessed? All of this information can be compiled into a training SOP. If in-house training is going to be conducted, then create the training material and assessments.

Q10) Make a plan for how activity-level evaluations will expand into the future in your organisation. This may be a plan for:
- Having additional people trained in the area,
- Expanding the use of activity-level thinking into other areas of the workflow (such as making triage decisions using CAI),
- Expanding the practice into other disciplines (and eventually combining the findings across disciplines),
- Expanding the practice to more complex evaluations, e.g. using DNA amounts instead of presence/absence,
- Developing an in-house research group that can assist in carrying out an experiment to help inform TPPR probability assignments.

Write these expansion ideas in a document and put a timeframe on them. This will give you something to work on over the years and ensure that there is continual improvement within your institution.

11.7 REFERENCES

1. J. de Zoete, M. Sjerps, R. Meester, Evaluating evidence in linked crimes with multiple offenders, *Science & Justice: Journal of the Forensic Science Society* 57(3) (2017) 228–238.
2. R. Wieten, J. de Zoete, B. Blankers, B. Kokshoorn, The interpretation of traces found on adhesive tapes, *Law, Probability and Risk* 14(4) (2015) 305–322.
3. J.A. de Koeijer, M.J. Sjerps, P. Vergeer, C.E.H. Berger, Combining evidence in complex cases: A practical approach to interdisciplinary casework, *Science & Justice* 60(1) (2020) 20–29.
4. I.W. Evett, P.D. Gill, G. Jackson, J. Whitaker, C. Champod, Interpreting small quantities of DNA: The hierarchy of propositions and the use of Bayesian networks, *Journal of Forensic Sciences* 47(3) (2002) 520–530.
5. B. Kokshoorn, B.J. Blankers, J. de Zoete, C.E.H. Berger, Activity level DNA evidence evaluation: On propositions addressing the actor or the activity, *Forensic Science International* 278 (2017) 115–124.
6. N. Fenton, M. Neil, Avoiding probabilistic reasoning fallacies in legal practice using Bayesian networks, *Australian Journal of Legal Philosophies* 36 (2011) 114–150.

7. P. Gill, T. Hicks, J.M. Butler, E. Connolly, L. Gusmão, B. Kokshoorn, N. Morling, R.A.H. van Oorschot, W. Parson, M. Prinz, P.M. Schneider, T. Sijen, D. Taylor, DNA commission of the International society for forensic genetics: Assessing the value of forensic biological evidence: Guidelines highlighting the importance of propositions. Part II: Evaluation of biological traces considering activity level propositions, *Forensic Science International: Genetics* 44 (2020) 102186.
8. R.D. Stoel, C. Berger, W. Kerkhoff, E.J.A.T. Mattijssen, I.E. Dror, M. Hickman, K. Strom, Minimizing contextual bias in forensic casework. Forensic science and the administration of justice, *Critical Issues and Directions* 67 (2014) 67–86.
9. I. Dror, W. Thompson, C. Meissner, I. Kornfield, D. Krane, M. Saks, M. Risinger, Letter to the editor-context management toolbox: A linear sequential unmasking (LSU) approach for minimizing cognitive bias in forensic decision making, *Journal of Forensic Sciences* 60 (2015) 1111–1112.
10. W.C. Thompson, Painting the target around the matching profile: The Texas sharpshooter fallacy in forensic DNA interpretation, *Law, Probability and Risk* 8 (2009) 257–276.
11. W. Thompson, What role should investigative facts play in the evaluation of scientific evidence?, *Australian Journal of Forensic Science* 43 (2011) 123–124.
12. I.E. Dror, G. Hampikian, Subjectivity and bias in forensic DNA mixture interpretation, *Science & Justice: Journal of the Forensic Science Society* 51(4) (2011) 204–208.
13. A. Jeanguenata, B. Budowle, I. Dror, Strengthening forensic DNA decision making through a better understanding of the influence of cognitive bias, *Science and Justice* 57 (2017) 415–420.
14. R. Stoel, C. Berger, W. Kerkhoff, E. Mattijssen, I. Dror, *Forensic Science and the Administration of Justice: Critical Issues and Directions*, N-America, Europe and Asia: SAGE Publishing, 2014.
15. Forensic Science Regulator, Guidance: Cognitive bias effects relevant to forensic science examinations, in: *FSR-G-217*, 2015. https://www.gov.uk/government/publications/cognitive-bias-effects-relevant-to-forensic-science-examinations [last accessed 10 January 2023]

12 Beyond Forensic Biology

Bas Kokshoorn

CONTENTS

12.1 INTRODUCTION

Throughout the book we have dealt with evaluation of biological traces analyses given activity-level propositions. The use of Bayesian statistics in forensic science is not restricted to this particular domain, as was mentioned already in Chapter 1. Maybe the first publication with an application of Bayesian statics to a forensic inference problem was written by Finkelstein and Fairley [1] in 1970. They work through an example case in which they show how the diagnostic value of a palm print on the handle of a knife may affect the posterior probability of the defendant being involved in the stabbing of his girlfriend. Lindley [2], in 1977, applied a likelihood ratio (or rather a Bayes factor) calculation to address the question whether fragments of glass could be attributed to the same source. This study was picked up by others, for instance, Seheult [3] compared the Lindley approach to the frequentist approaches that were in common use in most jurisdictions at the time. Evett published a paper in 1983 [4] describing a Bayesian approach to the issue of identification of the donor of a bloodstain left at a crime scene, and subsequently in 1984 a generalisation of the Bayesian approach to identification issues in forensic science [5].

What these studies show is that the same statistical framework can be used to address forensically relevant issues across the different fields of forensic science. The same holds for the principles

DOI: 10.4324/9781003273189-12

435

of evaluation of evidence and the hierarchy of propositions. These concepts are not unique to forensic biology and can (and should) be applied to the evaluation of forensic findings across the forensic disciplines. As such, the framework for the interpretation of evidence is a crucial aspect that binds these disciplines together.

This means that evaluating findings given activity-level propositions, as is discussed for forensic biology throughout this book, can be applied in the same way to forensic findings from other disciplines. And while the general statistical framework is the same, as are considerations on the TPPR of traces, the findings and the factors impacting on the probability of TPPR may differ substantially between the fields. Consider for instance the categorical nature of the number of glass particles and the way in which they transfer and persist on clothing compared to the amount of latent DNA on such clothing. Another example is that indirect transfer of biological material is something that frequently needs to be considered, while indirect transfer of fingermarks is hardly an issue (except when adhesive surfaces are involved) and the recovery of a fingermark from a surface is usually 'proof' of a direct contact between a hand and that surface.

In this chapter, the application of the principles of evidence evaluation to closely related fields like fingermarks analysis and bloodstain pattern analysis is explored in relation to forensic biology. Beyond that, we look at other transfer evidence fields, as well as touch on forensic disciplines like forensic medicine and digital forensics.

In the final section of this chapter, approaches to combining evidence from multiple cases and from different forensic disciplines to address issues at the activity level will be discussed.

12.2   ON TRANSFER OF MARKS, MATERIALS AND SIGNALS

As forensic scientists we are concerned with the application of our scientific methodology to issues of relevance to investigation, defence or prosecution of alleged criminal acts. The scientific methods applied are broad and encapsulate those developed in chemistry, physics, biology, mathematics and medical disciplines (this is a non-exhaustive list). All forensic disciplines apply their methods to the recovery, analysis and interpretation of traces created during activities of interest. These traces can be broadly grouped into three categories, or modes, which are marks, materials and signals.

Marks are basically a form of damage, deformation or marking of a surface or item. If we take an imaginary walk through a crime scene, we may encounter a number of marks that might warrant further investigation. Outside the house we find footprints in mud leading to a window. On the windowsill we observe scratches that could be related to the window being forced open and thereby broken. Inside we find the body of a deceased victim whose hands and feet are tied with duct tape. A roll of tape is found on the table. Physical comparison of the tape ends may tell us whether or not the tape around the wrists of the victim came from the roll of tape, and in what order the pieces of tape were applied. The victim has been shot and lines on the surface of the empty cartridges found on the floor may, sometime during the investigation, be linked to the inside of the barrel of a firearm. The medical examination of the body by a pathologist will also consider the marks left on the body by the bullets, for instance, to determine points of entry and exit.

At the same crime scene, we may encounter trace materials as well. We have, in this book, extensively discussed human biological traces for instance. DNA of the perpetrator may be found on the windowsill or on the duct tape. Gunshot residue may at some point be found on clothing of a suspect. Glass fragments of a broken window may have been transferred to the clothing of the perpetrator. Likewise, fibres from the gloves worn by the perpetrator may have been left on the sticky side of the duct tape.

Signal data may also come into the investigation when data from license plate registration cameras are analysed by the police to see which vehicles were near the scene at the time of the incident. CCTV cameras may have recorded the individuals who passed by the shops in the street leading to the victim's house. When a suspect is arrested, their smartphone may be examined for data stored in the device on its location at the time of the incident, and possibly whether it is connected to the Wi-Fi router in the victim's home.

Whether we are interested in marks, materials or signals, when we are addressing issues at the activity level, we need to consider aspects of transfer, persistence and recovery of the traces to some extent. Table 12.1 provides some examples of TPPR aspects that need to be considered for the three different modes.

Evidently, not all aspects of TPPR will impact on all modes of traces in the same way. Indirect transfer of materials may be a crucial aspect to consider, and we have given numerous examples throughout this book of indirect transfer of biological material. However, such indirect causation of marks is generally not possible as there needs to be a physical interaction between the surface and the object causing the marks. On the other hand, the factors impacting on the persistence of wounds in a decaying body (like ambient temperature, moisture and microbial activity) are likely to be very similar to those impacting on the persistence of biological traces from the offender on the skin of the same body as they will impact on the survivability of biological material. Therefore, even though the findings being evaluated are quite different in nature (physical marks vs the presence of foreign biological material), their common medium links them.

The three modes of traces each may contribute information that can distinguish between case-relevant scenarios and propositions. Different modes may be used to address a similar issue at the activity level, resulting in evaluations combining evidence from multiple forensic disciplines. We will discuss an example of such combined evaluations in Section 12.6.

The knowledgebase on factors impacting on TPPR, as well as data to support assignments of probabilities to TPPR events (the sources of which only partly overlap, as we discussed in Chapter 3), varies substantially between different fields. This is illustrated by an analysis of the data in the Transfer Traces Activity DataBase (TTADB) which is hosted and maintained by the University of Québec at Trois-Rivières (UQTR), Québec, Canada [17]. This database hosts records of scientific publications on the transfer, persistence, prevalence and recovery of forensically relevant trace materials. Each record is assigned to one or more main topic categories. An analysis of the number of records in each category is shown in Figure 12.1.

As can be taken from Figure 12.1, there are substantial differences in the number of publications supporting our knowledgebase in the different forensic disciplines. The total number of records in the database on 18 July 2021 was 2,208. As some records are assigned to more than one category,

TABLE 12.1
Examples of TPPR Aspects That Impact on the Three Modes of Traces

	Marks	Materials	Signals
Transfer	The force applied by a finger resulting in a broader, distorted fingermark [6]	Number of glass particles transferring to subjects from different distances [7]	Traces left in an android operating system database after connecting to a Wi-Fi router [8]
Persistence	Ligature marks on the skin of a victim fading over time [9]	The amount of organic gunshot residue on hands of a shooter decreasing over time [10]	User data retained in the physical memory of a Windows operating system [11]
Prevalence and background	Bruises on the skin of children who are not abused [12]	Common fibre types on bus seats [13]	The number of cell phones connecting to relay towers simultaneously [14]
Recovery	The methods used to visualise latent fingermarks [15]	Excluding a true donor from a mixture in DNA analysis [16]	Methods applied to unlock a smartphone to recover content [11]

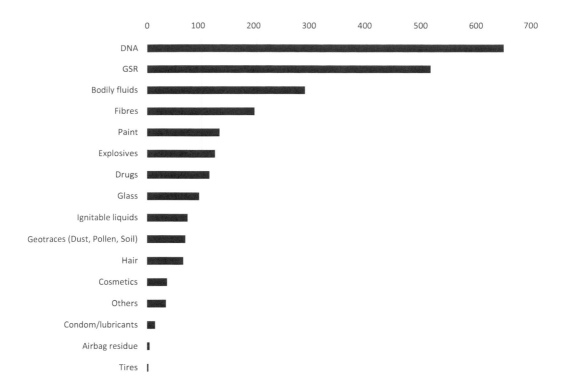

FIGURE 12.1 Number of records for each trace type category in the TTADB (as accessed on 18 July 2021). Records may be categorised in more than one category.

the total number of records per category sums up to 2,420. Overlap can for instance be found between DNA and bodily fluids, as well as between GSR and explosives. This overlap is not surprising, as factors impacting transfer or persistence are likely to be relevant for multiple disciplines. Analysis methods may also be highly similar between fields of forensic expertise, which is another reason why studies into recovery may also cover multiple categories. Differences and similarities between these aspects are explored further in the following sections.

12.3 WELCOME TO THE FORENSIC BIOLOGY FAMILY

As it is with family members, some characters are shared but there will also be marked differences between them. So, it is with what, for lack of a more appealing analogy, will be referred to as the 'forensic biology' family of forensic genetics, bloodstain pattern analysis, fingermark analysis and non-human forensic biology.

Forensic genetics and bloodstain pattern analysis share their interest in human blood, and they will use the same tests to determine if a stain or fluid is blood. With fingermarks there is the common forensically relevant issue of identification. Both disciplines have the ability, using extensive forensic databases, to identify potential donors to a trace using their respective biometric information. Further shared interests lie in the visualisation of latent biological deposits, and we see for instance that fingermarks that have been visualised may be sampled for DNA [18]. The fourth family member, which covers all non-human biological analyses, shares certain methodologies with forensic genetics. Non-human biological markers (i.e. microbiome) may also be used for human identification [19] or for identification of body fluid stains [20], amongst other applications [21].

12.3.1 BLOODSTAIN PATTERN ANALYSIS

'Bloodstain pattern analysis (BPA) refers to the collection, categorisation and interpretation of the shape and distribution of bloodstains connected with a crime' [22]. Bloodstain patterns can be documented at crime scenes or on items being examined in a laboratory setting. Such patterns will subsequently be analysed and interpreted to categorise (patterns of) stains based on their physical appearance and the mechanisms by which they are formed. Liquid blood has certain specific characteristics (i.e. viscosity, surface tension) and as a liquid has fluid dynamics that follow general laws of physics. Based on these defined characteristics, the mechanism that caused liquid blood to transfer to a surface may be predicted from visible characteristics of the resulting stains. Two main mechanisms are passive stains and projected patterns. Drops of blood falling due to gravitation only and resulting in a drip stain are an example of passive stains. Projected patterns result from other forces coming into play. Examples of such projected patterns are cast-off patterns from a bloodied object being swung or an arterial pattern caused by a severed arterial blood vessel. All patterns can subsequently be affected by secondary changes, for instance, due to capillary action or wicking in fabrics or by insect activity.

Within the main categories of bloodstain patterns, a number of subdivisions are made which encompass specific aspects of the mechanism that may have caused them. While the definition of the different pattern types has changed over time, current terminology is defined in standards provided by the American Academy of Forensic Science (AAFS) [23] based on work by the Organization of Scientific Area Committees (OSAC) and the FBI's SWGSTAIN working group. The international Association of Bloodstain Pattern Analysts (IABPA) has adopted this terminology.

BPA analysts generally provide categorical conclusions on the classification of bloodstains [24]. If a subcategory cannot be determined, a conclusion will be drawn on the main category. As this categorical determination of patterns is partly subjective, and affected by information on the case circumstances like the location of the stains and the scenarios that are being considered, the accuracy and reproducibility of bloodstain pattern classification vary between examiners [25].

As was discussed in Chapter 9, the strength of support of results from a test for a specific cell type can be assessed probabilistically. A similar approach has been advocated for inferences made on observed bloodstain patterns [26]. At this time, a likelihood ratio framework is not in common use in the field. As has been expressed by Attinger et al. [26], to attain this, advances need to be made in (1) research in the underlying physics, (2) data sharing and (3) training materials for analysts on the essential statistical background.

The observed patterns of bloodstains may also be evaluated given propositions at the activity level. While (from a statistical point of view) there are no reasons why a likelihood ratio framework should not be applied to the evaluation of bloodstain patterns given activity-level propositions, to date few publications have addressed the application of such a framework [27]. A reason for this may be the lack of systematic studies in TPPR of bloodstain patterns given forensically relevant scenarios. However, the lack of data in itself should not be a reason to abandon a logical framework for the interpretation of findings in a case [28]. Nonetheless, efforts are required by the field to extend the current knowledgebase on these issues.

Evidently, patterns of human bloodstains may be encountered in forensic casework together with other sources of biological material. While the areas of expertise are quite clearly defined (with interpretation of the type of stain or direction of stains being clearly in the domain of the BPA analyst), in some of these cases it may be unclear which expertise is required to evaluate the findings given activity-level propositions. Overlap in areas of expertise may be found between BPA analysts and DNA analysts when considering bloodstains given activity-level propositions. For instance, when the findings consist of one or only a few stains of which the pattern cannot be determined. The size and location of the stain may be the only aspects (other than the cell type test and DNA results) that can inform the likelihood ratio. In such cases, one will need to decide whether the findings are restricted to these observations (and hence may be best evaluated by a

DNA analyst), or whether the absence of identifiable patterns also needs to be considered. If so, the probability of finding a bloodstain pattern that cannot be categorised needs to be assigned. This probability would be in the domain of expertise of the BPA analyst. It may therefore be advisable that, if bloodstains may be part of the set of expected case findings, consultation between the DNA analyst and BPA analyst is part of the case pre-assessment. In this phase, it may be determined what the expectations are for the full set of human biological traces and whether BPA analysis should be part of the evaluation.

12.3.2 FINGERMARKS

Friction ridge skin impressions (or finger*marks*) may be found at a crime scene or on items of interest and are a somewhat exceptional trace with respect to the categorisation of traces discussed in the previous section. They consist of residues left by ridged skin from hands or feet on surfaces. From that perspective, fingermarks can be considered traces of materials. However, in common fingermarks examinations focused on identification of the source, it is not the residue itself we are concerned about. It is the impression of the ridge skin *pattern* that is left on the surface which is of interest. The examination of such patterns fits more closely with the forensic disciplines that study marks rather than traces of materials.

The residue itself has become more relevant in recent years for forensic investigations as the composition of the biological and chemical excretion materials in a fingermark may provide us with intelligence information on the donor. Studies of traces left by contact with friction ridge skin (and such a deposition of finger*traces* may result in simultaneous creation of finger*marks*) have shown that from proteomic, metabolomic and/or spectroscopic analyses we may deduce whether the donor is male or female [29], has a particular diet [30], has personal habits [31], is a smoker [32] or is an abuser of substances [33]. Studies have also addressed the age of the fingermark [34], the presence of condom lubricants [35] and the cell types present in the trace [36].

Nonetheless, while the focus is usually on the finger*mark*, the behaviour of the deposited material follows the general principles of transfer, persistence and recovery we find with trace materials. The residues that make up the fingermark may transfer directly, or indirectly, as we see with DNA. However, it is highly unlikely that the pattern itself transfers indirectly, as any distortions to the pattern will quickly render it useless for identification purposes. Exceptions to this have been nevertheless been found [37, 38]. An example is tape that is applied to a surface containing a fingermark. During this event, the mark may be transferred to the sticky side of the tape (or vice versa, as has been demonstrated with transfer of marks from self-adhesive stamps to paper [39]). For interpretation of fingermark findings given propositions suggesting an indirect transfer of the marks, it is therefore crucial to document whether or not the fingermark pattern is mirrored compared to a reference fingerprint.

An evaluation of findings from a fingermark examination given activity-level propositions may consider a number of factors that are similar to those considered in the domain of forensic biology. Such factors are for instance the location of the traces, or the quality and degradation of the trace. Specific aspects that may provide information that may support one proposition over another are the direction of the trace (which way do the marks point?), the part of the hand or finger of which marks are found, and their position relative to each other. For instance, the position of the thumb relative to the other fingers. Two studies have described mock case examples of fingermark evaluations given activity-level propositions [40, 41]. In one of these studies (De Ronde et al [41]), the position of marks from different parts of the hand on the handle of a knife was examined. Volunteers were asked to use a knife to either cut a piece of gingerbread or stab the knife into a foam figure. The positions of fingers, palm and thumb on the handle and blade of the knife were recorded. This study for instance showed that marks from the thumb were found on the back of the handle (the end which is perpendicular to the blade) after stabbing and never after cutting. Similarly, marks from the fingers on the (blunt) top of the blade were only found after cutting, not after stabbing. The use of this information in an evaluation given activity-level propositions were demonstrated in this study using Bayesian networks.

The network was constructed using the template that was developed by Taylor et al. [42] for use in forensic biology. This is the same BN architecture and construction as was explained in Chapter 6 of this book. As similar aspects of transfer, persistence and recovery need to be considered, the template can be applied to the evaluation of findings from fingermarks analyses as well as forensic biology findings. An adapted version of the network is shown in Figure 12.2 with the three result nodes instantiated to calculate a ratio of posterior probabilities (which equals the LR given that equal prior probabilities have been assigned) for the propositions (in which S denotes the suspect and V the victim):

H1. S stabbed the victim with the knife. S did not use the knife to cut food
H2. V was not stabbed with the knife. S only used the knife to cut food.

From the calculations of the probability of the observation of marks from the palm and fingers on the handle, none on the blade, and from the thumb on the back of the handle, given the propositions (and a number of assumptions which have been discussed by the authors, including assumptions on the relevance of the experimental data to a casework situation), the findings are approximately 30 times more probable if H1 is true rather than if H2 is true. This example demonstrates that the part of the hand that made contact with specific parts of the knife may be informative when addressing competing activities that have been proposed.

Examination for fingermarks may interact with examination of an item for biological traces. To visualise latent fingermarks, several different chemicals may be applied to a surface, some followed by one or more rinsing steps with water. These chemicals may interact with the biological material (fixating or degrading it), or the rinsing steps may remove any biological materials present. Several studies have shown the impact of latent fingermark detection on the persistence of biological traces [43–46]. Additionally, studies have shown a number of additional routes of indirect transfer (contamination) of biological materials when for instance fingerprint brushes [47] or superglue chambers [48] are used.

Interestingly, the impact of the chemical treatment on biological traces might also be a factor impacting on the relative contribution of donors to a mixed biological sample. Harush-Brosh et al. [49] applied amido black, which is a chemical commonly used to improve the visualisation of fingermarks that have been deposited in or with blood, to mixed samples containing blood from donor A and trace DNA from donor B. They tested samples in varying ratios of the amount of DNA from both donors and found that amido black consistently changed the mixture ratio of the sample. The reason was that less DNA of the blood donor was recovered from the sample compared to the donor of the trace DNA. The underlying cause for this was speculated on by the authors (differential degradation, fixation of blood cells), but the effect was pronounced as about half the DNA from blood was lost.

When assessing findings from forensic biology on an item given activity-level propositions in situations where such latent fingermark detection techniques have been applied, these issues need to be considered when assigning probabilities to the persistence and recovery of biological material. As with bloodstain pattern analysis, it is recommended to involve fingermark examiners in the case of pre-assessment consultation when fingermarks are expected to have a role in addressing the issue at the activity level.

12.3.3 NON-HUMAN BIOLOGICAL TRACES

The field of non-human biological traces is extremely broad, both in types of analyses as well as in the forensically relevant issues that are being considered. This ranges from the identification of species (for instance in the illegal trade of endangered species [50] or the identification of cannabis

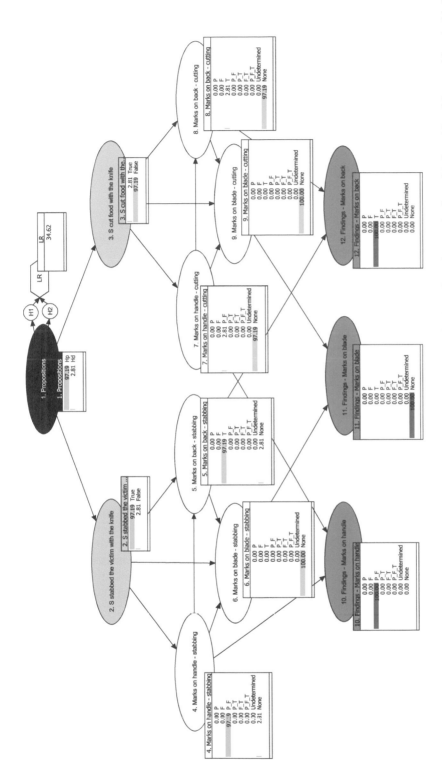

FIGURE 12.2 Bayesian network adapted from De Ronde et al. [40]. The results have been defined as observing marks from friction ridge skin on wither the blade, handle or back of the handle from the palm (P), fingers (F) and/or thumb (T). The state 'undetermined' reflects the observation of marks that cannot be attributed to either P, F or T.

[51]), to forensic palynology [52], forensic entomology [53] or the use of microbiomes [54]. Aspects of most of these subfields may interact with forensic human biology when addressing issues at the activity level in a case.

12.3.3.1 Species Identification

Traces from plants or animals may assist in linking a person or an item to a specific animal or location. Examples of examinations that are being performed in cases are the identification of a specific cat or dog that is the source of hairs found on clothing [55], or seeds from trees that can be linked to an individual tree or cluster of cloned individuals [56]. Such data can evidently be very useful to address issues at the activity level if the issue is for instance whether or not a suspect was present at the scene of a crime. When using this type of information, issues of transfer, persistence, prevalence and recovery will come into play. An example case (ECLI:NL:RBMNE:2018:3274) with a snail shell may serve to illustrate this.

A young woman was found dead in a ditch that holds water year-round. Autopsy revealed she may have been forcefully drowned. A suspect was apprehended who claimed that he met her earlier in the day but wasn't involved in her death, nor was he in water recently (particularly not the ditch in which she was found). In a pocket of his coat, a shell was found of a snail. The specimen was identified as *Anisus vortex* (Linnaeus, 1758), a freshwater snail that is common in ditches and streams in the Netherlands that harbour aquatic vegetation.

The prosecution took the position that the snail shell linked the suspect to the scene of the crime as the findings from the forensic examination of the shell and material from the scene were reported as:

> The vegetation in the victim's hair examined by the NFI consists mainly of rhizomes. In addition to the rhizomes, snail shells are also present. The snail shells in this vegetation have been identified on the basis of external features as, among other species, Whirlpool ramshorn snails. The snail shell from the left pocket of the jacket has been identified on the basis of its external features as a Whirlpool ramshorn snail. This species corresponds to one of the species in the vegetation in the hair [of the victim]. The Whirlpool ramshorn snail is a common freshwater snail found in stagnant to weakly flowing fresh or brackish surface water, such as a ditch, in which many plants are present.

Defence counsel disputed the stance of the prosecution and stated:

> There is no independent evidence of an offence committed by the defendant. The fact that a snail shell was found in the jacket of the defendant does not make this different. It has not been established conclusively that this snail shell came into the jacket pocket of the defendant at the location where the victim was found, nor that this snail shell came from this location or had not been in the jacket pocket for some time.

The disputed issue may be captured in propositions:

H1. The defendant wrestled with the victim in the ditch.
H2. Somebody other than the defendant wrestled with the victim in the ditch, and the defendant was not involved.

The defendant claimed not to have been in a ditch recently prior to or following the incident, and the prosecution did not question his statement. Hence, we can take this claim as undisputed case information (I).

If we now consider the findings (an intact shell of *A. vortex*) in a coat of his jacket given proposition H1, we must consider:

- *Transfer* of the shell to the pocket of the jacket. How probable is it when wrestling with somebody in a ditch that an object enters the pocket of a jacket? Here we would for instance need to consider the abundance of snails in the ditch, and how long the jacket would be submerged.

- *Persistence* of the shell in the pocket in the time between it entering the pocket and it being recovered at the laboratory. What is the probability of the snail shell staying in the pocket and remaining intact during this time? Here we would for instance need to consider the type of fabric (was it padded?), the fragility of the shell (was the animal alive at the time, or was it already an empty – and thus possibly degraded – shell?) and the activities performed with the coat (including how tight was it packaged during transport and storage after it was secured?)
- *Recovery* of the shell. What is the probability of recovering the intact shell during the examination? What was the examination strategy (were all pockets searched for traces?), would such a shell (of less than a centimetre in diameter and less than a millimetre high) be found during this search and would it be identified as *A. vortex* correctly by the analyst?

Some data are available that may be informative when assigning these probabilities, for instance from studies on attachment of snails (including *A. vortex*) to floating objects like mallards [57] or studies on erosion of snail shells [58, 59]. However, more relevant data may be obtained in situations like these by performing case-specific experiments.

Under proposition H2 similar aspects would need to be considered, depending on a reasonable alternative scenario that details how a freshwater snail shell could be found in the pocket of a jacket. In this case, such a scenario was not presented. The court considered as much when concluding:

> The defendant has not been able to give an explanation for the presence of the snail shell in his jacket pocket and diatoms on his clothes and on the pedals [of his bicycle] that is refutable or otherwise plausible for him, other than that he does not know, and he does not agree with it. Furthermore, the defendant has stated that he has not been in a ditch recently.

As can be seen from this last quote, diatoms have also been recovered and examined in this case. Diatoms are another example of non-human biological traces that may be recovered in forensic casework. They have been used in forensic science mainly to determine if a person drowned or if their body was disposed in water after death by examining the content of the lungs for their presence [60]. However, as species assemblages and relative species densities vary between locations [61], diatoms also have great potential to link objects to specific soils or surface waters with which they may have been in contact. Recent studies are also looking at the transfer and the persistence of such traces under different situations [62–64].

12.3.3.2 Microbiome

As with diatoms, other micro-organisms may also be recovered as trace material in forensic casework. Whether it is as species assemblages in soils [65], as microbial traces transferred from skin to surfaces [19], changes in species composition of microbiome used in estimation of post-mortem intervals [66] or identification of body fluid stains using microbial composition of the stains [20, 67].

Of particular interest to forensic biology are the microbiomes associated with humans. These microbial assemblages can be found on the skin as well as in bodily orifices and the intestinal tract. Descriptive studies of the human microbiome have shown that there is substantial intra- and inter-individual variation [68]. As these microbiomes may transfer to the environment, they potentially leave forensically relevant traces. Intra-individual variation is found on different areas of the skin and as such may be relevant to determine the part of the body that contacted a surface [69, 70]. This information may potentially be very relevant when addressing recovered biological traces given activity-level propositions. This potential is not yet fully realised as much and is still unknown about the mechanics of transfer (what part of our skin biota is transferred and how do factors like receiving surface impact this?), persistence (particularly the temporal shift in the composition of microbial communities; their ecology) and recovery (how do commonly used sampling methods affect the recovery of different microbial species?). Nor is it yet well understood to what extent indirect transfer of such microbial communities occurs and how that impacts on the species assemblage.

12.4   OTHER TRACES, MARKS AND SIGNAL DATA
GIVEN ACTIVITY-LEVEL PROPOSITIONS

There are numerous other areas of expertise focused on the forensic analysis of traces, marks or signal data. It would go well beyond the scope of this book to discuss them all. There are nonetheless a number of topics discussed in the scientific literature in relation to the evaluation of evidence that bear relevance to the interpretation of forensic biology findings given activity-level propositions. They either point to aspects relating to TPPR of traces that deserve further exploration, or they reinforce points made throughout this book on the evaluation of findings given activity-level propositions and on modelling decisions that need to be made. Some of these topics are reiterated in this section.

12.4.1   THE RELEVANT POPULATION

Throughout this book, the probability of finding background DNA in a sample from a surface has not been modelled specifically for each of the propositions. This means that it is generally assumed that the probability of finding background DNA is the same for each proposition. In the Bayesian network examples in Chapters 6–8, the background probability nodes have therefore been included as (grey) root nodes. A relevant question is whether this assumption holds or not.

Champod et al. [71] have shown that different population datasets may be relevant, depending on the case circumstances, when addressing source-level issues. They distinguish between 'Offender related', 'Innocent suspect related' and 'crime related' databases. The first type of database is relevant to inform the probability of the observation given that the defendant is the offender (generally proposed under the prosecution proposition). Such a database could be constructed of observations from individuals convicted for similar types of crimes. The second database would be used to inform the probability of the observations given that the defendant was not involved (and hence they or objects associated with them are not the source of the questioned material). Such a database might consist of observations from the appropriate population of individuals who have not been associated with relevant crimes. The third database may be used to inform the probability of the observation given potential sources for the questioned material that are found in relation to similar crime scenes. This database would be scene centred rather than centred around a suspect.

Curran et al. [72] in their book chapter on the interpretation of glass evidence discuss the relevant population in relation to the interpretation of findings given activity-level propositions. They suggest that different populations may be sampled to inform probabilities of finding background levels of glass given the alternate propositions. The probability of finding background levels of glass (which refers to the number of groups of different sources of glass as well as the number of particles in each group) may be assigned differently for persons suspected of breaking glass (e.g. suspects of burglary etc.) than for members of the general population. This means that data from surveys of casework items that have been submitted may be relevant to assign this probability under the prosecution proposition, while surveys of items from the general population may be more relevant to assign the probability given the defence proposition under which the defendant is innocent and thus a member of the general population.

While stressing that more research is needed to gain knowledge on background levels of glass in different population groups, they refer to the work of Coulson et al. [73] as a key study in this area. Coulson et al. have found in surveys they performed of clothing from members of the general population and of clothing of suspects that 'there is considerably more glass present on the clothing of people who are suspected of breaking crimes'.

Translating this discussion to forensic biology means that consideration must be made of the probability of finding background DNA (either the number of contributors or the amount of DNA from these unknown sources) given the prosecution proposition and the defence proposition separately. Would it, given the case circumstances, be reasonable to expect these probabilities to be different for samples taken from items related to the defendant? This may be the case if under both propositions there is an expectation of different behaviours or different levels of hygiene of the person of interest. In forensic biology too, more research in this area may be beneficial to expand the knowledge on the background and prevalence of DNA in different population groups. On a more general note, the relevance of the population groups sampled in published studies is a consideration that should be taken into account when assigning probabilities to DNA TPPR in a casework situation where the person of interest is not a member of those population groups. Many studies make use of for instance students or employees of a forensic laboratory. These groups may neither be representative of the general population, nor of the persons involved in the case at hand.

12.4.2   A Broader Scope in Scenario Testing

Human biological traces may be deposited by a person or persons involved in a series of events. Such traces may thus allow us to reconstruct elements of these events as we have been discussing throughout this book. Biological traces, however, may only address specific aspects of these events, while other types of traces, marks or signals may be able to provide information that assists in the reconstruction of other parts. While it is still relatively uncommon to see DNA and other types of traces being evaluated by a scientist given case-relevant scenarios, such evaluations are more common in other fields of forensic science. Examples of this can for instance be found in vehicle accident reconstructions where findings from analysis of marks (impact marks on vehicles and the road), traces (of paint on roads or objects) and signals (for instance from sensory data in the electronic circuits of the vehicle involved) are combined in the reconstruction of events. Another example is fire scene investigations where data from chemical analysis, electronics and construction are jointly evaluated to reconstruct the cause of the fire.

The importance of considering other evidence types in the case assessment process can therefore not be overstated. Some examples are given to illustrate how data from other traces, marks or signals may be combined with findings from forensic biology.

12.4.2.1   DNA and Other Particulate Traces

There are different situations in which examination for DNA as well as for other particulate evidence types may assist in addressing a common issue. We discuss three examples as follows.

- A victim was found sitting in his vehicle with a gun in his hand and a shot wound to the head. The issue in the case was whether this concerned a suicide or whether the situation was staged, and the defendant had shot the victim.
  While the examination initially focused on DNA examinations of the firearm, examination of the clothing of the defendant for gunshot residue proved potentially informative. Findings from both examinations could be evaluated given the propositions:
  H1. The defendant shot the victim and subsequently placed the firearm in the hand of the victim.
  H2. The victim shot himself, and the defendant was not present at this time.
     This case is an example of a situation where the combined forensic examination of items for DNA and other particulate trace evidence can be used as 'parallel' evidence to address the same disputed issue. The two evidence types may strengthen the overall weight of the evidence when both provide support for the same proposition or reduce it when they provide conflicting evidence.
- Linking a defendant to the breaking of a glass display cabinet during a robbery. In this example, the issue is whether it was the defendant that broke the cabinet or someone else. If the item under examination is a sweater left on the street, examination for DNA may link the sweater to the wearer, while examination for glass particles may link the sweater to the

broken cabinet. Both types of evidence are required to address the issue at hand (unless it can be assumed based on other evidence that the sweater was either worn by the defendant or that the sweater was worn by the perpetrator).

This example is a situation where the DNA findings provide only part of a chain of 'serial' evidence to link a person of interest to an item or activity. The findings from the glass analysis are needed to assess the strength of the entire chain that connects the defendant to the breaking of the glass cabinet. This overall strength is determined by the weakest link.

- The issue of whether a perpetrator wore gloves is commonly discussed in cases as this is an aspect of the case circumstances that is frequently unknown. In these instances, assumptions need to be made on whether or not the perpetrator wore gloves when performing the activities of interest.

In cases where adhesive tapes have been used to tie a person or object, these tapes are examined for DNA which could potentially provide information to address the issue of who tied the person or object. Relevant propositions in such an instance may be:

H1. The defendant tied the victim using duct tape.

H2. Somebody other than the defendant tied the victim using duct tape.

Results from the examination of fingermarks might also provide information to address this issue. However, if no marks are found, this is not necessarily uninformative as it might provide information to support the assumption that gloves were worn. Particularly the examination of the tape for (groups) of fibres or impressions of fabric in the glue might also provide such information and may even direct to a specific type of gloves being used. So, while not directly informative with regard to the issue presented in the propositions, such examinations may provide information on which crucial assumptions in the case need to be made.

12.4.2.2 DNA and Marks

Case types where multidisciplinary analyses to address the same issue are common are sexual assaults. Both forensic biology (seminal fluid or latent DNA traces in body cavities of the complainant) and forensic medicine (lacerations or other wounds in the genital area) may for instance assist in distinguishing between whether or not penetration of the body of the complainant has occurred. McGregor et al. [74] have shown in a retrospective study that there is a strong correlation between both the decision to prosecute and convictions, and forensic medical examinations resulting in documented injury of the victim. Patterns of injury have been suggested to be informative in distinguishing between consensual and non-consensual penetration [75], but this has been disputed [76]. Nonetheless, the findings from medical examinations of complainants of sexual assault may provide information that can be assessed probabilistically and combined with findings from forensic biology examinations.

Another example where the analysis of marks may successfully be combined with those from the analysis of biological traces is the forensic examination of adhesive tapes. Wieten et al. [77] have addressed the combination of findings from DNA analyses and physical end matching of duct tape. Their paper primarily addresses traces of DNA and their location in relation to the reconstructed order of pieces of tape and their position on the roll. The same has been done in the supplement to this paper for fingermarks and the combination of DNA and fingermarks. This publication is discussed in some more detail in Section 12.5.

The following case illustrates another example where the combination of findings from the analysis of marks with those from forensic biology may be relevant.

A complainant claimed they were attacked by the defendant. The defendant claimed no such thing happened and the complainant staged the attack. The clothing of the complainant was examined and damage to the clothing was found. The clothing was sampled for DNA near these damaged locations.

The propositions in the case that reflect the issue are:

H1. The defendant fought with the complainant before or during the stabbing incident. The T-shirt was damaged during this fight.
H2. The defendant did not fight with the complainant before or during the stabbing incident. The complainant damaged their own T-shirt afterwards.

The results from the DNA analyses as well as the examination of the damage in the T-shirt could be evaluated given these propositions. The examination of the damage in the textile for instance revealed that it was nearly impossible for the wearer to create these damages (location, direction, etc.) themselves.

12.4.2.3   DNA and Signals

Digital forensics is rapidly growing and quickly becoming one of the largest domains in forensic science. The use of electronic devices has grown in our daily lives, and so have the forensic opportunities that come with that. With most activities that we perform, we leave traces in the digital domain. Actively and consciously, we leave traces when posting on social media like Facebook, Instagram or WhatsApp, or when sending an e-mail. However, most of the traces we leave are not actively created by the user, nor are most of us conscious of the fact we leave those traces. The 'internet of things' connects our smartphone to our home appliances through the Wi-Fi network we setup at our homes, our smartphone tracks our movement through GPS and wireless connections to relay towers, but also through step counter apps. The computers on board our cars register sensory data from our trips, speed, acceleration, direction of movement, etc. Many modern vehicles contain 'black boxes' analogous to those on-board airplanes, to be recovered after an incident. They contain sensory data as well as data on activities performed by the driver in the moments prior to the incident and are very valuable in accident reconstruction. These examples of digital traces show that our daily activities in the physical world increasingly merge with those in 'cyber space'.

While digital forensics is traditionally aimed primarily at recovery of data and messages, there is an increasing interest in using digital data to reconstruct events. Such reconstructions effectively are evaluations of digital data given activity-level propositions. Recent studies have demonstrated the use of such data in crime reconstruction.

Bosma et al. [14] have addressed the issue of whether a 'burner phone' (a cell phone used solely for communication during a single criminal act and disposed immediately after) was used by a person of interest (PoI) or by another individual. They have done this by looking at location data for the private telephone of the PoI and for the burner phone and test for co-location of the two devices. They have shown that such data can distinguish between the proposition that both devices travelled with the same user and the proposition that both devices travelled with different users. While their paper addresses the analysis of the signal data only, one can imagine that the phone of interest will also be examined for fingermarks and DNA, opening the opportunity to combine these findings to address the issue of whether or not it was a defendant who used both phones.

Zandwijk and Boztas [78, 79] have shown that the Samsung Health app or WhatsApp app installed on iPhones record sensor data in log files. These data allow for differentiation between walking and running at a particular time, or movement by foot or vehicle. The data also show when a phone, stored in a backpack, is dropped to the floor.

If the issue is for instance whether the defendant was present in a vehicle at the time of an incident, the analysis of biological traces from the vehicle may be informative [80], but so may sensor data from the defendant's iPhone.

12.4.3   STANDARDISATION OF MODELLING ACROSS DISCIPLINES

Biedermann et al. [81] have explored the use of evidence from the analysis of ignitable liquids to address the (semi-activity-level) issue of the cause of a fire (natural cause, technical failure or through human action). They constructed a Bayesian network to evaluate the findings from such an analysis given propositions on these three causes (Figure 12.3). The structure of the network bears resemblance to the networks shown throughout this book for the evaluation of DNA evidence.

Given that the transfer, persistence, prevalence, background and recovery of traces need to be addressed for any particulate evidence to be evaluated at the activity level, it is not surprising that we find similar structures of the networks across disciplines.

Throughout this book, we have seen that the general structure of Bayesian networks that were developed for the case examples is very similar. Following the template proposed by Taylor et al. [42], the structure flows from the propositions node, through activity nodes, TPR nodes, to the findings or results nodes. This generalised structure is applicable to most casework situations as the same aspects need to be considered with the transfer of biological material through specific activities. We have seen, however, that this template can be applied to other types of particulate

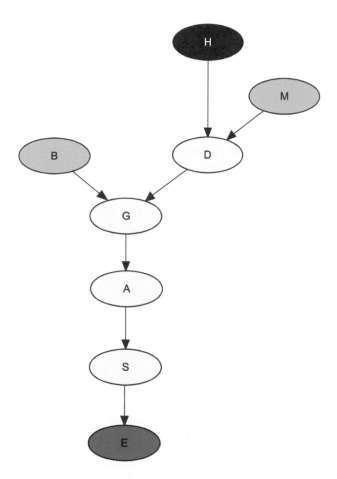

FIGURE 12.3   Bayesian network structure after Biedermann et al. [81]. The network allows for the calculation of the weight of the findings from the analysis of ignitable liquids in fire incident residues. Node letters refer to: propositions (H), relevance of sample (M), background levels (B), gasoline present (G), with persistence and recovery aspects covered in (A) and (S), resulting in the analysis result (E).

transfer evidence. The Bayesian network shown in Figure 12.2 which was constructed to evaluate the findings from a fingermarks analysis follows the same template structure. Examples from other fields can also be found. Uitdehaag et al. [82] have constructed a Bayesian network for the analysis of soil samples taken from a shovel and from the site of a clandestine grave to address the activity-level issue of whether or not the defendant dug and filled the grave with the shovel. The network has the same general structure as we are now familiar with, but has specific aspects modelled that need to be considered when with the transfer and persistence of the soil on the shovel (like the soil type and the wetness of the soil, as well as the subsequent use of the shovel after the alleged use with the digging of the grave).

Convergence of modelling decisions within forensic biology is a 'must' to progress the field, particularly in the implementation of the use of Bayesian networks in casework practice. The template proposed by Taylor et al. [42] has proven to work well in practice and allows for further harmonisation and standardisation, allowing for acceptance of the practice by the legal community and the courts. What can be seen from the examples presented here is that such harmonisation can also be achieved across forensic disciplines. Adoption of the template BN approach by other fields will also stimulate and facilitate the combination of evidence types in evaluations of findings given activity-level propositions.

12.4.4 BOUNDARIES OF EXPERTISE

In Chapter 3, the transfer, persistence and recovery of human hairs were briefly discussed. Historically the morphological analysis, classification and comparison of human hairs are part of the expertise linked with other types of fibres and textile materials. Currently, it is a common practice to use genetic analysis on DNA attached to the hair or present in the hair shaft to identify the donor of the hair. As such, we have seen that the forensic analysis of hairs for identification purposes has moved into the forensic biology domain. Consequently, questions on the relevance of hairs in relation to proposed scenarios are also addressed to the DNA scientist. As this issue is generally at the activity level, the question is who is best informed to address aspects of TPPR of human hairs. While the identification of the hairs (e.g. a source-level issue) is in the domain of the forensic biologist, knowledge on the TPPR of such traces may be more commonly available with experts on fibres and textiles as hairs 'behave' more like other natural fibres than like latent DNA traces or bodily fluids. In situations like these, an interdisciplinary evaluation of the findings may be required. In practice it may be difficult to assign this task to multiple experts due to operational or legal constraints.

Here we see that the definition of areas of expertise and the boundaries set by the technological developments rather than the forensically relevant issues may frustrate the process of properly addressing the issues raised by the criminal proceedings, thereby potentially limiting the impact that forensic science may make on the judicial process. Roux et al. [83] have recently highlighted this issue and concluded that:

> In (re)defining forensic science as a distinct discipline, studying traces (e.g. how, where and when material transfer) with an explicit and common purpose, we can potentially have a more impactful and long-lasting effect. This would further develop a forensic science culture around three primary identified purposes: the contribution of traces to (1) shed light on past events, (2) monitor criminality/security issues and (3) prevent future harms. Thus, we propose that shifting the forensic science focus from means to a purpose, i.e. reinitiating development on the discipline and its fundamental principles, will provide a basis on which organisation (s) and current practice(s) can more adequately evolve

Specifically for interdisciplinary evaluations of findings at the activity level, we must consider that the framework for the interpretation of the evidence is the same across forensic disciplines. The principles for evidence evaluation, the hierarchy of propositions, or Bayesian reasoning are the same and should be applied in the same way whether dealing with DNA evidence or with fibres, gunshot residue or fingermarks. A scientist that is properly trained in these principles should thus be able to apply them, regardless of the type of particulate evidence that is being considered. Where the differences appear is with the consideration of factors that may impact the TPPR of traces and the subsequent assignment of probabilities to these events. This knowledgebase is specific to the trace evidence type.

Different decisions can be made on the way laboratory procedures are set up. A practical approach may be to provide experts in all currently recognised fields training in case assessment and interpretation and statistical modelling to enable them to address their findings given activity-level propositions. This may be a significant investment in resources and may conflict with the recommendations by Roux et al. A path forward, and possibly the most efficient and effective way to deal with interdisciplinary evaluations, is to train a group of 'senior criminalists' within a laboratory who have state-of-the-art knowledge on case assessment and interpretation and statistical modelling. They may perform the formal evaluation with the support of experts from the relevant fields who point to relevant factors to be considered (and thus modelled) and who assign the conditional probabilities. A third approach could be experts from the relevant fields reporting on their findings given activity-level propositions, and a senior criminalist combining these findings given the core activity-level propositions. This option, while possibly the easiest and most convenient, will have disadvantages when conditional dependencies between findings from different fields of expertise need to be considered. This will effectively not be possible if such considerations have not been part of the individual evaluations of the findings.

12.5 COMBINING EVIDENCE

Findings from multiple samples or items of evidence need to be combined at one point to assist the decision-making process. While it may be argued that it is in the domain of the court to do so, there is an inherent risk in combining evidence when the potential conditional dependence between findings is not considered. This may result in an overstatement of the weight of the evidence. The case against Sally Clark has become a classic illustration of this risk. Sally Clark was suspected of murdering her two infant children who died in 1996 and 1998, respectively. The defendant claimed that both children died due to Sudden Infant Death Syndrome (SIDS).

Initially, an expert was consulted to calculate the probability of two SIDS cases occurring in the same family. The probability of a child dying due to SIDS was estimated to be 1 in 8,543 and the expert calculated the probability of two children in the same family dying of SIDS by multiplying these two probabilities, resulting in a probability of 1 in ~73 million. As was already discussed by Hill [84], the assumption of conditional independence between the two SIDS events does not hold. Fenton [85] discussed statistical issues with the interpretation of the forensic findings in this case. Although there were multiple issues (conditional dependence between the two incidents, lack of consideration of the prior probability of murder versus SIDS, and the selection of proper propositions), we will here focus on the first issue only. Fenton assigns a probability of 1 in 1,491 of SIDS occurring given a previous SIDS event in the same family. Multiplying this probability with the probability of the initial SIDS incident results in a probability of 1 in ~12.6 million, clearly demonstrating the risk of overstating the weight of the evidence (here by a factor of ~5.7) if conditional dependencies between findings are overlooked.

The Sally Clark case example shows an example of conditional dependence between findings *within the same field* of forensic science. We have discussed this for forensic biology in some detail in Chapter 5, and a further example Bayesian network for cross-transfer of DNA traces can be found in [86]. However, such conditional dependencies may also play a role when combining findings from multiple disciplines. As can be seen from the previous examples, addressing issues at the activity level is not restricted to forensic biology. Many disciplines may contribute to the reconstruction of events, and several may thus assist in distinguishing between the same set of propositions. Take, for instance, the propositions defined in example case 2 ('The three burglars'):

- H1: D held the knife and grabbed the hair of the victim.
- H2: AO held the knife and grabbed the hair of the victim, after wearing gloves of D.

The results from forensic biology examination have been evaluated given these propositions in Chapters 5 and 6. Let us now assume that the handle of the knife has also been examined for the presence of fingermarks. The presence of fingermarks on the knife may be highly informative and could assist in addressing the issue. Fingermarks on the handle of the knife corresponding to fingerprints taken from the defendant are highly improbable (if not impossible given the set of case circumstances) under H2. Conversely, the presence of fingermarks of an unknown person (with the exclusion of defendant, victim and witness) might provide support for H2 over H1.

When combining findings from DNA analyses and fingermarks given these propositions, conditional dependencies between the evidence types need to be considered. The evaluation of discipline-specific findings given activity-level propositions is part of the field-specific expertise and should thus be performed by properly trained experts in that field as was discussed in Chapter 11. In this example case with the three burglars, we would thus expect to see separate reports from a scientist on the weight of the forensic biology findings given the propositions and a report from a scientist on the weight of the fingermarks findings given the same set of propositions. The question remains who would be best informed to combine these reports into a multidisciplinary evaluation of the findings. The main issue that needs to be considered is whether or not there is a conditional dependence between the findings in both reports. One could argue that a solid statistical working knowledge is needed to address this issue, as well as a general understanding of the two fields of expertise. This knowledgebase is quite rare and may require specific training of the scientists involved. This was discussed in Chapter 11 as a natural extension to training provided to experts when evaluations given activity-level propositions are relatively mature within an organisation. Evidently, the task of assigning a prior probability (and subsequently combining the likelihood ratio with that into a posterior probability of the propositions, and ultimately making the leap of faith to decide which proposition is true) remains with the judiciary or jury.

Bayesian networks are by definition suitable for calculations with conditional probabilities. Hence, network structures have been published that deal with evidence from multiple disciplines in interdisciplinary evaluations. Two examples are given in the next section.

12.5.1 INTERDISCIPLINARY EVALUATIONS USING BAYESIAN NETWORKS

In practice we see that there are specific types of cases where an evaluation of forensic biology findings given propositions at the activity level is often combined with such an evaluation of findings from specific other disciplines. A commonly encountered example is the evaluation of results from a DNA analysis on traces taken from pieces of duct tape. Cases where duct tape has been used to tie up the victim of a robbery are quite common. As tape is usually rich in different types of traces, such items of evidence are commonly examined for DNA and fingermarks. When the issue is at the activity level, the experts may be requested to evaluate their findings given propositions as:

H1: The defendant tied the victim with the tape.
H2: Somebody other than the defendant tied the victim with the tape.

Both the experts in DNA and in fingermarks would be able to evaluate their findings given this set of propositions. Depending on further case information, the position and location of the traces on the pieces of tape in relation to the original roll of tape may become an important aspect to consider. For instance, when the defendant claims that he has used the roll of tape sometime prior to the incident, but somebody else has used it to tie up the victim. The presence of fingermarks or DNA of the defendant on the original end or outer layer of tape on the roll may be well explained by earlier use, while these traces may be less probable if they are found on tape that was several metres from the original end of the roll of tape.

If the position of the traces on the roll is relevant, physical end matching of the pieces of tape may assist in reconstructing the order in which they came from a roll. If a roll of tape is found at the scene, the position of the roll in relation to the pieces of tape may be reconstructed. Wieten et al. [77] have constructed a Bayesian network to combine findings from forensic biology and physical end matching of pieces of tape given a set of propositions at the activity level for a single trace found on a piece of tape used in taping an object or person. They addressed propositions:

Hp: The suspect taped the object.
Hd: An unknown person taped the object.

In the network the factors impacting on the TPPR of DNA were included, as was uncertainty inherent in the source attribution. For the location of the trace, the distance from the original end of the roll was included, as was the issue of whether the trace was located on the backing or on the sticky side of the piece of tape. For the roll of tape, the diameter of the roll is included as a factor as well as the position of the roll relative to the pieces of tape used to tape the object. As data to support probability assignments to DNA TPPR events to tape are scarce, the authors explored the effect of different probability density functions on the resulting LR. They concluded, based on sensitivity analyses, that future studies should primarily address the prevalence of DNA of legitimate users on rolls of duct tape, as the probability of finding DNA of a prior user further away from the original roll end will be the main parameter determining the LR given these propositions at the activity level.

Biedermann and Taroni [87] have addressed the combination of findings from the examination of marks on shells fired with firearms and traces of gunshot residues given *offense*-level propositions:

H1. The suspect if the offender.
H2. Some unknown person is the offender.

They constructed a Bayesian network that can evaluate the findings from both types of analyses given these propositions. While the network construct is not necessarily meant to calculate likelihood ratios (although it could be used as such), the authors show 'how an approach based on a probabilistic network environment can be used for the formal analysis and construction of arguments'.

Note that in both examples discussed here, the evidence types are considered conditionally independent. Nonetheless, Bayesian networks are excellent tools to properly combine conditionally dependent evidence types in an evaluation.

12.5.2 Interdisciplinary Evaluations Using Evidence Schemes

De Koeijer et al. [88] describe a generalised method to address the issue of combining *conditionally independent* evidence from multiple disciplines to address one or more sets of propositions describing one or more specific activities. They describe the use of 'evidence schemes' to graphically describe the relation between persons, items, locations and activities. To illustrate the concept, we will discuss their graphical abstract of the paper. The scheme is based on a fictive case example that could fit any number of shooting incidents, and we suggest the set of case circumstances is as follows.

A man was found dead in a field. Autopsy reveals that the man has been killed by a single shot within the last 24 hours. Next to the body tire tracks were found, but insufficient detail in the tracks was present for any comparison purposes.

A thorough search of the field results in a single bullet being found in the ground close to the body of the victim. The bullet is tested positive for the presence of blood and subsequently sampled for DNA. A single source DNA profile corresponding to that of the victim is recovered from the bullet.

A suspect is identified shortly after the body had been found and the investigation reveals that the suspect had rented a vehicle on the day the victim was shot. The vehicle is examined, and fibres are found on the driver seat, as well as traces of DNA on the steering wheel and gear shift. The DNA profile of the suspect corresponds to those traces.

In the wheel arches of the vehicle some plant remains are found as well as some Birch seeds. These remnants of vegetation correspond to plants found growing in and around the field in which the body was found.

The house of the suspect is searched, and clothing is seized. Fibres from a Jersey correspond to the fibres found in the vehicle. In the house a firearm is also found. The outside of the firearm is sampled for DNA analysis and test shots are fired and the bullets are compared to the one recovered from near the body of the victim.

At this point, the investigation is focused on the suspect. The scenario that the police consider most probable is one where the suspect drove to the field where he met the victim. In an argument the suspect shot the victim and fled the scene. In this scenario there are two main activities at issue (refer back to the discussion of 'package deal' propositions in Section 4.2.1.4; here we have proposition H1a versus proposition H2e1). These are whether (I) the suspect shot the victim, or if someone other than the suspect shot the victim, and (II) whether the suspect was at the scene in the rental car or not, which will be the 'core propositions' we consider. The combined evidence will thus be considered given the propositions:

$H_{scenario}$: The suspect shot the victim and fled the scene with his rental car.

$H_{alternative}$: Someone other than the suspect shot the victim. The suspect was not at the scene nor was his rental car.

The link between the suspect and the scenario with these two activities is formed by the findings from the examinations that were performed in the case. Figure 12.4 shows how these elements connect in the fictitious case.

Looking at Figure 12.4, we see two 'chains' connecting the suspect to the case-relevant scenario. One chain linking him to the firearm, the firearm to the bullet and the bullet to the shooting of the victim, and another chain linking him to the car through fibres and DNA and the car to the scene through the vegetation in the wheel arch.

The findings from the forensic examinations have been evaluated given activity-level 'sub'-propositions. The results of these evaluations are shown in Table 12.2.

We have discussed the concepts of parallel and serial evidence briefly in Chapter 9. In the evidence scheme, we find two sets of parallel evidence; the link between suspect and vehicle through both fibres and DNA, and the two chains linking the suspect to the scenario. There are also two serial evidence chains. The first is the one linking the suspect to the activity of shooting the victim by the firearm and bullet. The second one is the chain connecting the suspect to fleeing the scene through the vehicle. The parallel evidence is thus part of a serial evidence chain.

In their paper De Koeijer et al. [88] have shown that:

- If the evidence types are conditionally independent under both propositions, the strength of parallel evidence can be calculated by multiplying the individual LRs.
- If the evidence types are *not* conditionally independent under one or both propositions – and no data are available that can assist in assigning a weight to the combined evidence given the conditional dependence – the strength of the parallel evidence is *at least* as strong as the strongest of the parallel evidence types.
- Serial evidence is *at least* as strong as the weakest link in the chain.

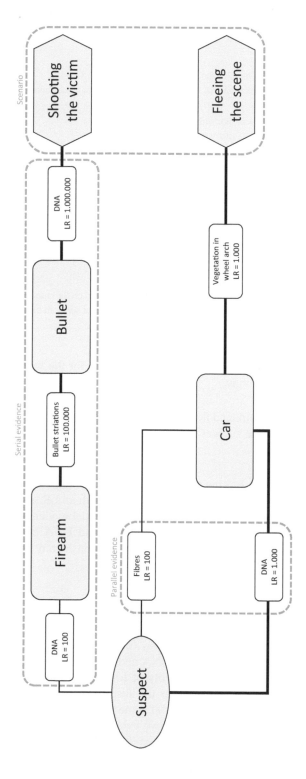

FIGURE 12.4 An evidence scheme reproduced from one constructed by De Koeijer et al. as a graphical abstract to [88]. Lines connect the person (suspect), items (firearm, bullet and car) and activities (shooting and fleeing based on the scenario) in this scheme. The thickness of the lines corresponds to the 'strength' of the connection (the LR). In the scheme, both 'serial evidence' and 'parallel evidence' are shown.

TABLE 12.2

Conclusions Based on the Evaluation of the Case Findings Given Activity-Level Sub-propositions

Evidence Type	Sub-propositions	LR
DNA	H_{s1}: The suspect handled the firearm when shooting the victim	100
	H_{s2}: Someone other than the suspect handled the firearm when shooting the victim	
	H_{s3}: The victim was shot with this bullet	1,000,000
	H_{s4}: The victim was shot with another bullet	
	H_{s5}: The suspect drove the car away from the field	1,000
	H_{s6}: Someone other than the suspect drove the car away from the field	
Firearm examination	H_{s7}: The victim was shot with a bullet that was fired from this firearm	100,000
	H_{s8}: The victim was shot with a bullet that was fired from another firearm	
Fibres	H_{s5}: The suspect drove the car away from the field	100
	H_{s6}: Someone other than the suspect drove the car away from the field	
Biology	H_{s9}: The car was driven in the field where the victim was shot	1,000
	H_{s10}: The car was not driven in the field where the victim was shot but in a different field at another location	

These sub-propositions follow from the core propositions that were formulated based on the case-relevant scenario. Note that any relevant assumptions and other task-relevant information are left out of scope in this example.

Given these three basic rules, we can calculate the 'lower bound' for the weight of the combined evidence given the scenario propositions:

$H_{scenario}$: The suspect shot the victim and fled the scene with his rental car.

$H_{alternative}$: Someone other than the suspect shot the victim. The suspect was not at the scene nor was his rental car.

First, we look at the serial evidence linking the suspect to the shooting. Of the three links in the chain, the one linking the suspect to the firearm is the weakest with an LR of 100. The strength of the total chain is thus at least 100. Secondly, we look at the parallel evidence of fibres and DNA linking the suspect to the vehicle. We can consider the fibre and DNA evidence conditionally independent under both propositions. Both evidence types are considered under the same set of sub-propositions:

H_{s5}: The suspect drove the car away from the field.

H_{s6}: Someone other than the suspect drove the car away from the field.

Under H_{s5} we know that the suspect drove the car away from the field. Under this proposition we have a reasonable expectation of finding both his DNA and fibres from his clothing. Finding the one will not change our assignment of the probability of finding the other. Under H_{s6} somebody else drove the car away from the field. However, as the suspect rented the vehicle on the day of the incident, it is a given that he has driven the car at some time prior to the incident. Again, we can assume conditional independence between the two evidence types as, given the case context and propositions, the presence of DNA of the suspect on the steering wheel and gear shift will not change our belief in the presence or absence of fibres from his clothing. As the fibre evidence and DNA evidence are conditionally independent, the combined findings will result in an LR of 100 × 1,000 = 100,000 supporting H_{s5} over H_{s6}.

In the serial evidence chain connecting the suspect to fleeing the scene with his car, the result from the vegetation remains analysis is the weakest link (1,000 versus 100,000 for the parallel fibres and DNA evidence). We can therefore conclude that this entire chain has a weight of at least 1,000.

The final step is to assess the parallel evidence chains for all evidence. The question is whether knowing any results in either chain will change our beliefs in any of the other evidence types given the propositions and other task-relevant information. Considering the three separate locations (field, car and house of the suspect), the fact that know where all items and traces were found, and that we know the relation of the suspect to those items and locations, we can conclude that there is no conditional dependence between the evidence types in the two parallel chains. We can therefore calculate the combined weight of the evidence as at least $100 \times 1,000 = 100,000$ in support of scenario proposition $H_{scenario}$ over $H_{alternative}$. In other words, the findings are at least *far more probable* (This verbal term relates to the ENFSI scale of verbal equivalents for numerical likelihood ratios which can be found at https://enfsi.eu/wp-content/uploads/2016/09/m1_guideline.pdf) if the suspect shot the victim and fled the scene with his car rather than if someone other than the suspect shot the victim.

This example shows how evidence schemes can be used to graphically convey the complexities of combining evidence. They can be used by experts but are equally useful for all parties involved in the case assessment and interpretation process.

12.5.3 MULTI-CASE EVALUATIONS

Occasionally multiple cases may be encountered as a series that might be linked to the same perpetrator. Examples are a series of burglaries in a particular neighbourhood, or multiple sexual assaults with a similar *modus operandi*. While as experts we are generally concerned with forensic examinations and evaluation of our findings within the boundaries of a single case, such boundaries are to a certain extent arbitrary, as they follow from police procedure and legal decisions that are likely to differ between jurisdictions.

When two crimes have been committed, the police may consider whether or not these two crimes are related. Aspects that may be relevant to consider are the relative locations where the crimes were committed (close together or far away?), the time interval between the crimes (shortly after each other or a long time apart?) and the *modus operandi*, the way in which the crime was committed (e.g. the way a burglar enters a building or the way in which victims of sexual assault are approached or treated by the perpetrator). The closer the crimes are together in time and space, and the more similar the modus operandi are, the stronger one would generally believe that both crimes were committed by the same perpetrator. This has been recently shown to be the case in a retrospective study of cases examined in Biology at Forensic Science SA, where crimes tended to be close together in time and space, and often would link to other crimes of similar types [89].

If there is a strong belief that two or more crimes are committed by the same (group of) perpetrators, the question becomes how evidence in one case may influence the weight of the evidence in another case. De Zoete et al. have studied the evaluation of evidence in crime-linkage scenarios using Bayesian networks. In a 2015 paper [90], they explored a relatively simple situation with two cases each having a single offender. This concept was extended to multiple offenders in a second paper [91]. In their papers the authors explored and suggested ways to deal with three evidence types across cases:

1. Evidence relevant to address the issue of who the offender is in a single case.
2. Evidence relevant to address the issue of whether two or more cases have the same offender.
3. Evidence relevant to address a combination of 1 and 2.

An example of the first category could be a DNA sample taken from the scene of case A which may assist in identifying the perpetrator of case A. An example of the second category could be a similar shoe mark beneath the entry window of the house that was burgled in both cases A and B. While such marks do not directly lead to the perpetrator in each case, they do provide evidence supporting the same perpetrator in both cases. If a suspect was identified and his house was searched resulting in a pair of shoes that could be compared to the shoe marks in both cases, this would constitute evidence of the third type. Each evidence type may be relevant when evaluating findings given activity-level propositions in one or more cases. We will discuss an example case from the Netherlands (ECLI:NL:RBROT:2017:4573 (trial verdict) and ECLI:NL:GHDHA:2019:2683 (appeal trial verdict)) in which the issue of multiple cases in the evaluation of forensic biology findings was addressed.

The Case of the Newspaper Delivery Lady

Around 5:15 am police get a call from a witness who sees a newspaper delivery lady covered in blood lying in the street. Police arrive a few minutes later and find the wounded victim (victim 1). She is unconscious and doesn't regain consciousness before she passes away in the hospital a few days later. A few minutes before the police were alerted by the witness, a man calls the police from the same area in which victim 1 was found. He claims he was a victim of an attempted robbery. This victim 2 states that a man approached him in the street, grabbed his vest by the arm and neck and demanded money and cigarettes. He fended the man off and ran away, after which he called the police. While both incidents took place only minutes and a short distance apart, they are treated by police as separate cases (although a link between them is considered).

Samples taken from the clothing of victim 1 resulted in a DNA profile searched against the national DNA database. This resulted in a suspect being identified. The room of the suspect was searched, and clothing was seized, including a hoody.

In their ruling the appeal court summarises the findings in the case:

> The court of appeal is of the opinion, all things considered and in summary, that the circumstances evident from the evidence are:
>
> - [Victim 2] was assaulted on [Location A] shortly before 05:14.
> - [victim 2] heard a bicycle fall shortly thereafter and at the same time heard screams from a woman, after which he called the police at 05:14.
> - [Victim 1] was subsequently found severely injured on [Location A] by [Witness 1] shortly before 05:16.
> - DNA traces of the defendant were found on the nails and the jacket of [victim 1], on a deodorant can found at the crime scene and on the vest of [victim 2].
> - DNA traces of [victim 1] were found on a hoody found in the defendant's rooming house […].

In the course of the appeal, the Netherlands Forensic Institute was requested to evaluate the DNA findings in the case given propositions:

> H1. The defendant stabbed the victim, while wearing the hoody.
> H2. Somebody other than the defendant stabbed the victim, while wearing the hoody.

The focus here is on the hoody that was found in the defendant's room. The hoody was extensively sampled but only DNA of the defendant was found. Except for one trace which resulted in a mixture of DNA of the defendant and Victim 1.

The prosecution position was that the defendant wore the hoody during the incident. Defense claimed that the hoody was lent to a friend of the defendant and that the defendant was not involved, neither in the attempted robbery nor in the stabbing incident.

For both cases, a Bayesian network was constructed following the template proposed by Taylor et al. [42]. Figure 12.5 shows part of the two networks. The findings nodes as well as the nodes containing probability assignments on DNA TPPR are left out as the focus in this example is not the evaluation of the findings themselves. In this example, we will not go into the formal evaluation that was performed but focus on the aspect where DNA of the suspect was found in two related cases and how this was used in the final report to the court.

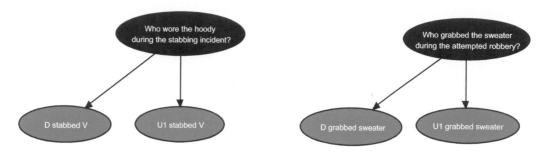

FIGURE 12.5   Part of the Bayesian networks constructed for the stabbing case (A; left) and the attempted robbery (B; right). The proposition nodes are shown in black and the activity nodes in blue.

There is a high probability that both cases are related, as both in time and place they occurred very close together. Both cases were also tried together. If it is assumed that the same perpetrator was involved in both cases, traces found in one case constitute findings that should also be evaluated in the other case. This is because under this assumption any findings in case A providing support for the defendant being the perpetrator will also provide support for him being the perpetrator in case B and vice versa. And, equally important, any findings providing support for him *not* being the perpetrator in one of the cases should also reduce the probability of him being the perpetrator in the other case.

The two networks in the two cases were connected following the principles as outlined in Evett et al. [92] and De Zoete et al. [90]. Figure 12.6 shows the structure linking the two cases.

A proposition node is added with propositions 'Same perpetrator' and 'different perpetrator', as based on the case files it was clear that only one perpetrator was involved in both cases. Both proposition nodes in the two cases also have a parent node which specifies the 'offender configuration' (which may be considered an 'ultimate' propositions node if the two cases are tried together). This node has a number of states which are listed in Table 12.3. In the configuration of perpetrators, the defendant can be involved in both incidents, in either one of them, or in neither. The same holds for the unknown perpetrator under the defence propositions. However, if there are different perpetrators in the two cases and the defendant is not involved, a second unknown perpetrator needs to be involved. This is why, in the network, U2 is introduced.

Figure 12.7 shows some different instantiations of node states to illustrate the logic behind this network structure. Figure 12.7A shows the network where the activity node D having grabbed victim 2 in the attempted robbery is instantiated to 'Yes'. We can see that this changes the posterior probabilities in the offender configuration node. All states where the defendant was not the one who grabbed the vest of victim 2 get a posterior probability of zero. When the node 'Same perpetrator?'

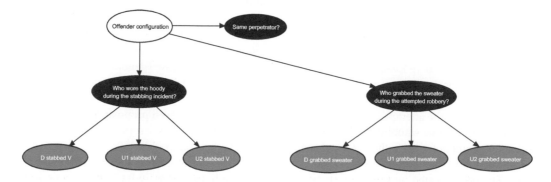

FIGURE 12.6   Bayesian network structure connecting the two cases.

TABLE 12.3

**States of the Offender
Configuration Node**

States

Defendant (D) both

D stabbing; U1 robbery

D stabbing; U2 robbery

U1 stabbing; D robbery

U2 stabbing; D robbery

U1 stabbing; U2 robbery

U2 stabbing; U1 robbery

U1 both

U2 both

is instantiated to 'Yes' (Figure 12.7B), we can see (as we would expect) that the only possible option is that the defendant is the perpetrator in both cases. Conversely, if we know that there are different perpetrators in both cases and the defendant is the perpetrator in one of them, he cannot be the perpetrator in the other case (Figure 12.7C). This setup of the network allows for exploration of the impact of findings in one case on the LR given a set of activity-level propositions in another case. The node 'Same offender?' may be used in different ways. Firstly, if results are available from forensic examinations that provide information on whether or not the same perpetrator was involved (as for example a similar shoe mark at both scenes), these may be evaluated in the network under this 'Same offender?' propositions node. Secondly, if no such evidence is available, prior probabilities other than equal prior probabilities may be assigned to the states if information is available in the case file that supports such assignments. Note that this prior probability assignment is in the domain of the court, but the scientist may facilitate a discussion on this issue. Thirdly, the node may be instantiated (as is shown in Figure 12.7B). By sequentially instantiating both states of the node, the impact on the likelihood ratio of the *assumption* that the same or a different perpetrator was involved can be explored.

In the discussed case, however, the weight of the findings was not calculated for the sample from the vest of victim 2. The sample consisted of a mixture of at least seven individuals and the statistical models available at the time were not able to deal with such a high number of contributors. At the time it was also unclear whether the court would consider both cases to have the same perpetrator or not.

Nonetheless, it was explored what the impact would be on the LR in the stabbing case if the sample from the vest was assumed to include DNA of the suspect. DNA of the suspect was found in one sample, while DNA of multiple unknowns was found on the vest. Detailed information about the interaction between the perpetrator and victim 2 was missing at the time, which made a proper evaluation of the findings given activity-level propositions complex. Calculations were made under different sets of assumptions, but the resulting LR remained quite low (generally in the range of 1–10). The LR for the findings in the stabbing case was about 240, which was reported as *appreciably more probable* (a verbal equivalent used with a numerical range of 100–10,000). The impact of the findings from the attempted robbery case on the LR of interest (given propositions on who wore the hoody during the stabbing incident) would be low and the conclusion presented (much more probable) would not alter as the LR would remain in this verbal 'ballpark').

This was why the findings from the attempted robbery case were not included in the final evaluation. The report stated on this (in a section discussing findings that were left out of scope in the evaluation):

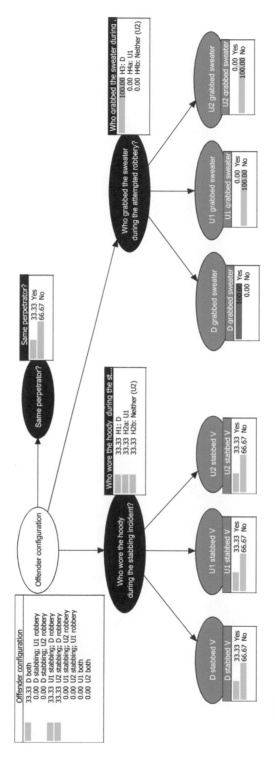

FIGURE 12.7A Bayesian network from Figure 12.6 with node 'D grabbed sweater' instantiated to 'Yes'.

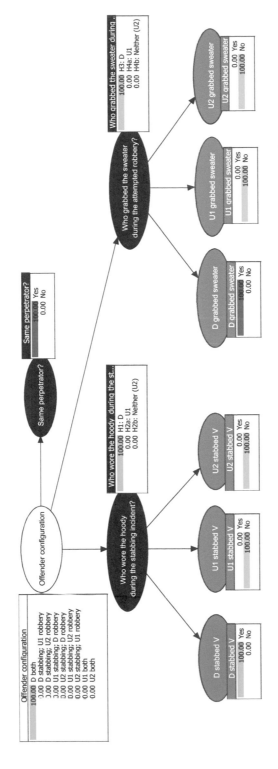

FIGURE 12.7B   Bayesian network from Figure 12.6 with nodes 'D grabbed sweater' and 'Same perpetrator?' instantiated to 'Yes'.

Beyond Forensic Biology

463

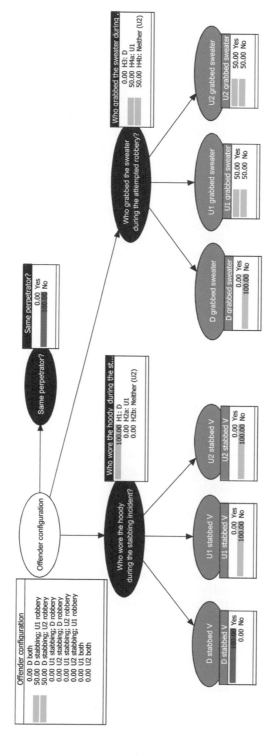

FIGURE 12.7C Bayesian network from Figure 12.6 with node 'D stabbed V' instantiated to 'Yes' and node 'Same perpetrator?' instantiated to 'No'.

Another result is the DNA analysis of the samples taken from the vest of [Victim 2] (see the [LABORATORY 1] reports on case examination requests x, y and z of this case, and the [LABORATORY 2] report). If the attempted robbery and the stabbing incident are assumed to have been carried out by the same perpetrator, finding DNA from the defendant on the vest of [Victim 2] provides information as to who wore the hoody. Because (1) I do not know whether it is reasonable to assume the perpetrator is the same in both cases, and (2) the evidential value of the match found between the defendant's DNA profile and the mixed DNA profile found on the vest has not been calculated, I did not consider this result in the evaluation.

The section was added to inform the court that combining evidence from the two cases could potentially be of interest if it can be assumed that both cases have the same perpetrator. It also highlighted again that no weight of evidence was calculated for the 'match' between the DNA profile of the defendant and the trace from the vest of victim 2.

Conversely, the findings from the stabbing case do have probative value for the attempted robbery case. The court concludes as much in their verdict on the attempted robbery:

> That the defendant was actually at the crime scene that early morning is also evident from the following. [Victim 2] states that only minutes before he heard a bicycle fall and a woman scream he was almost mugged by a man. The spot he describes where this allegedly took place is only about 20 meters from where [victim 1] was found. In addition, from a sample on the right sleeve of [victim 2]'s vest, a mixed DNA profile was obtained to which the DNA profile of the defendant matched. That is exactly the spot about which [victim 2] states that he was grabbed by the robber. The fact that no probative value can be calculated for this DNA trace because of the amount of cell donors that have contributed to this mixed DNA profile does not detract from this for the court. Viewed in conjunction with the defendant's DNA traces found on [victim 1], this also places the defendant at the crime scene.

The example case shows how the concepts developed by Evett et al. [92] and De Zoete et al. [90] may be used in casework practice. The case illustrates that findings from multiple cases may be relevant to address issues in one of those cases and that findings from one case may have an impact on the assessment of findings from another case. This is a natural extension of the duty of the scientist; all relevant findings need to be considered when addressing issues at the activity level. This means findings within a case and within the field of expertise of the scientist, findings within the case but outside their field of expertise (other scientists may need to be involved), but also findings from other, possibly related cases. The mandating authority should be made aware of the potential impact of findings from other cases on the case at issue.

12.6 REFERENCES

1. M.O. Finkelstein, W.B. Fairley, A Bayesian approach to identification evidence, *Harvard Law Review* 83(3) (1970) 489–517.
2. D.V. Lindley, A problem in forensic science, *Biometrika* 64(2) (1977) 207–213.
3. A. Seheult, On a problem in forensic science, *Biometrika* 65(3) (1978) 646–648.
4. I.W. Evett, What is the probability that this blood came from that person? A meaningful question, *Journal of the Forensic Science Society* 23 (1983) 35–39.
5. I.W. Evett, A quantitative theory for interpreting transfer evidence in criminal cases, *Applied Statistics* 33(1) (1984) 25–32.
6. N.D. Kalka, R.A. Hicklin, On relative distortion in fingerprint comparison, *Forensic Science International: Genetics* 244 (2014) 78–84.
7. T.J. Allen, J.K. Scranage, The transfer of glass—part 1: Transfer of glass to individuals at different distances, *Forensic Science International* 93(2) (1998) 167–174.
8. P. Andriotis, G. Oikonomou, T. Tryfonas, Forensic analysis of wireless networking evidence of Android smartphones, in: 2012 IEEE International Workshop on Information Forensics and Security (WIFS), 2012, 109–114.
9. L. Spagnoli, D. Mazzarelli, D. Porta, D. Gibelli, M. Grandi, A. Kustermann, C. Cattaneo, The persistence of ligature marks: Towards a new protocol for victims of abuse and torture, *International Journal of Legal Medicine* 128(1) (2014) 243–249.

10. M. Maitre, M. Horder, K.P. Kirkbride, A.-L. Gassner, C. Weyermann, C. Roux, A. Beavis, A forensic investigation on the persistence of organic gunshot residues, *Forensic Science International: Genetics* 292 (2018) 1–10.

11. V.V. Rao, A.S.N. Chakravarthy, Analysis and bypassing of pattern lock in android smartphone, in: 2016 IEEE International Conference on Computational Intelligence and Computing Research (ICCIC), 2016, pp. 1–3.

12. A.M. Kemp, S.A. Maguire, D. Nuttall, P. Collins, F. Dunstan, Bruising in children who are assessed for suspected physical abuse, *Archives of Disease in Childhood* 99(2) (2014) 108–113.

13. R. Palmer, E. Burnett, N. Luff, C. Wagner, G. Stinga, C. Carney, K. Sheridan, The prevalence of two 'commonly' encountered synthetic target fibres within a large urban environment, *Science & Justice: Journal of the Forensic Science Society* 55(2) (2015) 103–106.

14. W. Bosma, S. Dalm, E. van Eijk, R. El Harchaoui, E. Rijgersberg, H.T. Tops, A. Veenstra, R. Ypma, Establishing phone-pair co-usage by comparing mobility patterns, *Science & Justice: Journal of the Forensic Science Society* 60(2) (2020) 180–190.

15. M. Trapecar, Fingerprint recovery from wet transparent foil, *Egyptian Journal of Forensic Sciences* 2(4) (2012) 126–130.

16. C.C.G. Benschop, A. Nijveld, F.E. Duijs, T. Sijen, An assessment of the performance of the probabilistic genotyping software EuroForMix: Trends in likelihood ratios and analysis of Type I & II errors, *Forensic Science International: Genetics* 42 (2019) 31–38.

17. L. Cadola, M. Charest, C. Lavallée, F. Crispino, The occurrence and genesis of transfer traces in forensic science: A structured knowledge database, *Canadian Society of Forensic Science Journal* 54(2) (2021) 86–100.

18. J. Fraser, K. Sturrock, P. Deacon, S. Bleay, D.H. Bremner, Visualisation of fingermarks and grab impressions on fabrics. Part 1: gold/zinc vacuum metal deposition, *Forensic Science International: Genetics* 208(1–3) (2011) 74–78.

19. A. Neckovic, R.A.H. van Oorschot, B. Szkuta, A. Durdle, Challenges in human skin microbial profiling for forensic science: A review, *Genes* 11(9) (2020) 1015. doi:10.3390/genes11091015

20. E.N. Hanssen, E. Avershina, K. Rudi, P. Gill, L. Snipen, Body fluid prediction from microbial patterns for forensic application, *Forensic Science International: Genetics* 30 (2017) 10–17.

21. C. Díez López, A. Vidaki, M. Kayser, Integrating the human microbiome in the forensic toolkit: Current bottlenecks and future solutions, *Forensic Science International: Genetics* 56 (2022) 102627.

22. O. Peschel, S.N. Kunz, M.A. Rothschild, E. Mützel, Blood stain pattern analysis, *Forensic Science, Medicine, and Pathology* 7(3) (2011) 257–270.

23. ASB Technical Report 033, *Terms and Definitions in Bloodstain Pattern Analysis* (1st ed.), 2017.

24. A. Bettison, M.N. Krosch, J. Chaseling, K. Wright, Bloodstain pattern analysis: Does experience equate to expertise?, *Journal of Forensic Sciences* 66(3) (2021) 866–878.

25. R.A. Hicklin, K.R. Winer, P.E. Kish, C.L. Parks, W. Chapman, K. Dunagan, N. Richetelli, E.G. Epstein, M.A. Ausdemore, T.A. Busey, Accuracy and reproducibility of conclusions by forensic bloodstain pattern analysts, *Forensic Science International: Genetics* 325 (2021) 110856.

26. D. Attinger, K. De Brabanter, C. Champod, Using the likelihood ratio in bloodstain pattern analysis, *Journal of Forensic Sciences* (2021).

27. L. Meijrink, A logical framework approach for evaluating BPA evidence in casework, IABPA *Journal of Bloodstain Pattern Analysis* 34(3) (2019) 8–9.

28. A. Biedermann, C. Champod, G. Jackson, P. Gill, D. Taylor, J. Butler, N. Morling, T.H. Champod, J. Vuille, F. Taroni, Evaluation of forensic DNA traces when propositions of interest relate to activities: Analysis and discussion of recurrent concerns, *Frontiers in Genetics* 7 (2016) 215–220.

29. K.G. Asano, C.K. Bayne, K.M. Horsman, M.V. Buchanan, Chemical composition of fingerprints for gender determination, *Journal of Forensic Sciences* 47(4) (2002) 805–807.

30. K. Kuwayama, K. Tsujikawa, H. Miyaguchi, T. Kanamori, Y.T. Iwata, H. Inoue, Time-course measurements of caffeine and its metabolites extracted from fingertips after coffee intake: A preliminary study for the detection of drugs from fingerprints, *Analytical and Bioanalytical Chemistry* 405(12) (2013) 3945–3952.

31. C. Ricci, S. Kazarian, Collection and detection of latent fingermarks contaminated with cosmetics on nonporous and porous surfaces, *Surface and Interface Analysis* 42 (2010) 386–392.

32. W. van Helmond, A.W. van Herwijnen, J.J.H. van Riemsdijk, M.A. van Bochove, C.J. de Poot, M. de Puit, Chemical profiling of fingerprints using mass spectrometry, *Forensic Chemistry* 16 (2019) 100183.

33. G. Groeneveld, M. de Puit, S. Bleay, R. Bradshaw, S. Francese, Detection and mapping of illicit drugs and their metabolites in fingermarks by MALDI MS and compatibility with forensic techniques, *Scientific Reports* 5(1) (2015) 11716.

34. S. Oonk, T. Schuurmans, M. Pabst, L. de Smet, M. de Puit, Proteomics as a new tool to study fingermark ageing in forensics, *Scientific Reports* 8(1) (2018) 16425.

35. W. van Helmond, M.P.V. Begieneman, R. Kniest, M. de Puit, Classification of condom lubricants in cyanoacrylate treated fingerprints by desorption electrospray ionization mass spectrometry, *Forensic Science International: Genetics* 305 (2019) 110005.

36. S. Kamanna, J. Henry, N.H. Voelcker, A. Linacre, K. Paul Kirkbride, A mass spectrometry-based forensic toolbox for imaging and detecting biological fluid evidence in finger marks and fingernail scrapings, *International Journal of Legal Medicine* 131(5) (2017) 1413–1422.

37. R. Jabbal, R.E. Boseley, S. Lewis, Preliminary Studies into the Secondary Transfer of Undeveloped Latent Fingermarks Between Surfaces, *Journal of Forensic Identification,* 68(3) (2018) 421–437.

38. B. Geller, J. Almog, P. Margot, E. Springer, A CHRONOLOGICAL REVIEW OF FINGERPRINT FORGERY, *Journal of Forensic Sciences* 44 (1999) 963–968.

39. R. Ruprecht, R. Suter, M. Manganelli, A. Wehrli, M. Ender, B. Jung, Collection of evidence from the reverse side of self-adhesive stamps: A combined approach to obtain dactyloscopic and DNA evidence, *Forensic Science International: Genetics* 330 (2022) 111123.

40. A. de Ronde, B. Kokshoorn, C.J. de Poot, M. de Puit, The evaluation of fingermarks given activity level propositions, *Forensic Science International: Genetics* 302 (2019) 109904.

41. A. de Ronde, B. Kokshoorn, M. de Puit, C.J. de Poot, Using case specific experiments to evaluate fingermarks on knives given activity level propositions, *Forensic Science International: Genetics* 320 (2021) 110710.

42. D. Taylor, A. Biedermann, T. Hicks, C. Champod, A template for constructing Bayesian networks in forensic biology cases when considering activity level propositions, *Forensic Science International: Genetics* 33 (2018) 136–146.

43. P. Kumar, R. Gupta, R. Singh, O.P. Jasuja, Effects of latent fingerprint development reagents on subsequent forensic DNA typing: A review, *Journal of Forensic and Legal Medicine* 32 (2015) 64–69.

44. M.K. Balogh, J. Burger, K. Bender, P.M. Schneider, K.W. Alt, STR genotyping and mtDNA sequencing of latent fingerprint on paper, *Forensic Science International: Genetics* 137 (2003) 188–195.

45. B. Bhoelai, B.J.d. Jong, M. Puit, T. Sijen, Effect of common fingerprint detection techniques on subsequent STR profiling, *Forensic Science International: Genetics Supplement Series* 3(1) (2011) e429–e430.

46. J. Raymond, C. Roux, E.D. Pasquier, J. Sutton, C. Lennard, The Effect of Common Fingerprint Detection Techniques on the DNA Typing of Fingerprints Deposited on Different Surfaces, *Journal of Forensic Identification* 54(1) (2004) V23.

47. R.A. van Oorschot, S. Treadwell, J. Beaurepaire, N.L. Holding, R.J. Mitchell, Beware of the possibility of fingerprinting techniques transferring DNA, *Journal of Forensic Sciences* 50(6) (2005) 1417–1422.

48. C. Gibb, S.J. Gutowski, R.A.H. van Oorschot, Assessment of the possibility of DNA accumulation and transfer in a superglue chamber, *Journal of Forensic Sciences* 62(5) (2012) 409–424.

49. Y. Harush-Brosh, Y. Levy-Herman, R. Bengiat, C. Oz, M. Levin-Elad, M. Horowitz, M. Faerman, Back to Amido Black: Uncovering touch DNA in blood-contaminated fingermarks, *Journal of Forensic Sciences* 66(5) (2021) 1697–1703.

50. J.E. Huffman, J.R. Wallace, *Wildlife Forensics: Methods and Applications* (Vol. 6), John Wiley & Sons, Hoboken, 2012.

51. S. Gilmore, R. Peakall, J. Robertson, Short tandem repeat (STR) DNA markers are hypervariable and informative in Cannabis sativa: Implications for forensic investigations, *Forensic Science International: Genetics* 131(1) (2003) 65–74.

52. D.C. Mildenhall, P.E. Wiltshire, V.M. Bryant, Forensic palynology: Why do it and how it works, *Forensic Science International: Genetics* 163(3) (2006) 163–172.

53. J. Amendt, C.S. Richards, C.P. Campobasso, R. Zehner, M.J. Hall, Forensic entomology: Applications and limitations, *Forensic Science, Medicine, and Pathology* 7(4) (2011) 379–392.

54. J.L. Metcalf, Z.Z. Xu, A. Bouslimani, P. Dorrestein, D.O. Carter, R. Knight, Microbiome tools for forensic science, *Trends in Biotechnology* 35(9) (2017) 814–823.

55. C.R. Tarditi, R.A. Grahn, J.J. Evans, J.D. Kurushima, L.A. Lyons, Mitochondrial DNA sequencing of cat hair: An informative forensic tool*, *Journal of Forensic Sciences* 56(s1) Supplement 1 (2011) S36–S46.

56. M. Wesselink, A. Dragutinović, J.W. Noordhoek, L. Bergwerff, I. Kuiper, DNA typing of birch: Development of a forensic STR system for *Betula pendula* and *Betula pubescens*, *Forensic Science International: Genetics* 35 (2018) 70–81.

57. C.H.A. van Leeuwen, G. van der Velde, Prerequisites for flying snails: External transport potential of aquatic snails by waterbirds, *Freshwater Science* 31(3) (2012) 963–972.
58. G.C. CADÉE, Bioerosion of shells by terrestrial gastropods, *Lethaia* 32(3) (1999) 253–260.
59. T.A. Pearce, When a Snail Dies in the Forest, how Long will the Shell Persist? Effect of Dissolution and Micro-bioerosion, *American Malacological Bulletin* 26(1/2) (2008) 111–117, 7.
60. Y. Zhou, Y. Cao, J. Huang, K. Deng, K. Ma, T. Zhang, L. Chen, J. Zhang, P. Huang, Research advances in forensic diatom testing, *Forensic Sciences Research* 5(2) (2020) 98–105.
61. B.P. Horton, S. Boreham, C. Hillier, The development and application of a diatom-based quantitative reconstruction technique in forensic science, *Journal of Forensic Sciences* 51(3) (2006) 643–50.
62. K.R. Scott, R.M. Morgan, N.G. Cameron, V.J. Jones, Freshwater diatom transfer to clothing: Spatial and temporal influences on trace evidence in forensic reconstructions, *Science & Justice* 59(3) (2019) 292–305.
63. K.R. Scott, V.J. Jones, C. Ng, J.M. Young, R.M. Morgan, Freshwater diatom persistence on clothing I: A quantitative assessment of trace evidence dynamics over time, *Forensic Science International* (Online) 325 (2021) 110898.
64. K.R. Scott, V.J. Jones, N.G. Cameron, J.M. Young, R.M. Morgan, Freshwater diatom persistence on clothing II: Further analysis of species assemblage dynamics over investigative timescales, *Forensic Science International* 326 (2021) 110897.
65. S. Uitdehaag, F. Quaak, I. Kuiper, *Soil Comparisons Using Small Soil Traces, A Case Report*, 2016, pp. 61–69.
66. D. Roy, S. Tomo, P. Purohit, P. Setia, Microbiome in death and beyond: Current vistas and future trends, *Frontiers in Ecology and Evolution* 9 (2021) 630397.
67. F.C.A. Quaak, T. van Duijn, J. Hoogenboom, A.D. Kloosterman, I. Kuiper, Human-associated microbial populations as evidence in forensic casework, *Forensic Science International: Genetics* 36 (2018) 176–185.
68. V.K. Gupta, S. Paul, C. Dutta, Geography, Ethnicity or subsistence-specific variations in human microbiome composition and diversity, *Frontiers in Microbiology* 8(1162) (2017).
69. E.A. Grice, J.A. Segre, The skin microbiome, *Nature Reviews Microbiology* 9(4) (2011) 244–253.
70. P.A. Dimitriu, B. Iker, K. Malik, H. Leung, W.W. Mohn, G.G. Hillebrand, D.S. Guttman, E.A. Grice, P. Lee, New Insights into the Intrinsic and Extrinsic Factors That Shape the Human Skin Microbiome, *mBio* 10(4) (2019) e00839-19. https://doi.org/10.1128/mBio.00839-19
71. C. Champod, I.W. Evett, G. Jackson, Establishing the most appropriate databases for addressing source level propositions, *Science & Justice* 44(3) (2004) 153–164.
72. J. Curran, T. Hicks, T. Trejos, Interpretation of glass evidence, in: V.J. Desiderio, C.E. Taylor, N.N. Daéid (Eds.) *Handbook of Trace Evidence Analysis*, John Wiley & Sons Ltd, Hoboken, 2020, pp. 377–420.
73. S.A. Coulson, J.S. Buckleton, A.B. Gummer, C.M. Triggs, Glass on clothing and shoes of members of the general population and people suspected of breaking crimes, *Science & Justice* 41(1) (2001) 39–48.
74. M.J. McGregor, J.D. Mont, T.L. Myhr, Sexual assault forensic medical examination: Is evidence related to successful prosecution?, *Annals of Emergency Medicine* 39(6) (2002) 639–647.
75. S. Anderson, N. McClain, R.J. Riviello, Genital findings of women after consensual and nonconsensual intercourse, *Journal of Forensic Nursing* 2(2) (2006) 59–65.
76. G. Walker, The (in)significance of genital injury in rape and sexual assault, *Journal of Forensic and Legal Medicine* 34 (2015) 173–178.
77. R. Wieten, J. de Zoete, B. Blankers, B. Kokshoorn, The interpretation of traces found on adhesive tapes, *Law, Probability and Risk* 14(4) (2015) 305–322.
78. J.P. van Zandwijk, A. Boztas, The iPhone health app from a forensic perspective: Can steps and distances registered during walking and running be used as digital evidence?, *Digital Investigation* 28 (2019) S126–S133.
79. J.P. van Zandwijk, A. Boztas, The phone reveals your motion: Digital traces of walking, driving and other movements on iPhones, *Forensic Science International: Digital Investigation* 37 (2021) 301170.
80. T.R. De Wolff, L.H.J. Aarts, M. van den Berge, T. Boyko, R.A.H. van Oorschot, M. Zuidberg, B. Kokshoorn, Prevalence of DNA of regular occupants in vehicles, *Forensic Science International: Genetics* 320 (2021) 110713.
81. A. Biedermann, F. Taroni, O. Delemont, C. Semadeni, A.C. Davison, The evaluation of evidence in the forensic investigation of fire incidents (Part I): an approach using Bayesian networks, *Forensic Science International: Genetics* 147(1) (2005) 49–57.
82. S.C.A. Uitdehaag, T.H. Donders, I. Kuiper, F. Wagner-Cremer, M.J. Sjerps, Use of Bayesian networks in forensic soil casework, *Science & Justice* 62(2) (2022) 229–238.

83. C. Roux, S. Willis, C. Weyermann, Shifting forensic science focus from means to purpose: A path forward for the discipline?, *Science & Justice* 61(6) (2021) 678–686.
84. R. Hill, Reflections on the cot death cases, *Significance* 2(1) (2005) 13–16.
85. N. Fenton, Assessing evidence and testing appropriate hypotheses, *Science & Justice: Journal of the Forensic Science Society* 54(6) (2014) 502–504.
86. C.G.G. Aitken, F. Taroni, P. Garbolino, A graphical model for the evaluation of cross-transfer evidence in DNA profiles, *Theoretical Population Biology* 63 (2003) 179–190.
87. A. Biedermann, F. Taroni, A probabilistic approach to the joint evaluation of firearm evidence and gunshot residues, *Forensic Science International: Genetics* 163(1–2) (2006) 18–33.
88. J.A. de Koeijer, M.J. Sjerps, P. Vergeer, C.E.H. Berger, Combining evidence in complex cases: A practical approach to interdisciplinary casework, *Science & Justice* 60(1) (2020) 20–29.
89. D. Taylor, D. Abarno, Using big data from probabilistic genotyping to solve crime, *Forensic Science International: Genetics* 57 (2022) 102631.
90. J. de Zoete, M. Sjerps, D. Lagnado, N. Fenton, Modelling crime linkage with Bayesian networks, *Science & Justice* 55(3) (2015) 209–217.
91. J. de Zoete, M. Sjerps, R. Meester, Evaluating evidence in linked crimes with multiple offenders, *Science & Justice: Journal of the Forensic Science Society* 57(3) (2017) 228–238.
92. I. Evett, G. Jackson, D. Lindley, D. Meuwly, Logical evaluation of evidence when a person is suspected of committing two separate offences, *Science & Justice: Journal of the Forensic Science Society* 46 1 (2006) 25–31.

13 Looking to the Future

Duncan Taylor and Bas Kokshoorn

CONTENTS

13.1 AFTER READING THIS BOOK

We hope that you have enjoyed reading this book and that, if you are not already doing so, that it has inspired you to start evaluating forensic findings with activities in mind. For an organisation first starting the journey to implement activity-level reporting, the task ahead is large. However, the benefits to the justice system – given the increased helpfulness of opinions being provided – make all the effort worthwhile. The personal experience of the authors is one of professional pride and satisfaction. This comes from being able to assist with case investigations or court proceedings by providing the most helpful and relevant information and placing the biology findings into the framework of circumstances in which the case sits.

Like many forensic fields, research continues to be carried out to improve the practice of activity-level evaluation. Already in the time it has taken to write and publish this book, there have been a number of key articles published, and no doubt by the time you are reading this there will be many more. We briefly recall some of the emerging trends in the area and point to areas that require further study.

13.2 CURRENT TRENDS IN RESEARCH

There are research trends that can be seen heading off into a number of different branches, that all add, in their own way, to our overall understanding of issues relating to activity-level evaluation. This research is designed to add to the existing knowledge base on transfer, persistence, prevalence or recovery of biological material, to add to our knowledge of Bayesian network architecture and their application to case circumstances, to provide new types of data that can be incorporated into evaluations, or improve procedures and understanding of these issues throughout the forensic process.

13.2.1 ADVANCES IN TESTS PERFORMED AND DATA COLLECTED

Much research effort in the forensic biology field is devoted to the improvement of existing techniques and the development of new approaches for trace recovery and analysis. Examples of such developments are increased specificity in body fluid tests [1], Massively Parallel Sequencing (MPS) for better DNA profile resolution [2], sampling devices that can obtain DNA from surfaces in new

DOI: 10.4324/9781003273189-13

and potentially better ways (like spray-on rubber coating for bricks [3], or vacuuming of textiles M-Vac [4]), or that can target new types of samples altogether (e.g. surface sampling [5] or even air sampling (which has been unsuccessful in a forensic context in the past [6, 7] but may become possible with new sample collection technology [8]). These advances in methodologies applied to sample detection, collection, analysis and interpretation have consequences for the evaluation of findings, given activity-level propositions as they all impact the probability of recovery of relevant traces. These developments also increase the need for comparative DNA TPPR studies to compare such different methodologies and to assess their relative impact on the probability of recovery of biological materials. Such comparative studies are crucial for the application of DNA TPPR studies across laboratories and DNA analysis platforms.

13.2.2   ADVANCES IN TPPR STUDIES

The implementation of reporting given activity-level propositions and carrying out activity-level evaluations for casework often uncover gaps in knowledge around TPPR issues [9]. This means that casework is an ongoing source of inspiration for new TPPR research. With the rise in knowledge about TPPR issues (both in court and by the forensic community) and the increasing focus on reporting given activity-level propositions, there has been an explosion of TPPR studies in the last 5 years. These studies mimic either casework situations and generate data that may be relevant to assign probabilities to DNA TPPR events or address fundamental questions about aspects of TPPR such as the composition of transferred material [10–12], properties of surfaces that affect transfer [13, 14], and in-depth studies into shedder status of individuals [15]. We can expect to see this development continue with further harmonisation and standardisation of study design and data sharing [16, 17]. Such standardisation of experimental protocol can also be expected across disciplines with universal protocols for TPR studies have been proposed for trace evidence [18, 19].

13.2.3   ADVANCES IN EVALUATION TOOLS AND FRAMEWORK

The more studies that are conducted on TPPR issues, the greater our knowledge becomes as to how the world works, and what factors influence different events. With this increase in fundamental data-driven knowledge, there is also a need for the theoretical framework of evaluations to evolve and incorporate our new understanding. We have seen this occur already, with research into how existing literature can be pulled together in a Bayesian network to address issues of DNA transfer during direct contact [20]. We expect this trend to continue, and so would expect to see studies being published that give new insights into data modelling and how to incorporate that modelling within an evaluation. Some recent examples that have already shown this trend emerging are:

- Incorporating common unknowns into evaluation [21] – This study extends the idea of background DNA leading to unknown DNA, to consider the multiple mechanisms by which there could be common unknown DNA present on an item.
- Incorporating mixture-to-mixture comparison [22–25] – While Taylor et al. [21] included considerations of common unknowns in their analysis in a binary fashion (i.e. designating them as being the same or different), one could imagine the BN being extended to consider the probability of there being common unknown donors.
- Better and more standardised modelling of data [26] –We expect that research will begin to focus on standardised ways this data can be modelled, or on producing guidelines to ensure that data being generated and published is presented in a common format that is easily accessible to activity-level evaluation.
- More accessible sensitivity analyses on complex BN modelling [27] – As the use of activity-level evaluations increases and they become more prominent in court, we would expect that questions and challenges will arise around the robustness of the evaluation. There are

two areas (and probably others) where we see further research being carried out. The first of these is to make conducting a sensitivity analysis more formalised and more accessible. The second area of research we envisage is around communication.

- Machine learning dependencies from data [28] – Throughout this book, the process of designing the evaluation, identifying the important factors and assigning dependencies has been completely carried out by the analyst, using their knowledge and experience. An area of research that we see in the area of evidence evaluation in forensic science is to take the concepts of machine learning that are prominent in other applications of BNs. In a machine learning paradigm the response variable along with numerous other pieces of information about each datum is analysed by statistical algorithms so that the dependencies are learned from the data. These dependencies would then be incorporated into downstream evaluations.
- Combining results from different disciplines – We expect the use of activity-level evaluations to become more widespread, not only in forensic biology (as is the focus of our book) but also in other disciplines, such as many of those mentioned in Chapter 12. As this happens, there will be a greater drive from the legal community to combine the findings from different disciplines into a single evaluation for presentation in court. We already see examples of this type of evaluation occurring in the work of de Koeijer et al. [29].
- Increasing sophistication of evaluations – It is inevitable that as activity-level evaluations become more prominent, and the amount of data on DNA TPPR issues become more bountiful, there will be a drive to increase the sophistication of the evaluation framework. This increased sophistication may be to consider DNA amounts more often (rather than operating in a presence/absence paradigm) and more fine-scale events such as the bi-directional transfer of DNA in a contact, as demonstrated by Taylor et al. [30]. Through these types of studies, it may also become clearer what resolution evaluations can meaningfully occur.

13.2.4   IMPLICATIONS FOR CRIME SCENE PROCEDURES

When confronted with the scene of an incident, crime scene officers will formulate (often implicitly) scenarios that may explain their observations at the scene as well as fit other information that is provided to them at the time (like witness statements or CCTV imagery). Given these scenarios, they will decide what potentially relevant locations are to be sampled for biological traces, or what items should be secured for examination in the laboratory. To be able to select relevant traces or items, a thorough understanding of trace dynamics is needed. Given the growing body of knowledge on these topics, we expect that there will be several implications for crime scene procedures:

- The creation (or restoration) of a direct feedback loop between decisions made in sampling (and the methods applied) and the results from DNA analysis. For instance, through the creation of digital tools that may increase real-time access to such information on DNA typing success rates.
- The increased accessibility of knowledge on DNA TPPR for crime scene officers. As nearly all data on DNA TPPR are currently stored in the scientific literature, efforts will increase to enhance the availability of such data as well as the accessibility of the information (for instance through summaries of major findings presented in videos, infographics, etc.).
- Increased awareness of the need for and relevance of environmental samples to be collected at the scene. Given that activity-level reporting often focuses on the prevalence of DNA of persons of interest in an environment, samples to inform this prevalence, rather than base the probability assignments on generic studies, are crucial for robust evaluations. Examples of such environmental sampling strategies were discussed in Chapter 3 (Section 3.3.3).

- Increased attention to the documentation of crime scene decisions, sampling methods and the relation of traces and items collected to the scenarios being considered at the time. Such information is crucial for case pre-assessment as well as to determine the relevance of certain findings. An example of this comes from the field of fingermarks analysis. While it is common to properly document the mark itself (through photography and lifting of the mark) for identification purposes, the exact location of the mark at the scene, its position and direction are frequently not recorded, while such information may be crucial to evaluate the presence of the mark, given activity-level propositions [31].

13.2.5 ACCEPTANCE IN COURT

Reporting on the significance of forensic findings given activity-level propositions (using the formal framework that we recognise today) has been occurring for decades (since the formalisation of the activity-level evaluation framework by the Forensic Science Service in the 1990s). However, there has been a recent increase in interest and use of activity-level evaluation, in part to do with the increased sensitivity of typical DNA profiling systems, and in part to do with the realisation of issues of DNA transfer by members of the judiciary (for a more detailed discussion on these points, we point the reader to Chapters 1 and 2). Due to this relative newness of the field and activity-level evaluation, there is limited knowledge in the area of its use in legal settings (but see 9). There is very limited information on large-scale legal challenges to the process.

This situation leads to an area of study that is likely to emerge in the coming years, and that is the intersection of activity-level evaluations and their use in legal proceedings. There is a need for legal research into the admissibility of evaluations presented that consider activities:

- Whether there are bounds on the role of the scientist to make assumptions or the use of background information.
- Whether there are bounds on the types of prior probabilities that can be considered, and whether they stray into the duty of the court.
- Whether there is any place for the scientist to explain not only their evaluation, but the concepts of how it should be considered by the court (within the idea of base rates).
- How the results of testing the evaluation for robustness should be conveyed, or indeed whether it should be conveyed at all.
- The types of studies or theories that would satisfy the court that the process has been validated.
- How best to incorporate aspects of uncertainty about the scenario into the evaluation, how best to convey what has been done and whether there are bounds which the scientist should not cross in this aspect for fear of overruling the rights of the defendant.
- How best to deal with defence explanations of the data that are their right to put forward to the scientist on the stand, but from an evidence interpretation viewpoint, do not represent valid propositions.

We expect that as the use of activity-level evaluations becomes more widespread, particularly in countries that have adversarial legal systems, challenges to their introduction into court will occur. This challenge mechanism should be seen as a positive occurrence, as it will guide the areas of legal research that are most pressing. Even a legal challenge based on a misunderstanding of the theory of evaluation tells us that more communication is needed in those areas. Again, it could be that study from a legal viewpoint, and presentation of findings within legal literature is what is needed. This would be similar to challenges to a number of aspects in other forensic evidence evaluations, such as the evaluation of DNA evidence.

We see the legal studies into the legal admissibility and bounds of activity-level evaluations as being an inevitable area of research that will need to occur at some point.

13.3 REFERENCES

1. T. Sijen, S. Harbison, On the identification of body fluids and tissues: A crucial link in the investigation and solution of crime, *Genes* 12(11) (2021) 1728.
2. C. Børsting, N. Morling, Next generation sequencing and its applications in forensic genetics, *Forensic Science International: Genetics* 18 (2015) 78–89.
3. M.J. Cahill, B. Chapman, The novel use of a spray-on rubber coating to recover cellular material from the surface of bricks, *Forensic Science International: Genetics* 49 (2020) ISPM 102404.
4. J.M. McLamb, L.D. Adams, M.F. Kavlick, Comparison of the M-Vac® wet-vacuum-based collection method to a wet-swabbing method for DNA recovery on diluted bloodstained substrates, *Journal of Forensic Sciences* 65(6) (2020) 1828–1834.
5. L. Puliatti, O. Handt, D. Taylor, The level of DNA an individual transfers to untouched items in their immediate surroundings, *Forensic Science International: Genetics* 54 (2021) 102561.
6. M. Vandeowoestyne, D. van Hoofstat, S. De Groote, N. van Thuyne, S. Haerinck, F. van Nieuwerburgh, D. Deforce, Sources of DNA contamination and decontamination procedures in the forensic laboratory, *Journal of Forensic Research* (special issue 2) (2011).
7. N. Witt, G. Rodger, J. Vandesompele, V. Benes, A. Zumla, G.A. Rook, J.F. Huggett, An assessment of air as a source of DNA contamination encountered when performing PCR, *Journal of Biomolecular Techniques* 20(5) (2009) 236–240.
8. C. Fantinato, P. Gill, A.E. Fonneløp, Detection of human DNA in the air, *Forensic Science International: Genetics Supplement Series* 8 (2022) 282–284.
9. B. Kokshoorn, M. Luijsterburg, Reporting on forensic biology findings given activity level issues in the Netherlands, *Forensic Science International* (2022) 111545.
10. J. Burrill, B. Daniel, N. Frascione, A review of trace "Touch DNA" deposits: Variability factors and an exploration of cellular composition, *Forensic Science International: Genetics* 39 (2019) 8–18.
11. I. Quinones, B. Daniel, Cell free DNA as a component of forensic evidence recovered from touched surfaces, *Forensic Science International: Genetics* 6(1) (2012) 26–30.
12. L. Jansson, M. Swensson, E. Gifvars, R. Hedell, C. Forsberg, R. Ansell, J. Hedman, Individual shedder status and the origin of touch DNA, *Forensic Science International: Genetics* 56 (2022) 102626.
13. M. Goray, E. Eken, R.J. Mitchell, R.A.H. van Oorschot, Secondary DNA transfer of biological substances under varying test conditions, *Forensic Science International: Genetics* 4(2) (2010) 62–67.
14. M. Goray, R.J. Mitchell, R.A.H. van Oorschot, Investigation of secondary transfer of skin cells under controlled conditions, *Legal Medicine* 12 (2010) 117–120.
15. M. Goray, R.A.H. van Oorschot, Shedder status: Exploring means of determination, *Science & Justice* 61(4) (2021) 391–400.
16. G.E. Meakin, B. Kokshoorn, R.A.H. van Oorschot, B. Szkuta, Evaluating forensic DNA evidence: Connecting the dots, *WIREs Forensic Science* 3(4) (2021) e1404.
17. A. Gosch, C. Courts, On DNA transfer: The lack and difficulty of systematic research and how to do it better, *Forensic science international: Genetics* 40 (2019) 24–36.
18. H. Ménard, C. Cole, A. Gray, R. Mudie, J.K. Klu, N. Nic Daéid, Creation of a universal experimental protocol for the investigation of transfer and persistence of trace evidence: Part 1 - From design to implementation for particulate evidence, *Forensic Science International: Synergy* 3 (2021) 100165.
19. H. Ménard, C. Cole, R. Mudie, J.K. Klu, M. Lawson, S. Green, S. Doyle, E.H. MacNeill, B. Hamilton, K. Sheridan, N. Nic Daéid, Creation of a universal experimental protocol for the investigation of transfer and persistence of trace evidence: Part 2 – Implementation and preliminary data, *Forensic Science International: Synergy* 3 (2021) 100164.
20. D. Taylor, A. Biedermann, L. Samie, K.-M. Pun, T. Hicks, C. Champod, Helping to distinguish primary from secondary transfer events for trace DNA, *Forensic Science International: Genetics* 28 (2017) 155–177.
21. D. Taylor, L. Volgin, B. Kokshoorn, C. Champod, The importance of considering common sources of unknown DNA when evaluating findings given activity level propositions *Forensic Science International: Genetics* 53 (2021) 102518.
22. K. Slooten, Identifying common donors in DNA mixtures, with applications to database searches, *Forensic Science International: Genetics* 26 (2017) 40–47.
23. J.-A. Bright, D. Taylor, Z. Kerr, J. Buckleton, M. Kruijver, The efficacy of DNA mixture to mixture matching, *Forensic Science International: Genetics* 41 (2019) 64–71.
24. D. Taylor, M. Kruijver, Combining evidence across multiple mixed DNA profiles for improved resolution of a donor when a common contributor can be assumed, *Forensic Science International: Genetics* 49 (2020) 102375.

25. M. Kruijver, D. Taylor, J.-A. Bright, Evaluating DNA evidence possibly involving multiple (mixed) samples, common donors and related contributors, *Forensic Science International: Genetics in Review* 54 (2023) IPMS 102532.

26. P. Gill, Ø. Bleka, A. Roseth, A.E. Fonneløp, An LR framework incorporating sensitivity analysis to model multiple direct and secondary transfer events on skin surface, *Forensic Science International: Genetics* 53 (2021) 102509.

27. L. Samie, C. Champod, D. Taylor, F. Taroni, The use of Bayesian networks and simulation methods to identify the variables impacting the value of evidence assessed under activity level propositions in stabbing cases, *Forensic Science International: Genetics* 48 (2020) 102334.

28. S. Stahlschmidt, H. Tausendteufel, W.K. Härdle, Bayesian networks for sex-related homicides: Structure learning and prediction, *Journal of Applied Statistics* 40(6) (2013) 1155–1171.

29. J.A. de Koeijer, M.J. Sjerps, P. Vergeer, C.E.H. Berger, Combining evidence in complex cases: A practical approach to interdisciplinary casework, *Science & Justice* 60(1) (2020) 20–29.

30. D. Taylor, L. Samie, C. Champod, Using Bayesian networks to track DNA movement through complex transfer scenarios, *Forensic Science International: Genetics* 42 (2019) 69–80.

31. A. de Ronde, B. Kokshoorn, C.J. de Poot, M. de Puit, The evaluation of fingermarks given activity level propositions, *Forensic Science International* 302 (2019) 109904.

14 Answers to Practice Questions

Duncan Taylor and Bas Kokshoorn

CONTENTS

DOI: 10.4324/9781003273189-14

CHAPTER 1 – ANSWERS TO QUESTIONS

Q1

An exhaustive set of events is one where all possible outcomes are included. For the practise question parts:

> Part A – All values that are possible on a 6-sided die are listed and so this set is <u>exhaustive</u>.
>
> Part B – Days of the week that end in 'y' are Monday, Tuesday, Wednesday, Thursday, Friday, Saturday and Sunday. This is all possible days of the week and hence the set is <u>exhaustive</u>.
>
> Part C – Months of the year that end in 'y' are January, February, May and July. This is only a list of 4 out of the possible 12 months of the year and so the set is <u>not exhaustive</u>. Other eight months make up the rest of the exhaustive set of months.
>
> Part D – Even numbers on a 6-sided die are 2, 4 and 6. Prime numbers on a 6-sided die are 3 and 5. The value of 1 is not prime as the mathematical definition of a prime number is a positive integer that has exactly two positive divisors. The number 1 only has a single divisor and therefore is not prime. Therefore, as 1 is not included in the set, it is <u>not exhaustive</u>.
>
> Part E – Similarly to part D, number 1 will not be included in either set. Also 9 will not be included in either set (as it is neither even nor prime). Therefore, this set is <u>not exhaustive</u>.

Q2

Mutually exclusive sets are those where the values in one set have no overlap with the values in the other set. For the practice question parts:

> Part A – The visible colours of the rainbow are red, orange, yellow, green, blue, indigo and violet. Prime colours are red, yellow and blue. All the prime colours are present in the colours of the rainbow, and therefore the two sets are <u>not mutually exclusive</u>.
>
> Part B – These two sets each only describe one activity. It depends on the way you have interpreted the question as to whether you consider these to be mutually exclusive. If your

interpretation was that we are talking about a single glass of water that is being drunk in one moment in time, then the two activities cannot both occur, i.e. Mr Smith and someone else cannot both drink that glass of water at the same time and so they are <u>mutually exclusive</u>. However, if you interpret these statements as meaning different glasses of water, or multiple moments in time, then both Mr Smith and someone else can drink their own separate glasses of water or from the same glass at different moments and the sets would be <u>not mutually exclusive</u>.

Part C – Unlike the previous part, in this part there is no ambiguity about the main subject of Mr Smith drinking. Mr Smith cannot both drink a glass of water and not drink a glass of water at the same time and so the sets are <u>mutually exclusive</u>. They are also exhaustive as they list all possibilities.

Part D – The months of the year ending in 'y' are January, February, May and July. January, May and July also have 31 days and so the two sets are <u>not mutually exclusive</u>.

Part E – Even numbers on a 6-sided die are 2, 4 and 6. Prime numbers on a 6-sided die are 3 and 5. As there are no overlapping numbers, these sets are <u>mutually exclusive</u>.

Q3

Consider events A and B. The second law of probability states that 'If two events, A and B, are mutually exclusive then the probability of A or B occurring is the sum of the probability of A occurring and the probability of B occurring'. The factor that distinguishes the second law of probability from the generalised form is that the generalised form does not make the assumption that A and B are mutually exclusive, i.e. A and B can occur at the same time. For example, we could consider that A is that the day of the week is Tuesday and B is that it is raining. As both of these events can occur at the same time, if we want to determine the probability of the day being Tuesday or the weather being rainy, we need to add the probability of the day being Tuesday to the probability of the day being rainy and then subtract the probability of it being both Tuesday and rainy.

The third law of probability states that 'The probability of two independent events occurring is the product of the probability of each event occurring'. The factor that differentiates the third law of probability from the generalised third law of probability is that the generalised third law of probability does not assume independence between events A and B. For example, imagine that A is getting a head in a coin toss and B is not getting a tail in a coin toss. These two are clearly dependent (in fact they are the same thing!). So, the generalised third law of probability tells us that the probability of both of these events is the probability of the first event (getting a head in a coin toss) multiplied by the probability of the second event <u>given the first event</u> (in this case the probability of not getting a tail given that you have gotten a head). In this simplistic example, the probability of B given A is 1 and so the probability of both occurring is the same as the probability of just the first one occurring (which makes sense as they are the same event).

Q4

Mr Smith's hat could have been knocked off by (not exhaustive, but we are not considering very remote possibilities like rabbits jumping out of the hat, a hunter shooting the hat off or a small alien spaceship bumping into the hat):

The wind – evidence of this would be the tree branches swaying or rustling.

An apple falling and knocking his hat off – evidence of this would be an apple landing on the ground nearby.

A bird flying past and knocking his hat off – evidence of this may be a bird flying off past him, or feathers floating down, or bird poo nearby or on him).

A person knocking his hat off – evidence of this may be the person being close by, or the person appearing mischievous.

Walking under a low branch which knocks his hat off – evidence of this may be low branches on the apple trees.

As we are listing out possibilities, we are using abductive reasoning. But by listing out the observations, we expect from each possibility we are using deductive reasoning.

Q5

The bird poo on his shoulder suggests that a bird may have flown over or near him recently and so increases his belief in his hat having been knocked off by a bird flying past. The fact that the person behind him is 4 m away would make it difficult to have knocked off Mr Smith's hat and so decreases his belief in the possibility that his hat has been knocked off by a person.

As we are now using the observations to consider the probabilities of potential causes of Mr Smith's hat being knocked off, we are using inductive reasoning.

Q6

Probabilities (P) are converted to odds (O) by the equation $O = P/(1 - P)$. Therefore, the conversions are:

A: 9:1 for
B: 1:1 for
C: ~0.0101:1 for (or 99:1 against)
D: ~0.00050025:1 for (or 1999: against)

Odds are converted to probabilities by the equation $P = O/(1 + O)$, where odds are given in the form O:1. Therefore, the conversions are:

E: 0.5
F: ~0.982
G: We first convert this to 1.25:1 and then calculate probability as ~0.556
H: We first convert to 0.05:1 and calculate probability as ~0.048. Alternatively, we could have calculated the probability for the odds of 20:1 (where this is now shown as odds against an event occurring), which is ~0.952 (which is the probability of the event occurring) and then converted it to the probability of the event not occurring by subtracting it from 1 to get ~0.048

Q7

Define the term 'D' as having the disease, 'not-D' as not having the disease, 'P' as having a positive test result, and 'N' as having a negative test result. We are given the false positive rate of 0.02%, therefore $\Pr(P|not\text{-}D) = 0.0002$. We are also given the false negative rate of 2%, therefore $\Pr(N|D) = 0.02$. We are also given the base rate of the disease in the population of $\Pr(D) = 1/15000 \sim 0.000067$. In this question, we seek the probability of having the disease if you get a negative test result, i.e. $\Pr(D|N)$. Using the laws of probability, we know:

$$\Pr(D \mid N) = \frac{\Pr(N \mid D)\Pr(D)}{\Pr(N)} = \frac{\Pr(N \mid D)\Pr(D)}{\Pr(N \mid D)\Pr(D) + \Pr(N \mid not - D)\Pr(not - D)}$$

$$\Pr(D \mid N) = \frac{(0.02)(0.000067)}{(0.05)(0.000067) + (1 - 0.0002)(1 - 0.000067)}$$

$$\Pr(D \mid N) \sim 0.00000133$$

Q8

We now seek Pr($D|P$). Again, using the laws of probability:

$$\Pr(D \mid P) = \frac{\Pr(P \mid D)\Pr(D)}{\Pr(P)} = \frac{\Pr(P \mid D)\Pr(D)}{\Pr(P \mid D)\Pr(D) + \Pr(P \mid not - D)\Pr(not - D)}$$

$$\Pr(D \mid P) = \frac{(1 - 0.02)(0.000067)}{(1 - 0.02)(0.000067) + (0.0002)(1 - 0.000067)}$$

$$\Pr(D \mid P) \sim 0.247$$

Q9

One of the main reasons for considering two propositions is that it is difficult to understand the significance of a result when only considering one possibility for the observation. The example was given in Chapter 1 of a rocking child. While the frequency of rocking in an abused child seemingly made the fact that a child was rocking very significantly in support of abuse, it in fact may not be significant at all. Whether or not the fact that a child is rocking, and is significant in its support for that child's abuse, can only be determined when we also have information about the frequency of rocking in non-abused children.

A second reason that considering two propositions is important is that the effect of irrelevant information does not affect the overall evaluation, i.e. while incorporating irrelevant information will affect the probability of the findings given one proposition, it would affect the probability of the findings given an alternate proposition to the exact same relative amount. This was shown in the mathematical justification given for the use of the *LR*.

Q10

There are many advantages of evaluating forensic findings in light of activity-level propositions, rather than only providing the evaluation considering propositions at lower levels in the hierarchy. In general, the end consumer of the evaluation (i.e. police or courts) will be interested in whether an offence has occurred. If a scientist can address these issues, i.e. place the findings in a wider case context of the potential activities that lead to their deposition (or absence), then they will be providing the most assistance to the consumer. They will be helping the consumer to directly address the activity-level component of their question. All of this hopefully shows that the process of carrying out an activity-level evaluation is not only advantageous but also necessary in order for the significance of the forensic findings to be properly understood.

It is important to have moved up the hierarchy of issues to activity level in order to assist the consumer with understanding the significance of the forensic findings in a case context. However, caution (or avoidance) should be exercised when considering offence-level propositions. While it is possible to evaluate the forensic findings given offence-level propositions, actually all the information that is being used from the proposition is the activity component, i.e., if the offence-level proposition was that the defendant raped the complainant, then DNA results can be evaluated considering this; however, all the information that will really be used from the proposition is that the defendant and complainant had sexual intercourse. The legal component (i.e. whether the intercourse was consensual) is not being considered within the forensic evaluation of the findings, and including it in the proposition may give the false impression to the consumer that it somehow has been. We note that there may be some instances where a forensic examination will comment on what seems to be the ultimate issue, such as the presence of vaginal tissue damage in an alleged rape. It may be that the presence of this damage is evaluated by a medical professional with regard to whether it is more probable if the sexual intercourse had been consensual or non-consensual. However, the lines between activity and offence still exist if the question is whether the intercourse was non-consensual, or consensual but rough. Viewed in this light the medical profession could be seen to still be commenting on activities of 'standard sexual intercourse' vs 'rough sexual intercourse', and the issue of consent (and hence offence) is again left to the consumer.

CHAPTER 2 – ANSWERS TO QUESTIONS

Q1

The issue with the statement 'the absence of evidence is evidence of absence' is that it is often misunderstood as to what 'absence of evidence' means. If the absence of evidence means that there is simply no information available at all, then the statement is not true, because without any information there can be no support for any activity (or lack of activity). In this situation, a better phrase would be 'the absence of evidence is uninformative'.

If the absence of evidence means that there is evidence of a lack of something, i.e. someone's DNA on an item, then this may well provide some support to an activity having not occurred. Whether this is the case and to what extent depend on the propositions and the framework of case circumstances. In this situation, a better phrase would be 'the absence of evidence may support an absence depending on the case circumstances' (but we note that this is not as catchy).

Q2

The potential for contamination can be considered within every evaluation, and usually there are multiple types of contamination that could be considered. Whether or not contamination should be taken into account will depend on the circumstances of the case. In many instances, the presence of a person's DNA on an object or in an area is explained by competing activities in the two propositions (for example sexual intercourse vs social interaction). Contamination is taken into account as a route by which the DNA of someone could have come to be in a forensic sample. When the scenario is one of competing activities (and not competing actors, i.e. the person who undertook the activity), it is common not to take contamination into account as the probability of contamination will be very low compared to the probability of DNA transfer in the activity specified under Hd, i.e. the vast majority of any power in the results to distinguish between propositions comes from the TPPR considerations and not the probability of contamination (which is generally very low in comparison). In a similar vein, it may be unnecessary to consider contamination in the case of a disputed actor if there is an expected prevalence of the suspect's DNA. For example, consider a case of an assault where the defendant and complainant live together. Regardless of whether the assault is in dispute, or the identity of the assailant is in dispute, there is a probability that the defendant's DNA will be present on the complainant (or the complainant's possessions) from cohabitation that is likely to be much greater than the probability of contamination.

There are circumstances where contamination will be important to incorporate into the evaluation. It may be that contamination is the most reasonable alternate explanation for the DNA of a person being on an item, particularly when the findings are from a single sample only. For example, if the defence states that the defendant had no direct or indirect contact with the complainant prior to an assault, and DNA matching the defendant is found on the complainant, then the scientist will need to consider events such as crime scene or laboratory contamination, or an alternate offender with a chance matching reference DNA profile. If these considerations are not made, then the presence of DNA matching the defendant will lead to certainty that *Hp* has occurred.

Also, there may be modes of contamination that can be considered which are outside the standard manner in which contamination is often thought. For example, if the location of DNA on an item is important and DNA is found in several locations on that item, then the potential for site-to-site transfer on the object within an exhibit package may be included in the model. This is a form of contamination (i.e. the placement of DNA on an area or object which is not associated with the activities being considered).

Q3

There are several different sources of unknown DNA that can be considered within an evaluation:

a) Unknown DNA that has come from an alternate offender (if one is being posited by the defence)
b) Unknown DNA that is in the form of background DNA
c) Unknown DNA that has come from an individual expected to be on an item, but who has not been profiled. For example, if the owner of an object or habitant in a residence is unavailable for a reference DNA profile to be taken, then we may account for an expected presence of unknown DNA

Q4

Part A: This statement is correct. The term 'match' is used, which is not required and may not necessarily be appropriate (particularly if a mixture of DNA is being evaluated) but is commonly used. An alternate phrasing that does not require the word 'match', or the implication that matching has occurred would be:

The DNA profile from the evidence has been compared to the reference DNA profile from the defendant. The probability of these profiles occurring is 1 million times higher if the defendant is the source of DNA than if someone else is the source of the DNA.

Part B: This statement is also correct; however, it gives a probability (in this case a random match probability) and it is not clear whether the rarity of the profile given refers to the evidence profile or the reference of the defendant (or to the shared loci in both profiles only). A better formulation to present the evaluation is in a likelihood ratio form
Part C: This statement is incorrect as it is making a statement about the probability of the defendant being the source of the DNA and not a statement about the DNA profile. This has made the classic error of a transposed conditional if the prior probability of the defendant being the source of the DNA is not taken into account. A better formulation is that given in part A
Part D: This statement is ambiguous, but probably correct. We are not sure of what the 'it' is referring to at the beginning of the second sentence, but it appears to be the matching profiles. The statement could be made clearer if the term 'it' is explicitly stated as 'The match of these profiles …'

Q5

There are many advantages of evaluating forensic findings in light of activity-level propositions, rather than only providing the evaluation considering propositions at lower levels in the hierarchy.

In general, the end consumer of the evaluation (i.e. police or courts) will be interested in whether an offence has occurred. This offence will typically have two components: some activity that has occurred and a legal component such as a state of mind, or an intent behind carrying out that activity. When forensic biology findings are evaluated considering sub-source (or source)-level propositions, there is a large inferential gap that must be made by the consumer in order to place the findings in the context of their ultimate question. Often this gap will involve matters of DNA transfer, persistence, prevalence or recovery and so best sit with the forensic scientists to address. If scientists can address these issues, i.e. place the findings in a wider case context of the potential activities that lead to their deposition (or absence), then they will be providing the most assistance to the consumer. They will be helping the consumer to directly address the activity-level component of their question. This activity-level evaluation should be carried out by the scientist even when it is difficult to do so (i.e. there is a sparsity of data available to inform the evaluation) because the inferential leap between sub-source and activity will still need to take place and if the scientist doesn't assist then the consumer must do so unaided (and how can they be expected to do so without all the knowledge on DNA that a forensic scientist has).

If a formal evaluation of the findings given activity-level propositions is not performed by the scientist for whatever reason (lack of task-relevant information, for instance), the customer needs to be made aware of the risk of carrying over the *LR* from the sub-source level to the activity level or offence level.

Q6

This is a criticism of the *LR* that is common but comes from a misunderstanding of the way in which an evaluation is carried out. The response is to explain that the evaluation of the evidence does not assume that the defendant has contributed DNA, it considers two possibilities, as stated in the propositions, and evaluates the evidence given those two propositions. Whether or not the defendant has contributed DNA is indeed up to the court to determine given the case circumstances, and the evaluation of the DNA profile evidence you have provided. The lawyer does not have a valid point here, and simply needs to be educated on how forensic evidence evaluation is carried out.

Q7

Persistence can be considered in a number of circumstances but is particularly important to do so when the timeframe, or more generally the framework of circumstances surrounding an item, is different between the two propositions. For example, consider a case of alleged sexual assault, where there has been a social encounter between the defendant and complainant some days prior. If clothing from the complainant is seized and tested for the presence of DNA of the defendant, then under *Hp* the DNA could have been transferred during the assault. But under *Hd* the DNA could have transferred during the social interaction and would then have had to persist for the days between the social interaction and when the top was sampled as a result of the assault occurring. The example just given is one of timeframe, but more generally persistence should be considered whenever there is a difference in the framework of circumstances between the propositions. For instance, if the timeframe is the same under both propositions, but perhaps the item was washed in one scenario and not the other.

From the above description, it should then be clear that when the framework of circumstances surrounding an item is the same (or close enough that they can be approximated to be the same), then persistence will not materially affect the evaluation. For example, if the case of alleged sexual assault occurs during a social interaction (such as a party), then we would expect persistence to act in the same manner on the item under both *Hp* and *Hd* and so it will have no effect on the resulting *LR*.

Q8

Some of this was covered in the previous question but types of persistence may be:

a) Persistence over time, with the removal of DNA coming from natural loss with time and movement
b) Persistence through a washing event such as handwashing, clothes washing in a washing machine, deliberate attempts at evidence removal
c) Persistence over time being on or in a body, such as on hands, or in a cavity (oral, anal, vaginal)
d) Persistence through hostile environments such as burning or bleaching
e) Persistence of the DNA in DNA extracts over years being stored in a laboratory fridge or freezer (particularly important if dealing with a cold-case scenario where items have been tested many years after they were first examined)

Q9

Some of this was already covered in the answer to Question 2. The types of contamination that could be considered are:

a) Contamination of an exhibit by the (reference) DNA of an individual (most critically DNA of one of the individuals already suspected to be involved in the case through non-DNA means)
b) Contamination of an exhibit by DNA from another exhibit in the case
c) Contamination of the exhibit through a biological fluid (semen, blood, saliva, etc.) from another exhibit in the case
d) Contamination of one area of an exhibit from DNA or biological material from another area of the same exhibit (typically through being folder or stored in packaging)
e) Contamination of the exhibit at the crime scene by it coming into contact with other surfaces, or being in close proximity to someone at the scene

The types of contamination being considered at sub-source level will necessarily be those that consider only the movement of DNA. These will typically be considered as occurring in the forensic laboratory at the end stages of sample processing. When contamination is being considered during an activity-level evaluation, it may also be required to model how that contamination has occurred. For example, at sub-source it is common to incorporate the possibility of contamination as a single event acting on the DNA sample with a probability of occurring. However, if evaluating a case at the activity level, if it is alleged that the police have contaminated an exhibit with the DNA of the defendant as they didn't change gloves between arresting the defendant and picking up an exhibit, then it is likely the contamination route would be modelled through transfer events (a more complex consideration than a single probability as is used in sub-source).

Q10

In Section 2.4.1.5 we explained the birthday problem and gave an example of how to calculate the probability of seeing two people with the same birthday in a room with varying numbers of people. In general, if we have N objects the number of pairwise comparisons between them can be calculated by $N(N-1)/2$. The probability of seeing two matching objects, $Pr(M)$ if there are X different states that each object can take (with an equal probability of $1/X$, although this assumption may be questionable with birthdays) is:

$$\Pr(M) = 1 - \left(1 - \frac{1}{X}\right)^{\frac{N(N-1)}{2}}$$

Therefore, if we are interested in determining the value of N required to obtain a specific $\Pr(M)$, we rearrange this to:

$$\log\left[1 - \Pr(M)\right] = \frac{N(N-1)}{2}\log\left[1 - \frac{1}{X}\right]$$

$$N^2 - 2N + 1 - \frac{2\log\left[1 - \Pr(M)\right]}{\log\left[1 - \frac{1}{X}\right]} = 0$$

If we define:

$$C = 1 - \frac{2\log\left[1 - \Pr(M)\right]}{\log\left[1 - \frac{1}{X}\right]}$$

Then we have $N^2 - 2N + C = 0$ and using the quadratic equation:

$N = \dfrac{2 \pm \sqrt{4 - 4C}}{2} = 1 \pm \sqrt{1 - C}$, of which only $N = 1 + \sqrt{1 - C}$ makes sense in this context (i.e. the number has to be positive).

Therefore, if we are interested in a 75% chance of two people having the same birthday, $\Pr(M) = 0.75$, and we know there are 365 possible birthdays, $X = 365$, then we can calculate $C = -1{,}009.61$ and $N = 32.79$. To make sure that we have at least 75% chance would require 33 people.

The alternative method could have been to use the original formula $\Pr(M) = 1 - \left(1 - \frac{1}{X}\right)^{\frac{N(N-1)}{2}}$

and graph $\Pr(M)$ for each value of N between 1 and 100 and determine at what value of N a value of greater than 0.75 was reached. This would have yielded the graph below, where we can again see the answer is 33.

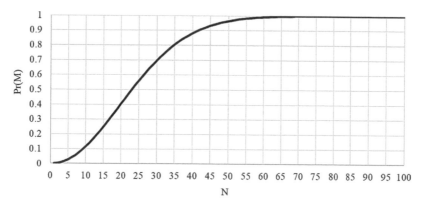

CHAPTER 3 – ANSWERS TO QUESTIONS

Q1

To answer this question, we first need to define what we mean by 'unknown DNA'. As it is a given that a reference DNA profile of you is available, we can eliminate your own contribution to samples as being a 'known' contributor. If no other reference samples are available, any other source of DNA than your own will by definition be considered 'unknown DNA'. This would include DNA from family members or other house mates, friends who visited, etc.

If all these individuals are considered unknown DNA, then we would consider the following potential routes for transfer of unknown DNA to the handle of the knife:

Prior to 'incident'
- DNA being transferred from other objects inside the drawer,
- DNA being transferred from the inside of the drawer itself,
- DNA being transferred to the handle by the person handling it prior to it having been placed in the drawer (for instance when it was taken from the dishwasher).

During 'incident'
- DNA from your hands that was transferred to the handle when you took it from the drawer,
- DNA from your hands that was transferred to the handle when using it to remove gum.

We can take this further and consider routes of transfer of unknown DNA to your hands:

- DNA being transferred to your hands from the outside front door handle,
- DNA being transferred from your clothing or other objects that you may touch while walking through the hallway,
- DNA being transferred to your hands from the kitchen door handle,
- DNA being transferred to your hands from the kitchen drawer handle,
- DNA being transferred to your hands from the inside of the kitchen drawer,
- DNA being transferred to your hands from objects inside the drawer (while rummaging through to grab the knife),
- DNA being transferred to your hands from the face of the front door while removing the gum,
- DNA being transferred to your hands from the gum while removing it.

Note that we focused on the transfer of the unknown DNA only and did not consider persistence. As the activities listed are sequential, each contact following another possibly affects the persistence of DNA that was acquired from the previous activity. In casework, it is therefore crucial to make these activities (also those between the activities of focus, which are the grabbing of the knife and the removing of the gum in this example) explicit and consider how each subsequent activity may impact on the transfer as well as persistence of DNA.

If we would have had the availability of more reference samples of people inhabiting the house or frequently visiting, as well as from other associates that you interacted with recently, some of the routes of transfer of unknown DNA may be considered highly unlikely and therefore obsolete. The point here is that the probability of recovering background DNA or prevalent DNA in a case is crucially dependent on the extent to which witnesses, and other associates have provided reference samples.

Q2

If you washed your hands prior to opening the kitchen drawer, any unknown DNA present on your hands from prior contacts may be removed. From the answer list under Q1, two routes for transfer of unknown DNA to your hands may be removed (but note that the manner of washing may impact on the amount of non-self DNA that is actually removed):

- DNA being transferred to your hands from the outside front door handle,
- DNA being transferred from your clothing or other objects that you may touch while walking through the hallway,
- DNA being transferred to your hands from the kitchen door handle.

However, the washing process may introduce new sources of unknown DNA, like:

- DNA being transferred to your hands from the handle on the faucet on the sink,
- DNA being transferred to your hands from the towel used to dry them after.

To summarise, the routes of transfer of unknown DNA to the handle of the knife remain the same, regardless of whether you wash your hands prior to opening the drawer or not. The sources of unknown DNA on your hands, however, may change.

Q3

From the case description in Chapter 2, we take it that the perpetrator grabbed the hair of the complainant (C) with their bare hands. We need to consider potential routes for transfer of DNA of the defendant (D) to the hair of C under all scenarios that have been presented in the case. The first is the scenario that is likely put forward by the prosecution:

- DNA of D transferred to the hair of C when D grabbed the hair of C.

D suggested a specific alternative route through the hands of AO:

- DNA of D transferred to the inside of his gloves while wearing them. The unknown 'alternative offender' (AO) wore these gloves subsequently at which time DNA of D transferred to the hands of AO. AO took off the gloves and grabbed the hair of C, transferring DNA of D to the hair of C.

However, other implicit routes may also need to be considered, depending on details surrounding the case circumstances. Depending on when D last met AO prior to the incident, one may need to consider:

- DNA of D transferring to the hands of AO when shaking hands,
- DNA of D transferring to clothing of AO (say a coat), being 'parked' there for some time, after which DNA of D transferred to the hands of AO when they put on their coat prior to the incident,
- Etc.

Q4

From the case description in Chapter 2, we take it that the complainant (C) stayed in the house of the defendant (D) (her brother) for a week. She claims at some point he bit her over the underwear.

We need to consider possible routes of transfer of DNA of D to the underwear of C under all case-relevant scenarios. One obvious route for transfer is through the biting event:

- DNA of D transferred to the underwear at the time D bit the victim over her underwear.

Other relevant routes for transfer relate to D and C co-habiting for a week. Possible routes to consider may be dictated by the case circumstances that have been provided or may be based on common knowledge. These routes may for instance include DNA of D transferring to the underwear:

- by C placing clean folded underwear on a shelf in a cabinet containing DNA of D,
- by C picking up DNA from D from surfaces in the house and transferring it to the underwear when handling it,

- by C wearing the underwear and sitting on a surface (say a duvet on the bed) containing DNA of D,
- by C using the bathroom, sitting on the toilet seat, picking up DNA from D who used the toilet prior to C, putting on the underwear and transferring DNA from D from her legs/ buttocks to the underwear,
- by D washing, drying and/or folding the underwear,
- etc.

Q5

The short answer is: All of them.

- Donning laboratory gloves may lead to the outside of the gloves being contaminated with DNA from the wearer (as well as with any non-self DNA that was present on the hands). DNA may also already have been present on the gloves due to exposure to the environment prior to them being worn.
- Another aspect to consider is the handling of the bag itself; any movement may cause friction and compression of the item of evidence in the bag, resulting in material being lost to the (inside of the) bag and relocated on the item. Both aspects are considered types of contamination as well.
- Opening the door while wearing gloves (like contacting any surface) may result in any DNA present on the door handle being transferred to the outside of the gloves.
- The same holds for handling the evidence bag while wearing the gloves. The outside of the bag is likely to contain DNA from a number of sources that may transfer to the outside surface of the gloves.
- By placing the evidence bag on the side table without cleaning the table top prior to doing so, any DNA present on that surface may transfer to the outside of the bag. This biological material may in subsequent steps be transferred inadvertently to other surfaces, including to your gloves or to the item of evidence.
- By opening the bag, biological material on the outside surface may 'flake' resulting in aerial transfer to other surfaces. Subsequent manipulations of the bag like letting the item slide out and moving the bag to the side table may continue this process.
- Folding open the clothing means manipulating it with your hands or with tools like tweezers. Any such manipulation may cause traces present on the item to be removed or relocated. Also, if the gloves or tools used are not DNA free, biological materials may be transferred to the item during this process.
- After manipulating the item, any surface that is contacted may be contaminated with DNA that was picked up from the item. In this instance, biological material may be transferred to the camera. This causes the camera to become a potential vector for future contamination events.

This question serves to illustrate that the forensic process is an integral part of the factors that need to be considered when assessing the relevance of traces recovered (or their absence) in the context of a case. Crucially, information on procedures followed at the scene of the crime, packaging, transport and handling of items of evidence in the laboratory needs to be shared with the scientist to be able to include these aspects in an evaluation of the findings given activity-level propositions.

Q6

The case circumstances as given in Chapter 2 provide us with information that the complainant (C) was grabbed by the arm (presumably covered by part of the top C was wearing). The top was secured for forensic examination one day after the incident.

When considering persistence of traces that were transferred to the top during the incident, we would need to consider any activities, contacts and exposures of the top prior to it being sampled for DNA using tape lifts. As no details are provided on what happened between the incident and the securing of the top, assumptions may need to be made on these circumstances. Examples of factors that could impact on the persistence of the traces are:

- Clothing put on over the top (like a coat), resulting in DNA being lost to the inside of the coat,
- Subsequent touching of the surface that was grabbed by the perpetrator by the victim or other individuals,
- Hugging or consoling of C by others,
- Manipulation of the clothing, for instance by taking it off, throwing it on the floor, folding it, placing it in a laundry bag, washing it, etc.,
- Manipulation of the clothing during the forensic process, for instance by handling of the item by a police officer, packaging it, transporting it and examining it in the lab.

Q7

We are considering a single route of TPR of DNA of the defendant (D) only; from the laundry bag and items therein, through the process of securing the gun, up to sampling and subsequent DNA analyses.

Regarding transfer we might consider as relevant factors:

- The way in which the gun was placed in the bag (e.g. were items loosely draped over the gun or was pressure and friction caused by stuffing the bag?),
- The number of clothing items in contact with the gun,
- The history of the clothing items (duration of wear, and type of biological material expected on the items; wet or dry body fluids may transfer differently and possibly more readily than latent DNA traces),
- The type of fabric of the clothing (as cotton or wool tends to have a higher retention of biological material than for instance polyester clothing),
- The surface of the gun (clean or oily, smooth surfaces or rough, etc.),

Regarding persistence:

- The way in which the gun was recovered (how was it handled, packaged?).

Regarding recovery:

- The sampling strategy (were smooth or rough surfaces targeted, or only the inside surfaces?),
- The sampling method,
- DNA analysis methods (e.g. direct PCR, DNA extraction method, amplification, CE, etc.),
- DNA interpretation methods (what thresholds for interpretation were applied?).

Q8

There are a number of aspects to consider when deciding whether or not to use a study to assign probabilities to TPR events. One will first need to assess to what extent the study design aligns with the case circumstances, and secondly whether or not this 'alignment' meets some arbitrary threshold for use.

Regarding the first aspect, one will need to consider:

- The activities as performed in the study versus those postulated in the case.

    In this instance, the activity in the case that is postulated is the grabbing and pulling of the hair of the victim for some time until she is locked in her bathroom. In the study, volunteers were asked to hold pieces of glass, wood or fabric (without high pressure or friction) for one minute.
- The experimental conditions.

    In the case the hair on the scalp of the victim was sampled (likely containing DNA of the victim herself given the sebaceous nature of the scalp), while in the study the surface that is sampled was pre-cleaned to reduce the presence of background DNA.
- The methods for sampling, analysis and interpretation.

    In the study mini-tapes were used for sampling versus moistened cotton swabs in the case. DNA analysis in the study was performed with the AmpFlSTR® SGM Plus™ Amplification Kit, while the Next-Generation Multiplex (NGM) was used in the case. In the study the number of observed alleles was reported, while a formal interpretation and evaluation of the DNA findings were performed, including calculation of the weight of the findings using LRmixStudio software.

Other aspects that may be considered relevant are the number of repeat experiments (n) in a study or the accessibility of the relevant data in the way that they are presented in the publication (can crucial information be obtained from the study or supplementary data, for instance the link between the DNA-profiling result and the DNA concentration in a sample?)

The second aspect, the decision to use the study or not, will depend on the extent to which the study aligns with the case circumstances (in an extreme example; data on DNA transfer with SGMPlus in an experiment where volunteers touched the handle of a knife may bear little relevance when one needs to assign a probability to DNA TPR given Y-STR DNA typing on a high vaginal swab from the victim of a sexual assault). Other aspects that will need to be considered are whether other studies are available that better align with the circumstances of the case, and whether or not several studies together provide information on reasonable probabilities that may be assigned given slightly different (unknown or uncertain) case circumstances.

Importantly, it is up to the scientist to decide what data are used to support probability assignments. This decision needs to be motivated, regardless of whether or not a potentially relevant study is used.

Q9

A person's shedder status is the amount of DNA a person consistently leaves on a surface after a skin-to-surface contact, relative to an average individual in the relevant population after the same type of contact with the same type of surface.

In this sense, people could be good shedders (if they consistently leave a higher amount of DNA than an average individual from the same population), or bad shedders (if they consistently leave a lower amount of DNA than an average individual from the same population).

Whether or not this status is consistent over longer periods of time, and how this status is best measured is still uncertain. Future research may provide answers to these questions, as well as provide guidance on how to apply such shedder status testing to forensic casework.

Q10

Shedder status is still poorly understood as a factor at this time. Studies have shown that some factors (like skin conditions) may consistently impact on the amount of DNA an individual transfers.

When such specific conditions are known to have affected a person of interest in the case at the time of any postulated activities, this may be accounted for in the assignment of the probabilities of TPR of that individual to the sampled substrate. For (unknown) individuals where no specific conditions are known that may have impacted the transfer of biological material through skin contact to a substrate, probabilities are best assigned based on data from studies with volunteers that are representative of the general population.

CHAPTER 4 – ANSWERS TO QUESTIONS

Q1

Of the five bits of information, the following may be considered task relevant:

- Fibres were found clustered on the sticky side of the tapes, suggesting gloves were worn by the perpetrator.
  This information may support an assumption that is made on whether or not the perpetrator wore gloves. Information may be available from the type of fibres on the type of gloves worn. This information is task relevant as it will impact on the assignment of a probability to transfer of DNA from perpetrator to tape.
- The tape was wrapped around the wrists of the victim.
  This information may be considered task relevant as it may inform the assignment of a probability to the levels of prevalent DNA from the victim or background DNA from unknowns. The latter may be different for wrists of a victim being bound rather than, for instance, tape being wrapped around an object.
- The victim has bitten in the tape to remove it from their wrists.
  This information is task relevant as it may impact on the assignment of a probability to the persistence of DNA of the perpetrator on the tape (as well as on prevalent DNA of the victim).

Two bits of information should be considered task irrelevant:

- The fibres 'match' those from the fabric of gloves that were seized at the home of the suspect.
  This information follows from a comparison of fibres from the tape to that of gloves from the suspect. This information may be highly relevant to address the issue of whether the suspect was the one who bound the victim. However, this would require an evaluation of the findings from the fibres examination given propositions (and very likely the same propositions as for the DNA findings) at the activity level. This evaluation is however outside the field of expertise of the DNA scientist. Access to this task-irrelevant information may (un)consciously bias the DNA scientist as they may increase their expectation of finding DNA of the suspect as fibres from his gloves were also found.
- A witness claims to have seen the defendant near the scene at the time of the incident.
  This information is task irrelevant. The propositions will already specify the presence or absence of the suspect at the scene. Such a witness statement may increase the belief of the scientist in one proposition over the other, possibly resulting in an unconscious bias towards one of the propositions.

Q2

This is a situation that is occasionally encountered in casework. It is a disputed activity in which defence claims no sexual assault took place but that the complainant staged the incident. Propositions could be formulated as:

H1: D had sexual intercourse with C at the time of the incident
H2: No sexual intercourse took place between D and C after the end of their relation. C
   planted biological traces of D on her body.

The probability of the findings under H2 will depend strongly on the type of biological material
that the complainant had access to, and their knowledge of criminal investigation procedures and
sampling and analysis strategies. This information would be required for the scientist to make any
meaningful assessment of the findings under H2.

Q3

There are two aspects to this set of propositions that make them less useful for an evaluation of
findings.
   First, they appear to be explanations for results that were found (saliva from D is present on
the balaclava). The results should be omitted from the propositions and the focus should be on the
activities that are case relevant.
   This brings us to the second point; under H2 it is unclear whether the balaclava was worn during
the robbery or not (and if so, by whom). The propositions focus on the presence of saliva from D,
while the presence or absence of DNA of others may also be informative.
   A more useful set of propositions may therefore be:

H1: D wore the balaclava during the robbery
H2: AO wore the balaclava during the robbery, after which AO threw the balaclava in a puddle
   of spit from D

Q4

There are two aspects to these propositions to be considered.
   Firstly, the term 'assault' is a legal term. Using it in the propositions may suggest that the propositions are set at the offence level rather than the activity level. While the evaluation or its outcome
is unlikely to be affected by this, using the propositions like this may mislead the court in their
considerations.
   Secondly 'assault' is not very specific. This could mean any form of physical attack, including
acts of a sexual nature. The activity needs to be specified in more detail. This may be done in the
propositions themselves or in the undisputed contextual information.
   Propositions in a situation like this may be formulated as:

H1: D hit the victim in the face and put his hand down the front of her trousers
H2: AO hit the victim in the face and put his hand down the front of her trousers

In this way, the focus is again on the activities and the disputed actor. Whether or not these activities
are considered an assault is for the court to decide.

Q5

For the answer to this question, please refer back to Figure 4.1. Following this flowchart, we need to
answer the following questions:

1. Are either of the activities disputed in H2?
      Yes. There are two questioned activities in this case. First is the forced kiss, which D
   states he did not do, and second is the hitting over the head, again something D disputes

has happened. There is no suggestion that an alternate offender is involved, hence both activities are disputed to have occurred.

2. Are the findings relevant to, or impacted by, both activities?

No. As it is highly unlikely that biological traces from the alleged kiss two weeks prior to the forensic examination of the victim would have persisted, any findings from these samples bear no relation to this activity.

At this point, the formulation of the propositions is suggested to follow H1C and H2D, which are given as:

H1c1: D did X
H2d1: X did not occur
H1c2: D did Y
H2d2: Y did not occur

In other words, the activities should be considered separately under separate sets of propositions:

H1c1: D kissed C on the mouth
H2d1: D did not kiss C on the mouth
    and
H1c2: D hit C over the head
H2d2: D did not hit C over the head

It may be considered that, as the findings provide no information with respect to the first set of propositions (the kissing), it is not needed to report on this set of propositions and leave them out of an evaluation based on a case pre-assessment. Although the findings will not be able to distinguish between H1c1 and H2d1, it is still sensible to communicate to the court that this is the case. Relevant literature on the persistence of biological traces may be presented and it reduces the risk of the findings being misinterpreted as providing support for an activity that has not been reported on.

Q6

In this case prosecution states that D stabbed CA after he took his own knife out of his own pocket. The defence counsel claims that the defendant was not involved, and that D lost the knife one week prior to the incident. Under the prosecution scenario there is thus one unknown offender (O2), under the defence scenario there are two (O1 and O2).

- H1: D stabbed CA with his own knife after which O2 stabbed CB with the same knife
- H2: O1 stabbed CA with a knife that D lost one week prior, after which O2 stabbed CB with that knife. D was not involved in the stabbing.

The order of the activities is important as the sequence of stabbing is likely to affect the persistence of DNA of the first stabber. Additionally, persistence of DNA of the owner of the knife may need to be taken into account.

Q7

In this case the relevant activity (which is not disputed by the defence to have occurred) is described in some detail by the complainant. As the order of the actions performed by the offender is important to assess the TPR of body fluids to different locations and orifices of the body and clothing of the victim, these need to be specified. One option is to do this in the propositions. Another is to leave it in the undisputed case information. The latter may be preferred as including it in the propositions makes them very long and difficult to use and repeat in a report. These actions and their sequence are not disputed by the defence (they only dispute D being the offender and suggest an alternative offender, AO). Hence propositions may be formulated as:

- H1: D attacked C and had sexual intercourse with her at the time of the incident
- H2: AO attacked C and had sexual intercourse with her at the time of the incident
- I: The activities consisted of (in order):
    - The perpetrator grabbing the shirt of C,
    - The perpetrator forcing C to the ground,
    - The perpetrator removing the skirt and underwear of C,
    - The perpetrator penetrating C's vagina with his finger,
    - C performing fellatio on the perpetrator,
    - The perpetrator penetrating C's vagina with his penis,
    - The perpetrator ejaculating over the shirt of C.

Q8

Under both propositions, we need to consider an indirect route of transfer of DNA from D, via the backpack and the hands of the witness to the shoulder of the shirt of C. Under H1 D brought his own backpack, under H2 AO brought the backpack from D that was stolen from him three months prior to the incident.

We are faced with two options.

We can assume under both propositions that the backpack belonged to D and that the inside and outside surfaces contained DNA from D. This DNA may have subsequently transferred indirectly to the hands of W and subsequently to the shirt of C.

Alternatively, we can examine the backpack to see what the current situation is. While the traces present before W searching the backpack may have been affected by this activity, it is unlikely that all traces of DNA (if present) would have been removed. Such an examination of the backpack may therefore assist in the evaluation of the findings from the shirt.

One aspect to consider is whether the weight of the evidence will be impacted by the examination of the backpack.

If we assume that DNA of D was present on the backpack, under H2 DNA of D would have transferred from the backpack to the hands of W, and from the hand of W to the shirt of C. No DNA of W is found on the shirt of C. The probability of such a result would be low under this proposition. Under H1 the probability of finding DNA of D and not of W on the shirt of C is (much) higher than under H2.

If we examine the backpack for traces from D, finding DNA of D on the backpack will not change this as we already assumed the presence of his DNA. Not finding his DNA would further lower the probability (it is not zero, as DNA may have been lost in the process of searching the bag by W and the subsequent packaging and transport of the item to the lab) of recovering DNA from D from the shirt of C, resulting in an increase in the weight of the evidence. Whether or not this increase in the weight of the evidence is useful to the deliberation in court is something that the mandating authority needs to decide. A case assessment prior to performing the evaluation may assist in informing the mandating authority on the merits of examination of the backpack.

Q9

About uncertain or unknown aspects of the case circumstances the scientist will need to make assumptions. Ideally these assumptions are discussed with the mandating authority prior to the evaluation. This may prevent evaluations being performed under a set of assumptions that may not be case relevant.

In an instance where it is uncertain whether gloves were worn, one could assume that the perpetrator wore no gloves. Such an assumption should be motivated in the report and may be reasonable if the incident took place under circumstances where one wouldn't normally wear gloves (summer at the beach, indoors, etc.) and there is no pre-meditation (a fight after a spilled drink in a bar, road rage incident, etc.). Under other circumstances (Alaska in winter or a robbery), this assumption may not be reasonable.

If there is a dispute about the assumption being made, additionally the evaluation may be performed under the assumption that the perpetrator did wear gloves. This assumption is more complex, as one would need additionally to make a series of assumptions about the type of gloves (latex gloves, winter gloves or other) and their life history prior to the incident (frequency and duration of wear, by which individual(s), history of washing, etc).

Such detailed assumptions may not be reasonable and can likely not be motivated. It may therefore suffice to explore two extreme situations:

1. Clean latex gloves without a history of prior use were worn by the perpetrator, and
2. Unwashed winter gloves that were owned (and frequently worn) by the POI were worn by the perpetrator.

Q10

The case information that was provided in Chapter 2 can be structured as follows. First, propositions are formulated following the three-step approach.

1. Is it disputed that an activity took place?
   No. The gun was found in the laundry basket. There is no dispute that somebody put it there.
2. If the activity itself is not disputed, who performed the activity?
   The prosecution position is that the defendant (D) placed the gun in the laundry basket (after he committed the robbery). Defence claims that this did not happen because the defendant has never seen the firearm. They state that somebody else (AO) has placed the gun in the basket within 24 hours prior to it being found by police and seized.

Propositions could be formulated as:

- H1: D placed the gun in the laundry basket
- H2: AO placed the gun in the laundry basket

3. What alternative routes for TPR or prevalence of materials are being disputed?

As the gun was found in the laundry basket which contained unwashed clothing of D, indirect transfer of D's DNA through contact of the gun with the clothing should be considered. As it is not in dispute that the gun was retrieved from the basket, this route of indirect transfer needs to be considered under both propositions.

Propositions (H)

- H1. D placed the gun in the laundry basket
- H2. AO placed the gun in the laundry basket

Undisputed case information (I)

- The gun was taken from a basket containing unwashed laundry of D.

The moment in which the gun was placed in the basket may differ between the propositions. The defendant claims that 24 h prior to the gun being seized he did his laundry. In the defence scenario the gun must therefore have been placed in the basket within 24 h of it being seized. However, the robbery took place 48 h prior to the gun being seized. It is not specified at which time the gun was placed in the basket. Assuming that prosecution does not dispute the defendant doing his laundry at the time specified, the gun will have been placed in the basket after this time. This may be part of the assumptions or of the undisputed case information, depending on the position the prosecution takes.

Assumptions (A)

Assumptions need to be made under both propositions about what happened with the gun in the first 24 h after the incident, as well as some other topics:

- The gun was placed in the basket within 24 h before it being seized.
- The basket contained only clothing of D, which he wore for some time.
- There was no extreme pressure or friction between the gun and the clothing in the basket. The clothing was draped over the gun loosely.
- The gun was not handled or cleaned after the incident, apart from it being placed in the laundry basket after one day.
- The person who placed the gun in the basket did not wear gloves while doing so.
- The way in which the gun was seized, packaged and transported to the laboratory followed existing procedures and the risk of any loss or redistribution of traces was minimised.
- The DNA in the sample *is* derived from the individuals to which it was associated based on the sub-source-level evaluations.

CHAPTER 5 – ANSWERS TO QUESTIONS

Q1

If we are considering the results in a binary fashion (i.e. DNA present or absent), then we need to convert the three categories of the table into 2. Although we are only interested in the friction category, the table below collapses the high and low amount categories into a single 'transfer' category. Also, in the table below an equal prior count of 1 is applied to the two categories and provided the posterior probabilities, 'p'.

Transfer Category	Type of Contact		
	Passive	Pressure	Friction
DNA transfer	$N = 2+7 = 9$	$N = 9+15 = 24$	$N = 13+0 = 13$
	$p = (9+1)/(20+2) = 0.45$	$p = (24+1)/(27+2) = 0.86$	$p = (13+1)/(13+2) = 0.93$
No DNA transfer	$N = 11$	$N = 3$	$N = 0$
	$p = (11+1)/(20+2) = 0.55$	$p = (3+1)/(27+2) = 0.14$	$p = (0+1)/(13+2) = 0.07$
total	20	27	13

Therefore, in a friction scenario, for the probability of DNA transfer a probability of 0.93 and for the probability of no DNA transfer a probability of 0.07 would be assigned.

Q2

In this question we need to consider the possibility of a chance matching profile between C's friend and AO. The pathway for *Hp* does not change as this is only a consideration under *Hd*:

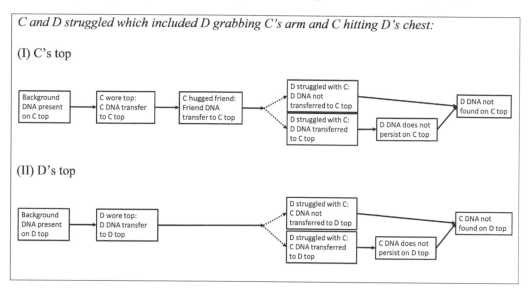

The pathway under *Hd* will change, but only for C's top (as under *Hd* D had nothing to do with the alleged incident). The new pathways for *Hd* would be:

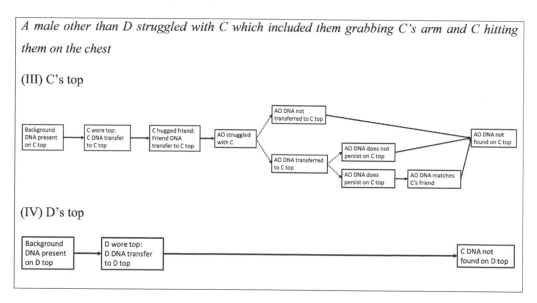

Although note that the results listed as 'AO DNA not found on C top' is no longer really an adequate description of the result we are considering (as it may have been found, but just is the same as C's friend). A better description may have been 'no unknown male DNA found on C top'.

Q3

The *LR* derived directly from the pathways for the Drummond case in Chapter 5 was

$$LR = \frac{\left[B_C \times T_{C \to C} \times T_{F \to C} \left(\overline{T_{D \to C}} + T_{D \to C} \times \overline{P_{D \to C}} \right) \right] \left[B_D \times T_{D \to D} \left(\overline{T_{C \to D}} + T_{C \to D} \times \overline{P_{C \to D}} \right) \right]}{\left[B_C \times T_{C \to C} \times T_{F \to C} \left(\overline{T_{AO \to C}} + T_{AO \to C} \times \overline{P_{AO \to C}} \right) \right] \left[B_D \times T_{D \to D} \right]}$$

If we carry out the same process of *LR* derivation for the pathways in Question 2, only the denominator component $\left(\overline{T_{AO \to C}} + T_{AO \to C} \times \overline{P_{AO \to C}} \right)$ will change. By adding in the consideration of a matching profile between C's friend and AO (which is designated as 'M'), we end up with:

$$LR = \frac{\left[B_C \times T_{C \to C} \times T_{F \to C} \left(\overline{T_{D \to C}} + T_{D \to C} \times \overline{P_{D \to C}} \right) \right] \left[B_D \times T_{D \to D} \left(\overline{T_{C \to D}} + T_{C \to D} \times \overline{P_{C \to D}} \right) \right]}{\left[B_C \times T_{C \to C} \times T_{F \to C} \left(\overline{T_{AO \to C}} + T_{AO \to C} \left(\overline{P_{AO \to C}} + P_{AO \to C} \times M_{AO=F} \right) \right) \right] \left[B_D \times T_{D \to D} \right]}$$

The difference between the two formulae is that:

$$\overline{P_{AO \to C}}$$

Has been replaced by:

$$\overline{P_{AO \to C}} + P_{AO \to C} \times M_{AO=F}$$

If we consider that the probability of matching profiles is in the order of 0.000000001 (or 1×10^{-9}), and the probability of no persistence in this case, while small, is likely to be in the order of 0.01 (given the number of experiments that we could be informed by), then

$$\overline{P_{AO \to C}} \approx \overline{P_{AO \to C}} + P_{AO \to C} \times M_{AO=F}$$

And therefore, the two *LR*s would be roughly equivalent.

Q4

If we consider that under *Hd* no attack occurred, then this will change the pathway that would apply under *Hd* for C's top (i.e. replacing pathway III in Figure 5.1). The new pathway for this item would be:

Again, 'AO DNA not found on C top' is no longer really an adequate description of the result we are considering (there is no AO). A better description may have been 'no unknown male DNA found on C top'.

Q5

We wish to derive an *LR* formula for the defence scenario that an attack did not occur. The scenario put forward by prosecution has not changed and so the derivation for Pr(*E*|*Hp*) will also not have changed, i.e.:

$$\Pr(E \mid Hp) = \left[B_C \times T_{C \to C} \times T_{F \to C} \left(\overline{T_{D \to C}} + T_{D \to C} \times \overline{P_{D \to C}} \right) \right] \left[B_D \times T_{D \to D} \left(\overline{T_{C \to D}} + T_{C \to D} \times \overline{P_{C \to D}} \right) \right]$$

Under *Hd* there has been no attack and so using the pathways above we only end up modelling the transfers that are also accepted under *Hp*:

$$\Pr(E \mid Hd) = \left[B_C \times T_{C \to C} \times T_{F \to C} \right]\left[B_D \times T_{D \to D} \right]$$

And the *LR* is

$$LR = \frac{\left[B_C \times T_{C \to C} \times T_{F \to C} \left(\overline{T_{D \to C}} + T_{D \to C} \times \overline{P_{D \to C}} \right) \right]\left[B_D \times T_{D \to D} \left(\overline{T_{C \to D}} + T_{C \to D} \times \overline{P_{C \to D}} \right) \right]}{\left[B_C \times T_{C \to C} \times T_{F \to C} \right]\left[B_D \times T_{D \to D} \right]}$$

$$= \left(\overline{T_{D \to C}} + T_{D \to C} \times \overline{P_{D \to C}} \right)\left(\overline{T_{C \to D}} + T_{C \to D} \times \overline{P_{C \to D}} \right)$$

And if we make the assumption (as was made in Chapter 5) that given the short timeframe between the alleged offence and the sampling of exhibits means that DNA (if transferred) would persist, then the *LR* simplifies to:

$$= \left(\overline{T_{D \to C}} \right)\left(\overline{T_{C \to D}} \right) = \overline{T}^2$$

The main effect on the evaluation that can be seen is that the *LR* when an AO was considered was $\overline{T}$, and when an alternate offender is not considered it is $\overline{T}^2$. In other words, the *LR* is squared in size when an AO is not considered, and this is because we do not cancel one of the non-transfers in the numerator with a non-transfer from the AO to C's top in the denominator as we previously did.

Q6

Considering D has a skin condition will mean that we are not able to make the assumption that the probability of transfer (or non-transfer) from D is the same as that of C or an AO. Start with the original formula derived for the propositions:

Hp: D struggled with C
Hd: an AO struggled with C

which was

$$LR = \frac{\left[B_C \times T_{C \to C} \times T_{F \to C} \left(\overline{T_{D \to C}} + T_{D \to C} \times \overline{P_{D \to C}} \right) \right]\left[B_D \times T_{D \to D} \left(\overline{T_{C \to D}} + T_{C \to D} \times \overline{P_{C \to D}} \right) \right]}{\left[B_C \times T_{C \to C} \times T_{F \to C} \left(\overline{T_{AO \to C}} + T_{AO \to C} \times \overline{P_{AO \to C}} \right) \right]\left[B_D \times T_{D \to D} \right]}$$

Some cancellations can still occur, leading to:

$$LR = \frac{\left(\overline{T_{D \to C}} + T_{D \to C} \times \overline{P_{D \to C}} \right)\left(\overline{T_{C \to D}} + T_{C \to D} \times \overline{P_{C \to D}} \right)}{\overline{T_{AO \to C}} + T_{AO \to C} \times \overline{P_{AO \to C}}}$$

If we can still make the assumption that the probability of transfer from C to D is the same as the probability of transfer from AO to C (i.e. we are not taking into account a unique transfer probability for each person), then the *LR* can still cancel to:

$$LR = \overline{T_{D \to C}} + T_{D \to C} \times \overline{P_{D \to C}}$$

And again, if we assume that DNA persists if transferred (as in the answer to the last question):

$$LR = \overline{T_{D \to C}}$$

So, the *LR* formula does not change, but the *LR* value will change as we will be using a different value for the probability of a non-transfer by Drummond. In the question given, where Drummond has a skin condition meaning that he is more likely to transfer DNA, then the *LR* will decrease (become more in support of *Hd* compared to *Hp*) compared to the *LR* assigned if Drummond did not have a skin condition.

Q7

Figure 5.5 is shown below.

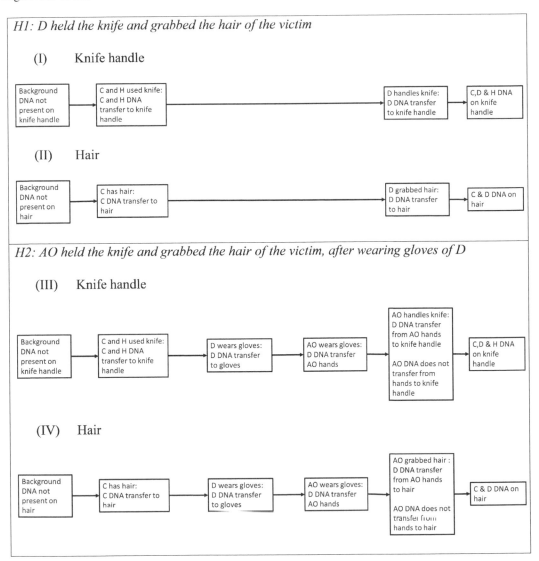

Carrying out the same process of conversion from the pathway to a formula gives the following:

First, note that there are two events that occur in the final step of the pathways under section (III) and (IV), that DNA of the AO does not transfer and that DNA of D does transfer. We place these in the box together as they have both occurred from the one action (the AO grabbing the knife in (III) or the hair in (IV), and so are dependent. This gives rise to the conditional probability term.

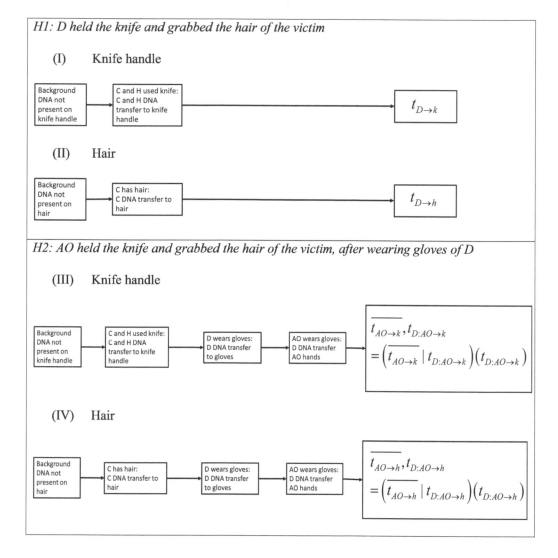

We now use the simplification that if an event occurs in the stream of both *Hp* and *Hd* (so that all terminal points must have passed through this event) that it can be cancelled, which we show by crossing out the boxes (paired in *Hp* and *Hd* by colour).

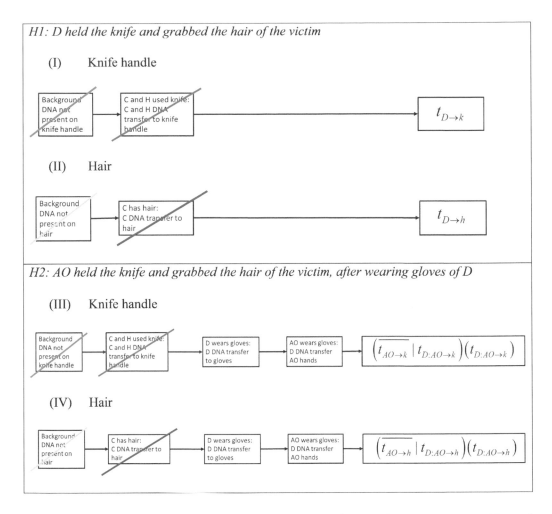

There are no further terms to consider under paths (I) and (II) and two terms to consider under (III) and (IV). Note that if the same event occurs in multiple pathways within a proposition, then it is only considered once probabilistically, and we symbolise this by crossing out the instances that we are not considering probabilistically.

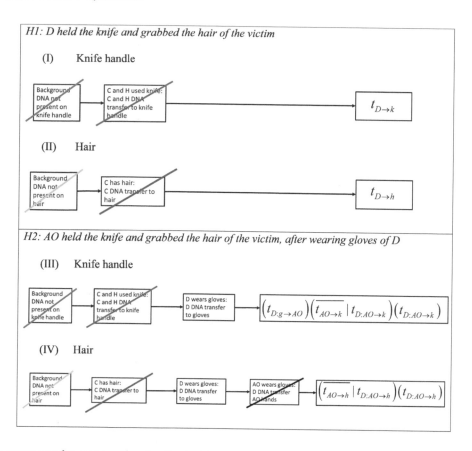

We carry out the same action for the last term:

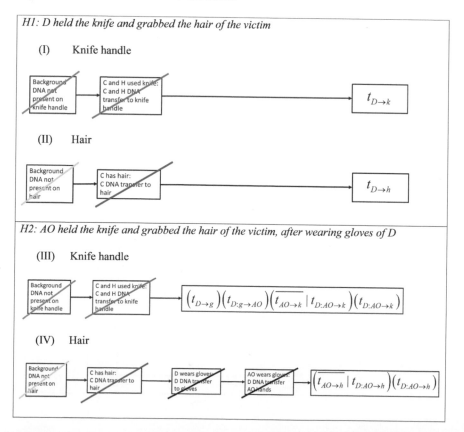

We then obtain the final terms for Pr(E|Hp) by multiplying the terms for each exhibit (I) and (II). We do the same to obtain Pr(E|Hd) by multiplying terms (III) and (IV). Dividing the two terms derived above leads to the *LR* formula given in Chapter 5:

$$LR = \frac{\left(t_{D \to k}\right)\left(t_{D \to h}\right)}{\left(t_{D \to g}\right)\left(t_{D:g \to AO}\right)\left(t_{D:AO \to k}\right)\left(\overline{t_{AO \to k}} \mid t_{D:AO \to k}\right)\left(t_{D:AO \to h}\right)\left(\overline{t_{AO \to h}} \mid t_{D:AO \to h}\right)}$$

Figure 5.8 is shown below.

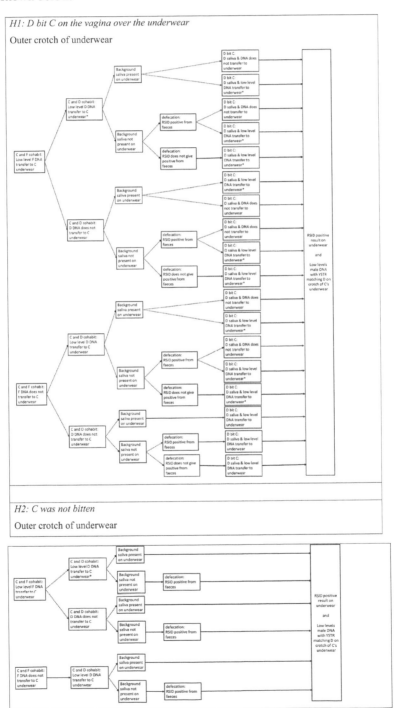

Carrying out the same process of conversion from the pathway to a formula gives the following:

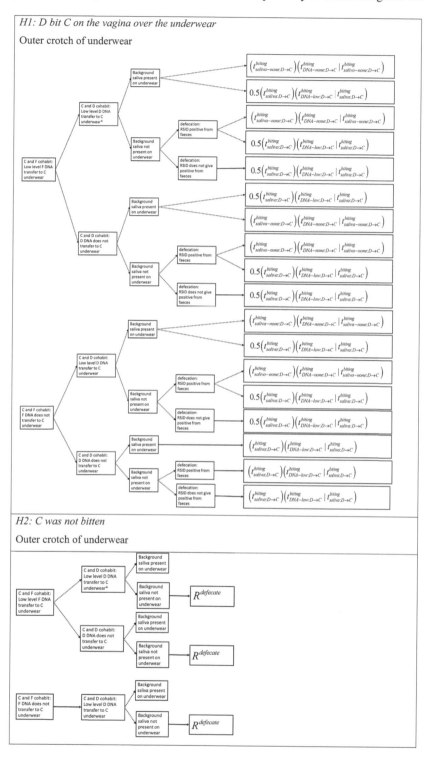

We have shown the terms in this first step in their conditional form. Note that as these two factors (the deposition of saliva and DNA) are in the same box as they are dependent on each other, i.e. if there was a high level of saliva deposition, then this is expected to be associated with a high level of DNA deposition also. We make the simplifying assumption that if low levels of saliva are transferred, then so too will be low levels of DNA and if saliva is not transferred then DNA will not be transferred. We mildly abuse terminology here and use the ambiguous term $t^{biting}_{none:D \to C}$ to denote no transfer of saliva and DNA, and $t^{biting}_{low:D \to C}$ to denote low levels of transfer of saliva and DNA. We incorporate this into the next step, along with the definition that $t^{bite} = t^{biting}_{none:D \to C} + 0.5\left(t^{biting}_{low:D \to C}\right)$.

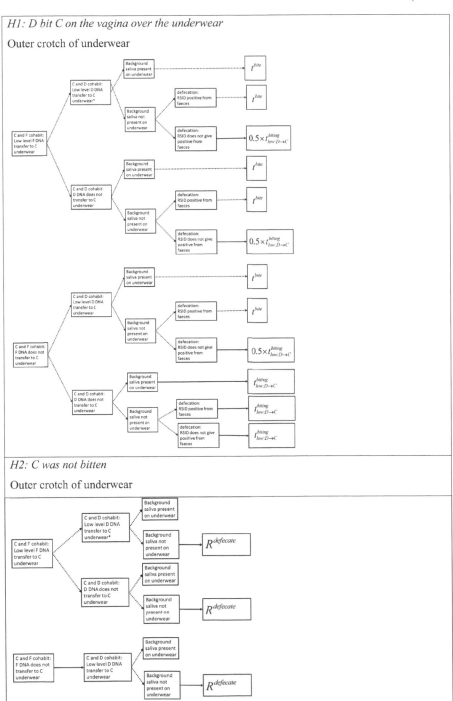

Now going back to the next term:

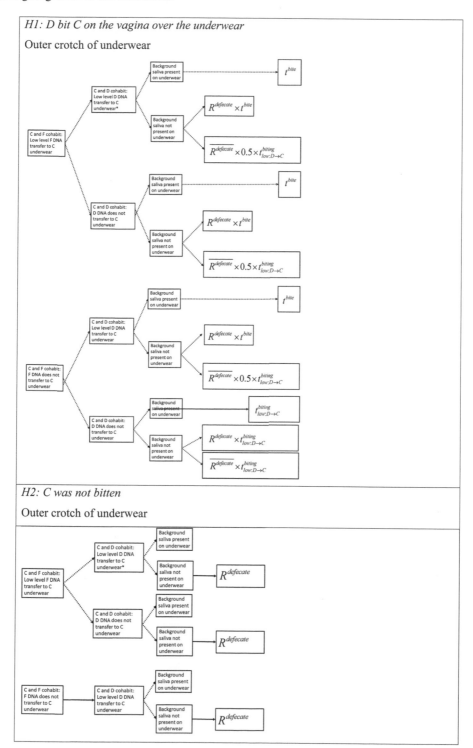

And using the simplification that $t^{bite/defecate} = R^{defecate} \times t^{bite} + \overline{R^{defecate}} \times 0.5\left(t^{biting}_{low:D \to C}\right)$ gives:

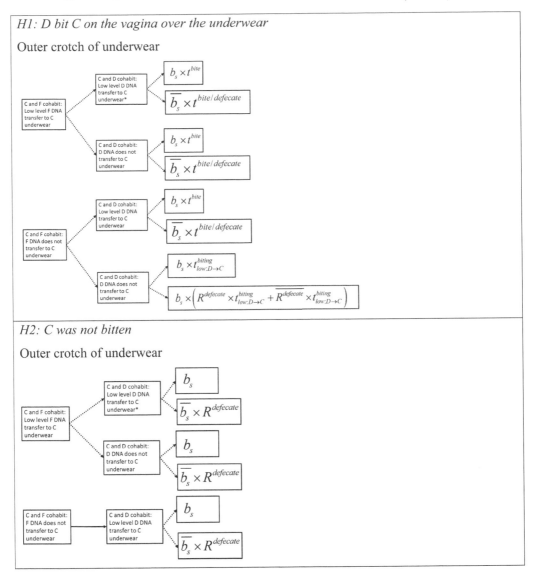

And using the simplification that $t^{bite/defecate/background} = b_s \times t^{bite} + \overline{b_s} \times t^{bite/defecate}$ gives:

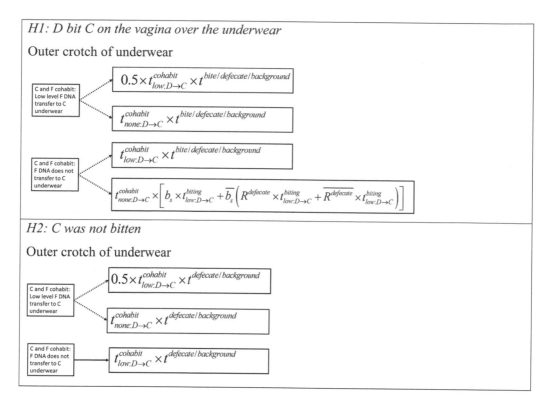

And the final steps are:

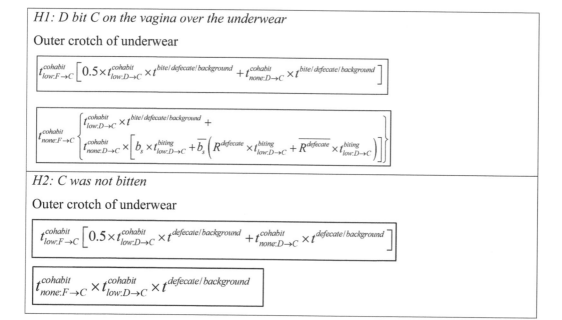

and

H1: D bit C on the vagina over the underwear

Outer crotch of underwear

$$t_{low:F\to C}^{cohabit}\left[0.5\times t_{low:D\to C}^{cohabit}\times t^{bite/defecate/background}+t_{none:D\to C}^{cohabit}\times t^{bite/defecate/background}\right]+$$

$$t_{none:F\to C}^{cohabit}\left(t_{low:D\to C}^{cohabit}\times t^{bite/defecate/background}+t_{none:D\to C}^{cohabit}\times t_{low:D\to C}^{biting}\right)$$

H2: C was not bitten

Outer crotch of underwear

$$t_{low:F\to C}^{cohabit}\left[0.5\times t_{low:D\to C}^{cohabit}\times t^{defecate/background}+t_{none:D\to C}^{cohabit}\times t^{defecate/background}\right]+$$

$$t_{none:F\to C}^{cohabit}\times t_{low:D\to C}^{cohabit}\times t^{defecate/background}$$

Dividing the two terms derived above leads to the *LR* formula given in the text of Chapter 5:

$$LR=\frac{\left\{\begin{array}{l}t_{low:F\to C}^{cohabit}\left[0.5\times t_{low:D\to C}^{cohabit}\times t^{bite/defecate/background}+t_{none:D\to C}^{cohabit}\times t^{bite/defecate/background}\right]+\\[2mm]t_{none:F\to C}^{cohabit}\left\{\begin{array}{l}t_{low:D\to C}^{cohabit}\times t^{bite/defecate/background}+\\[2mm]t_{none:D\to C}^{cohabit}\times\left[b_s\times t_{low:D\to C}^{biting}+\overline{b_s}\left(R^{defecate}\times t_{low:D\to C}^{biting}+\overline{R^{defecate}}\times t_{low:D\to C}^{biting}\right)\right]\end{array}\right\}\end{array}\right\}}{\left\{\begin{array}{l}t_{low:F\to C}^{cohabit}\left[0.5\times t_{low:D\to C}^{cohabit}\times t^{defecate/background}+t_{none:D\to C}^{cohabit}\times t^{defecate/background}\right]+\\[2mm]t_{none:F\to C}^{cohabit}\times t_{low:D\to C}^{cohabit}\times t^{defecate/background}\end{array}\right\}}$$

Which, with some simplification gives:

$$LR=\frac{\left\{\begin{array}{l}t_{low:F\to C}^{cohabit}\left[0.5\times t_{low:D\to C}^{cohabit}\times t^{bite/defecate/background}+t_{none:D\to C}^{cohabit}\times t^{bite/defecate/background}\right]+\\[2mm]t_{none:F\to C}^{cohabit}\left(t_{low:D\to C}^{cohabit}\times t^{bite/defecate/background}+t_{none:D\to C}^{cohabit}\times t_{low:D\to C}^{biting}\right)\end{array}\right\}}{\left\{\begin{array}{l}t_{low:F\to C}^{cohabit}\left[0.5\times t_{low:D\to C}^{cohabit}\times t^{defecate/background}+t_{none:D\to C}^{cohabit}\times t^{defecate/background}\right]+\\[2mm]t_{none:F\to C}^{cohabit}\times t_{low:D\to C}^{cohabit}\times t^{defecate/background}\end{array}\right\}}$$

Q8

Table 5.4 is a table that laid out all the combinations of events (and their respective probabilities) for Case 3 under H2 that could lead to the findings. This table was:

Events That Account for the Low Levels of Make DNA		Events That Account for the Positive RSID Test		
DNA Transfer FC from Cohabitation	DNA Transfer DC from Cohabitation	Background Saliva Present on C Underwear	Faecal Staining	Product
$t_{low.F\to C}^{cohabit}$	$0.5 \times t_{low.D\to C}^{cohabit}$	b_s	$R^{defecate}$	$t_{low.F\to C}^{cohabit} \times 0.5 \times t_{low.D\to C}^{cohabit} \times b_s \times R^{defecate}$
$t_{low.F\to C}^{cohabit}$	$0.5 \times t_{low.D\to C}^{cohabit}$	b_s	$\overline{R^{defecate}}$	$t_{low.F\to C}^{cohabit} \times 0.5 \times t_{low.D\to C}^{cohabit} \times b_s \times \overline{R^{defecate}}$
$t_{low.F\to C}^{cohabit}$	$0.5 \times t_{low.D\to C}^{cohabit}$	$\overline{b_s}$	$R^{defecate}$	$t_{low.F\to C}^{cohabit} \times 0.5 \times t_{low.D\to C}^{cohabit} \times \overline{b_s} \times R^{defecate}$
$t_{low.F\to C}^{cohabit}$	$t_{none.D\to C}^{cohabit}$	b_s	$R^{defecate}$	$t_{low.F\to C}^{cohabit} \times t_{none.D\to C}^{cohabit} \times b_s \times R^{defecate}$
$t_{low.F\to C}^{cohabit}$	$t_{none.D\to C}^{cohabit}$	b_s	$\overline{R^{defecate}}$	$t_{low.F\to C}^{cohabit} \times t_{none.D\to C}^{cohabit} \times b_s \times \overline{R^{defecate}}$
$t_{low.F\to C}^{cohabit}$	$t_{none.D\to C}^{cohabit}$	$\overline{b_s}$	$R^{defecate}$	$t_{low.F\to C}^{cohabit} \times t_{none.D\to C}^{cohabit} \times \overline{b_s} \times R^{defecate}$
$t_{none.F\to C}^{cohabit}$	$t_{low.D\to C}^{cohabit}$	b_s	$R^{defecate}$	$t_{none.F\to C}^{cohabit} \times t_{low.D\to C}^{cohabit} \times b_s \times R^{defecate}$
$t_{none.F\to C}^{cohabit}$	$t_{low.D\to C}^{cohabit}$	b_s	$\overline{R^{defecate}}$	$t_{none.F\to C}^{cohabit} \times t_{low.D\to C}^{cohabit} \times b_s \times \overline{R^{defecate}}$
$t_{none.F\to C}^{cohabit}$	$t_{low.D\to C}^{cohabit}$	$\overline{b_s}$	$R^{defecate}$	$t_{none.F\to C}^{cohabit} \times t_{low.D\to C}^{cohabit} \times \overline{b_s} \times R^{defecate}$

And in the text of Chapter 5, it was shown how this came from the formula:

$$\Pr(E \mid H2) = t_{low.F\to C}^{cohabit} \times 0.5 \times t_{low.D\to C}^{cohabit} \times t^{defecate/background} +$$
$$t_{low.F\to C}^{cohabit} \times t_{none.D\to C}^{cohabit} \times t^{defecate/background} +$$
$$t_{none.F\to C}^{cohabit} \times t_{low.D\to C}^{cohabit} \times t^{defecate/background}$$

We now wish to do the same for the combinations of events (and their respective probabilities) for Case 3 under H1 that could lead to the findings. However, instead of going from the formula to the table, we will list out all possible combinations of events in a table and show (through, summation, multiplication and simplification) that they result in the same *LR* formula as produced by the pathway method in the previous question. Setting up a similar table to that produced for Pr(*E*|H2) gives:

Number	Event That Account for the Low Levels of Male DNA		Events That Account for the Positive RSID Test		Events That Account for the Low Levels of Male DNA and the Positive RSID Test	Product
	DNA Transfer FC from Cohabitation	DNA Transfer DC from Cohabitation	Background Saliva Present on C Underwear	Faecal Staining	Saliva and DNA Transfer DC from Biting	
1	$t_{low:F\rightarrow C}^{cohabit}$	$0.5 \times t_{low:D\rightarrow C}^{cohabit}$	b_s	$R^{defecate}$	$t_{none:D\rightarrow C}^{biting}$	$t_{none:D\rightarrow C}^{biting} \times t_{low:F\rightarrow C}^{cohabit} \times 0.5 \times t_{low:D\rightarrow C}^{cohabit} \times b_s \times R^{defecate}$
2	$t_{low:F\rightarrow C}^{cohabit}$	$0.5 \times t_{low:D\rightarrow C}^{cohabit}$	b_s	$\overline{R^{defecate}}$	$t_{none:D\rightarrow C}^{biting}$	$t_{none:D\rightarrow C}^{biting} \times t_{low:F\rightarrow C}^{cohabit} \times 0.5 \times t_{low:D\rightarrow C}^{cohabit} \times b_s \times \overline{R^{defecate}}$
3	$t_{low:F\rightarrow C}^{cohabit}$	$0.5 \times t_{low:D\rightarrow C}^{cohabit}$	$\overline{b_s}$	$R^{defecate}$	$t_{none:D\rightarrow C}^{biting}$	$t_{none:D\rightarrow C}^{biting} \times t_{low:F\rightarrow C}^{cohabit} \times 0.5 \times t_{low:D\rightarrow C}^{cohabit} \times \overline{b_s} \times R^{defecate}$
4	$t_{low:F\rightarrow C}^{cohabit}$	$t_{none:D\rightarrow C}^{cohabit}$	b_s	$R^{defecate}$	$t_{none:D\rightarrow C}^{biting}$	$t_{none:D\rightarrow C}^{biting} \times t_{low:F\rightarrow C}^{cohabit} \times t_{none:D\rightarrow C}^{cohabit} \times b_s \times R^{defecate}$
5	$t_{low:F\rightarrow C}^{cohabit}$	$t_{none:D\rightarrow C}^{cohabit}$	b_s	$\overline{R^{defecate}}$	$t_{none:D\rightarrow C}^{biting}$	$t_{none:D\rightarrow C}^{biting} \times t_{low:F\rightarrow C}^{cohabit} \times t_{none:D\rightarrow C}^{cohabit} \times b_s \times \overline{R^{defecate}}$
6	$t_{low:F\rightarrow C}^{cohabit}$	$t_{none:D\rightarrow C}^{cohabit}$	$\overline{b_s}$	$R^{defecate}$	$t_{none:D\rightarrow C}^{biting}$	$t_{none:D\rightarrow C}^{biting} \times t_{low:F\rightarrow C}^{cohabit} \times t_{none:D\rightarrow C}^{cohabit} \times \overline{b_s} \times R^{defecate}$
7	$t_{none:F\rightarrow C}^{cohabit}$	$t_{low:D\rightarrow C}^{cohabit}$	b_s	$R^{defecate}$	$t_{none:D\rightarrow C}^{biting}$	$t_{none:D\rightarrow C}^{biting} \times t_{none:F\rightarrow C}^{cohabit} \times t_{low:D\rightarrow C}^{cohabit} \times b_s \times R^{defecate}$
8	$t_{none:F\rightarrow C}^{cohabit}$	$t_{low:D\rightarrow C}^{cohabit}$	b_s	$\overline{R^{defecate}}$	$t_{none:D\rightarrow C}^{biting}$	$t_{none:D\rightarrow C}^{biting} \times t_{none:F\rightarrow C}^{cohabit} \times t_{low:D\rightarrow C}^{cohabit} \times b_s \times \overline{R^{defecate}}$
9	$t_{none:F\rightarrow C}^{cohabit}$	$t_{low:D\rightarrow C}^{cohabit}$	$\overline{b_s}$	$R^{defecate}$	$t_{none:D\rightarrow C}^{biting}$	$t_{none:D\rightarrow C}^{biting} \times t_{none:F\rightarrow C}^{cohabit} \times t_{low:D\rightarrow C}^{cohabit} \times \overline{b_s} \times R^{defecate}$
10	$t_{low:F\rightarrow C}^{cohabit}$	$0.5 \times t_{low:D\rightarrow C}^{cohabit}$	b_s	$R^{defecate}$	$0.5 \times t_{low:D\rightarrow C}^{biting}$	$0.5 \times t_{low:D\rightarrow C}^{biting} \times t_{low:F\rightarrow C}^{cohabit} \times 0.5 \times t_{low:D\rightarrow C}^{cohabit} \times b_s \times R^{defecate}$
11	$t_{low:F\rightarrow C}^{cohabit}$	$0.5 \times t_{low:D\rightarrow C}^{cohabit}$	b_s	$\overline{R^{defecate}}$	$0.5 \times t_{low:D\rightarrow C}^{biting}$	$0.5 \times t_{low:D\rightarrow C}^{biting} \times t_{low:F\rightarrow C}^{cohabit} \times 0.5 \times t_{low:D\rightarrow C}^{cohabit} \times b_s \times \overline{R^{defecate}}$
12	$t_{low:F\rightarrow C}^{cohabit}$	$0.5 \times t_{low:D\rightarrow C}^{cohabit}$	$\overline{b_s}$	$R^{defecate}$	$0.5 \times t_{low:D\rightarrow C}^{biting}$	$0.5 \times t_{low:D\rightarrow C}^{biting} \times t_{low:F\rightarrow C}^{cohabit} \times 0.5 \times t_{low:D\rightarrow C}^{cohabit} \times \overline{b_s} \times R^{defecate}$
13	$t_{low:F\rightarrow C}^{cohabit}$	$t_{none:D\rightarrow C}^{cohabit}$	b_s	$R^{defecate}$	$0.5 \times t_{low:D\rightarrow C}^{biting}$	$0.5 \times t_{low:D\rightarrow C}^{biting} \times t_{low:F\rightarrow C}^{cohabit} \times t_{none:D\rightarrow C}^{cohabit} \times b_s \times R^{defecate}$
14	$t_{low:F\rightarrow C}^{cohabit}$	$t_{none:D\rightarrow C}^{cohabit}$	b_s	$\overline{R^{defecate}}$	$0.5 \times t_{low:D\rightarrow C}^{biting}$	$0.5 \times t_{low:D\rightarrow C}^{biting} \times t_{low:F\rightarrow C}^{cohabit} \times t_{none:D\rightarrow C}^{cohabit} \times b_s \times \overline{R^{defecate}}$
15	$t_{low:F\rightarrow C}^{cohabit}$	$t_{none:D\rightarrow C}^{cohabit}$	$\overline{b_s}$	$R^{defecate}$	$0.5 \times t_{low:D\rightarrow C}^{biting}$	$0.5 \times t_{low:D\rightarrow C}^{biting} \times t_{low:F\rightarrow C}^{cohabit} \times t_{none:D\rightarrow C}^{cohabit} \times \overline{b_s} \times R^{defecate}$
16	$t_{none:F\rightarrow C}^{cohabit}$	$t_{low:D\rightarrow C}^{cohabit}$	b_s	$R^{defecate}$	$0.5 \times t_{low:D\rightarrow C}^{biting}$	$0.5 \times t_{low:D\rightarrow C}^{biting} \times t_{none:F\rightarrow C}^{cohabit} \times t_{low:D\rightarrow C}^{cohabit} \times b_s \times R^{defecate}$
17	$t_{none:F\rightarrow C}^{cohabit}$	$t_{low:D\rightarrow C}^{cohabit}$	b_s	$\overline{R^{defecate}}$	$0.5 \times t_{low:D\rightarrow C}^{biting}$	$0.5 \times t_{low:D\rightarrow C}^{biting} \times t_{none:F\rightarrow C}^{cohabit} \times t_{low:D\rightarrow C}^{cohabit} \times b_s \times \overline{R^{defecate}}$

18	$t^{cohabit}_{none:F\to C}$	$t^{cohabit}_{low:D\to C}$	$\bar{b}_s$	$R^{deficate}$	$0.5\times t^{biting}_{low:D\to C}$	$0.5\times t^{biting}_{low:D\to C}\times t^{cohabit}_{none:F\to C}\times t^{cohabit}_{low:D\to C}\times \bar{b}_s \times R^{deficate}$
19	$t^{cohabit}_{low:F\to C}$	$0.5\times t^{cohabit}_{low:D\to C}$	$\bar{b}_s$	$\overline{R^{deficate}}$	$0.5\times t^{biting}_{low:D\to C}$	$0.5\times t^{biting}_{low:D\to C}\times t^{cohabit}_{low:F\to C}\times 0.5\times t^{cohabit}_{low:D\to C}\times \bar{b}_s \times \overline{R^{deficate}}$
20	$t^{cohabit}_{low:F\to C}$	$t^{cohabit}_{none:D\to C}$	$\bar{b}_s$	$\overline{R^{deficate}}$	$0.5\times t^{biting}_{low:D\to C}$	$0.5\times t^{biting}_{low:D\to C}\times t^{cohabit}_{low:F\to C}\times t^{cohabit}_{none:D\to C}\times \bar{b}_s \times R^{deficate}$
21	$t^{cohabit}_{none:F\to C}$	$t^{cohabit}_{low:D\to C}$	$\bar{b}_s$	$\overline{R^{deficate}}$	$0.5\times t^{biting}_{low:D\to C}$	$0.5\times t^{biting}_{low:D\to C}\times t^{cohabit}_{none:F\to C}\times t^{cohabit}_{low:D\to C}\times \bar{b}_s \times \overline{R^{deficate}}$
22	$t^{cohabit}_{none:F\to C}$	$t^{cohabit}_{none:D\to C}$	b_s	$R^{deficate}$	$t^{biting}_{low:D\to C}$	$t^{biting}_{low:D\to C}\times t^{cohabit}_{none:F\to C}\times t^{cohabit}_{none:D\to C}\times b_s \times R^{deficate}$
23	$t^{cohabit}_{none:F\to C}$	$t^{cohabit}_{none:D\to C}$	b_s	$\overline{R^{deficate}}$	$t^{biting}_{low:D\to C}$	$t^{biting}_{low:D\to C}\times t^{cohabit}_{none:F\to C}\times t^{cohabit}_{none:D\to C}\times b_s \times \overline{R^{deficate}}$
24	$t^{cohabit}_{none:F\to C}$	$t^{cohabit}_{none:D\to C}$	$\bar{b}_s$	$R^{deficate}$	$t^{biting}_{low:D\to C}$	$t^{biting}_{low:D\to C}\times t^{cohabit}_{none:F\to C}\times t^{cohabit}_{none:D\to C}\times \bar{b}_s \times R^{deficate}$
25	$t^{cohabit}_{none:F\to C}$	$t^{cohabit}_{none:D\to C}$	$\bar{b}_s$	$\overline{R^{deficate}}$	$t^{biting}_{low:D\to C}$	$t^{biting}_{low:D\to C}\times t^{cohabit}_{none:F\to C}\times t^{cohabit}_{none:D\to C}\times \bar{b}_s \times \overline{R^{deficate}}$

Note that with the exception of the column 'Events that account for the low levels of male DNA and the positive RSID test', rows 1–9 are the same as those in the table for Pr(E|H2). Rows 10–28 are the same again, but with considering the transfer of low amounts of DNA and saliva through biting. The formula for Pr(E|H1) is then obtained by the sum of all 25 terms in the product column:

$$\Pr(E \mid H1) = t_{none:D \to C}^{biting} \times t_{low:F \to C}^{cohabit} \times 0.5 \times t_{low:D \to C}^{cohabit} \times b_s \times R^{defecate}$$

$$+ t_{none:D \to C}^{biting} \times t_{low:F \to C}^{cohabit} \times 0.5 \times t_{low:D \to C}^{cohabit} \times b_s \times \overline{R^{defecate}}$$

$$+ t_{none:D \to C}^{biting} \times t_{low:F \to C}^{cohabit} \times 0.5 \times t_{low:D \to C}^{cohabit} \times \overline{b_s} \times R^{defecate}$$

$$+ t_{none:D \to C}^{biting} \times t_{low:F \to C}^{cohabit} \times t_{none:D \to C}^{cohabit} \times b_s \times R^{defecate}$$

$$+ t_{none:D \to C}^{biting} \times t_{low:F \to C}^{cohabit} \times t_{none:D \to C}^{cohabit} \times b_s \times \overline{R^{defecate}}$$

$$+ t_{none:D \to C}^{biting} \times t_{low:F \to C}^{cohabit} \times t_{none:D \to C}^{cohabit} \times \overline{b_s} \times R^{defecate}$$

$$+ t_{none:D \to C}^{biting} \times t_{none:F \to C}^{cohabit} \times t_{low:D \to C}^{cohabit} \times b_s \times R^{defecate}$$

$$+ t_{none:D \to C}^{biting} \times t_{none:F \to C}^{cohabit} \times t_{low:D \to C}^{cohabit} \times b_s \times \overline{R^{defecate}}$$

$$+ t_{none:D \to C}^{biting} \times t_{none:F \to C}^{cohabit} \times t_{low:D \to C}^{cohabit} \times \overline{b_s} \times R^{defecate}$$

$$+ 0.5 \times t_{low:D \to C}^{biting} \times t_{low:F \to C}^{cohabit} \times 0.5 \times t_{low:D \to C}^{cohabit} \times b_s \times R^{defecate}$$

$$+ 0.5 \times t_{low:D \to C}^{biting} \times t_{low:F \to C}^{cohabit} \times 0.5 \times t_{low:D \to C}^{cohabit} \times b_s \times \overline{R^{defecate}}$$

$$+ 0.5 \times t_{low:D \to C}^{biting} \times t_{low:F \to C}^{cohabit} \times 0.5 \times t_{low:D \to C}^{cohabit} \times \overline{b_s} \times R^{defecate}$$

$$+ 0.5 \times t_{low:D \to C}^{biting} \times t_{low:F \to C}^{cohabit} \times t_{none:D \to C}^{cohabit} \times b_s \times R^{defecate}$$

$$+ 0.5 \times t_{low:D \to C}^{biting} \times t_{low:F \to C}^{cohabit} \times t_{none:D \to C}^{cohabit} \times b_s \times \overline{R^{defecate}}$$

$$+ 0.5 \times t_{low:D \to C}^{biting} \times t_{low:F \to C}^{cohabit} \times t_{none:D \to C}^{cohabit} \times \overline{b_s} \times R^{defecate}$$

$$+ 0.5 \times t_{low:D \to C}^{biting} \times t_{none:F \to C}^{cohabit} \times t_{low:D \to C}^{cohabit} \times b_s \times R^{defecate}$$

$$+ 0.5 \times t_{low:D \to C}^{biting} \times t_{none:F \to C}^{cohabit} \times t_{low:D \to C}^{cohabit} \times b_s \times \overline{R^{defecate}}$$

$$+ 0.5 \times t_{low:D \to C}^{biting} \times t_{none:F \to C}^{cohabit} \times t_{low:D \to C}^{cohabit} \times \overline{b_s} \times R^{deficate}$$

$$+ 0.5 \times t_{low:D \to C}^{biting} \times t_{low:F \to C}^{cohabit} \times 0.5 \times t_{low:D \to C}^{cohabit} \times \overline{b_s} \times \overline{R^{deficate}}$$

$$+ 0.5 \times t_{low:D \to C}^{biting} \times t_{low:F \to C}^{cohabit} \times t_{none:D \to C}^{cohabit} \times \overline{b_s} \times \overline{R^{deficate}}$$

$$+ 0.5 \times t_{low:D \to C}^{biting} \times t_{none:F \to C}^{cohabit} \times t_{low:D \to C}^{cohabit} \times \overline{b_s} \times \overline{R^{deficate}}$$

$$+ t_{low:D \to C}^{biting} \times t_{none:F \to C}^{cohabit} \times t_{none:D \to C}^{cohabit} \times b_s \times R^{defecate}$$

$$+ t_{low:D \to C}^{biting} \times t_{none:F \to C}^{cohabit} \times t_{none:D \to C}^{cohabit} \times b_s \times \overline{R^{defecate}}$$

$$+ t_{low:D \to C}^{biting} \times t_{none:F \to C}^{cohabit} \times t_{none:D \to C}^{cohabit} \times \overline{b_s} \times R^{defecate}$$

$$+ t_{low:D \to C}^{biting} \times t_{none:F \to C}^{cohabit} \times t_{none:D \to C}^{cohabit} \times \overline{b_s} \times \overline{R^{defecate}}$$

It is then a matter of algebraically collecting like terms to obtain the *LR* formula from the text:

$$\Pr(E \mid H1) = t_{low:F \to C}^{cohabit} \left\{ \begin{array}{l} t_{none:D \to C}^{biting} \times 0.5 \times t_{low:D \to C}^{cohabit} \times b_s \times R^{defecate} + t_{none:D \to C}^{biting} \times 0.5 \times t_{low:D \to C}^{cohabit} \times b_s \times \overline{R^{defecate}} + \\ t_{none:D \to C}^{biting} \times 0.5 \times t_{low:D \to C}^{cohabit} \times \overline{b_s} \times R^{defecate} + t_{none:D \to C}^{biting} \times t_{none:D \to C}^{cohabit} \times b_s \times R^{defecate} + \\ t_{none:D \to C}^{biting} \times t_{none:D \to C}^{cohabit} \times b_s \times \overline{R^{defecate}} + t_{none:D \to C}^{biting} \times t_{none:D \to C}^{cohabit} \times \overline{b_s} \times R^{defecate} + \\ 0.5 \times t_{low:D \to C}^{biting} \times 0.5 \times t_{low:D \to C}^{cohabit} \times b_s \times R^{defecate} + 0.5 \times t_{low:D \to C}^{biting} \times 0.5 \times t_{low:D \to C}^{cohabit} \times b_s \times \overline{R^{defecate}} + \\ 0.5 \times t_{low:D \to C}^{biting} \times 0.5 \times t_{low:D \to C}^{cohabit} \times \overline{b_s} \times R^{defecate} + 0.5 \times t_{low:D \to C}^{biting} \times 0.5 \times t_{low:D \to C}^{cohabit} \times \overline{b_s} \times \overline{R^{defecate}} + \\ 0.5 \times t_{low:D \to C}^{biting} \times t_{none:D \to C}^{cohabit} \times b_s \times R^{defecate} + 0.5 \times t_{low:D \to C}^{biting} \times t_{none:D \to C}^{cohabit} \times b_s \times \overline{R^{defecate}} + \\ 0.5 \times t_{low:D \to C}^{biting} \times t_{none:D \to C}^{cohabit} \times \overline{b_s} \times R^{defecate} + 0.5 \times t_{low:D \to C}^{biting} \times 0.5 \times t_{low:D \to C}^{cohabit} \times \overline{b_s} \times \overline{R^{defecate}} + \\ 0.5 \times t_{low:D \to C}^{biting} \times t_{none:D \to C}^{cohabit} \times \overline{b_s} \times R^{defecate} \end{array} \right\} +$$

$$t_{none:F \to C}^{cohabit} \left\{ \begin{array}{l} t_{none:D \to C}^{biting} \times t_{low:D \to C}^{cohabit} \times b_s \times R^{defecate} + t_{none:D \to C}^{biting} \times t_{low:D \to C}^{cohabit} \times b_s \times \overline{R^{defecate}} + \\ t_{none:D \to C}^{biting} \times t_{low:D \to C}^{cohabit} \times \overline{b_s} \times R^{defecate} + 0.5 \times t_{low:D \to C}^{biting} \times t_{low:D \to C}^{cohabit} \times b_s \times R^{defecate} + \\ 0.5 \times t_{low:D \to C}^{biting} \times t_{low:D \to C}^{cohabit} \times b_s \times \overline{R^{defecate}} + 0.5 \times t_{low:D \to C}^{biting} \times t_{low:D \to C}^{cohabit} \times \overline{b_s} \times R^{defecate} + \\ 0.5 \times t_{low:D \to C}^{biting} \times t_{low:D \to C}^{cohabit} \times \overline{b_s} \times R^{defecate} + t_{low:D \to C}^{biting} \times t_{none:D \to C}^{cohabit} \times b_s \times R^{defecate} + \\ t_{low:D \to C}^{biting} \times t_{none:D \to C}^{cohabit} \times b_s \times \overline{R^{defecate}} + t_{low:D \to C}^{biting} \times t_{none:D \to C}^{cohabit} \times \overline{b_s} \times R^{defecate} + \\ t_{low:D \to C}^{biting} \times t_{none:D \to C}^{cohabit} \times \overline{b_s} \times R^{defecate} \end{array} \right\}$$

$$\Pr(E \mid H1) = t_{low:F \to C}^{cohabit} \left\{ \begin{array}{l} t_{low:D \to C}^{cohabit} \left[\begin{array}{l} t_{none:D \to C}^{biting} \times 0.5 \times b_s \times R^{defecate} + t_{none:D \to C}^{biting} \times 0.5 \times b_s \times \overline{R^{defecate}} + \\ t_{none:D \to C}^{biting} \times 0.5 \times \overline{b_s} \times R^{defecate} + 0.5 \times t_{low:D \to C}^{biting} \times 0.5 \times b_s \times R^{defecate} + \\ 0.5 \times t_{low:D \to C}^{biting} \times 0.5 \times b_s \times \overline{R^{defecate}} + 0.5 \times t_{low:D \to C}^{biting} \times 0.5 \times \overline{b_s} \times R^{defecate} + \\ 0.5 \times t_{low:D \to C}^{biting} \times 0.5 \times \overline{b_s} \times R^{defecate} + 0.5 \times t_{low:D \to C}^{biting} \times 0.5 \times \overline{b_s} \times \overline{R^{defecate}} \end{array} \right] + \\ \\ t_{none:D \to C}^{cohabit} \left[\begin{array}{l} t_{none:D \to C}^{biting} \times b_s \times R^{defecate} + t_{none:D \to C}^{biting} \times b_s \times \overline{R^{defecate}} + \\ t_{none:D \to C}^{biting} \times \overline{b_s} \times R^{defecate} + 0.5 \times t_{low:D \to C}^{biting} \times b_s \times R^{defecate} + \\ 0.5 \times t_{low:D \to C}^{biting} \times b_s \times \overline{R^{defecate}} + 0.5 \times t_{low:D \to C}^{biting} \times \overline{b_s} \times R^{defecate} + \\ 0.5 \times t_{low:D \to C}^{biting} \times \overline{b_s} \times \overline{R^{defecate}} \end{array} \right] \end{array} \right\} +$$

$$t_{none:F \to C}^{cohabit} \left\{ \begin{array}{l} t_{low:D \to C}^{cohabit} \left[\begin{array}{l} t_{none:D \to C}^{biting} \times b_s \times R^{defecate} + t_{none:D \to C}^{biting} \times b_s \times \overline{R^{defecate}} + \\ t_{none:D \to C}^{biting} \times \overline{b_s} \times R^{defecate} + 0.5 \times t_{low:D \to C}^{biting} \times b_s \times R^{defecate} + \\ 0.5 \times t_{low:D \to C}^{biting} \times b_s \times \overline{R^{defecate}} + 0.5 \times t_{low:D \to C}^{biting} \times \overline{b_s} \times R^{defecate} + \\ 0.5 \times t_{low:D \to C}^{biting} \times \overline{b_s} \times R^{defecate} \end{array} \right] + \\ \\ t_{none:D \to C}^{cohabit} \left[\begin{array}{l} t_{low:D \to C}^{biting} \times b_s \times R^{defecate} + t_{low:D \to C}^{biting} \times b_s \times \overline{R^{defecate}} + \\ t_{low:D \to C}^{biting} \times \overline{b_s} \times R^{defecate} + t_{low:D \to C}^{biting} \times \overline{b_s} \times \overline{R^{defecate}} \end{array} \right] \end{array} \right\}$$

$$\Pr(E\mid H1)=t^{cohabit}_{low:F\to C}\left\{\begin{array}{l} 0.5\times t^{cohabit}_{low:D\to C}\left[\begin{array}{l} b_s\left(\begin{array}{l} t^{biting}_{none:D\to C}\times R^{defecate}+t^{biting}_{none:D\to C}\times\overline{R^{defecate}}+0.5\times t^{biting}_{low:D\to C}\times R^{defecate}+\\ 0.5\times t^{biting}_{low:D\to C}\times\overline{R^{defecate}}+\end{array}\right)+\\ \overline{b_s}\left(\begin{array}{l} t^{biting}_{none:D\to C}\times R^{defecate}+0.5\times t^{biting}_{low:D\to C}\times R^{defecate}+0.5\times t^{biting}_{low:D\to C}\times R^{defecate}+\\ 0.5\times t^{biting}_{low:D\to C}\times\overline{R^{deficate}}\end{array}\right)\end{array}\right]+\\[2em] t^{cohabit}_{none:D\to C}\left[\begin{array}{l} b_s\left(\begin{array}{l} t^{biting}_{none:D\to C}\times R^{defecate}+t^{biting}_{none:D\to C}\times\overline{R^{defecate}}+0.5\times t^{biting}_{low:D\to C}\times R^{defecate}+\\ 0.5\times t^{biting}_{low:D\to C}\times\overline{R^{defecate}}\end{array}\right)+\\ \overline{b_s}\left(t^{biting}_{none:D\to C}\times R^{defecate}+0.5\times t^{biting}_{low:D\to C}\times R^{defecate}+0.5\times t^{biting}_{low:D\to C}\times\overline{R^{defecate}}\right)\end{array}\right]\end{array}\right\}+$$

$$t^{cohabit}_{none:F\to C}\left\{\begin{array}{l} t^{cohabit}_{low:D\to C}\left[\begin{array}{l} b_s\left(t^{biting}_{none:D\to C}\times R^{defecate}+t^{biting}_{none:D\to C}\times\overline{R^{defecate}}+0.5\times t^{biting}_{low:D\to C}\times R^{defecate}+0.5\times t^{biting}_{low:D\to C}\times\overline{R^{defecate}}\right)+\\ \overline{b_s}\left(t^{biting}_{none:D\to C}\times R^{deficate}+0.5\times t^{biting}_{low:D\to C}\times R^{deficate}+0.5\times t^{biting}_{low:D\to C}\times\overline{R^{deficate}}\right)\end{array}\right]+\\[1.5em] t^{cohabit}_{none:D\to C}\left[b_s\left(t^{biting}_{low:D\to C}\times R^{defecate}+t^{biting}_{low:D\to C}\times\overline{R^{defecate}}\right)+\overline{b_s}\left(t^{biting}_{low:D\to C}\times R^{defecate}+t^{biting}_{low:D\to C}\times\overline{R^{defecate}}\right)\right]\end{array}\right\}$$

$$\Pr(E\mid H1)=t^{cohabit}_{low:F\to C}\left\{\begin{array}{l} 0.5\times t^{cohabit}_{low:D\to C}\left[\begin{array}{l} b_s\left(R^{defecate}\left(t^{biting}_{none:D\to C}+0.5\times t^{biting}_{low:D\to C}\right)+\overline{R^{defecate}}\left(t^{biting}_{none:D\to C}+0.5\times t^{biting}_{low:D\to C}\right)\right)+\\ \overline{b_s}\left(R^{defecate}\left(t^{biting}_{none:D\to C}+0.5\times t^{biting}_{low:D\to C}\right)+\overline{R^{defecate}}\left(0.5\times t^{biting}_{low:D\to C}\right)+R^{defecate}\left(0.5\times t^{biting}_{low:D\to C}\right)\right)\end{array}\right]+\\[1.5em] t^{cohabit}_{none:D\to C}\left[\begin{array}{l} b_s\left(R^{defecate}\left(t^{biting}_{none:D\to C}+0.5\times t^{biting}_{low:D\to C}\right)+\overline{R^{defecate}}\left(t^{biting}_{none:D\to C}+0.5\times t^{biting}_{low:D\to C}\right)\right)+\\ \overline{b_s}\left(R^{defecate}\left(t^{biting}_{none:D\to C}+0.5\times t^{biting}_{low:D\to C}\right)+\overline{R^{defecate}}\left(0.5\times t^{biting}_{low:D\to C}\right)\right)\end{array}\right]\end{array}\right\}+$$

$$t^{cohabit}_{none:F\to C}\left\{\begin{array}{l} t^{cohabit}_{low:D\to C}\left[\begin{array}{l} b_s\left(R^{defecate}\left(t^{biting}_{none:D\to C}+0.5\times t^{biting}_{low:D\to C}\right)+\overline{R^{defecate}}\left(t^{biting}_{none:D\to C}+0.5\times t^{biting}_{low:D\to C}\right)\right)+\\ \overline{b_s}\left(R^{defecate}\left(t^{biting}_{none:D\to C}+0.5\times t^{biting}_{low:D\to C}\right)+\overline{R^{defecate}}\left(0.5\times t^{biting}_{low:D\to C}\right)\right)\end{array}\right]+\\[1.5em] t^{cohabit}_{none:D\to C}\left[b_s\left(t^{biting}_{low:D\to C}\times R^{defecate}+t^{biting}_{low:D\to C}\times\overline{R^{defecate}}\right)+\overline{b_s}\left(t^{biting}_{low:D\to C}\times R^{defecate}+t^{biting}_{low:D\to C}\times\overline{R^{defecate}}\right)\right]\end{array}\right\}$$

$$\Pr(E\mid H1)=t^{cohabit}_{low:F\to C}\left\{\begin{array}{l} 0.5\times t^{cohabit}_{low:D\to C}\left[b_s\left(t^{biting}_{none:D\to C}+0.5\times t^{biting}_{low:D\to C}\right)+\overline{b_s}\left(R^{defecate}\left(t^{biting}_{none:D\to C}+0.5\times t^{biting}_{low:D\to C}\right)+0.5\times t^{biting}_{low:D\to C}\right)\right]+\\ t^{cohabit}_{none:D\to C}\left[b_s\left(t^{biting}_{none:D\to C}+0.5\times t^{biting}_{low:D\to C}\right)+\overline{b_s}\left(R^{defecate}\left(t^{biting}_{none:D\to C}+0.5\times t^{biting}_{low:D\to C}\right)+\overline{R^{deficate}}\left(0.5\times t^{biting}_{low:D\to C}\right)\right)\right]\end{array}\right\}+$$

$$t^{cohabit}_{none:F\to C}\left\{\begin{array}{l} t^{cohabit}_{low:D\to C}\left[b_s\left(t^{biting}_{none:D\to C}+0.5\times t^{biting}_{low:D\to C}\right)+\overline{b_s}\left(R^{deficate}\left(t^{biting}_{none:D\to C}+0.5\times t^{biting}_{low:D\to C}\right)+\overline{R^{deficate}}\left(0.5\times t^{biting}_{low:D\to C}\right)\right)\right]+\\ t^{cohabit}_{none:D\to C}\left[b_s\left(t^{biting}_{low:D\to C}\right)+\overline{b_s}\left(t^{biting}_{low:D\to C}\right)\right]\end{array}\right\}$$

Now we apply the simplifying definitions. First, $t^{bite}=t^{biting}_{none:D\to C}+0.5\left(t^{biting}_{low:D\to C}\right)$:

$$\Pr(E\mid H1)=t_{low:F\rightarrow C}^{cohabit}\left\{\begin{array}{l}0.5\times t_{low:D\rightarrow C}^{cohabit}\left[b_s\left(t^{bite}\right)+\overline{b_s}\left(R^{defecate}\left(t^{bite}\right)+0.5\times t_{low:D\rightarrow C}^{biting}\right)\right]+\\[2mm]t_{none:D\rightarrow C}^{cohabit}\left[b_s\left(t^{bite}\right)+\overline{b_s}\left(R^{defecate}\left(t^{bite}\right)+\overline{R^{defecate}}\left(0.5\times t_{low:D\rightarrow C}^{biting}\right)\right)\right]\end{array}\right\}+$$

$$t_{none:F\rightarrow C}^{cohabit}\left\{t_{low:D\rightarrow C}^{cohabit}\left[b_s\left(t^{bite}\right)+\overline{b_s}\left(R^{deficate}\left(t^{bite}\right)+\overline{R^{deficate}}\left(0.5\times t_{low:D\rightarrow C}^{biting}\right)\right)\right]+t_{none:D\rightarrow C}^{cohabit}\left[b_s\left(t_{low:D\rightarrow C}^{biting}\right)+\overline{b_s}\left(t_{low:D\rightarrow C}^{biting}\right)\right]\right\}$$

Then $t^{bite/defecate}=R^{defecate}\times t^{bite}+\overline{R^{defecate}}\times 0.5\left(t_{low:D\rightarrow C}^{biting}\right)$:

$$\Pr(E\mid H1)=t_{low:F\rightarrow C}^{cohabit}\left\{0.5\times t_{low:D\rightarrow C}^{cohabit}\left[b_s\left(t^{bite}\right)+\overline{b_s}\left(t^{bite/defecate}\right)\right]+t_{none:D\rightarrow C}^{cohabit}\left[b_s\left(t^{bite}\right)+\overline{b_s}\left(t^{bite/defecate}\right)\right]\right\}+$$

$$t_{none:F\rightarrow C}^{cohabit}\left\{t_{low:D\rightarrow C}^{cohabit}\left[b_s\times t^{bite}+\overline{b_s}\left(t^{bite/defecate}\right)\right]+t_{none:D\rightarrow C}^{cohabit}\times t_{low:D\rightarrow C}^{biting}\right\}$$

And finally $t^{bite/defecate/background}=b_s\times t^{bite}+\overline{b_s}\times t^{bite/defecate}$:

$$\Pr(E\mid H1)=t_{low:F\rightarrow C}^{cohabit}\left\{0.5\times t_{low:D\rightarrow C}^{cohabit}\times t^{bite/defecate/background}+t_{none:D\rightarrow C}^{cohabit}\times t^{bite/defecate/background}\right\}+$$

$$t_{none:F\rightarrow C}^{cohabit}\left\{t_{low:D\rightarrow C}^{cohabit}\times t^{bite/defecate/background}+t_{none:D\rightarrow C}^{cohabit}\times t_{low:D\rightarrow C}^{biting}\right\}$$

As derived in the previous question and Chapter 5.

Q9

There are several ways in which a formula could be derived to evaluate the finding of Ms B's DNA on Mr A's hands, i.e. by conceptual derivation, or by using pathways, or by listing out all possibilities in a table. This scenario is simple enough that it is possible to consider it without pathways or tables, which we demonstrate below (although you may choose to do either of these other two if they are your preference).

If we recover DNA from Ms B on Mr A's hands, then, given prosecution's propositions this can be explained by three events:

(1) There was transfer of DNA from the dental check and not from handling the wallet, $t(1-s)$.
(2) There was both transfer from the dental check and handling the wallet, ts.
(3) There was no transfer from the dental check but there was transfer from handling the wallet, $(1-t)s$.

Under the defence proposition, there was only transfer from the dental check, t. Taking these mechanisms into account, the LR in our example is:

$$LR=\frac{t(1-s)+ts+(1-t)s}{t}$$

simplifying to:

$$LR=1+\frac{(1-t)s}{t}$$

which would be slightly greater than one. For practical purposes, the presence of Ms B's DNA on Mr A's hands does not provide much support to either proposition, and this is because $t \sim 1$. Or more generally, it is because the probability of detecting DNA given that only the undisputed activity took place will be comparable to that which is expected given that both the undisputed and the disputed activity took place.

Q10

Let us say that we derived an LR formula that only considered the activity associated with the offence under prosecution's view of events, and that we did not account for part of the information. In this instance, we end up with:

$$LR = s / t$$

which clearly does not reflect the case circumstances, most demonstrably because if we detected DNA matching Ms B, then we would end up with $LR < 1$ (coming from the fact that $s << t$). But of course, this does not make sense as detecting Ms B's DNA on Mr A's hands should be uninformative (or slightly supportive of the prosecution proposition) and thus support neither of the propositions if the dental checkup is not disputed. This expected behaviour is seen in the previous formula derived in Question 9, which was:

$$LR = 1 + \frac{(1-t)s}{t}$$

CHAPTER 6 – ANSWERS TO QUESTIONS

Q1

We provide the basic multiplications in the table below. When the genotype of the child cannot be produced by the genotype of the parent, then 0 is given without further explanation.

Genotype of Parent		XX	XY	XZ	YY	YZ	ZZ
Genotype of Child	XX	1×0.5 $= 0.5$	0.5×0.5 $= 0.25$	0.5×0.5 $= 0.25$	0	0	0
	XY	1×0.49 $= 0.49$	$0.5 \times 0.5 + 0.5 \times 0.49$ $= 0.495$	0.5×0.49 $= 0.245$	1×0.5 $= 0.5$	0.5×0.5 $= 0.25$	0
	XZ	1×0.01 $= 0.01$	0.5×0.01 $= 0.005$	$0.5 \times 0.5 + 0.5 \times 0.01$ $= 0.255$	0	0.5×0.5 $= 0.25$	1×0.5 $= 0.5$
	YY	0	0.5×0.49 $= 0.245$	0	1×0.49 $= 0.49$	0.5×0.49 $= 0.245$	0
	YZ	0	0.5×0.01 $= 0.005$	0.5×0.49 $= 0.245$	1×0.01 $= 0.01$	0.5×0.5 $= 0.25$	1×0.49 $= 0.49$
	ZZ	0	0	0.5×0.01 $= 0.005$	0	0.5×0.01 $= 0.005$	1×0.01 $= 0.01$

Q2

The probability of a child having genotype [X,X] was derived in Chapter 6 and shown to be 0.25, but we include it here as well.

$$\Pr(G_C = [X,X]) =$$

$$\Pr(G_C = [X,X] \mid G_P = [X,X]) \Pr(G_P = [X,X]) +$$

$$\Pr(G_C = [X,X] \mid G_P = [X,Y]) \Pr(G_P = [X,Y]) +$$

$$\Pr(G_C = [X,X] \mid G_P = [X,Z]) \Pr(G_P = [X,Z]) +$$

$$\Pr(G_C = [X,X] \mid G_P = [Y,Y]) \Pr(G_P = [Y,Y]) +$$

$$\Pr(G_C = [X,X] \mid G_P = [Y,Z]) \Pr(G_P = [Y,Z]) +$$

$$\Pr(G_C = [X,X] \mid G_P = [Z,Z]) \Pr(G_P = [Z,Z])$$

$$\Pr(G_C = [X,X]) = 0.5 \times 0.25 + 0.25 \times 0.49 + 0.25 \times 0.01 + 0 \times 0.2401 + 0 \times 0.0098 + 0 \times 0.0001$$

$$= 0.25$$

Similar:

$$\Pr(G_C = [X,Y]) =$$

$$\Pr(G_C = [X,Y] \mid G_P = [X,X]) \Pr(G_P = [X,X]) +$$

$$\Pr(G_C = [X,Y] \mid G_P = [X,Y]) \Pr(G_P = [X,Y]) +$$

$$\Pr(G_C = [X,Y] \mid G_P = [X,Z]) \Pr(G_P = [X,Z]) +$$

$$\Pr(G_C = [X,Y] \mid G_P = [Y,Y]) \Pr(G_P = [Y,Y]) +$$

$$\Pr(G_C = [X,Y] \mid G_P = [Y,Z]) \Pr(G_P = [Y,Z]) +$$

$$\Pr(G_C = [X,Y] \mid G_P = [Z,Z]) \Pr(G_P = [Z,Z])$$

$$\Pr(G_C = [X,X]) = 0.49 \times 0.25 + 0.495 \times 0.49 + 0.254 \times 0.01 + 0.5 \times 0.2401 + 0.25 \times 0.0098 + 0 \times 0.0001$$

$$= 0.49$$

$$\Pr(G_C = [X,Z]) =$$

$$\Pr(G_C = [X,Z] \mid G_P = [X,X]) \Pr(G_P = [X,X]) +$$

$$\Pr(G_C = [X,Z] \mid G_P = [X,Y]) \Pr(G_P = [X,Y]) +$$

$$\Pr(G_C = [X,Z] \mid G_P = [X,Z]) \Pr(G_P = [X,Z]) +$$

$$\Pr(G_C = [X,Z] \mid G_P = [Y,Y]) \Pr(G_P = [Y,Y]) +$$

$$\Pr(G_C = [X,Z] \mid G_P = [Y,Z]) \Pr(G_P = [Y,Z]) +$$

$$\Pr(G_C = [X,Z] \mid G_P = [Z,Z]) \Pr(G_P = [Z,Z])$$

$$\Pr(G_C = [X,X]) = 0.01 \times 0.25 + 0.005 \times 0.49 + 0.255 \times 0.01 + 0 \times 0.2401 + 0.25 \times 0.0098 + 0.5 \times 0.0001$$

$= 0.01$

$$\Pr(G_C = [Y,Y]) =$$
$$\Pr(G_C = [Y,Y] \mid G_P = [X,X])\Pr(G_P = [X,X]) +$$
$$\Pr(G_C = [Y,Y] \mid G_P = [X,Y])\Pr(G_P = [X,Y]) +$$
$$\Pr(G_C = [Y,Y] \mid G_P = [X,Z])\Pr(G_P = [X,Z]) +$$
$$\Pr(G_C = [Y,Y] \mid G_P = [Y,Y])\Pr(G_P = [Y,Y]) +$$
$$\Pr(G_C = [Y,Y] \mid G_P = [Y,Z])\Pr(G_P = [Y,Z]) +$$
$$\Pr(G_C = [Y,Y] \mid G_P = [Z,Z])\Pr(G_P = [Z,Z])$$

$$\Pr(G_C = [X,X]) = 0 \times 0.25 + 0.245 \times 0.49 + 0 \times 0.01 + 0.49 \times 0.2401 + 0.245 \times 0.0098 + 0 \times 0.0001$$

$= 0.2401$

$$\Pr(G_C = [Y,Z]) =$$
$$\Pr(G_C = [Y,Z] \mid G_P = [X,X])\Pr(G_P = [X,X]) +$$
$$\Pr(G_C = [Y,Z] \mid G_P = [X,Y])\Pr(G_P = [X,Y]) +$$
$$\Pr(G_C = [Y,Z] \mid G_P = [X,Z])\Pr(G_P = [X,Z]) +$$
$$\Pr(G_C = [Y,Z] \mid G_P = [Y,Y])\Pr(G_P = [Y,Y]) +$$
$$\Pr(G_C = [Y,Z] \mid G_P = [Y,Z])\Pr(G_P = [Y,Z]) +$$
$$\Pr(G_C = [Y,Z] \mid G_P = [Z,Z])\Pr(G_P = [Z,Z])$$

$$\Pr(G_C = [X,X]) = 0 \times 0.25 + 0.005 \times 0.49 + 0.245 \times 0.01 + 0.01 \times 0.2401 + 0.25 \times 0.0098 + 0.49 \times 0.0001$$

$= 0.0098$

$$\Pr(G_C = [Z,Z]) =$$
$$\Pr(G_C = [Z,Z] \mid G_P = [X,X])\Pr(G_P = [X,X]) +$$
$$\Pr(G_C = [Z,Z] \mid G_P = [X,Y])\Pr(G_P = [X,Y]) +$$
$$\Pr(G_C = [Z,Z] \mid G_P = [X,Z])\Pr(G_P = [X,Z]) +$$
$$\Pr(G_C = [Z,Z] \mid G_P = [Y,Y])\Pr(G_P = [Y,Y]) +$$
$$\Pr(G_C = [Z,Z] \mid G_P = [Y,Z])\Pr(G_P = [Y,Z]) +$$
$$\Pr(G_C = [Z,Z] \mid G_P = [Z,Z])\Pr(G_P = [Z,Z])$$

$$\Pr(G_C = [X,X]) = 0 \times 0.25 + 0 \times 0.49 + 0.005 \times 0.01 + 0 \times 0.2401 + 0.005 \times 0.0098 + 0.01 \times 0.0001$$

$= 0.0001$

Q3

A patient gives a positive result to a test for a disease that has a prevalence in the population of 1 in 1 million people. The test is known to have 99.9% true positive rate and a 0.001% false positive rate. There are two factors that need to be considered within a BN; the disease status of a person and the disease test result for that person. This means the BN will have two nodes:

- Disease state, with states of diseased (D) and not diseased ($\overline{D}$)
- Test result, with states of positive (P) and negative (N)

The disease state is the parent of the test result node and will have prior probabilities of:

Disease state		
	Diseased	$\mathrm{Pr}(D) = 0.000001$
	Not diseased	$\mathrm{Pr}(\overline{D}) = 0.999999$

The test result node will have four conditional probabilities:

	Disease State	**Diseased**	**Not diseased**
Test result	**Positive**	$\mathrm{Pr}(P\,|\,D) = 0.999$	$\mathrm{Pr}(P\,|\,\overline{D}) = 0.00001$
	Negative	$\mathrm{Pr}(N\,|\,D) = 0.001$	$\mathrm{Pr}(N\,|\,\overline{D}) = 0.99999$

Below is a screen capture of the BN set up with these probabilities, showing the probability of having the disease given a test result (left).

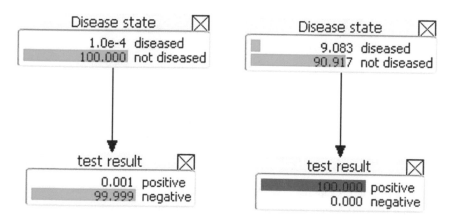

When a positive test result is instantiated (right), it can be seen that the positive test result means that the patient has an approximately 9.1% chance of being truly diseased (just as shown in Chapter 1).

Q4

In this question, we consider the architecture of a BN for Case 1 (the Drummond case) considering the defence proposition that there was no attacker at all, i.e. the complainant made a false statement.

Taking the BN seen in Figure 6.22, we delete the branch that relates to the AO struggling with C, i.e. nodes 3, 7, 13 and 15. This gives the BN architecture of:

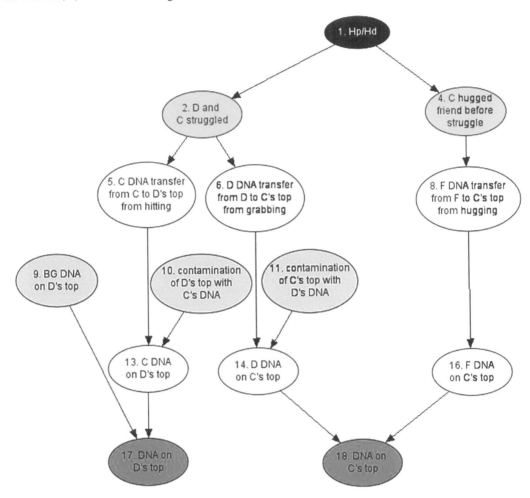

Where the only difference to conditional probability tables is in node 18 (noting that the numbering is disjointed in the BN given above, but retains node numbering as it is in Figure 6.22 for the remaining nodes), which is now:

16. F DNA on C's Top		Yes		No	
14. D DNA on C's Top		Yes	No	Yes	No
18. DNA on C's top	D+F	1	0	0	0
	F	0	1	0	0
	D	0	0	1	0
	None	0	0	0	1

Q5

In order to extend the BN for Case 1 (the attempted kidnapping case) if C and D had both been present at a party together the night before the alleged assault, we could add in nodes for a new activity 'D and C social interaction', and then two new transfers, one from C to D's top from the social interact and one from D to C's top form the social interaction (assuming they wore the same clothes during the party and at the time of the incident). Taking Figure 6.22 as a starting point, these two new transfer nodes would then be parents of nodes 13 and 14, as shown in the BN below:

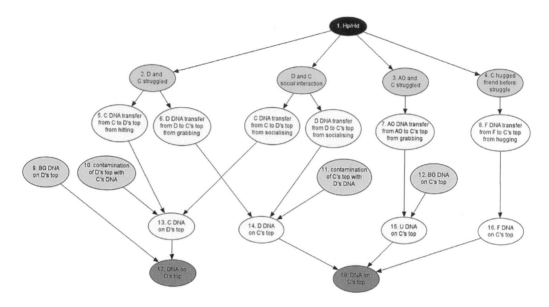

Note that this makes the assumption that the clothing worn by C and D is the same at the party as on the day of the alleged attack. It also assumes that the DNA has persisted from the social interaction the night before to the time of the alleged attack. If we didn't make the assumption of persistence, then the BN architecture could be extended further to:

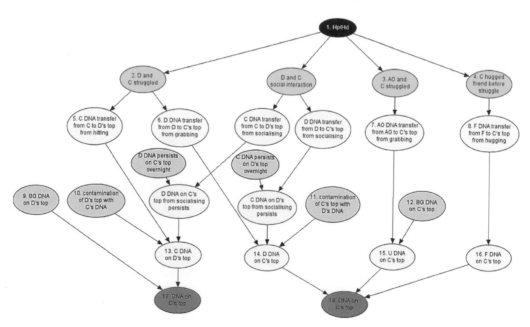

Q6

In the evaluation framework for Case 2, it was considered that if the AO had DNA from D on his hands, then there was a dependency between the transfer of AO DNA and the transfer of D DNA from AO's hands. The arc in a BN represents the dependencies of the different factors. The dependency between AO's DNA and D's DNA being transferred from AO's hands to the knife handle (when held by AO) is modelled by the arc from node 11 to node 13 in the BN from Figure 6.33. The dependency between AO's DNA and D's DNA being transferred from AO's hands to the hair (when grabbed by AO) is modelled by the arc from node 12 to node 14 in the BN from Figure 6.33.

If these dependencies were not modelled, then the model assumes that when someone who has multiple different sources of DNA on their hands touches an item that they can independently transfer the different sources of DNA. In the evaluation of Case 2, the findings were that D's DNA was found on the knife and the hair and no unknown DNA (that could have come from a potential AO). If an AO had grabbed these items, and not transferred their own DNA, having the dependency that we have just been discussing in place means that we would also expect that the AO would transfer none of D's DNA. The modelled dependencies mean that the absence of unknown DNA coupled with the presence of D's DNA provide support to *Hp* over *Hd*. If the dependencies were removed then the model allows the transfer of D's DNA from AO's hands coupled with no transfer of AO's DNA without any consideration other than independent, hand-to-object transfer probabilities. We would therefore expect that the findings would be more acceptable under *Hd* without the dependencies in place and so the *LR* would decrease when the case results were instantiated. The BN below shows this effect (where the conditional probabilities in nodes 13 and 14 are the same as those in 9 and 10).

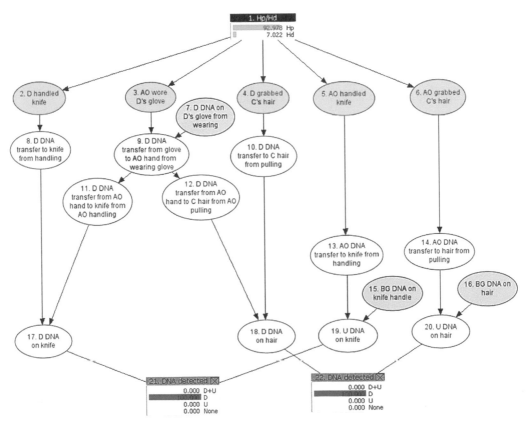

We can see that the predictions were correct, with the *LR* dropping from approximately 140 using the original BN seen in Figure 6.36 to approximately 13 using the BN above.

Q7

Nodes 15 and 16 from the BN constructed for Case 2 (the three burglars) required probability for background DNA being present on items, specifically a knife handle and hair. There are various factors that could be important to consider when assigning a probability of background. For the knife we may consider how often it was used and cleaned. We consider that it is a knife in a domestic kitchen environment rather than a commercial kitchen or one found somewhere outside and accessible to the public. If it has been recently washed, we could consider the type of washing (i.e. a rinse under cold water, or a hot dishwasher machine wash). For the hair we may consider again, how recently it has been washed and what type of washing was done. We could also consider how much time the person (whose hair we are interested in) spent out of the house, and whether they came into contact or close proximity to other people during that time or used public transport.

It is unlikely that all of the specific factors in the case will ever be represented by published literature. One possible source of data would be to carry out bespoke experiments within the forensic laboratory that matched, as closely as possible, the known details of the case. This will be the most directly informative when assigning the probability but may not be possible if time and resources are limited. In-house validation studies may hold some information, for example the validation of a tapelift, swab or extraction technique may have had some samples taken from knives to mimic real casework-type samples. It may be that published literature is the only source of information and that some adjustment of literature values can be made in order to represent the specific case details. If this is the case, then the limitation is that the literature data may be only partly applicable, and that the amount of adjustment (or how the adjustment should be made) is not immediately apparent. Any divergences from the factors surrounding the source of data used to assign a probability, and the factors in the framework of circumstances in the case where the probability is being applied, should be made transparent in the report and the evaluation.

We do not provide specific literature sources in our answer as newly published material comes out on almost a daily basis. However, we suggest you try to find some potential sources of data that would apply to the transfers in this case and comment on any differences between the literature and the case, i.e. such as differences in the published work compared to the circumstances of the case.

Q8

We wish to consider Case 3 and the alternate proposition being that someone other than D bit C. Starting with the BN in Case 3 (Figure 6.41), the BN structure would need to be changed by adding in an additional activity node for the AO biting C as shown below:

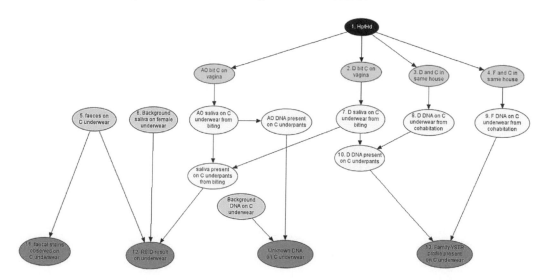

We also consider now whether we detect unknown DNA present on the underwear by adding a new results node. We could have chosen instead to alter the states of node 13 to 'Y-STR profile present on C underwear' and had combinations of the 'family' Y-STR profile, and unknown male DNA. We also add in the possibility that unknown DNA could be present due to background DNA on C's underwear (otherwise we would find that instantiating a result that showed the presence of unknown DNA would lead to complete support for *Hd*). In the above BN we assume that the AO DNA does not match the family Y-STR profile. We may wish to consider a chance match with the family Y-STR profile. This could be added to the BN architecture as shown below:

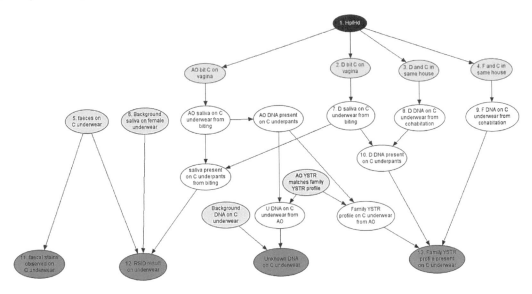

Where we can now see that the presence of AO DNA will either lead to the presence of unknown DNA or the family DNA depending on whether it matches.

Q9

Again, we consider the BN in Case 3 (Figure 6.41) and add into the BN the possibility of laboratory contamination of the underwear of C with the DNA of D. We consider that the probability of contamination is 1 in 1,000, but we must consider the probability of contamination by high amounts of DNA, low amounts of DNA or now DNA. We do so by adding a root node 'C underwear contaminated by D DNA' with probabilities:

C underwear contaminated by D DNA		
	Yes (high)	0.0001
	Yes (low)	0.0009
	No	0.999

Note that the combined probability of contamination is still 1 in 1,000, but we have split it into high and low, with high having lower probability than low. The BN structure is:

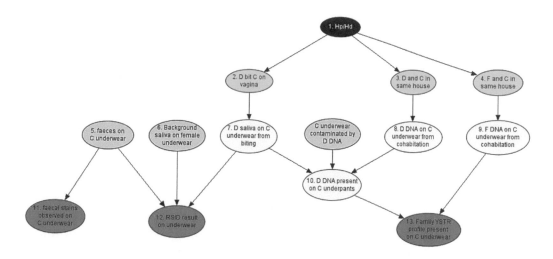

With the conditional probability table in node 10 being populated considering the same assumptions as before (i.e. that two low amounts of DNA have a probability of 0.5 of leading to high DNA and a probability of 0.5 of leading to low DNA). Recall that the instantiation of the original BN gave an *LR* of 0.19. Doing so for the BN shown above gives:

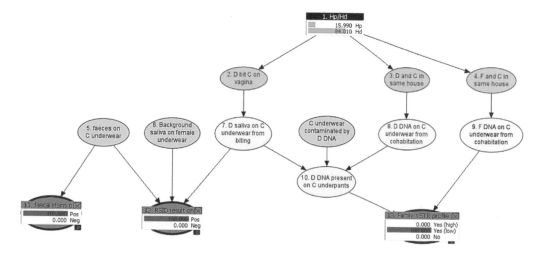

which is still approximately 0.19.

We now add in the possibility that the inheritance of the family DNA profile between F and D had a possibility of mutation (1 in 1,000). Including this into the BN requires us to add a mutation node with two children; any DNA present from F being the family Y-STR profile or any DNA present from F being an unknown Y-STR profile. As a result, we must also add in a results node that can handle the presence of unknown DNA (which we show in the BN below, but we could have also expanded the existing node 13 to include combinations of family and unknown profile). We also add in the possibility of background DNA being present on the underwear and leading to unknown DNA (otherwise the presence of unknown DNA would only be explainable by a mutation from DNA transferred by F to C's underwear). The resulting BN is shown below:

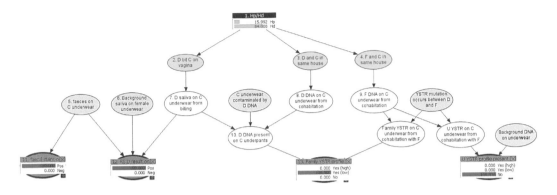

Again, we see that instantiating the results for this case, the *LR* is still approximately 0.19. The reason for the *LR* not changing is that the presence of the family Y-STR profile, either from the cohabitation or assault of C by D, or the cohabitation of C with F is a much more probable explanation than contamination or mutation. Therefore, in this instance, it would be acceptable to leave these factors out of the evaluation, and hence BN without risk that the result would be misleading.

Q10

In Chapter 5, in the supplemental information for Case 3 (Section 5.4.8) the following scenario was given:

> Mr A has been invited over to Ms B's home to perform an amateur dental check-up, which Mr A does without wearing any gloves. This is accepted by both prosecution and defence as having occurred legally and consensually. After the exam it is alleged that Mr A picked up Ms B's wallet and took out some money (without permission) and left the house. Swabs are taken of Mr A's hands two hours after the incident (before he washed his hands) to help with the issue of whether or not he has picked up Ms B's wallet. The DNA profile obtained from the swabs of Mr A's hand can be considered as a mixture of two persons corresponding to both DNA profiles of A and B.

A BN, using the seven steps in Chapter 6 could be constructed such as that shown below:

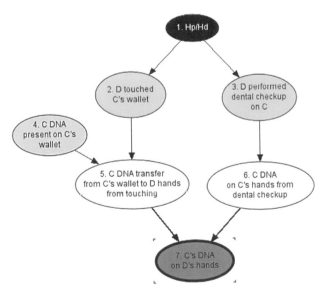

The BN that has been constructed is relatively simple and only contains seven nodes. There are only three probabilities that need to be assigned:

- The probability that C's DNA will be on his own wallet (for node 4).
- The probability that C's DNA will transfer from his wallet to the hands of D when touched (for node 5).
- The probability that C's DNA will transfer to the hands of D from D giving C a dental checkup (for node 6).

We do not cite specific literature for these probabilities (indeed, the rate at which new literature on DNA TPPR is published, the choice is likely to be outdated soon after this book has been published). We provide exemplar probability tables below:

1. Hp/Hd

Hp	0.5
Hd	0.5

1. Hp/Hd		Hp	Hd
2. D touched C's wallet	Yes	1	0
	No	0	1

1. Hp/Hd		Hp	Hd
3. D performed dental checkup on C	Yes	1	1
	No	0	0

2. D touched C's wallet		Yes		No	
4. C's DNA present on C's wallet		Yes	No	Yes	No
5. C DNA transfer from C's wallet to D hands from touching	Yes	0.2	0.0	0.0	0.0
	No	0.8	1.0	1.0	1.0

4. C's DNA present on C's wallet		
	Yes	0.99
	No	0.01

3. D performed dental checkup on C		Yes	No
6. C DNA on C's hands from dental checkup	Yes	0.99	0
	No	0.01	1

5. C DNA transfer from C's wallet to D's hands from touching		Yes		No	
6. C DNA on C's hands from dental checkup		Yes	No	Yes	No
7. C's DNA on D's hands	Yes	1	1	1	0
	No	0	0	0	1

When we instantiate the finding of C's DNA on D's hands, the resulting *LR* is approximately 1 (in fact very slightly in favour of *Hp* over *Hd*) as seen in the BN below, which is what we would expect from the circumstances of the case.

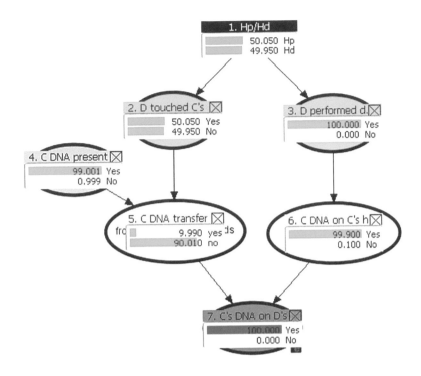

CHAPTER 7 – ANSWERS TO QUESTIONS

Q1

We are interested in the total number of fish caught on a fishing trip, where:

- The father catches 0, 1, 2, 3 or 4 fish with probabilities 0.1, 0.3, 0.3, 0.2 and 0.1
- The son catches 0 or 1 fish with probabilities of 0.5 and 0.5
- The friend catches 5, 6, 7 or 8 fish with probabilities of 0.2, 0.3, 0.3 and 0.2

The BN below uses numbered nodes to calculate the total number of fish expected to be caught from this fishing party.

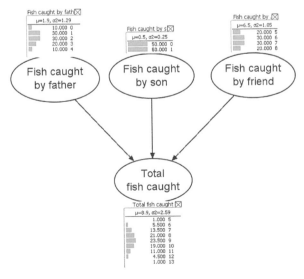

The total fish caught now ranges from 5 to 13 with probabilities calculated using the formula:
Total fish caught = Fish caught by father + Fish caught by son + Fish caught by friend

Q2

Each time a person touches a drawer handle, we expect them to transfer an amount of DNA with the associated probabilities given in the table below.

DNA Transferred (ng)	Probability of DNA Transfer
0–1	0.50
1–2	0.30
2–3	0.15
3–4	0.05

When trying to open a locked cabinet, a person is expected to contact the handle between 1 and 5 times with equal probability. In the BN construction, we will need a node for the amount of DNA transferred from someone's hands to a door handle from touching it. This node will use the DNA amounts given in the table for this question, and will be an interval node. The BN will also need a node for the number of touches, which will range from 1 to 5 and will be a numbered node. The final node will be a total DNA amount transferred node which will also need to be an interval node that will have probabilities assigned via the formula:
   Total DNA = DNA transferred per touch x number of touches
   The BN is shown below:

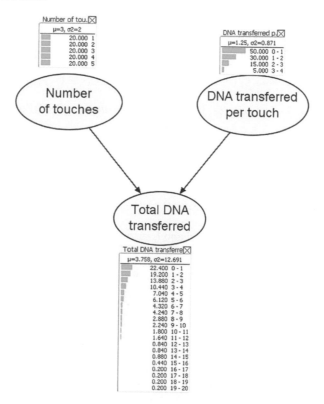

The total DNA must range up to 20 ng (which covers the highest number of touches multiplied by the highest transfer per touch $4 \times 5 = 20$).

Q3

In order to show how the BN is calculating the total amount of expected DNA in Question 2, we will use the value from the 'Total DNA transferred' node of 4–5 ng. Recall from earlier chapters the law of total probability. If we consider the number of touches as 't' and the DNA transferred per touch as 'd', then the probability of between 4 and 5 ng of total DNA being transferred can be calculated using the formula:

$$\Pr(total\ DNA = 4-5ng) = \sum_t \sum_d \Pr(total\ DNA = 4-5ng\,|\,d,t)\Pr(DNA\ transfer = d)\Pr(touches = t)$$

Given five different states for the touch node and four different states of the amount transferred node, there are 20 combinations of states and hence 20 elements to the sum of the equation above. Many of these can be discounted as the combination of amount and number of touches leads to $total\ DNA = 4-5ng\,|\,d,t) = 0$, for example consider 1 touch and 0–1 ng transferred, given these combinations of states it is impossible for 4–5 ng of total DNA to be transferred. In fact there are only four combinations of states for d and t that lead to non-zero values for $total\ DNA = 4-5ng\,|\,d,t)$, which are:

- $d = 0–1$ and $t = 5$
- $d = 1–2$ and $t = 3$
- $d = 1–2$ and $t = 4$
- $d = 2–3$ and $t = 2$

So that:

$$\Pr(total\ DNA = 4-5ng) =$$

$$\Pr(total\ DNA = 4-5ng\,|\,d = 0-1, t = 5)\Pr(DNA\ transfer = 0-1)\Pr(touches = 5) +$$

$$\Pr(total\ DNA = 4-5ng\,|\,d = 1-2, t = 3)\Pr(DNA\ transfer = 1-2)\Pr(touches = 3) +$$

$$\Pr(total\ DNA = 4-5ng\,|\,d = 1-2, t = 4)\Pr(DNA\ transfer = 1-2)\Pr(touches = 4) +$$

$$\Pr(total\ DNA = 4-5ng\,|\,d = 2-3, t = 2)\Pr(DNA\ transfer = 2-3)\Pr(touches = 2)$$

Each of the above leads to a uniform distribution, similar to the one seen in Figure 7.5 of Chapter 7. The first combination of states leads to a uniform distribution from 0 to 5, with a height of 0.2 (so that the area under the distribution is 1). Therefore:

$$\Pr(total\ DNA = 4-5ng\,|\,d = 0-1, t = 5) = (5-4) \times \frac{1}{5-0} = 0.2$$

And we know the priors for these states of d and t, respectively, have prior probabilities of 0.5 and 0.2. In a similar manner:

$$\Pr(total\ DNA = 4-5ng) =$$

$$0.2 \times 0.5 \times 0.2 +$$

$$\left[(5-4) \times \frac{1}{6-3}\right] \times 0.3 \times 0.2 +$$

$$\left[(5-4) \times \frac{1}{8-4}\right] \times 0.3 \times 0.2 +$$

$$\left[(5-4) \times \frac{1}{6-4}\right] \times 0.15 \times 0.2$$

~ 0.07 as shown in the BN image above (note that there may be slight differences in probabilities in further decimal places due to the algorithms used in software to calculate probabilities within a BN).

Q4

The amount of DNA expected to be on a firearm is given by the table below:

DNA on Firearm (ng)	Probability of DNA Amount
0–1	0.50
1–2	0.25
2–3	0.15
3–4	0.10

Each time someone wipes down a firearm, they are expected to remove one quarter of the DNA that is on it. We can construct a BN with three nodes, one interval node for the amount of DNA that is on the firearm to begin with, one numbered node for the number of times someone wipes the firearm, and one node for the remaining DNA (which will be the child of the two other nodes) and would best be represented by an interval node. The remaining DNA node will have conditional probabilities filled using the formula:

$$\text{Remaining DNA} = \text{starting DNA on firearm} \times (0.75)^{\text{wipes}}$$

The BN is shown below:

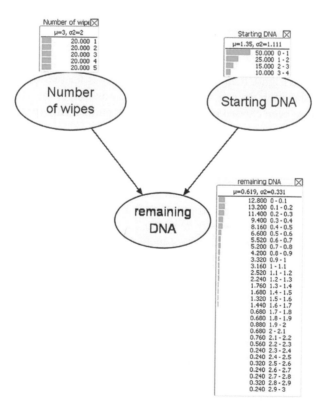

Note that there is no need for the remaining DNA node to consider ranges higher than 3 ng as this is not possible given the states of its parents. We have chosen to use a finer scale of states in the remaining DNA node than in previous examples (brackets of 0.1 ng rather than brackets of 1 ng).

We were asked how many wipe-downs a gun is most likely to have been subjected to if there was 1.6 ng of DNA detected on the firearm. This falls on a boundary between two brackets. In this case we will show the results for both below:

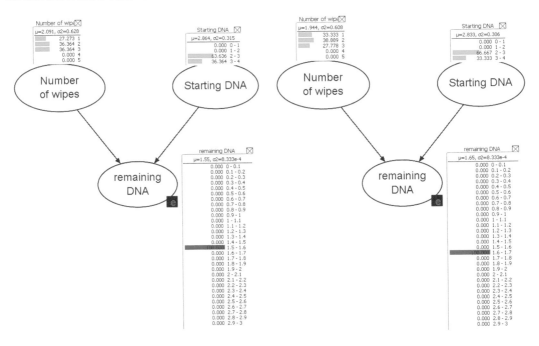

When the interval 1.5–1.6 ng is instantiated in the remaining DNA node (left), then the highest probability is for two to three wipes to have been carried out. When 1.6–1.7 ng is instantiated in the remaining DNA node (right), then the highest probability is for two wipes to have been carried out. Therefore, in combination we can say that if 1.6 ng was found, then two wipes are the most probable number to have been carried out (assuming each number of wipes to have the same prior probability).

Q5

Using the example from Section 7.2.2.1, we want to calculate the probabilities for:

Pr(4 ng < total DNA < 5 ng | transfer from object 1 = 2–4 ng, transfer from object 2 = 1–2 ng)
Pr(5 ng < total DNA < 6 ng | transfer from object 1 = 2–4 ng, transfer from object 2 = 1–2 ng)

We will be using the cumulative density function for the addition of two uniform random variables:

$$cdf(z) = \begin{cases} 0 & z \leq a+c \\ (z-a-c)^2 / [2(b-a)(d-c)] & a+c < z < b+c \\ (2z-a-b-2c)/[2(d-c)] & b+c \leq z < d+a \\ 1-(b+d-z)^2 / [2(b-a)(d-c)] & d+a \leq z < b+d \\ 1 & z \geq b+d \end{cases}$$

To calculate:

Pr(4 ng < total DNA < 5 ng | transfer from object 1 = 2–4 ng, transfer from object 2 = 1–2 ng)

The intervals we are interested in are [2, 4] and [1, 2], we need to satisfy the inequality $b + c \leq d + a$ and therefore we need to order the intervals [a = 1, b = 2] and [c = 2, d = 4]. We are interested in calculating the probability of 4–5 ng and so need:

$$cdf(z = 5) - cdf(z = 4)$$

5 is greater than or equal to $d + a$ and less $b + d$. 4 is greater than or equal to $b + c$ and less than $d + a$. Therefore:

$$cdf(z = 5) - cdf(z = 4) =$$

$$1 - \left(b + d - \{z = 5\}\right)^2 / \left[2(b-a)(d-c)\right] - (2\{z = 4\} - a - b - 2c) / [2(d-c)]$$

$$= 1 - \left(2 + 4 - 5\right)^2 / \left[2(2-1)(4-2)\right] - (2 \times 4 - 1 - 2 - 2 \times 2) / [2(4-2)]$$

- $= 0.5$

which is the same as the value shown in the corresponding cell of Table 7.2.

Next to calculate:

Pr(5 ng < total DNA < 6 ng | transfer from object 1 = 2–4 ng, transfer from object 2 = 1–2 ng)

The intervals we are interested in are [2, 4] and [1, 2], we need to satisfy the inequality $b + c \leq d + a$ and therefore we need to order the intervals [a = 1, b = 2] and [c = 2, d = 4]. We are interested in calculating the probability of 5–6 ng and so need:

$$= cdf(z = 6) - cdf(z = 5)$$

6 is greater than or equal to $b + d$. 5 is greater than or equal to $d + a$ and less $b + d$. Therefore:

$$cdf(z = 6) - cdf(z = 5) = 1 - \left\{ 1 - \left(b + d - \{z = 5\}\right)^2 / \left[2(b-a)(d-c)\right]\right\}$$

$$= 1 - \left\{ 1 - \left(2 + 4 - 5\right)^2 / \left[2(2-1)(4-2)\right]\right\}$$

$$= 0.25$$

which is the same as the value shown in the corresponding cell of Table 7.2.

Q6

In Section 7.2.2.2 the derivation of:

Pr(2 ng < transferred DNA < 3 ng | DNA on object 1 = 4–6 ng, transfer proportion = 0.66–1)

Using the cdf function:

$$
cdf(z) = \begin{cases}
0 & z < ac \\[4pt]
 & ac \leq z < bc \\[4pt]
\dfrac{z\left[\ln(z) - \ln(ac) - 1\right] + ac}{(b-a)(d-c)} & \\[10pt]
\Pr(Z \leq bc) + \dfrac{(z - bc)\ln\left(\dfrac{b}{a}\right)}{(b-a)(d-c)} & bc \leq z < ad \\[14pt]
\Pr(Z \leq ad) + \dfrac{z\left[1 - \ln\left(\dfrac{z}{bd}\right)\right] - ad\left[1 - \ln\left(\dfrac{ad}{bd}\right)\right]}{(b-a)(d-c)} & ad \leq z < bd \\[16pt]
1 & z \geq bd
\end{cases}
$$

was shown to equal 0.034, which was one of the values in Table 7.3 (the far-right column).
   Using the same method:

Pr(3 ng < transferred DNA < 4 ng | DNA on object 1 = 4–6 ng, transfer proportion = 0.66–1)

The intervals we are interested in are [4, 6] and [0.66, 1], we need to satisfy the inequality $ad \leq bc$, and therefore we need to order the intervals [$a = 4$, $b = 6$] and [$c = 0.66$, $d = 1$]. We are interested in calculating the probability of 3–4 ng and so need:

$$cdf(z = 4) - cdf(z = 3)$$

3 is greater than or equal to ac and less than bc. 4 is greater than or equal to ad and less than bd. Therefore:

$$cdf(z = 4) - cdf(z = 3)$$

$$
= \Pr(Z \leq ad) + \frac{\{z = 4\}\left[1 - \ln\left(\dfrac{\{z = 4\}}{bd}\right)\right] - ad\left[1 - \ln\left(\dfrac{ad}{bd}\right)\right]}{(b-a)(d-c)} - \frac{\{z = 3\}\left[\ln(\{z = 3\}) - \ln(ac) - 1\right] + ac}{(b-a)(d-c)}
$$

For these calculations we need to know the value of $\Pr(Z \leq z)$, at various critical points of z. These are:

$$\Pr(Z \leq bc) = \frac{bc\left[\ln(bc) - \ln(ac) - 1\right] + ac}{(b-a)(d-c)} \quad \text{(as given in Chapter 7)}$$

$$\Pr(Z \leq bc) = \frac{6 \times 0.66\left[\ln(6 \times 0.66) - \ln(4 \times 0.66) - 1\right] + 4 \times 0.66}{(6-4)(1-0.66)} \approx 0.42$$

And $\Pr(Z \leq ad)$, which can be calculated as the area under the cdf curve between $z = ac$ and $z = ad$, i.e.

$$Pr(Z \le ad) = Pr(Z \le bc) + \frac{(\{z = ad\} - bc)\ln\left(\frac{b}{a}\right)}{(b-a)(d-c)}$$

$$= 0.42 + \frac{(4 \times 1 - 6 \times 0.66)\ln\left(\frac{6}{4}\right)}{(6-4)(1-0.66)} = 0.42 + 0.02385 \approx 0.4439$$

Therefore, back to the calculation of $cdf(z = 4) - cdf(z = 3)$

$$= 0.4439 + \frac{4\left[1 - \ln\left(\frac{4}{6 \times 1}\right)\right] - 4 \times 1\left[1 - \ln\left(\frac{4 \times 1}{6 \times 1}\right)\right]}{(6-4)(1-0.66)} - \frac{3[\ln(3) - \ln(4 \times 0.66) - 1] + 4 \times 0.66}{(6-4)(1-0.66)}$$

$$= 0.040935$$

which is the value seen in the corresponding cell of Table 7.3 (far-right column, shown to two significant figures as 0.41).

Pr(4 ng < transferred DNA < 5 ng | DNA on object 1 = 4–6 ng, transfer proportion = 0.66–1)
We are interested in calculating the probability of 4–5 ng and so need:

$$cdf(z = 5) - cdf(z = 4)$$

4 is greater than or equal to ad and less than bd. 5 is also greater than or equal to ad and less than bd. Therefore:

$$cdf(z = 5) - cdf(z = 4)$$

$$= \left\{ Pr(Z \le ad) + \frac{\{z = 5\}\left[1 - \ln\left(\frac{\{z = 5\}}{bd}\right)\right] - ad\left[1 - \ln\left(\frac{ad}{bd}\right)\right]}{(b-a)(d-c)} \right\}$$

$$- \left\{ Pr(Z \le ad) + \frac{\{z = 4\}\left[1 - \ln\left(\frac{\{z = 4\}}{bd}\right)\right] - ad\left[1 - \ln\left(\frac{ad}{bd}\right)\right]}{(b-a)(d-c)} \right\}$$

$$= \frac{5\left[1 - \ln\left(\frac{5}{bd}\right)\right] - ad\left[1 - \ln\left(\frac{ad}{bd}\right)\right] - 4\left[1 - \ln\left(\frac{4}{bd}\right)\right] + ad\left[1 - \ln\left(\frac{ad}{bd}\right)\right]}{(b-a)(d-c)}$$

$$= \frac{5\left[1 - \ln\left(\frac{5}{bd}\right)\right] - 4\left[1 - \ln\left(\frac{4}{bd}\right)\right]}{(b-a)(d-c)}$$

$$= \frac{5\left[1 - \ln\left(\dfrac{5}{6 \times 1}\right)\right] - 4\left[1 - \ln\left(\dfrac{4}{6 \times 1}\right)\right]}{(6-4)(1-0.66)}$$

$$\sim 0.43$$

and the value seen in the corresponding cell of Table 7.3 (far-right column, shown to two significant figures as 0.44). Note that this slight difference comes from the fact that HUGIN has been used to create the values in Table 7.3, and HUGIN does not carry out the algorithmic calculations in the same way as described in Chapter 7. Specifically, HUGIN breaks interval nodes into a number (default 25) sub-intervals. From the HUGIN API reference manual

> For a given interval of the parent (i.e., for a specific parent state configuration), we compute probability distributions for the child, each distribution being obtained by instantiating the parent to a value in the interval under consideration. The average of these distributions is used as the conditional probability distribution for the child given the parent is in the interval state considered.

Pr(5 ng < transferred DNA < 6 ng | DNA on object 1 = 4–6 ng, transfer proportion = 0.66–1)

We are interested in calculating the probability of 5–6 ng and so need:

$$cdf(z = 6) - cdf(z = 5)$$

5 is also greater than or equal to ad and less than bd. 6 is greater than or equal to bd. Therefore:

$$cdf(z = 6) - cdf(z = 5)$$

$$= 1 - \left\{ \Pr(Z \le ad) + \frac{\{z = 5\}\left[1 - \ln\left(\dfrac{\{z = 5\}}{bd}\right)\right] - ad\left[1 - \ln\left(\dfrac{ad}{bd}\right)\right]}{(b-a)(d-c)} \right\}$$

$$= 1 - \left\{ 0.4439 + \frac{5\left[1 - \ln\left(\dfrac{5}{6 \times 1}\right)\right] - 4 \times 1\left[1 - \ln\left(\dfrac{4 \times 1}{6 \times 1}\right)\right]}{(6-4)(1-0.66)} \right\}$$

$$\sim 0.13$$

and the value seen in the corresponding cell of Table 7.3 (far-right column, shown to two significant figures as 0.12, which again is slightly to the value in Table 7.3 for the reasons previously mentioned).

Q7

An experiment to see how much DNA can be recovered from struck matches was performed. In ten experiments, the amounts of DNA obtained (in ng) were 0.094, 0.111, 759.455, 15.397, 602.348, 3.002, 0.727, 0.609, 0.002 and 875.931.

The first point to consider is what distribution should be used to model the data. Typical for DNA is to model the log transformation of the DNA amount with a normal distribution. If the data is not as well studied as DNA amounts, then some investigation into the best transformation or distribution will need to be carried out. Taking the values above, their log10 transformation values are:

$$-1.0268721, -0.9546770, 2.8805020, 1.1874361, 2.7798475, 0.4774107, -0.1384656, -0.2153827,$$
$$-2.6989700 \text{ and } 2.9424699$$

The density of these values can be seen to be approximately normal:

start bracket	end bracket	probability
-6	-5	0.00161
-5	-4	0.007061
-4	-3	0.023682
-3	-2	0.060774
-2	-1	0.119344
-1	0	0.179355
0	1	0.206294
1	2	0.181607
2	3	0.12236
3	4	0.063093
4	5	0.024895
5	6	0.007515

The mean of the values is 0.5233299 and the standard deviation is 1.9119. If we plot a normal distribution using these values over the density, we can see a reasonably close alignment:

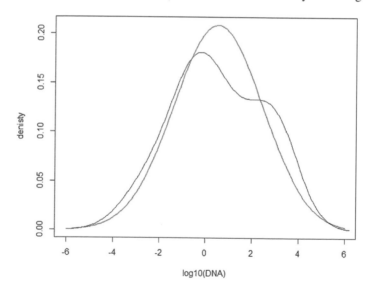

In R the commands to obtain these values are:

```
DNA <- c(0.094, 0.111, 759.455, 15.397, 602.348, 3.002, 0.727, 0.609,
0.002, 875.931)
logDNA <- log10(DNA)
mean(logDNA)
sd(logDNA)
```

```
plot(x=seq(from=-6, to=6, by=0.1), y=dnorm(x=seq(from=-6, to=6, by=0.1),
mean=mean(logDNA), sd=sd(logDNA)), type='l', col="red", ylab="denisty",
xlab="log10(DNA)", main="")
lines(density(logDNA))
```

In order to use these values in a table of a BN, they need to be discretised. The choice of bracket sizes is somewhat arbitrary, but we will use brackets for log(DNA) of width 1, and range from −6 to 6. The probability that is assigned to each bracket is the area under the normal distribution between the bracket boundaries. Some BN construction software will automatically be able to calculate the probability for intervals nodes with just the mean and standard deviation of the normal distribution provided. However, these values can also be calculated manually. For example, in R we can obtain them by commands:

```
start_brackets <- seq(from=-6, to=5, by=1)
end_brackets <- seq(from=-5, to=6, by=1)
prob_brackets <- pnorm(end_brackets, mean=mean(logDNA), sd=sd(logDNA))-
pnorm(start_brackets, mean=mean(logDNA), sd=sd(logDNA))
node_table <- cbind(start_brackets, end_brackets, prob_brackets)
colnames(node_table) <- c("bracket start", "bracket end", "probability")
node_table
```

and these produce the table:

Start Bracket	End Bracket	Probability
−6	−5	0.00161
−5	−4	0.007061
−4	−3	0.023682
−3	−2	0.060774
−2	−1	0.119344
−1	0	0.179355
0	1	0.206294
1	2	0.181607
2	3	0.12236
3	4	0.063093
4	5	0.024895
5	6	0.007515

One thing to note is that because we have started at −6 and ended at 6 (and the normal distribution goes beyond these points), the probabilities do not sum to 1. Because we have chosen bracket extremes that have a low probability, the sum of the probabilities is close to 1 (it is in fact in this instance approximately 0.998) and is not likely to have a practical effect on the result.

The other point to note is that the extremes of the brackets stray into areas that are quite unlikely to be reached in real life. For example, the upper bracket ends in 6, which equates to 1 million ng of DNA. If we wished to address both these issues, we could consider restricting the range of the brackets to values that are sensible and extend the extremes to infinity.

```
start_brackets <- c(-Inf, -3, -2, -1, 0, 1, 2)
end_brackets <- c(-3, -2, -1, 0, 1, 2, Inf)
prob_brackets <- pnorm(end_brackets, mean=mean(logDNA), sd=sd(logDNA))-
pnorm(start_brackets, mean=mean(logDNA), sd=sd(logDNA))
node_table <- cbind(start_brackets, end_brackets, prob_brackets)
```

```
colnames(node_table) <- c("bracket start", "bracket end", "probability")
node_table
```

which produces the table:

Start Bracket	End Bracket	Probability
−Inf	−3	0.032676
−3	−2	0.060774
−2	−1	0.119344
−1	0	0.179355
0	1	0.206294
1	2	0.181607
2	Inf	0.219951

Where the first bracket can be thought of s representing <0.001 ng and the final bracket can be thought of as representing >100 ng.

Q8

A person (D) is accused of attempting to break into someone's (C's) home. D went around and tried to open each of the three doors by the handle, but all were locked and so D left and is alleged to have spray-painted the wall as he did. C calls the police, who attend the next day. Each door handle is swabbed and submitted for DNA profiling. Overnight it has rained, and it is known from studies that there is a probability that this may wash away the DNA. It is also known that door handles have a high probability of possessing background DNA. Police recognise the spray-paint tag of D and arrest him. D denies ever having been near the house and through discussions you determine that there is no explainable route for indirect transfer of D's DNA. In your evaluation you wish to consider that DNA matching D's profile could come to be on the door handles either through contamination of the door handle samples with D's reference or because an alternate offender has a matching DNA profile.

The BN constructed to evaluate this scenario is shown below:

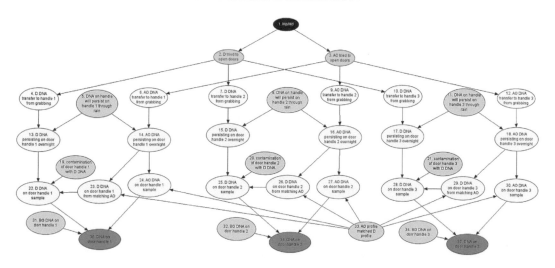

Some points to note about this BN are:

- By having a separate contamination node for each handle sample, we assume that contaminations are an independent event (i.e. as opposed to a single large event that would contaminate all the samples at once)
- We assume that the probability of DNA transfer to each handle is independent

The BN structure given may look complex (it does have 37 nodes!) but is actually a relatively simple BN structure that is repeated three times (once for each handle). The repeated structure is shown below, circled in red:

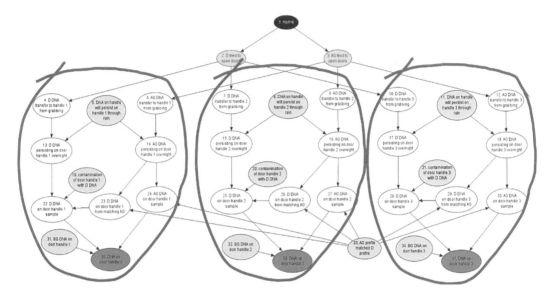

The class network that could be designed to replace these repeated structures is shown below. It has three input nodes:

- D tried to open doors
- AO tried to open doors
- AO profile matched D profile

And one output node:

- DNA on door handle

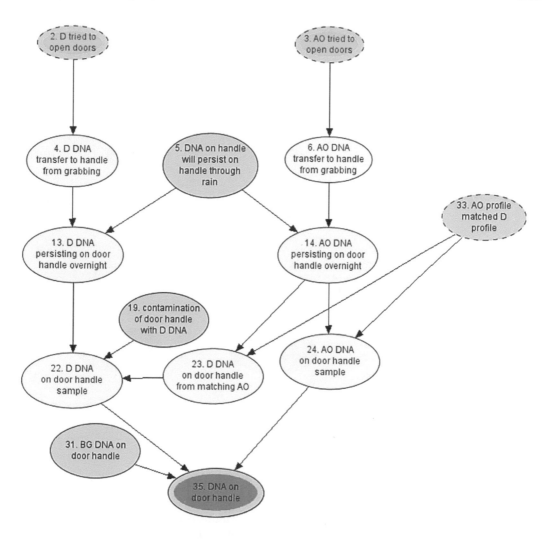

By setting up a class network in this way, we make the assumptions that:

- The contamination probabilities between all three samples are the same (this is a common assumption, but may not be the case if different sampling devices were used on the different handles, or they underwent different extraction techniques)
- The probability of DNA persistence on all three doors is the same (again, this may not be the case if the doors were in different environmental conditions)

Q9

Constructing the BN from Q8 with class network structure creates the following OOBN:

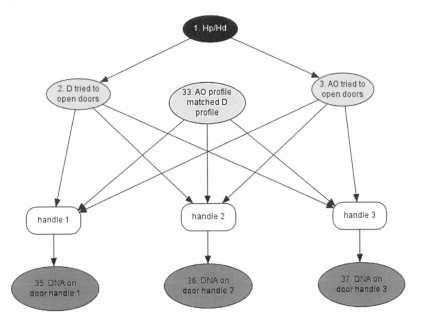

Note the very much visually simplified BN structure, as all the repeated structures in the BN have been encapsulated into a handle class network.

Q10

In the OOBN shown in Q9, the probability of matching profiles between D and AO is on the main layer of the OOBN and passed into the handle class networks. The reason for this is that DNA on a handle from an AO has a probability of matching D, but this is not independent for each handle, i.e. if AO's DNA on handle 1 matched D then so too would AO's DNA on handle 2 match D. Using a match probability for D's profile of 1 in 1 million, if the node for the probability of matching AO and D profiles were within each class network then for the AO to match D on all three handles, the probability would be 1 in 1 million million million.

Contamination on the other hand is within the handle class network, because we do treat each contamination as an independent event, i.e. if the laboratory rate of contaminating samples was 1 in 1,000, then for all three samples to be contaminated we would expect this to occur with a probability of 1 in 1 billion.

CHAPTER 8 – ANSWERS TO QUESTIONS

Q1

A particular transfer event was experimentally found to occur in 3 of 100 instances. We wish to use a Dirichlet distribution to resample the counts for this experiment and graph the distribution. This can be achieved using R to first sample 100 values from a Gamma(3, 1) and adding one to each (for the prior count):

```
positive <- rgamma(100,3,1) + 1
```

then sample 100 values from a Gamma(97, 1) and adding one to each (for the prior count):

```
negative <- rgamma(100,97,1) + 1
```

then normalising to create probabilities that have been drawn from a Dirichlet distribution with a uniform prior:

```
resampled_probabilities <- positive/(positive + negative)
```

the histogram can then be generated with command:

```
hist(resampled_probabilities)
```

which produces:

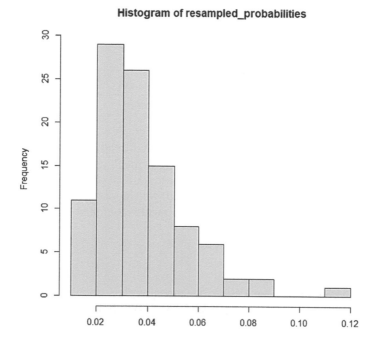

Q2

Now we imagine that the transfer experiment in Question 1 was able to carry out 1,000 tests and found 30 transfers occur. Using a Dirichlet distribution to resample the counts for this new number of experiments produces the following distribution (with R code shown prior):

```
positive <- rgamma(100,30,1) + 1
negative <- rgamma(100,970,1) + 1
resampled_probabilities <- positive/(positive + negative)
hist(resampled_probabilities, xlim=c(0, 0.12))
```

Histogram of resampled_probabilities

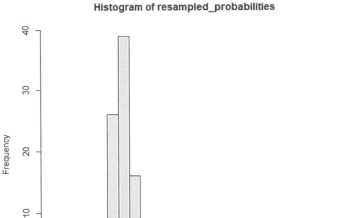

Note that in the last line of code given just before the graph we have specified the *x*-axis range to be the same as it was in Q1 so that the graphs are more directly comparable. By doing this, we can see that the distribution of resampled values in Q2 is much tighter around the experimentally observed probability than those for Q1. The reason for this is that while the ratio of positive to negative results is the same between Q1 and Q2, there are 10-fold more samples in Q2 to inform this ratio and so we have more belief that repeating the experiment would lead to the same result. In other words, there is less sampling variability.

Q3

In Chapter 6, a BN was developed for Case 2: The three burglars (shown in Figure 6.33). Adding in sensitivity nodes (as discussed in Section 8.3) leads to the BN shown below:

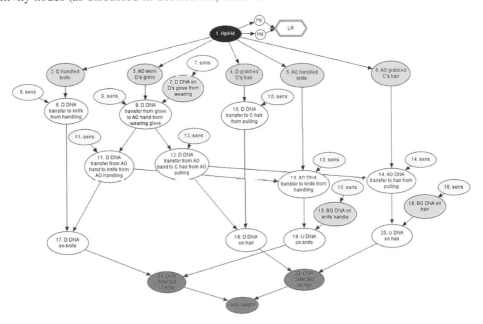

A small point of complexity arises when considering how the sensitivity should be applied to nodes 13 and 14. We have made the choice that the conditional probability for when D DNA has been transferred to nodes 11 and 12 is not altered (i.e. it remains as 0.9 for yes and 0.1 for no) and the sensitivity applies only when D DNA was not transferred to nodes 11 and 12.

Q4

Using the sensitivity prepared BN in Question 3, the results of carrying out a sensitivity analysis on the data underlying the different nodes in the BN is shown below in a series of graphs. The case values are marked with a blue dashed line.

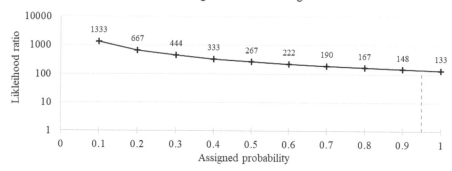

node 7. D DNA on D's glove from wearing

node 8. D DNA transfer to knife from handling
node 11. D DNA transfer from AO hand to knife from AO handling

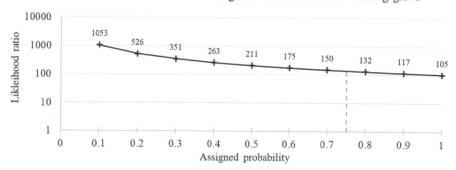

node 9. D DNA transfer from glove to AO hand from wearing glove

Note that node 8 and node 11 use the same underlying data and so have to be changed together. Doing so means that the *LR* does not change. This makes sense as either D will leave DNA or AO will transfer D's DNA with the same probability. Hence, the change in one is cancelled out by the same change in the other. We show below what the result would be if only the probability I node 8 were varied, which does not make sense to do, but importantly incorrectly represents the sensitivity of the evaluation to this underlying data.

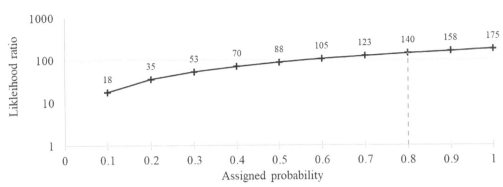

node 8. D DNA transfer to knife from handling

The same dual use of data occurs in nodes 10 and 12, and similarly the *LR* is completely insensitive (graph not shown). Also the *LR* is completely insensitive (graphs not shown) to the data underlying nodes 13, 14, 15 and 16 due to the fact that instantiation means that their probabilities are not part of the final evaluation (or better put, they are cancelled out in calculation of the *LR*).

Q5

Having completed the sensitivity analysis from Question 4, the *LR* is most sensitive to the data underlying nodes 7 and 9 approximately equally. This makes sense because in both cases, the evaluation critically relies on these events occurring in order for *Hd* to be possible. For example, if DNA from D has a very low probability of being present on his own gloves to begin with (node 8), then *Hd* is a very bad explanation of the observations. Similarly, if D's DNA has a very low probability of being transferred to AO hands from D's gloves, then again *Hd* is a very bad explanation of the observations.

For this evaluation we believe the *LR* is robust because while we have varied the probabilities across the range of [0.1, 1] a sensible range for the probabilities in nodes 7 and 9 is much tighter around the values suggested by the literature, i.e. we would expect a reasonably high probability of transfer in both cases, and it is not until the probability gets low that the *LR* rises dramatically from the reported value.

Q6

In Section 6.8, a BN was constructed for Case 3, the family assault. The figure below shows the BN set up for a sensitivity analysis.

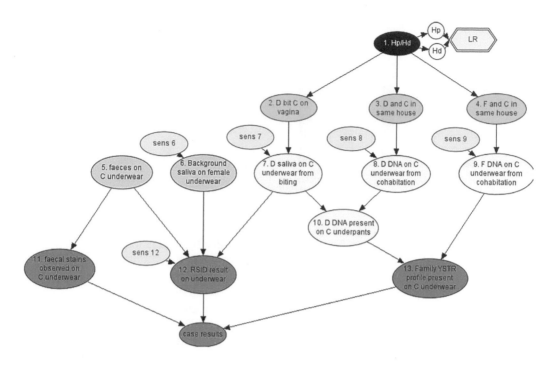

Note that the sensitivity nodes for nodes 6 and 12 are in the form of taking values from 0 to 1 (as these nodes are binary), while the sensitivity nodes for nodes 7, 8 and 9 have the extreme left to extreme right states so that they can handle the multi-state data in the node to which they apply.

Q7

Using the BN from Question 6, we wish to explore the sensitivity of the *LR* to the amount of saliva transferred during biting and from cohabitation. We use the following probabilities for the states of the sensitivity nodes 7, 8 and 9:

Distribution		Beta (1,5)	Beta (2,5)	Beta (1,1)	Beta (5,2)	Beta (5,1)
Category		Extreme Left	Left	Uniform	Right	Extreme Right
State	High	0.86831	0.64883	0.33333	0.01783	0.00412
	Low	0.12757	0.33333	0.33333	0.33333	0.12757
	None	0.00412	0.01783	0.33333	0.64883	0.86831

The graphs below show the sensitivity of the *LR* to the data in nodes 7, 8 and 9 (remembering that the same data underlies nodes 8 and 9 and so these are changed together). As before the case *LR* is marked with a blue dashed line.

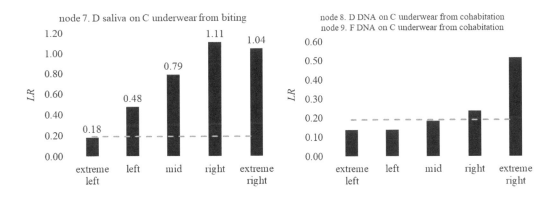

node 7. D saliva on C underwear from biting

node 8. D DNA on C underwear from cohabitation
node 9. F DNA on C underwear from cohabitation

In this case the *LR* is not very sensitive to either sets of data, i.e. the *LR* does not change much over the full range of values trialled. Therefore, we can say that the evaluation is robust.

Q8

In Figures 7.24 and 7.25, a BN was shown that demonstrated an object-oriented architecture to evaluate the findings in the Drummond case. We wish to use this OOBN and set it up for a sensitivity analysis to the data underlying the 'DNA transfer to top' node within the class network (shown in Figure 7.24). There are different ways in which this can be done. One is to add a sensitivity node to the 'DNA transfer to top' node within the class network (shown in Figure 7.24) and then instantiate this (in each instance of the class network) to carry out the sensitivity analysis. The class network with the sensitivity node would become:

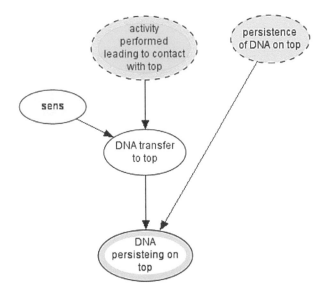

If there are numerous different class networks (or nodes on the main layer of the OOBN), then it may be desirable to make the sensitivity nodes within the class networks input nodes, and have sensitivity nodes on the main layer of the OOBN, so that all sensitivity nodes can be instantiated from the main layer, without the need for expanding individual class networks and instantiating nodes within them. This would be achieved with the following structure:

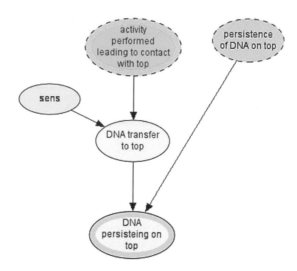

A final option that could be used in the case of an OOBN that had many instances of a class network would be to change the probabilities assigned within the class network. Changing these probability values once in the class network would then apply them to each instance within the OOBN. This method would only become preferable if the number of instances of the class network meant that it was quicker to modify conditional probability tables for each probability trialled rather than instantiating multiple instances of sensitivity nodes.

Regardless of which method is used, the sensitivity of the *LR* to the probability assignment is shown in the graph below:

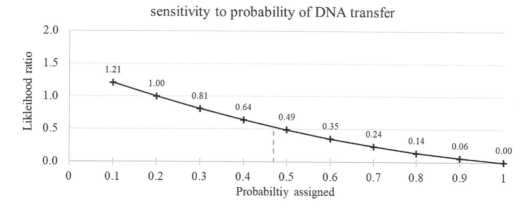

The *LR* is not very sensitive to the data except when the probability of transfer approaches 1, which is not a realistic value for it to take. Therefore, we feel the evaluation is robust.

Q9

Using the BN for Case 2 (in Figure 6.33, the three burglars), we wish to explore how sensitive is the *LR* to the presence of unknown DNA on the samples taken from the case. This can be achieved by instantiating different results for the knife and hair sample, with or without unknown DNA present. We will assume that D is still present on both items. Such a sensitivity analysis may be considered if there were some ambiguities around the presence of unknown DNA (i.e. peaks that are difficult to determine as to whether they are artefactual or allelic on the electropherogram, sub-analytical information that suggests an additional contributor, large peak imbalances that suggest an additional contributor, etc.)

The table below shows the effect of these instantiations on the result:

State of Node 21 (DNA Detected on Knife)	State of Node 22 (DNA Detected on Hair)	LR
D	D	140
D + U	D	7
D	D + U	7
D + U	D + U	0.4

We can see above that the *LR* is quite sensitive to the presence of unknown DNA. If unknown DNA was detected on either of the items then the *LR* drops from 140 to 7, and if unknown DNA was present on both items then the *LR* favours *Hd* over *Hp*. It may be decided that the BN needs to be altered to account for this sensitivity. For example, two ways in which we may wish to expand the BN are:

1) Take into account whether the same unknown DNA was present on both items (so in the case that D + U was found on the knife and the hair the BN can consider whether this is the same person such as an AO)
2) Take into account DNA amounts. In reality we know that the absence of any unknown DNA and the presence of a very small amount of unknown DNA should make very little difference to the evaluation. Part of the reason for the sensitivity seen in this evaluation is that it is dealing with the presence or absence of unknown DNA in a binary fashion, i.e. a single small unknown peak, just above the analytical threshold has the same effect as if the profiles were dominated by unknown DNA.

Q10

We wish to expand the BN in Figure 6.33 (for Case 2) to take into account the fact that unknown DNA might be present in the samples, but undetected by the DNA-profiling system (through stochastic effects). One way this could be achieved is to add in a node between nodes 19 and 21 that considers whether any unknown DNA on the knife will be detectable. Similarly adding a node between nodes 20 and 22 that takes into account whether any unknown DNA on the hair would be detectable. The architecture of this solution is shown below:

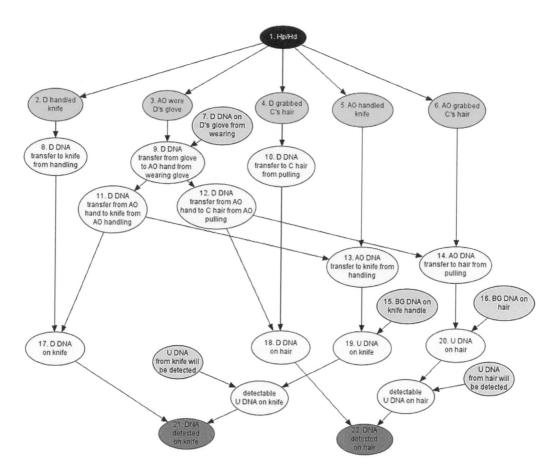

Adding these nodes specifically relates to recovery, i.e. the R in TPPR. Another alternative could be to adjust the transfer probabilities in the original BN for nodes 19 and 20 to reflect a lower probability of the DNA being transferred, persisting and being recovered.

Note that if both new root nodes ('U DNA from knife will be detected' and 'U DNA from hair will be detected') are instantiated to yes, then this returns the original *LR* of 140. If we were to provide prior probabilities for these two root nodes that there is a probability of 0.9 that unknown DNA would be detected if present, then the *LR* (when instantiating the state 'D' for both results nodes) leads to an *LR* of approximately 20. Lower than 140 due to the fact that there is additional uncertainty being taken into account in the evaluation around the ability to observe the results.

CHAPTER 9 – ANSWERS TO QUESTIONS

Q1

While the observation of spermatozoa on the microscope slide may provide strong support for the presence of semen in the sample, there are several aspects that need to be considered in the interpretation of the result. For example:

- Contamination of the slide. Anecdotally, spermatozoa have been transferred to the microscope objective through careless handling of the slides. Subsequent slides may be scored positive while no semen is present in the sample.

- Type of staining used. Some staining provides less resolution between (for instance) yeast cells and spermatozoa, allowing positive scoring in situations where no spermatozoa are present.
- Inter-observer variation. Depending on the level of training and experience, as well as other human factor aspects, the probability of spermatozoa being scored or missed during the examination may vary (see S. Tobe, L. Dennany, M. Vennemann, An assessment of the subjectivity of sperm scoring, *Forensic Science International* 251 (2015) 83–86).
- Trace context. Anecdotally samples have been scored positive for human spermatozoa in animal abuse cases. Swabs taken from the animals have been shown to contain non-human spermatozoa, which were not identified as such by the analyst.

These examples of false positives may occur. However, depending on the case circumstances, analysis methods used, laboratory procedures and knowledge on training and expertise of the analyst, one may *decide* that the findings confirm the presence of human semen. This leap of faith may be justified given a specific set of case circumstances. The decision is however personal and based on case context. The test itself can therefore not be considered confirmatory, the *results* for the test and the *interpretation* thereof may be considered confirmatory.

Q2

There are effectively three options to report the test result. One can decide to provide a *technical*, *investigative* or *evaluative* opinion. A technical statement could read: 'A test for PSA was performed. The test result was positive'. No interpretation of these findings is provided. An investigative opinion would provide some interpretation of the findings: 'A test for PSA was performed. The test result was positive. This is an indication for the presence of seminal fluid in the sample'. The *indication for the presence of seminal fluid* is an interpretation. Based on the case context and positive test result, the presence of seminal fluid is considered the most probable *explanation* for the findings.

The third option would be to properly evaluate the findings given propositions at the source level. For instance:

H1: The sample contains seminal fluid
H2: The sample does not contain seminal fluid
I: Any other cell type or body fluid may be present in the sample

I consider the findings 50 times more probable if the sample contains seminal fluid than if it does not.

The support for the probability assignments needs to be transparent. Hence the report would need to contain a description of the sources of information used (e.g. literature, experiments performed, personal expertise, etc.)

Q3

The statement is an investigative opinion. The scientist has communicated that, based on the information available to them at the time and on the test result, they concluded that the presence of saliva in the sample was the best explanation for the findings.

As reviews are generally performed in the evaluative phase of a case, such a statement may no longer be opportune. If the presence or absence of saliva in the sample is a crucial piece of information for the court, a proper evaluation of the findings may be considered relevant. This may be commented on (and – depending on the questions put to you – also performed).

Alternatively, as the presence or absence of a cell type is generally only relevant if specific scenarios at the activity level are being discussed, one may comment that an evaluation of the findings given propositions at the activity level is the proper way to address the findings.

Q4

There are two paths that could lead us to the conclusion that the sample contains human blood. The first is that the test is conclusive in that a positive result can only be explained by the presence of human blood (to the exclusion of all other possibilities). The second path, if we cannot reasonably exclude other causes for a positive test result, would be to also consider the prior probability that human blood is present in the sample. Given this prior and the *LR* for the test result, we may assign a posterior probability to the presence of human blood and, based on that and considerations of 'utilities' (see: Biedermann, A., Bozza, S., & Taroni, F. (2008). Decision theoretic properties of forensic identification: underlying logic and argumentative implications. *Forensic Science International*, 177(2–3), 120–132) like for instance the impact on the case (how bad is it if we are wrong? E.g. What is the cost of being wrong?), we reach a decision on the presence of human blood in the sample.

Walking the first path we thus need to consider if we can exclude (based on an extremely low prior probability) other causes for a positive test result. The *HemaTrace* test has been shown to give a positive result in the presence of blood of human origin, but also of other higher primates and ferrets (see Johnston, S., Newman, J., & Frappier, R. (2003). Validation study of the Abacus Diagnostics ABAcard® HemaTrace® membrane test for the forensic identification of human blood. *Canadian Society of Forensic Science Journal*, 36(3), 173–183). The first assumption that needs to be made is that the validation of the system is sufficient, in that it covers all possible sources for false positive results (e.g. only blood from ferrets or higher primates may result in a positive test with the exclusion of all other species). Have for instance all species of *mustelids* (the group of species to which the ferret belongs), and all other species groups that conceivably result in a positive test been tested? Secondly, we will need to assume that the sample cannot contain blood from one or more ferrets or higher primates (and by extension from other species that have not been included in the validation which could potentially return a positive test result).

A third assumption is that we can exclude causes for a false positive result (like observer error, sample mix-up, contamination, etc.). If these three assumptions can be made, we may report that *given these assumptions* we conclude that the positive test result confirms the presence of human blood in the sample. It would evidently depend on the case circumstances if such assumptions were valid.

If any of these assumptions on other sources or causes for a positive test cannot reasonably be excluded, we need to extend our probabilistic assessment of the findings. In such instances, we may need to consider case information to assign a prior probability to the presence of human blood and other sources of blood in the sample. As was discussed in Section 9.3.2 of this chapter, assigning a prior (and thus also the decision on the presence of blood in the sample) should be left to the court. The scientist should thus report on the probability of their findings (the positive test result) given case-relevant propositions.

Q5

Assumptions would relate to the prior probability of blood from the defendant being present (did the defendant have open wounds on their hand or other parts of their body at the time?). Similar considerations need to be made for the female contributor. As these are outside the domain of expertise of the forensic biologist, such considerations are best left up to the court. However, the findings need not be evaluated at the source level before progressing to evaluating the findings given case-relevant propositions at the activity level.

Q6

One could either consider these findings non-informative with regard to the issue of whether or not spermatozoa are present in the sample, or informative in that they support the absence of spermatozoa. If the findings are considered uninformative one may consider that the sample is degraded (e.g. spermatozoa may have been damaged and cells may have lysed with the epithelial fraction during the differential extraction) or with only few spermatozoa (hence no enrichment is expected). If one considers the findings informative, one would expect based on the sample and case context that enrichment is observed if spermatozoa are present. Again, these decisions are very much based on case context.

Q7

The node 'C faeces on C underwear from wearing' has a single parent node ('C wore C underwear') which has two states ('Yes' and 'No'). The probability table would look like this:

'C faeces on C underwear from wearing'

'C wore C Underwear'	Yes	No
Yes (High)	*P(faeces high|worn by C)*	0
Yes (Low)	*P(faeces low|worn by C)*	0
None	*P(no faeces|worn by C)*	1

Factors to consider would be:

- Duration of wear of the underwear by C,
- Age of C (both for very young or very old individuals our expectations of finding faeces in the underwear may be different compared to adolescents or other adults),
- Specific medical conditions of C,
- etc.

Such information may be obtained from documentation in the case file. Specific aspects (like interpretation of medical conditions) may be outside of the field of expertise of the forensic biologist and may require consultation with experts in such fields.

Q8

The findings from an RSID saliva test and from DNA testing are dependent on each other (as saliva contains DNA). To determine if they are *conditionally* dependent, however, requires careful consideration of the conditioning information.

Would our expectation of finding DNA of C or the family Y-STR profile change if we have knowledge of the RSID saliva test result, *given* the conditioning information? As under both H1 and H2 we have modelled more than one specific route for transfer, persistence and recovery of both amylase and DNA, it may be argued that our expectation does not change, regardless of whether the RSID saliva test is positive or negative. Hence, the test results may be considered conditionally *independent*.

Q9

We may consider two routes for contamination of the underwear;

- The use of the toilet in the facility during which both latent DNA and body fluids from other sources may be transferred to the underwear
- The speaking by the police officer over the underwear during which saliva of this officer may transfer to the underwear through aerial transfer

As the toilet in the facility is used by all staff, we are unlikely to be able to obtain reference samples from all those that may have deposited their biological material there. Hence, we may consider this aspect to increase the probability of background biological material (including saliva, faeces, vaginal mucosa, etc.) transferring to the underwear. As a consequence, we may assign a higher probability of finding background amylase on C's underwear (node 23).

With regard to the potential contamination of the underwear with saliva from the police officer, our decision will depend on whether or not a reference DNA profile from this officer is available.

If no such reference is available we may consider, like with the toilet visit, a higher probability of finding background amylase. If a reference DNA profile is available, we may model the transfer, persistence and recovery of the contaminating saliva specifically.

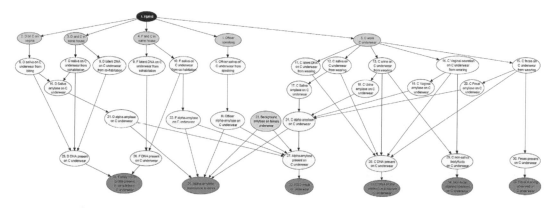

Nodes I–III have been added which model this route of transfer of saliva specifically. Node III connects to node 27 ('Alpha-amylase present on C underwear') as well as to the association test result node 36 ('Alpha-amylase association to donor'). The latter will thus gain a number of additional states which include the police officer.

Q10

Contamination may occur at any stage of the forensic process. In Question Q9 we discussed two routes for contamination of the underwear at the police station. Others may occur through improper packaging (like packaging the underwear with other clothing from the complainant, or improperly sealed packaging which may expose the underwear to outside sources of biological material during storage and transportation). Further sources may occur at the laboratory when the item is unpacked and examined (for instance contamination through contact with the outside of the packaging, or improper cleaning of the examination table, improper use of personal protection equipment like gloves or facemasks).

In this evaluation, given the disputed activity of biting by the defendant and the results (family Y-STR profile and positive amylase test), we are primarily concerned with the contamination of the underwear with other sources of amylase and DNA from sources that result in the family Y-STR profile. Such routes that are considered may be modelled specifically (as was done for the speaking by the police officer while securing the underwear in Q9) or may be introduced as 'root nodes' to the relevant TPR or findings nodes. Whether or not these factors impact on the outcome of the evaluation may be explored with sensitivity analyses. However, as for both the presence of amylase on the underwear and the presence of the family Y-STR profile several routes have been modelled and (generally) the probability for contamination to occur is quite low (at least some orders of magnitude lower than the other routes of TPR considered), one may decide to exclude other sources for contamination from the model. Such simplifying assumptions were discussed in Chapter 5 (Section 5.2.2).

CHAPTER 10 – ANSWERS TO QUESTIONS

We suggest that to prepare for court you compile a list of court-style questions and answers. In Chapter 10, we gave ten examples of questions that the authors have faced in court. We provide here some pointers to consider when formulating full answers to the example questions.

Q1)

- Assumptions need to be made on aspects of the case circumstances that are crucial to the evaluation, but about which no information is available (or is uncertain).
- Wearing of gloves by the perpetrator is such an aspect.
- In general, wearing of gloves may reduce the probability of transfer of DNA of the wearer through their hands.
- If in this case, it is relevant to assume that gloves were worn, additional assumptions need to be made on the type of gloves, their history, etc.
- Based on a different set of conditioning information I will be able to re-evaluate the findings and calculate the weight of evidence.

Q2)

- All assumptions need to be considered by the court. If any of them are considered not valid, this may affect the weight of the evidence in my conclusion.
- I made this assumption as I understood that the defendant did not dispute the presence of their DNA in the samples, but rather the mechanism by which it got where it was recovered from.
- If the presence of DNA is disputed, I can re-evaluate the findings under different assumptions. I may also consider the weight of the evidence supporting the presence of their DNA in my evaluation.

Q3)

- A BN is a graphical representation of statistical formulae.
- These formulae are based on the generalised laws of probability and can be used to calculate with conditional probabilities.
- BNs are widely used for this purpose in the forensic domain and, beyond that, in a.o. medicine, risk assessment, etc.

Q4)

- I have used software to construct the BN.
- This software is a commercial product and has been validated by the developer.
- In our laboratory I have performed a number of tests to check whether the software performs correctly. This I have done by creating frequently used network structures (idioms) and manually checking the calculations that were made.
- The model I constructed for this case is based on a proposed template that is used in forensic biology.
- The model has been checked by a second expert/statistician, as is the standard operating procedure in our laboratory.

Q5)

- In my evaluation of the findings, I have considered the case circumstances as they have been made known to me.
- Based on this information, I considered all relevant routes for direct and indirect transfer of DNA.
- I also used this information in my assignment of probabilities to those DNA TPPR events.
- In the construction of the model and the assignment of the probabilities, I have made assumptions on uncertain or unknown aspects of the case circumstances. These are listed in my report.
- I am happy to discuss these (simplifying) assumptions, and the reasons for making them, in more detail.

Q6)

- In general, different aspects of the DNA analysis results may be informative (cell type, location, amount of DNA, etc.)
- In this case I have focused on the presence or absence of DNA since I have found no relevant scientific literature to support probability assignments on the amount of DNA.
- It is my expectation that the amount of DNA does (not) affect the weight of the evidence in this case, because …
- If so required I am able to perform bespoke experiments to obtain data on TPPR of amounts of DNA.

Q7)

- I have used these data as I consider them reliable and informative for the issue in this case.
- Other studies support the general outcome of this particular study, namely.
- Based on sensitivity analyses that I have performed I consider the weight of the evidence robust for this probability assignment.

Q8)

- X number
- Prior to this, I have retrospectively worked through X number of cases from our casework workflow to gain experience in these matters.

Q9)

- Provide CV with training received.
- Gain experience with DNA TPPR studies, possibly by collaboration with others.
- Perform case file studies on success rates.

Q10)

- I have reported on the probability of the findings given the propositions.
- I have not commented on the probability of the propositions themselves.
- The latter is indeed in the domain of the court.

CHAPTER 11 – ANSWERS TO QUESTIONS

The questions given at the end of Chapter 11 were a developmental guide, specific to the laboratory. There are no answers provided.

Index